Text-to-Speech Synthesis

Text-to-Speech Synthesis provides a complete, end-to-end account of the process of generating speech by computer. Giving an in-depth explanation of all aspects of current speech synthesis technology, it assumes no specialised prior knowledge.

Introductory chapters on linguistics, phonetics, signal processing and speech signals lay the foundation, with subsequent material explaining how this knowledge is put to use in building practical systems that generate speech. The very latest techniques such as unit selection, hidden-Markov-model synthesis and statistical text analysis are covered, and explanations of the more-traditional techniques such as format synthesis and synthesis by rule are also provided.

By weaving together the various strands of this multidisciplinary field, this book is designed for graduate students in electrical engineering, computer science and linguistics. It is also an ideal reference for existing practitioners in the fields of human communication interaction and telephony.

PAUL TAYLOR is a Visiting Lecturer in Engineering at the University of Cambridge and Chief Executive Officer of Phonetic Arts. He received his Ph.D. in Linguistics from Edinburgh University, where he went on to serve as a lecturer and Director in the Centre for Speech Technology Research. He was also a co-founder and Chief Technical Officer at Rhetorical Systems.

Text-to-Speech Synthesis

PAUL TAYLOR
University of Cambridge

CAMBRIDGE
UNIVERSITY PRESS

University Printing House, Cambridge CB2 8BS, United Kingdom

Cambridge University Press is part of the University of Cambridge.

It furthers the University's mission by disseminating knowledge in the pursuit of education, learning and research at the highest international levels of excellence.

www.cambridge.org
Information on this title: www.cambridge.org/9780521899277

© Cambridge University Press 2009

This publication is in copyright. Subject to statutory exception and to the provisions of relevant collective licensing agreements, no reproduction of any part may take place without the written permission of Cambridge University Press.

First published 2009

A catalogue record for this publication is available from the British Library

ISBN 978-0-521-89927-7 Hardback

Cambridge University Press has no responsibility for the persistence or accuracy of URLs for external or third-party internet websites referred to in this publication, and does not guarantee that any content on such websites is, or will remain, accurate or appropriate.

This book is dedicated to the technical team at Rhetorical Systems

Summary of contents

	Foreword	xxiii
	Preface	xxvii
1	Introduction	1
2	Communication and language	8
3	The text-to-speech problem	26
4	Text segmentation and organisation	52
5	Text decoding	78
6	Prosody prediction from text	111
7	Phonetics and phonology	146
8	Pronunciation	192
9	Synthesis of prosody	225
10	Signals and filters	262
11	Acoustic models of speech production	309
12	Analysis of speech signals	341
13	Synthesis techniques based on vocal-tract models	387
14	Synthesis by concatenation and signal-processing modification	412
15	Hidden-Markov-model synthesis	435
16	Unit-selection synthesis	474
17	Further issues	517
18	Conclusions	533
	Appendix A	540
	Appendix B	553
	References	556
	Index	583

Contents

Foreword		*page* xxiii
Preface		xxvii

1	**Introduction**	1
	1.1 What are text-to-speech systems for?	2
	1.2 What should the goals of text-to-speech system development be?	3
	1.3 The engineering approach	4
	1.4 Overview of the book	5
	1.4.1 Viewpoints within the book	5
	1.4.2 Readers' backgrounds	6
	1.4.3 Background and specialist sections	7

2	**Communication and language**	8
	2.1 Types of communication	8
	2.1.1 Affective communication	8
	2.1.2 Iconic communication	9
	2.1.3 Symbolic communication	10
	2.1.4 Combinations of symbols	11
	2.1.5 Meaning, form and signal	12
	2.2 Human communication	13
	2.2.1 Verbal communication	14
	2.2.2 Linguistic levels	16
	2.2.3 Affective prosody	17
	2.2.4 Augmentative prosody	18
	2.3 Communication processes	18
	2.3.1 Communication factors	19
	2.3.2 Generation	20
	2.3.3 Encoding	21
	2.3.4 Decoding	22
	2.3.5 Understanding	22
	2.4 Discussion	23
	2.5 Summary	24

3 The text-to-speech problem — 26

- 3.1 Speech and writing — 26
  - 3.1.1 Physical nature — 27
  - 3.1.2 Spoken form and written form — 28
  - 3.1.3 Use — 29
  - 3.1.4 Prosodic and verbal content — 30
  - 3.1.5 Component balance — 31
  - 3.1.6 Non-linguistic content — 32
  - 3.1.7 Semiotic systems — 33
  - 3.1.8 Writing systems — 34
- 3.2 Reading aloud — 35
  - 3.2.1 Reading silently and reading aloud — 35
  - 3.2.2 Prosody in reading aloud — 36
  - 3.2.3 Verbal content and style in reading aloud — 37
- 3.3 Text-to-speech system organisation — 37
  - 3.3.1 The common-form model — 38
  - 3.3.2 Other models — 39
  - 3.3.3 Comparison — 40
- 3.4 Systems — 41
  - 3.4.1 A simple text-to-speech system — 41
  - 3.4.2 Concept to speech — 42
  - 3.4.3 Canned speech and limited-domain synthesis — 43
- 3.5 Key problems in text-to-speech — 44
  - 3.5.1 Text classification with respect to semiotic systems — 44
  - 3.5.2 Decoding natural-language text — 46
  - 3.5.3 Naturalness — 47
  - 3.5.4 Intelligibility: encoding the message in signal — 48
  - 3.5.5 Auxiliary generation for prosody — 49
  - 3.5.6 Adapting the system to the situation — 50
- 3.6 Summary — 50

4 Text segmentation and organisation — 52

- 4.1 Overview of the problem — 52
- 4.2 Words and sentences — 53
  - 4.2.1 What is a word? — 53
  - 4.2.2 Defining words in text-to-speech — 55
  - 4.2.3 Scope and morphology — 59
  - 4.2.4 Contractions and clitics — 60
  - 4.2.5 Slang forms — 61
  - 4.2.6 Hyphenated forms — 61
  - 4.2.7 What is a sentence? — 62
  - 4.2.8 The lexicon — 63
- 4.3 Text segmentation — 63

		4.3.1	Tokenisation	64
		4.3.2	Tokenisation and punctuation	65
		4.3.3	Tokenisation algorithms	66
		4.3.4	Sentence splitting	67
	4.4	Processing documents		68
		4.4.1	Markup languages	68
		4.4.2	Interpreting characters	70
	4.5	Text-to-speech architectures		71
	4.6	Discussion		75
		4.6.1	Further reading	75
		4.6.2	Summary	76

5 Text decoding: finding the words from the text — 78

	5.1	Overview of text decoding		78
	5.2	Text-classification algorithms		79
		5.2.1	Features and algorithms	79
		5.2.2	Tagging and word-sense disambiguation	82
		5.2.3	Ad-hoc approaches	83
		5.2.4	Deterministic rule approaches	83
		5.2.5	Decision lists	85
		5.2.6	Naive Bayes classifier	86
		5.2.7	Decision trees	87
		5.2.8	Part-of-speech tagging	88
	5.3	Non-natural-language text		92
		5.3.1	Semiotic classification	92
		5.3.2	Semiotic decoding	95
		5.3.3	Verbalisation	95
	5.4	Natural-language text		97
		5.4.1	Acronyms and letter sequences	99
		5.4.2	Homograph disambiguation	99
		5.4.3	Non-homographs	101
	5.5	Natural-language parsing		102
		5.5.1	Context-free grammars	102
		5.5.2	Statistical parsing	105
	5.6	Discussion		105
		5.6.1	Further reading	108
		5.6.2	Summary	109

6 Prosody prediction from text — 111

	6.1	Prosodic form		111
	6.2	Phrasing		112
		6.2.1	Phrasing phenomena	112
		6.2.2	Models of phrasing	113

	6.3	Prominence	115
		6.3.1 Syntactic prominence patterns	116
		6.3.2 Discourse prominence patterns	118
		6.3.3 Prominence systems, data and labelling	119
	6.4	Intonation and tune	121
	6.5	Prosodic meaning and function	122
		6.5.1 Affective prosody	123
		6.5.2 Suprasegmentality	124
		6.5.3 Augmentative prosody	125
		6.5.4 Symbolic communication and prosodic style	126
	6.6	Determining prosody from the text	127
		6.6.1 Prosody and human reading	127
		6.6.2 Controlling the degree of augmentative prosody	128
		6.6.3 Prosody and synthesis techniques	128
	6.7	Phrasing prediction	129
		6.7.1 Experimental formulation	129
		6.7.2 Deterministic approaches	130
		6.7.3 Classifier approaches	132
		6.7.4 HMM approaches	133
		6.7.5 Hybrid approaches	135
	6.8	Prominence prediction	136
		6.8.1 Compound-noun phrases	136
		6.8.2 Function-word prominence	138
		6.8.3 Data-driven approaches	138
	6.9	Intonational-tune prediction	139
	6.10	Discussion	139
		6.10.1 Labelling schemes and labelling accuracy	139
		6.10.2 Linguistic theories and prosody	141
		6.10.3 Synthesising suprasegmental and true prosody	142
		6.10.4 Prosody in real dialogues	143
		6.10.5 Conclusion	144
		6.10.6 Summary	144
7	**Phonetics and phonology**		**146**
	7.1	Articulatory phonetics and speech production	146
		7.1.1 The vocal organs	147
		7.1.2 Sound sources	147
		7.1.3 Sound output	150
		7.1.4 The vocal-tract filter	150
		7.1.5 Vowels	151
		7.1.6 Consonants	153
		7.1.7 Examining speech production	155
	7.2	Acoustics, phonetics and speech perception	156

		7.2.1	Acoustic representations	156
		7.2.2	Acoustic characteristics	159
	7.3	The communicative use of speech		160
		7.3.1	Communicating discrete information with a continuous channel	161
		7.3.2	Phonemes, phones and allophones	162
		7.3.3	Allophonic variation and phonetic context	166
		7.3.4	Coarticulation, targets and transients	168
		7.3.5	The continuous nature of speech	169
		7.3.6	Transcription	170
		7.3.7	The distinctiveness of speech in communication	171
	7.4	Phonology: the linguistic organisation of speech		172
		7.4.1	Phonotactics	172
		7.4.2	Word formation	179
		7.4.3	Distinctive features and phonological theories	181
		7.4.4	Syllables	184
		7.4.5	Lexical stress	186
	7.5	Discussion		189
		7.5.1	Further reading	189
		7.5.2	Summary	190
8	**Pronunciation**			192
	8.1	Pronunciation representations		192
		8.1.1	Why bother?	192
		8.1.2	Phonemic and phonetic input	193
		8.1.3	Difficulties in deriving phonetic input	194
		8.1.4	A structured approach to pronunciation	195
		8.1.5	Abstract phonological representations	196
	8.2	Formulating a phonological representation system		197
		8.2.1	Simple consonants and vowels	197
		8.2.2	Difficult consonants	199
		8.2.3	Diphthongs and affricates	201
		8.2.4	Approximant–vowel combinations	201
		8.2.5	Defining the full inventory	203
		8.2.6	Phoneme names	204
		8.2.7	Syllabic issues	206
	8.3	The lexicon		207
		8.3.1	Lexicon and rules	208
		8.3.2	Lexicon formats	210
		8.3.3	The offline lexicon	213
		8.3.4	The system lexicon	214
		8.3.5	Lexicon quality	215
		8.3.6	Determining the pronunciations of unknown words	216
	8.4	Grapheme-to-phoneme conversion		218

	8.4.1	Rule-based techniques	218
	8.4.2	Grapheme-to-phoneme alignment	219
	8.4.3	Neural networks	219
	8.4.4	Pronunciation by analogy	220
	8.4.5	Other data-driven techniques	221
	8.4.6	Statistical techniques	221
8.5	Further issues		222
	8.5.1	Morphology	222
	8.5.2	Language origin and names	223
	8.5.3	Post-lexical processing	223
8.6	Summary		224

9 Synthesis of prosody 225

9.1	Intonation overview		225
	9.1.1	F0 and pitch	226
	9.1.2	Intonational form	226
	9.1.3	Models of F0 contours	227
	9.1.4	Micro-prosody	229
9.2	Intonational behaviour		229
	9.2.1	Intonational tune	229
	9.2.2	Downdrift	230
	9.2.3	Pitch range	233
	9.2.4	Pitch accents and boundary tones	234
9.3	Intonation theories and models		236
	9.3.1	Traditional models and the British school	236
	9.3.2	The Dutch school	237
	9.3.3	Autosegmental–metrical and ToBI models	237
	9.3.4	The INTSINT model	239
	9.3.5	The Fujisaki model and superimpositional models	239
	9.3.6	The Tilt model	242
	9.3.7	Comparison	244
9.4	Intonation synthesis with AM models		245
	9.4.1	Prediction of AM labels from text	246
	9.4.2	Deterministic synthesis methods	246
	9.4.3	Data-driven synthesis methods	247
	9.4.4	Analysis with autosegmental models	248
9.5	Intonation synthesis with deterministic acoustic models		248
	9.5.1	Synthesis with superimpositional models	249
	9.5.2	Synthesis with the Tilt model	249
	9.5.3	Analysis with Fujisaki and Tilt models	250
9.6	Data-driven intonation models		250
	9.6.1	Unit-selection-style approaches	251
	9.6.2	Dynamic-system models	252

	9.6.3 Hidden Markov models	253
	9.6.4 Functional models	254
9.7	Timing	254
	9.7.1 Formulation of the timing problem	255
	9.7.2 The nature of timing	255
	9.7.3 Klatt rules	256
	9.7.4 The sums-of-products model	257
	9.7.5 The Campbell model	258
	9.7.6 Other regression techniques	259
9.8	Discussion	259
	9.8.1 Further reading	260
	9.8.2 Summary	260

10 Signals and filters 262

10.1 Analogue signals 262
    10.1.1 Simple periodic signals: sinusoids 263
    10.1.2 General periodic signals 265
    10.1.3 Sinusoids as complex exponentials 266
    10.1.4 Fourier analysis 269
    10.1.5 The frequency domain 270
    10.1.6 The Fourier transform 275

10.2 Digital signals 278
    10.2.1 Digital waveforms 279
    10.2.2 Digital representations 280
    10.2.3 The discrete-time Fourier transform 280
    10.2.4 The discrete Fourier transform 281
    10.2.5 The z-transform 282
    10.2.6 The frequency domain for digital signals 283

10.3 Properties of transforms 284
    10.3.1 Linearity 284
    10.3.2 Time and frequency duality 284
    10.3.3 Scaling 285
    10.3.4 Impulse properties 285
    10.3.5 Time delay 286
    10.3.6 Frequency shift 286
    10.3.7 Convolution 287
    10.3.8 Analytical and numerical analysis 287
    10.3.9 Stochastic signals 288

10.4 Digital filters 288
    10.4.1 Difference equations 289
    10.4.2 The impulse response 289
    10.4.3 The filter convolution sum 292
    10.4.4 The filter transfer function 293

	10.4.5 The transfer function and the impulse response	293
10.5	Digital filter analysis and design	294
	10.5.1 Polynomial analysis: poles and zeros	294
	10.5.2 Frequency interpretation of the z-domain transfer function	297
	10.5.3 Filter characteristics	298
	10.5.4 Putting it all together	304
10.6	Summary	305

11 Acoustic models of speech production — 309

11.1	The acoustic theory of speech production	309
	11.1.1 Components in the model	309
11.2	The physics of sound	311
	11.2.1 Resonant systems	311
	11.2.2 Travelling waves	313
	11.2.3 Acoustic waves	315
	11.2.4 Acoustic reflection	317
11.3	The vowel-tube model	318
	11.3.1 Discrete time and distance	319
	11.3.2 Junction of two tubes	320
	11.3.3 Special cases of junctions	322
	11.3.4 The two-tube vocal-tract model	323
	11.3.5 The single-tube model	325
	11.3.6 The multi-tube vocal-tract model	327
	11.3.7 The all-pole resonator model	329
11.4	Source and radiation models	330
	11.4.1 Radiation	330
	11.4.2 The glottal source	330
11.5	Model refinements	333
	11.5.1 Modelling the nasal cavity	333
	11.5.2 Source positions in the oral cavity	334
	11.5.3 Models with vocal-tract losses	335
	11.5.4 Source and radiation effects	336
11.6	Discussion	336
	11.6.1 Further reading	339
	11.6.2 Summary	339

12 Analysis of speech signals — 341

12.1	Short-term speech analysis	341
	12.1.1 Windowing	342
	12.1.2 Short-term spectral representations	343
	12.1.3 Frame lengths and shifts	345
	12.1.4 The spectrogram	348
	12.1.5 Auditory scales	351

	12.2	Filter-bank analysis	352
	12.3	The cepstrum	353
		12.3.1 Cepstrum definition	353
		12.3.2 Treating the magnitude spectrum as a signal	353
		12.3.3 Cepstral analysis as deconvolution	355
		12.3.4 Cepstral analysis discussion	356
	12.4	Linear-prediction analysis	357
		12.4.1 Finding the coefficients: the covariance method	358
		12.4.2 The autocorrelation method	360
		12.4.3 Levinson–Durbin recursion	361
	12.5	Spectral-envelope and vocal-tract representations	362
		12.5.1 Linear-prediction spectra	362
		12.5.2 Transfer-function poles	364
		12.5.3 Reflection coefficients	366
		12.5.4 Log area ratios	367
		12.5.5 Line-spectrum frequencies	367
		12.5.6 Linear-prediction cepstra	369
		12.5.7 Mel-scaled cepstra	370
		12.5.8 Perceptual linear prediction	370
		12.5.9 Formant tracking	370
	12.6	Source representations	372
		12.6.1 Residual signals	372
		12.6.2 Closed-phase analysis	374
		12.6.3 Open-phase analysis	377
		12.6.4 Impulse/noise models	378
		12.6.5 Parameterisation of glottal-flow signals	379
	12.7	Pitch and epoch detection	379
		12.7.1 Pitch detection	379
		12.7.2 Epoch detection: finding the instant of glottal closure	381
	12.8	Discussion	384
		12.8.1 Further reading	385
		12.8.2 Summary	386
13	**Synthesis techniques based on vocal-tract models**		387
	13.1	Synthesis specification: the input to the synthesiser	387
	13.2	Formant synthesis	388
		13.2.1 Sound sources	389
		13.2.2 Synthesising a single formant	390
		13.2.3 Resonators in series and parallel	391
		13.2.4 Synthesising consonants	392
		13.2.5 A complete synthesiser	394
		13.2.6 The phonetic input to the synthesiser	394
		13.2.7 Formant-synthesis quality	397

	13.3	Classical linear-prediction synthesis	399
		13.3.1 Comparison with formant synthesis	399
		13.3.2 The impulse/noise source model	400
		13.3.3 Linear-prediction diphone-concatenative synthesis	401
		13.3.4 A complete synthesiser	403
		13.3.5 Problems with the source	404
	13.4	Articulatory synthesis	405
	13.5	Discussion	407
		13.5.1 Further reading	409
		13.5.2 Summary	410
14	**Synthesis by concatenation and signal-processing modification**		**412**
	14.1	Speech units in second-generation systems	413
		14.1.1 Creating a diphone inventory	414
		14.1.2 Obtaining diphones from speech	414
	14.2	Pitch-synchronous overlap and add (PSOLA)	415
		14.2.1 Time-domain PSOLA	416
		14.2.2 Epoch manipulation	417
		14.2.3 How does PSOLA work?	421
	14.3	Residual-excited linear prediction (RELP)	421
		14.3.1 Residual manipulation	423
		14.3.2 Linear-prediction PSOLA	423
	14.4	Sinusoidal models	424
		14.4.1 Pure sinusoidal models	425
		14.4.2 Harmonic/noise models	426
	14.5	MBROLA	429
	14.6	Synthesis from cepstral coefficients	429
	14.7	Concatenation issues	431
	14.8	Discussion	433
		14.8.1 Further reading	433
		14.8.2 Summary	433
15	**Hidden-Markov-model synthesis**		**435**
	15.1	The HMM formalism	435
		15.1.1 Observation probabilities	436
		15.1.2 Delta coefficients	438
		15.1.3 Acoustic representations and covariance	439
		15.1.4 States and transitions	440
		15.1.5 Recognising with HMMs	440
		15.1.6 Language models	443
		15.1.7 The Viterbi algorithm	444
		15.1.8 Training HMMS	447
		15.1.9 Context-sensitive modelling	451

		15.1.10 Are HMMs a good model of speech?	454
	15.2	Synthesis from hidden Markov models	456
		15.2.1 Finding the likeliest observations given the state sequence	457
		15.2.2 Finding the likeliest observations and state sequence	459
		15.2.3 Acoustic representations	459
		15.2.4 Context-sensitive synthesis models	463
		15.2.5 Duration modelling	464
		15.2.6 Signal Processing in HMM synthesis	464
		15.2.7 HMM synthesis systems	465
	15.3	Labelling databases with HMMs	467
		15.3.1 Determining the word sequence	467
		15.3.2 Determining the phone sequence	468
		15.3.3 Determining the phone boundaries	468
		15.3.4 Measuring the quality of the alignments	470
	15.4	Other data-driven synthesis techniques	471
	15.5	Discussion	471
		15.5.1 Further reading	471
		15.5.2 Summary	472
16	**Unit-selection synthesis**		**474**
	16.1	From concatenative synthesis to unit selection	474
		16.1.1 Extending concatenative synthesis	475
		16.1.2 The Hunt and Black algorithm	477
	16.2	Features	479
		16.2.1 Base types	479
		16.2.2 Linguistic and acoustic features	480
		16.2.3 Choice of features	481
		16.2.4 Types of features	482
	16.3	The independent-feature target-function formulation	484
		16.3.1 The purpose of the target function	484
		16.3.2 Defining a perceptual space	485
		16.3.3 Perceptual spaces defined by independent features	486
		16.3.4 Setting the target weights using acoustic distances	488
		16.3.5 Limitations of the independent-feature formulation	491
	16.4	The acoustic-space target-function formulation	493
		16.4.1 Decision-tree clustering	494
		16.4.2 General partial-synthesis functions	496
	16.5	Join functions	497
		16.5.1 Basic issues in joining units	497
		16.5.2 Phone-class join costs	498
		16.5.3 Acoustic-distance join costs	499
		16.5.4 Combining categorical and and acoustic join costs	500
		16.5.5 Probabilistic and sequence join costs	501

		16.5.6 Join classifiers	502
	16.6	Searching	504
		16.6.1 Base types and searching	505
		16.6.2 Pruning	508
		16.6.3 Pre-selection	508
		16.6.4 Beam pruning	509
		16.6.5 Multi-pass searching	509
	16.7	Discussion	510
		16.7.1 Unit selection and signal processing	511
		16.7.2 Features, costs and perception	511
		16.7.3 Example unit-selection systems	512
		16.7.4 Further reading	514
		16.7.5 Summary	515
17	**Further issues**		**517**
	17.1	Databases	517
		17.1.1 Unit-selection databases	517
		17.1.2 Text materials	518
		17.1.3 Prosody databases	518
		17.1.4 Labelling	519
		17.1.5 What exactly is hand labelling?	519
		17.1.6 Automatic labelling	521
		17.1.7 Avoiding explicit labels	521
	17.2	Evaluation	522
		17.2.1 System testing: intelligibility and naturalness	523
		17.2.2 Word-recognition tests	523
		17.2.3 Naturalness tests	524
		17.2.4 Test data	525
		17.2.5 Unit or component testing	525
		17.2.6 Competitive evaluations	526
	17.3	Audio-visual speech synthesis	527
		17.3.1 Speech control	528
	17.4	Synthesis of emotional and expressive speech	529
		17.4.1 Describing emotion	529
		17.4.2 Synthesising emotion with prosody control	529
		17.4.3 Synthesising emotion with voice transformation	530
		17.4.4 Unit selection and HMM techniques	531
	17.5	Summary	531
18	**Conclusion**		**533**
	18.1	Speech technology and linguistics	533
	18.2	Future directions	536
	18.3	Conclusion	539

Appendix A Probability — 540

A.1 Discrete probabilities — 540
    A.1.1 Discrete random variables — 540
    A.1.2 Probability mass functions — 541
    A.1.3 Expected values — 541
    A.1.4 Moments of a PMF — 542
A.2 Pairs of discrete random variables — 542
    A.2.1 Marginal distributions — 543
    A.2.2 Independence — 543
    A.2.3 Expected values — 543
    A.2.4 Moments of a joint distribution — 544
    A.2.5 Higher-order moments and covariance — 544
    A.2.6 Correlation — 544
    A.2.7 Conditional probability — 545
    A.2.8 Bayes' rule — 545
    A.2.9 Sum of random variables — 545
    A.2.10 The chain rule — 546
    A.2.11 Entropy — 546
A.3 Continuous random variables — 547
    A.3.1 Continuous random variables — 548
    A.3.2 Expected values — 548
    A.3.3 The Gaussian distribution — 549
    A.3.4 The uniform distribution — 549
    A.3.5 Cumulative density functions — 549
A.4 Pairs of continuous random variables — 550
    A.4.1 Independent versus uncorrelated — 551
    A.4.2 The sum of two random variables — 551
    A.4.3 Entropy — 552
    A.4.4 Kullback–Leibler distance — 552

Appendix B Phone definitions — 553

References — 556
Index — 583

Foreword

Speech-processing technology has been a mainstream area of research for more than 50 years. The ultimate goal of speech research is to build systems that mimic (or potentially surpass) human capabilities in understanding, generating and coding speech for a range of human-to-human and human-to-machine interactions.

In the area of speech coding a great deal of success has been achieved in creating systems that significantly reduce the overall bit rate of the speech signal (from of the order of 100 kilobits per second to rates of the order of 8 kilobits per second or less), while maintaining speech intelligibility and quality at levels appropriate for the intended applications. The heart of the modern cellular industry is the 8 kilobit per second speech coder, embedded in VLSI logic on the more than two billion cellphones in use worldwide at the end of 2007.

In the area of speech recognition and understanding by machines, steady progress has enabled systems to become part of everyday life in the form of call centres for the airlines, financial, medical and banking industries, help desks for large businesses, form and report generation for the legal and medical communities, and dictation machines that enable individuals to enter text into machines without having to type the text explicitly. Such speech-recognition systems were made available to the general public as long as 15 years ago (in 1992 AT&T introduced the Voice Recognition Call Processing system which automated operator-assisted calls, handling more than 1.2 billion requests each year with error rates below 0.5%) and have penetrated almost every major industry since that time. Simple speech-understanding systems have also been introduced into the marketplace and have had varying degrees of success for help desks (e.g., the How May I Help You system introduced by AT&T for customer-care applications) and for stock-trading applications (IBM system), among others.

It has been the area of speech generation that has been the hardest speech-technology area in which to obtain any viable degree of success. For more than 50 years researchers have struggled with the problem of trying to mimic the physical processes of speech generation via articulatory models of the human vocal tract, or via terminal analogue synthesis models of the time-varying spectral and temporal properties of speech. In spite of the best efforts of some outstanding speech researchers, the quality of synthetic speech generated by machine was unnatural most of the time and has been unacceptable for human use in most real-world applications. In the late 1970s the idea of generating speech by concatenating basic speech units (in most cases diphone units that represented pieces comprising pairs of phonemes) was investigated and shown to be practical once researchers had learned how to excise diphones from human speech reliably. After more than a decade of investigation into how to concatenate diphones optimally, the resulting synthetic speech was often highly intelligible (a big improvement over earlier systems) but regrettably remained highly unnatural. Hence concatenative speech-synthesis systems remained lab curiosities and were not employed in real-world applications such as reading email, user interactions in dialogue systems, etc. The really big breakthrough in speech synthesis came in the late 1980s when Yoshinori Sagisaka at ATR in Japan made the leap from single-diphone tokens as the basic unit set for speech synthesis to multiple-diphone tokens, extracted from carefully designed and read speech databases. Sagisaka realized that, in the limiting case, where you had thousands of tokens of each possible diphone of the English language, you could literally concatenate the "correct" sequence of diphones and produce natural-sounding human speech. The new problem that arose was that of deciding exactly which of the thousands of diphones should be used at each diphone position in the speech being generated. History has shown that, like most large-scale computing problems, there are solutions that make the search for the optimum sequence of diphones (from a large, virtually infinite database) possible in reasonable time and memory. The rest is now history as a new generation of speech researchers investigated virtually every aspect of the so-called "unit-selection" method of concatenative speech synthesis, showing that high-quality (both intelligibility and naturalness) synthetic speech could be obtained from such systems for virtually any task application.

Once the problem of generating natural-sounding speech from a sequence of diphones was solved (in the sense that a practical demonstration of the feasibility of such high-quality synthesis was made with unit-selection synthesis systems), the remaining long-standing problem was the conversion from ordinary printed text to the proper sequence of diphones, along with associated prosodic information about sound duration, loudness, emphasis, pitch, pauses, and other so-called suprasegmental aspects of speech. The problem of converting from text to a complete linguistic description of associated sound was one that has been studied for almost as long as synthesis itself; much progress had been made in almost every aspect of the linguistic description of speech as in the acoustic generation of high-quality sounds.

It is the success of unit-selection speech-synthesis systems that has motivated the research of Paul Taylor, the author of this book on text-to-speech synthesis systems. Paul Taylor has been in the thick of the research in speech-synthesis systems for more than 15 years, having worked at ATR in Japan on the CHATR synthesiser (the system that actually demonstrated near-perfect speech quality on some subset of the sentences that were input), at the Centre for Speech Technology research at the University of Edinburgh on the Festival synthesis system, and as Chief Technical Officer of Rhetorical Systems, also in Edinburgh.

Taylor has put together a book based on decades of research and the extraordinary progress over the past decade, which attempts to tie it all together and to document and explain the processes involved in a complete text-to-speech synthesis system. The first nine chapters of the book address the problem of converting printed text to a sequence of sound units (which characterise the acoustic properties of the resulting synthetic sentence) and an accompanying description of the associated prosody which is most appropriate for the sentence being spoken. The remaining eight chapters (not including the conclusion) provide a review of the associated signal-processing techniques for representing speech units and for seamlessly tying them together to form intelligible and natural speech sounds. This is followed by a discussion of the three generations of synthesis methods, namely articulatory and terminal analogue synthesis methods, simple concatenative methods using a single representation for each diphone unit and the unit-selection method based on multiple representations for each diphone unit. There is a single chapter devoted to a promising new synthesis approach, namely a statistical method based on the popular hidden-Markov-model (HMM) formulation used in speech-recognition systems.

According to the author, "Speech synthesis has progressed remarkably in recent years, and it is no longer the case that state-of-the-art systems sound overtly mechanical and robotic". Although this statement is true, there remains a great deal more to be accomplished before speech-synthesis systems are indistinguishable from a human speaker. Perhaps the most glaring need is expressive synthesis that imparts not only the message corresponding to the printed text input but also the emotion associated with the way a human might speak the same sentence in the context of a dialogue with another human being. We are still a long way from such emotional or expressive speech-synthesis systems.

This book is a wonderful addition to the literature in speech processing and will be a must-read for anyone wanting to understand the blossoming field of text-to-speech synthesis.

Lawrence Rabiner
August 2007

Preface

I'd like to say that *Text-to-Speech Synthesis* was years in the planning but nothing could be further from the truth. In mid 2004, as Rhetorical Systems was nearing its end, I suddenly found myself with spare time on my hands for the first time since ..., well, ever to be honest. I had no clear idea of what to do next and thought it might "serve the community well" to jot down a few thoughts on TTS. The initial idea was a slim "hardcore" technical book explaining the state of the art, but as writing continued I realised that more and more background material was needed. Eventually the initial idea was dropped and the more-comprehensive volume that you now see was written.

Some early notes were made in Edinburgh, but the book mainly started during a summer I spent in Boulder, Colorado. I was freelancing at that point but am grateful to Ron Cole and his group for accommodating me in such a friendly fashion. A good deal of work was completed outside the office and I am eternally grateful to Laura Michalis for putting me up, putting up with me and generally being there and supportive during that phase.

While the book was in effect written entirely by myself, I should pay particular thanks to Steve Young and Mark Gales as I used the HTK book and lecture notes directly in Chapter 15 and the appendix. Many have helped by reading earlier drafts, especially Bob Ladd, Keiichi Tokuda, Richard Sproat, Dan Jurafsky, Ian Hodson, Ant Tomlinson, Matthew Aylett and Rob Clark.

From October 2004 to August 2006 I was a visitor at the Cambridge University Engineering department, and it was there that the bulk of the book was written. During this time, I was sponsored by the Royal Society, via a fellowship that allowed individuals to transfer between industrial and academic research. This was a great scheme, and I am indebted to the Royal Society for making such a scheme open to me. This scheme allowed me to pursue open-ended research with enough time left over for me to write this book. I am very grateful to Steve Young in particular for taking me in and making me feel part of what was a very well-established group. Without Steve's help Cambridge wouldn't have happened for me and the book probably would have not been written. Within the department the other faculty staff made me feel very welcome. Phil Woodland and Mark Gales always showed an interest and were of great help on many issues. Finally, Bill Byrne arrived at Cambridge at the same time as me, and proved a great comrade in our quest to understand how on all Earth Cambridge University actually worked. Since, he has become a good friend and was a great encouragment in the writing of this book. All these and the other members of the Machine Intelligence lab have become good friends as well as respected colleagues. To my long suffering room mate Gabe Brostow I simply say thanks for being a good friend. Because of the strength of our friendship I know he won't mind me having the last word in a long-running argument and saying that Ireland really is a better place than Texas.

It is customary in acknowledgments such as these to thank one's wife and family; unfortunately, while I may have some skill in TTS, this has (strangely) not transfered into skill in holding down a relationship. Nonetheless, no man is an island and I'd like to thank Laura, Kirstin and Kayleigh for being there while I wrote the book. Final and special thanks must go to Maria Founda, who has a generosity of spirit second to none.

For most of my career I have worked in Edinburgh, first at the University's Centre for Speech Technology Research (CSTR) and latterly at Rhetorical Systems. The book reflects the climate in both those organisations, which, to my mind, had a healthy balance between eclectic knowledge of engineering, computer science and linguistics on the one hand and robust system building on the other. There are far too many individuals there to thank personally, but I should mention Bob Ladd, Alan Wrench and Steve Isard as being particularly influential, especially in the early days. The later days in the university were mainly spent on Festival and the various sub-projects it included. My long-term partner in crime Alan Black deserves particular thanks for all the long conversations about virtually everything that helped form that system and gave me a better understanding on so many issues in computational linguistics and computer science in general. The other main author in Festival, Richard Caley, made an enormous and often unrecognised contribution to Festival. Tragically Richard died in 2005 before his time. I would of course like to thank everyone else in CSTR and the general Edinburgh University speech and language community for making it such a great place to work.

Following CSTR, I co-founded Rhetorical Systems in 2000 and for 4 years had a great time. I think the team we assembled there was world class and for quite a while we had (as far as our tests showed) the highest-quality TTS system at that time. I thought the technical environment there was first class and helped finalise my attitudes towards a no-nonsense approach to TTS systems. All the techical team there were outstanding but I would like in particular to thank the members of the R&D team who most clearly had an impact on my thoughts and knowledge on TTS. They were Kathrine Hammervold, Wojciech Skut, David McKelvie, Matthew Aylett, Justin Fackrell, Peter Rutten and David Talkin. In recognition of the great years at Rhetorical and to thank all those who had to have me as a boss, this book is dedicated to the entire Rhetorical technical team.

Paul Taylor
Cambridge
19 November 2007

1 Introduction

This is a book about getting computers to read out loud. It is therefore about three things: the process of reading, the process of speaking, and the issues involved in getting computers (as opposed to humans) to do this. This field of study is known both as **speech synthesis**, that is the "synthetic" (computer) generation of speech, and as **text-to-speech** or **TTS**; the process of converting written text into speech. It complements other language technologies such as **speech recognition**, which aims to convert speech into text, and **machine translation**, which converts writing or speech in one language into writing or speech in another.

I am assuming that most readers have heard some synthetic speech in their life. We experience this in a number of situations; some telephone information systems have automated speech response, speech synthesis is often used as an aid to the disabled, and Professor Stephen Hawking has probably contributed more than anyone else to the direct exposure of (one particular type of) synthetic speech. The *idea* of artificially generated speech has of course been around for a long time – hardly any science-fiction film is complete without a talking computer of some sort. In fact science fiction has had an interesting effect on the field and our impressions of it. Sometimes (less technically aware) people believe that perfect speech synthesis exists because they "heard it on *Star Trek*".[1] Often makers of science-fiction films fake the synthesis by using an actor, although usually some processing is added to the voice to make it sound "computerised". Some actually use real speech-synthesis systems, but interestingly these are usually not state-of-the-art systems, since these sound too natural, and may mislead the viewer.[2] One of the genuine attempts to predict how synthetic voices will sound is the computer HAL in the film *2001: A Space Odyssey* [265]. The fact that this computer spoke with a calm and near-humanlike-voice gave rise to the sense of genuine intelligence in the machine. While many parts of this film were wide of the mark (especially the ability of HAL to understand, rather than just recognise, human speech), the makers of the film just about got it right in predicting how good computer voices would be in the year in question.

Speech synthesis has progressed remarkably in recent years, and it is no longer the case that state-of-the-art systems sound overtly mechanical and robotic. That said, it

---

[1] Younger readers please substitute the in-vogue science-fiction series of the day.
[2] In much the same way, when someone types the wrong password on a computer, the screen starts flashing and saying "access denied". Some even go so far as to have a siren sounding. Those of us who use computers know this never happens, but in a sense we go along with the exaggeration as it adds to the drama.

is normally fairly easy to tell that it is a computer talking rather than a human, so substantial progress has still to be made. When assessing a computer's ability to speak, one fluctuates between two judgments. On the one hand, it is tempting to paraphrase Dr Johnson's famous remark [61] "Sir, a talking computer is like a dog's walking on his hind legs. It is not done well; but you are surprised to find it done at all." Indeed, even as an experienced text-to-speech researcher who has listened to more synthetic speech than could be healthy in one life, I find that sometimes I am genuinely surprised and thrilled in a naive way that here we have a talking computer: "like wow! it talks!". On the other hand, it is also possible to have the impression that computers are quite dreadful at the job of speaking; they make frequent mistakes, drone on, and just sound plain *wrong* in many cases. These impressions are all part of the mysteries and complexities of speech.

1.1 What are text-to-speech systems for?

Text-to-speech systems have an enormous range of applications. Their first real use was in reading systems for the blind, where a system would read some text from a book and convert it into speech. These early systems of course sounded very mechanical, but their adoption by blind people was hardly surprising because the other options of reading braille or having a real person do the reading were often not available. Today, quite sophisticated systems exist that facilitate human–computer interaction for the blind, in which the TTS can help the user navigate around a windows system.

The mainstream adoption of TTS has been severely limited by its quality. Apart from users who have little choice (as is the case with blind people), people's reaction to old-style TTS is not particularly positive. While people may be somewhat impressed and quite happy to listen to a few sentences, in general the novelty of this soon wears off. In recent years, the considerable advances in quality have changed the situation such that TTS systems are more common in a number of applications. Probably the main use of TTS today is in call-centre automation, where a user calls to pay an electricity bill or book some travel and conducts the entire transaction through an automatic dialogue system. Beyond this, TTS systems have been used for reading news stories, weather reports, travel directions and a wide variety of other applications.

While this book concentrates on the practical, engineering aspects of text-to-speech, it is worth commenting that research in this field has contributed an enormous amount to our general understanding of language. Often this has been in the form of "negative" evidence, meaning that when a theory thought to be true was implemented in a TTS system it was shown to be false; in fact, as we shall see, many linguistic theories have fallen when rigorously tested in speech systems. More positively, TTS systems have made good testing grounds for many models and theories, and TTS systems are certainly interesting in their own terms, without reference to application or use.

1.2 What should the goals of text-to-speech system development be?

One can legitimately ask, regardless of what application we want a talking computer for, is it really necessary that the quality needs to be high and that the voice needs to sound like a human? Wouldn't a mechanical-sounding voice suffice? Experience has shown that people are in fact very sensitive, not just to the words that are spoken, but to the *way* they are spoken. After only a short while, most people find highly mechanical voices irritating and discomforting to listen to. Furthermore, tests have shown that user satisfaction increases dramatically the more "natural" sounding the voice is. Experience (and particularly commercial experience) shows that users clearly want natural-sounding (that is human-like) systems.

Hence our goals in building a computer system capable of speaking are to build a system that first of all clearly gets across the message and secondly does this using a human-like voice. Within the research community, these goals are referred to as **intelligibility** and **naturalness**.

A further goal is that the system should be able to take any written input; that is, if we build an English text-to-speech system, it should be capable of reading any English sentence given to it. With this in mind, it is worth making a few distinctions about computer speech in general. It is of course possible simply to record some speech, store it on a computer and play it back. We do this all the time; our answering machine replays a message we have recorded, the radio plays interviews that were previously recorded and so on. This is of course simply a process of playing back what was originally recorded. The idea behind text-to-speech is to "play back" messages that weren't originally recorded. One step away from simple playback is to record a number of common words or phrases and recombine them, and this technique is frequently used in telephone dialogue services. Sometimes the result is acceptable, sometimes not, since often the artificially joined speech sounds stilted and jumpy. This allows a certain degree of flexibility, but falls short of open-ended flexibility. Text-to-speech, on the other hand, has the goal of being able to speak anything, regardless of whether the desired message was originally spoken or not.

As we shall see in Chapter 13, there are various techniques for actually generating the speech. These generally fall into two camps, which we can call bottom-up and concatenative. In the bottom-up approach, we generate a speech signal "from scratch", using our knowledge of how the speech-production system works. We artificially create a basic signal and then modify it, in much the same way as the larynx produces a basic signal that is then modified by the mouth in real human speech. In the concatenative approach, there is no bottom-up signal creation per se; rather we record some real speech, cut this up into small pieces, and then recombine these to form "new" speech. Sometimes one hears the comment that concatenative techniques aren't "real" speech synthesis in that we aren't generating the signal from scratch. This point is debatable, but it turns out that at present concatenative techniques outperform other techniques, and for this reason concatenative techniques currently dominate in commercial applications.

1.3 The engineering approach

In this book, we take what is known as an **engineering approach** to the text-to-speech problem. The term "engineering" is often used to mean that systems are simply bolted together, with no underlying theory or methodology. Engineering is of course much more than this, and it should be clear that great feats of engineering such as the Brooklyn bridge were not simply the result of some engineers waking up one morning and banging some bolts together. So by "engineering" we mean that we are tackling this problem in the best traditions of other engineering; these include working with the materials available and building a practical system that doesn't, for instance, take days to produce a single sentence. Furthermore, we don't use the term engineering to mean that this field is only relevant or accessible to those with (traditional) engineering backgrounds or education. As we explain below, TTS is a field relevant to people from many different backgrounds.

One point of note is that we can contrast the engineering approach with the scientific approach. Our task is to build the best possible text-to-speech system and in doing so we will use any model, mathematics, data, theory or tool that serves our purpose. Our main job is to build an *artefact* and we will use any means possible to do so. All artefact creation can be called engineering, but *good* engineering involves more: often we wish to make good use of our resources (we don't want to use a hammer to crack a nut); we also in general want to base our system on solid principles. This is for several reasons. First, using solid (say mathematical) principles assures us that we are on well-tested ground; we can trust these principles and don't have to verify experimentally every step we take. Second, we are of course not building the last ever text-to-speech system; our system is one step in a continual development; by basing our system on solid principles we hope to help others to improve and build on our work. Finally, using solid principles has the advantage of helping us diagnose the system, for instance to help us find why some components do perhaps better than expected, and allows the principles on which these components are based to be used for other problems.

Speech synthesis has also been approached from a more scientific aspect. Researchers who pursue this approach are not interested in building systems for their own sake, but rather as models that will shine light on human speech and language abilities. Thus, the goals are different, and, for example, it is important in this approach to use techniques that are at least plausible possibilities for how humans would handle this task. A good example of the difference is in the concatenative waveform techniques which we will use predominantly; recording large numbers of audio waveforms, chopping them up and gluing them back together can produce very-high-quality speech. It is of course absurd to think that this is how humans do it. We bring this point up because speech synthesis is often used (or was certainly used in the past) as a testing ground for many theories of speech and language. As a leading proponent of the scientific viewpoint states, so long as the two approaches are not confused, no harm should arise (Huckvale [225]).

1.4 Overview of the book

I must confess to generally hating sections entitled "how to read this book" and so on. I feel that, if I bought it, I should be able to read it any way I damn well please! Nevertheless, I feel some guidelines may be useful.

This book is what one might call an *extended text book*. A normal text book has the job of explaining a field or subject to outsiders and this book certainly has that goal. I qualify this by using the term "extended" for two reasons. Firstly, the book contains some original work and is not simply a summary, collection or retelling of existing ideas. Secondly, the book aims to take the reader right up to the current state of the art. In reality this can never be fully achieved, but the book is genuinely intended to be an "all you ever need to know". More modestly, it can be thought of as "all that I know and can explain". In other words, this is it: I certainly couldn't write a second book that dealt with more advanced topics.

Despite these original sections, the book is certainly not a monograph. This point is worth reinforcing: because of my personal involvement in many aspects of TTS research over the last 15 years, and specifically because of my involvement in the development of many well-known TTS systems, including CHATR [53], Festival [55], and rVoice, many friends and colleagues have asked me whether this is a book about those systems or the techniques behind those systems. Let me clearly state that this is not the case; *Text-to-Speech Synthesis* is not a system book that describes one particular system; rather I aim for a general account that describes current techniques without reference to any particular single system or theory.

1.4.1 Viewpoints within the book

That said, this book aims to provide a single, coherent picture of text-to-speech, rather than simply a list of available techniques. While not being a book centred on any one system, it is certainly heavily influenced by the general philosophy that I have been using (and evolving) over the past years, and I think it is proper at this stage to say something about what this philosophy is and how it may differ from other views. In the broadest sense, I adopt what is probably the current mainstream view in TTS, namely that this is an engineering problem, which should be approached with the aim of producing the best possible system, rather than with the aim of investigating any particular linguistic or other theory.

Within the engineering view, I again have taken a more specialist view in posing the text-to-speech problem as one where we have a single integrated text-analysis component followed by a single integrated speech-synthesis component. I have called this the **common-form model** (this and other models are explained in Chapter 3). While the common-form model differs significantly from the usual "pipelined" models, most work that has been carried out in one framework can be used in the other without too much difficulty.

In addition to this, there are many parts that can be considered original (at least to the best of my knowledge) and in this sense the book diverges from being a pure text book at these points. Specifically, these parts are

1. the common-form model itself,
2. the formulation of text analysis as a decoding problem,
3. the idea that text analysis should be seen as a semiotic classification and verbalisation problem,
4. the model of reading aloud,
5. the general unit-selection framework and
6. the view that prosody is composed of the functionally separate systems of affective, augmentative and suprasegmental prosody.

With regard to the last topic, I should point out that my views on prosody diverge considerably from the mainstream. My view is that mainstream linguistics, and as a consequence much of speech technology, has simply got this area of language badly wrong. There is a vast, confusing and usually contradictory literature on prosody, and it has bothered me for years why several contradictory competing theories (of, say, intonation) exist, why no-one has been able to make use of prosody in speech-recognition and -understanding systems, and why all prosodic models that I have tested fall far short of the results their creators say we should expect. This has led me to propose a completely new model of prosody, which is explained in Chapters 3 and 6.

1.4.2 Readers' backgrounds

This book is intended for both an academic and a commercial audience. Text-to-speech or speech synthesis does not fall neatly into any one traditional academic discipline, so the level and amount of background knowledge will vary greatly depending on a particular reader's background. Most TTS researchers I know come from an electrical engineering, computer science or linguistics background. I have aimed the book at being directly accessible to readers with these backgrounds, but the book should in general be accessible to those from other fields.

I assume that all readers are computer literate and have some experience in programming. To this extent, concepts such as algorithm, variable, loop and so on are assumed. Some areas of TTS are mathematical, and here I have assumed that the entry level is that of an advanced high-school or first-year university course in maths. While some of the mathematical concepts are quite advanced, these are explained in full starting with the entry-level knowledge. For those readers with little mathematical knowledge (or inclination!), don't worry; many areas of TTS do not require much maths. Even for those areas which do, I believe a significant understanding can still be achieved by reading about the general principles, studying the graphs and, above all, trying the algorithms in practice. Digital filters can seem like a complicated and abstract subject to many; but I have seen few people fail to grasp its basics when give the opportunity to play around with filters in a GUI package.

My commercial experience made it clear that it was difficult to find software developers with any knowledge of TTS. It was seen as too specialist a topic and even for those who were interested in the field, there was no satisfactory introduction. I hope this book will help solve this problem, and I have aimed it at being accessible to software engineers (regardless of academic background) who wish to learn more about this area. While this book does not give a step-by-step recipe for building a TTS system, it does go significantly beyond the theory, and tries to impart a feel for the subject and to pass on some of the "folklore" that is necessary for successful development. I believe the book covers enough ground that a good software engineer should not have too much difficulty with implementation.

1.4.3 Background and specialist sections

The book contains a significant amount of background material. This is included for two reasons. Firstly, as just explained, I wanted to make the book seem complete to any reader outside the field of speech technology. I believe it is to the readers' benefit to have introductory sections on phonology or signal processing in a single book, rather than having to resort to the alternative of pointing the reader to other works.

There is a second reason, however, which is that I believe that the traditional approach to explaining TTS is too disjointed. Of course TTS draws upon many other disciplines, but the differences between these, to me, are often overstated. Too often, it is believed that only "an engineer" (that is someone who has a degree in engineering) can understand the signal processing, only "a linguist" (again, a degree in linguistics) can understand the phonetics and so on. I believe that this view is very unhelpful; it is ridiculous to believe that someone with the ability to master signal processing isn't able to understand phonetics and vice versa. I have attempted to bridge these gaps by providing a significant amount of background material, but in doing so have tried to make this firstly genuinely accessible and secondly focused on the area of text-to-speech. I have therefore covered topics found in introductory texts in engineering and linguistics, but tried to do so in a novel way that makes the subject matter more accessible to readers with different backgrounds. It is difficult to judge potential readers' exposure to the fundamentals of probability as this is now taught quite extensively. For this reason, I have assumed a knowledge of this in the body of the book, and have included a reference section on this topic in the appendix.

The book is written in English and mostly uses English examples. I decided to write the book and focus on one language rather than make constant references to the specific techniques or variations that would be required for every language of interest to us. Many newcomers (and indeed many in the field who don't subscribe to the data-driven view) believe that the differences between languages are quite substantial and that what works for English is unlikely to work for French, Finnish, or Chinese. While languages obviously do differ, in today's modern synthesisers these differences can nearly all be modelled by training and using appropriate data; the same core engine suffices in all cases. Hence concentrating on English does not mean that we are building a system that will work on only one language.

2 Communication and language

Before delving into the details of how to perform text-to-speech conversion, we will first examine some of the fundamentals of communication in general. This chapter looks at the various ways in which people communicate and how communication varies depending on the situation and the means which are used. From this we can develop a general model, which will then help us specify the text-to-speech problem more exactly in the following chapter.

2.1 Types of communication

We experience the world though our senses and we can think of this as a process of gaining **information**. We share this ability with most other animals: if an animal hears running water it can infer that there is a stream nearby; if it sees a ripe fruit it can infer that there is food available. This ability to extract information from the world via the senses is a great advantage in the survival of any species. Animals can, however, cause information to be created: many animals make noises, such as barks or roars, or gestures such as flapping or head nodding, which are intended to be interpreted by other animals. We call the process of *deliberate creation* of information with the *intention that it be interpreted* **communication**.

The prerequisites for communication are an ability to create information in one being, an ability to transmit this information and an ability to perceive the created information by another being. All three of these prerequisites strongly influence the nature of communication; for example, animals that live in darkness or are blind would be unlikely to use a visual system. Despite these restrictions, it is clear that there are still many possible ways to make use of the possibilities of creation, medium and perception to communicate. We will now examine the three fundamental communication techniques that form the basis for human communication.

2.1.1 Affective communication

The most basic and common type of communication is **affective** communication, where we express a primary emotional state with external means. A good example of this is the expression of pain, where we might let out a yell or cry upon hurting ourselves. A defining characteristic of this type of communication is that the intensity of the external

Figure 2.1 A (British) road sign, indicating a slippery road and high chance of skidding.

form is clearly a function of the the intensity of feeling; the more intense the pain the louder the yell. Other primary mental states such as happiness, anger and sadness can be expressed in this way. This type of communication is one that is common to most higher animals. While the *ability* to express these affective states is common among animals, the precise means by which these are expressed is not universal or always obvious. A high-pitched yell, squeal or cry often means pain, but it is by no means obvious that a dog's wagging tail and a cat's purring are expressions of happiness.

2.1.2 Iconic communication

Though fundamental and powerful, affective communication is severely limited in the range of things it can be used to express. Happiness can readily be conveyed, but other simple mental states such as hunger or tiredness are significantly more difficult to convey. To express more complex messages, we can make use of a second communication mechanism known as **iconic** communication. An iconic system is one where the created **form** of the communication somehow resembles the intended **meaning**. We use the term "form" here in a technical sense that allows us to discuss the common properties of communication systems: in acoustic communication, "form" can be thought of as a type of sound; in visual communication form might be types of hand signals, facial expressions and so on. For example, it is common to communicate tiredness iconically by the "sleeping gesture", whereby someone closes her eyes, puts her hands together and places her head sideways on her hands. The person isn't really asleep – she is using a gesture that (crudely) mimics sleep to indicate tiredness. Another good example of iconic communication is the road sign shown in Figure 2.1. In this case, the form is the diagram, and the meaning is *slippery road*, and the fact that the form visually resembles what can happen when a road is slippery means that this communication is iconic. Note that, just as with the sleep example, the form isn't a particularly accurate picture of a car, road, skid or so on; the idea is to communicate the essence of the meaning and little else.

Anecdotally, we sometimes think of pre-human communication also working like this, and, in such a system, the presence of a sabre-tooth tiger might be communicated by imitating its growl, or by acting like a tiger and so on. Such systems have a certain advantage of transparency, in that, if the intended recipient does not know what you mean, a certain amount of mimicry or acting may get the point across. In essence, this is just like the road-sign example, in that the form and meaning have a certain direct connection. When travelling in a country where we don't speak the language, we often resort to miming some action with the hope that the meaning will be conveyed.

Iconic systems have several drawbacks, though. One is that, while it may be easy to imitate a tiger, it is less easy to imitate more abstract notions such as "nervous" or "idea". More importantly though, iconic systems can suffer from a lack of precision: when a first caveman imitates the tiger, a second caveman might not get this reference exactly – he might be sitting there thinking "well, it could be a tiger, or perhaps a lion, or maybe a large dog". By the time the first caveman has mimed his way through this and the action of "creeping up behind", both have probably departed to the great cave in the sky. While useful and important, iconic communication clearly has its limits.

2.1.3 Symbolic communication

In contrast to iconic and affective communication, we also have **symbolic** communication in which we give up the idea that the form must indicate the meaning. Rather we use a series of correspondences between form and meaning, in which the relationship is not direct. In symbolic systems, a tiger might be indicated by waving the left hand, a lion by waving the right. There is no reason why these forms should indicate what they do; it is merely a matter of convention. The advantage is that it is easier to devise a system where the forms are clear and distinct from one another – less confusion will arise. The disadvantage is that the fact that left-arm-wave means tiger, whereas right-arm-wave means lion, has to be *learned*; if you don't know, seeing the movement in isolation won't give a clue to the meaning. Despite the disadvantage of needing to learn, using conventions rather than icons can be hugely advantageous in that the conventions can be relatively brief; one noise or gesture may represent the tiger and this need not be acted out carefully each time. This brevity and clarity of form leads to a swiftness and precision seldom possible with iconic communication.

Once a convention-based communication system is used, it soon becomes clear that it is the notion of **contrast** in form that is the key to success. To put it another way, once the form no longer needs to resemble the meaning, the communication system gains benefit from making the forms as distinct from one another as possible. To show this point, consider the following experiment.

Eight subjects were grouped into pairs, and each pair was asked, in isolation, to design a communication system based on colour cards. The premise was that the subjects were in a noisy pub, with one of the pair at the bar while the other was sitting down at a table. We said that there were four basic concepts to communicate: "I would like a drink of water", "I would like some food", "I would like a beer" and "I would like to listen to some music". Each pair was given a set of 100 differently coloured cards arranged across the spectrum, and asked to pick one card for each concept and memorise which colour was associated with which concept. In a simple test, it was clear that the pairs could remember the colours and their meanings, and could effectively communicate these simple concepts with this system. The real purpose of the experiment was different though. While three of the groups chose a blue colour for the concept "I want a drink of water", for the other categories there was no commonality in the choice. So "I want a beer" had two pairs choose green, one yellow and one black. Interestingly, most gave some justification as to why each colour was chosen (green means "yes, let's have another") and so on; but it is clear that this justification was fairly superficial in that each group chose quite different colours. This demonstrates one of the main principles of a symbolic communication system, namely that the relationship between the form of the symbol (in this case the colour) and its meaning is in general **arbitrary**. These form/meaning pairs must be learned, but, as we shall see, in many cases the "cost" of learning the pairing to an individual is less than the "cost" of potential confusion and imprecision that can arise from having the form resemble the meaning. This simple study demonstrates another vital point. In each case, the pairs chose quite different colours; so group A chose a blue, a red, a yellow and a green, while group B chose a red, a yellow, a light blue and a black. In no case did a group choose, say, a green, a light blue, a medium blue and a dark blue for instance. In other words, the groups chose a set of colours that had clear contrast with each other. In a way this makes obvious sense; picking colours that are significantly different from one another lessens the possibility that they will be confused, but it does demonstrate a crucial point in symbolic systems, namely that the systems are based on the notion of contrast between forms rather than the absolute "value" of each form.

2.1.4 Combinations of symbols

This simple communication system works well because we have only four concepts we wish to convey. What happens when we need to communicate more concepts? We can see that the simple colour contrast will eventually reach its limits. If, say, 20 concepts are required, it soon becomes a difficult task to associate each with just one colour. Setting aside the issue of whether the individuals can remember the colours (let us assume they have a crib sheet to remind them), a major problem arises because, as more concepts get added, a new colour is associated with each one, and as the range of colours is fixed, each new added concept increases the likelihood that it will get confused with another concept that has been assigned a similar colour. While it may be possible for an individual to distinguish 20 colours, asking them to do so for 1000 colours seems unlikely to meet with success. We therefore need a separate solution. The obvious thing to do is use **combinations** of colour cards. For instance, we could associate the colour black with the concept *not*, pink with *large*, and then use these and one of the original colours to indicate a much larger range of concepts (so holding a black card, pink card and blue card would say "I don't want a large drink of water"). Note that this is different from holding a single card that is a mixed combination of black, pink and blue (which might be a murky brown). The reason why the combination system works is that each colour card on its own is clear, distinct and unconfusable, but, by combining them, we can generate

Figure 2.2 With three symbols in isolation only three meanings are possible. When we combine the symbols in sequences the number of meanings exponentially increases, while the ability to perceive the symbols is unaffected.

a large number of concepts from a small number of form/meaning pairs (Figure 2.2). The power of combination is considerable; in the limit, the number of complex forms exponentially multiplies, such that, if we have N simple forms and M instances of them, we have a total of N^M complex forms; so 10 forms combined in 10 different ways would produce over 10 billion different complex forms.

We call communication systems that use a fixed number of simple forms that can be combined into a much larger number of forms a **symbolic language**. Here we use the term "language" in both a very precise and a general sense. It is precise, in that it we use it only to mean this principle of combination, but general in the sense that all human languages make use of this principle in this sense, as do other systems, such as written numbers (we can create any number from a combination of the 10 digits), mathematics and even email addresses. Human language often goes under the description of **natural language** and we shall use that term here to distinguish it from the principle of languages in general. One of the main distinctive properties of natural language is the sheer scale of things that it can express. While we can talk of other animals having sophisticated communication systems (bee dancing etc.), none has anything like the scale of natural language, where a person might know, say, 10 000 words, and be able to combine these into trillions and trillions of sentences.

2.1.5 Meaning, form and signal

In our discussion so far, we have been somewhat vague and a little informal about what we really mean by "meaning" and "form". The study of meaning (often called semantics) is notoriously difficult, and has been the subject of debate since the earliest philosophy. It is an inherently tricky subject, because firstly it requires an understanding of the world (what the world contains, how the things in the world relate to one another and so on) and secondly, from a cognitive perspective, it is exceedingly difficult to peer inside someone's brain and find out exactly how they represent the concept of "warm" or "drink" and so on. As we shall see, developing a precise model of semantics does not seem to be necessary for text-to-speech, so we shall largely bypass this topic in a formal sense.

Our notion of "form" also needs some elaboration. A proper understanding of this part of communication *is* vital for our purposes, but thankfully we can rely on an established framework to help us. The main issue when discussing form is to distinguish the general nature of the form from any specific instantiation it may have. Going back to our simple colour example; it is important to realise that when one individual holds up a red card, it isn't the red card itself that is the important point, rather that red was used rather than the other choices of green, blue and yellow. So, for example, a different red card could be used successfully, and probably any red card would do, since it is not the particular red card itself which is the form, but rather that the card has the property "red" and it is this property and its contrasting values that is the basis of the communication system. In some accounts, the term **message** is used instead of form.

We use the term **signal** to refer to the physical manifestation of the form. Here signal is again used in a specific technical sense: it can be thought of as the thing which is stored or the thing which is transmitted during communication. The idea of "red" (and the idea that it contrasts with other colours) is the form; the physical light waves that travel are the signal. The above example also demonstrates that the same form (red) can have a number of quite different signals, each dependent on the actual mode of transmission.

It is important to see that the relationships between meaning and form and between form and signal are quite separate. For instance, in our example we have associated the colours red, yellow, blue and green each with a concept. Imagine now the sender holds up an orange card. Most probably the receiver will see this signal and correctly interpret the physical light waves into a form "orange". After this, however, she is stuck because she can't determine what concept is matched with this orange form. The signal/form relationship has worked well, but it has not been possible to determine the correct concept. This situation is similar to that when we read a new word – while we are quite capable of using our eyes to read the physical pattern from the page and create an idea of the letters in our head, if we don't know the form/meaning mapping of the word, we cannot understand it.

Finally, we use the term **channel** or **medium** to refer to the means by which the message is converted and transmitted as a signal. In this book, we deal mainly with the **spoken channel** and the **written channel**.

2.2 Human communication

The above examination of communication in the abstract has of course been leading us to an explanation of how human communication operates. Taking spoken communication first, let us make a primary distinction between the two main aspects of this, which we shall call the **verbal** component and the **prosodic** component. We use the term "verbal" to mean the part of human communication to do with **words**. In some books, the terms "verbal" and "language" are synonymous, but because "language" has so many other, potentially confusable meanings, we will stick with the term "verbal" or "verbal component" throughout. The verbal component is a symbolic system in that we have a discrete set of words that can be arranged in sequences to form a very large number of sentences.

In contrast, the prosodic component (often just called **prosody**) is different in that it is not primarily symbolic. We use prosody for many purposes: to express emotion or surprise, to emphasise a word, to indicate where a sentence ends and so on. It is fundamentally different in that it is not purely discrete and does not generally have units that combine in sequences and thus has a much more limited range of meanings that can be expressed. That said, prosody should not be taken as a poor relative of the verbal component; if we say "it's wonderful" with words, but use a tone of voice that indicates disgust, the listener will guess that we are being sarcastic. If the prosodic and verbal components disagree, the prosodic information wins. To a certain extent, these systems can be studied separately; in fact nearly all work in the field of linguistics is interested solely in the verbal component. The two do interact, though; first in the obvious sense that they share the same signal (speech), but also in more complex ways in that prosody can be used to emphasise one particular word in a sentence.

2.2.1 Verbal communication

The symbolic model of the verbal component states that the principal unit of form is the **phoneme**. Phonemes operate in a contrastive system of sounds, in which a fixed number of phonemes can combine to a much larger number of **words** (a vocabulary of more than 10 000 words is common, but many millions of words are possible). Next, these words are taken as the basic units, and they further combine into **sentences**. Verbal language is therefore seen as a combination of two systems: one that makes words from phonemes and another that makes sentences from words. Since the word/sentence system has quite a large set of basic units to start with, the number of combined units is vast. The two systems have different properties with regard to meaning. In the first system, the basic building blocks, the phonemes, have no direct (arbitrary, iconic or other) meaning; they are simply building blocks. These are combined to form words, and these do have meaning. The relation between a word's meaning and the phonemes which comprise it is completely arbitrary and so the meaning and pronunciation of each word has to be learned. The second system operates differently, in that the basic building blocks (the words) do have inherent meaning, and the meaning of the sentence is related to the meaning of the components. Exactly how the sentence meaning relates to the word meaning is still an issue of considerable debate; on the one hand there is considerable evidence that it is a function, in the sense that if we have JOHN KICKED MARY[1] we can substitute different people for JOHN and MARY (e.g. ANN KICKED GEORGE), or different verbs (e.g. JOHN KISSED MARY) and predict the sentence meaning from the parts. On the other hand, there is also considerable evidence that sentence meaning is not a simple function, in that JOHN KICKED THE BUCKET means John has died, and has nothing to do with kicking or buckets.

---

[1] From here we will use a small-caps typeface when we refer to form/message such as JOHN and a courier typeface for the written signal, e.g. john.

From this basic model, we can now explore some of the fundamental features of language. In doing so, we follow Hockett [214], who stated the principal defining properties of verbal language. Among these are the following.

Arbitrariness
Phonemes do not have inherent meaning but words do; a word is in effect a form/meaning pair in which the correspondence between the meaning and the form (i.e. the sequence of phonemes) is arbitrary. Because of this, there is no way to guess the meaning of a word if it is not learned; so, as a reader, you won't know the meaning of the word PANDICULATION unless you are informed of its meaning. From this, it follows that words of similar form can have completely different meanings; so BIT and PIT mean quite different things despite having only a slight sound difference, whereas DOG and WOLF are similar in meaning, but share no sounds.

Duality
The duality principle states that verbal language is not one system but two; the first system uses a small inventory of forms, phonemes, which sequentially combine to form a much larger inventory of words. The phonemes don't carry meaning but the words do, and this association is arbitrary. The second system combines the large number of words into an effectively limitless number of sentences.

Productiveness
The productiveness property means that there is an effectively unlimited number of things that one can say in a verbal language. It is not the case that we have a fixed number of messages and in communication simply choose one from this selection; rather many of the sentences we produce are unique, that is, the particular combinations of words we use have never been used before. Of course many of the sentences we say are conventionalised ("How do you do?" etc.), but most are unique. It might seem questionable whether this is really the case, since most people don't feel they are being particularly "original" when they speak. However, a simple examination of the combinatorics of combining even a modest vocabulary of 10 000 words shows that a vast number of possibilities exists (certainly larger than the number of atoms in the Universe).

Discreteness
Both the basic types of linguistic unit, the phoneme and the word, are **discrete**. Phonemes are distinct from one another and form a finite and fixed set. The set of words is less fixed (we can create new words), but nevertheless all words are distinct from one another and are countable. What is remarkable about the discreteness property of language is that the primary signal of language, speech, is not discrete and hence there is a mismatch between form and signal. Speech signals are continuous functions of pressure over time (see Chapter 10), so speaking is a process of converting a discrete phoneme representation into a continuous acoustic one. This process is reversed during listening, where we recover a discrete representation from the continuous signal. As we shall see, this is a remarkably

complex process, which is one of the main reasons why automatic speech synthesis and speech recognition are difficult.

2.2.2 Linguistic levels

Another common way to look at verbal language is as a series of so-called linguistic **levels**. So far we have talked about the duality of form such that we have phonemes, which combine to form words, which combine to form sentences. In our model, sentences, words and phonemes are considered the primary units of form, and suffice to describe the basic properties of verbal language. It is, however, extremely useful to make use of a number of secondary units, which we shall now introduce. **Morphemes** are units used to describe word formation or the internal structure of words. This describes for instance the connection between the three English words BIG, BIGGER, BIGGEST; and the study of this is called **morphology**. Next we have **syntax**, often defined as the study of the patterns of words within a sentence. Syntax describes the process of how nouns, verbs, prepositions and so on interact, and why some sequences of words occur but not others. The fundamental units of syntax are the **syntactic phrase** and sentence. With these levels, we can for a given sentence describe a **syntactic hierarchy** whereby a sentence comprises a number of phrases, each of which comprises a number of words, which then comprise morphemes, which then comprise phonemes.

We can also consider the **phonological hierarchy**, which is a different, parallel hierarchy that focuses on the sound patterns in a sentence. In this, we have the additional units of **syllables**, which are structural sound units used to group phonemes together, and **phonological phrases**, another type of structural sound unit that groups words together within the sentence. Representations using these units can also be expressed in a hierarchy of sentences, prosodic phrases, words, syllables and phonemes.

The difference between the two hierarchies is that the phonological one concentrates on the sound patterns alone, and is not concerned with any aspect of meaning, whereas the syntactic one ignores the sound patterns, and concentrates on the grammatical and semantic relationships between the units.

Next we consider levels between the form and the signal, which in spoken communication is an acoustic waveform. Here we have **speech acoustics**, which involves working with the acoustic speech sound signal. Next we have **phonetics**, which is often described as the study of the physical speech sounds in a language. Phonetics is often studied independently of other linguistic areas, since it is seen as part of the system which links form and signal rather than as part of the meaning/form verbal linguistic system. **Phonology** is the study of sound patterns as they relate to linguistic use; in contrast to phonetics, this *is* considered part of the core linguistic system because phonology describes the form in the meaning/form system. That said, phonology and phonetics are closely related and phonology is sometimes described as the organisational system of phonetic units.

Semantics is sometimes given a broad definition as the study of "meaning" in the general sense, but is sometimes given a much narrower definition concerned only with certain aspects of meaning, such as analytic truth, reference and logic. Finally, **pragmatics** concerns a number of related areas including the aspects of meaning not covered by semantics, the interaction of language, a speaker's knowledge and the environment and the use of language in conversations and discourses. Sometimes pragmatics is described as being "parallel" to semantics/syntax/phonology etc. since it is the study of how language is used.

It is important to realise that these terms are used to refer both to the linguistic level and to the area of study. In traditional linguistics these terms are often taken as absolute levels that exist in language, such that one sometimes sees "turf wars" between phoneticians and phonologists about who best should describe a particular phenomenon. From our engineering perspective, we will find it more convenient to adopt these terms, but not worry too much about the division between them; rather it is best to view these levels as salient points in a continuum stretching from pragmatics to acoustics.

2.2.3 Affective prosody

We now turn to a different component of human communication. Broadly speaking, we can state that **prosody** has two main communicative purposes, termed **affective** and **augmentative**. The affective use of prosody can be used to convey a variety of meanings: perhaps the most straightforward case is when it is used to indicate a primary emotion such as anger or pain. Prosody is particularly good at this; when we hurt ourselves and want to tell someone we usually let out a yell rather than say "look, I've cut off my fingers" in a calm measured tone. The use of prosody to convey primary emotion is also largely universal; when listening to someone who speaks a different language we find they use more or less the same form to indicate the same meanings. In fact this extends to animals too; we are often in little doubt as to what an animal snarl or growl is trying to convey. Beyond primary emotions we can consider a whole range of meanings that are expressible via prosody. These include secondary emotions such as angst or distrust, speech acts such as question, statement and imperative; modes of expression such as sarcasm or hesitancy and so on.

Many of these phenomena are language-specific (in contrast to primary emotion), and have a conventional or arbitrary character such that a speaker of that language has to learn the appropriate prosodic pattern in the same way as they learn the form/meaning correspondences for words in verbal language. In fact we can see these effects as lying on a scale with universal systems lying at one end and conventional arbitrary systems lying at the other. In many languages, a rising pitch pattern at the end of a sentence indicates a question; but this is not universally so. Ladd [269] shows that the pattern is reversed in Hungarian such that certain questions are indicated by a falling pitch.

One of the central debates in the field of prosodic linguistics is just how far this scale extends. Crudely speaking, the argument concerns whether prosody is *mainly* a universal, non-arbitrary system, with a few strange effects such as Hungarian question intonation; or whether it is in fact much more like verbal language in that it is *mainly* arbitrary, and has rules of syntax and so on.

2.2.4 Augmentative prosody

The second aspect of prosody is called **augmentative prosody**. The basic purpose of this is to *augment* the verbal component. Taking a simple example first, we see that many sentences are highly confusable with respect to syntactic structure. In sentences such as

(1) BILL DOESN'T DRINK BECAUSE HE'S UNHAPPY

we see that there are two completely different interpretations, which depend on which syntactic structure is used. The speaker can choose to insert a pause, and if she does so, she will bias the listener's choice as to which interpretation is intended and hence facilitate understanding. In addition to helping with phrasing structure, this component of prosody is also used to draw attention to words by emphasising them.

This use of prosody is called augmentative because it augments the verbal component and therefore helps the listener to find the intended form and meaning from the signal. In contrast to affective prosody, the use of augmentative prosody here does not add any new meaning or extra aspect to the existing meaning, it is simply there to help find the verbal content. One of the most interesting aspects of augmentative prosody is that speakers show considerable variation in how and when they use it. Whereas one speaker may pause at a certain point, another speaker will speak straight on, and we even find that the same speaker will quite happily use a pause on one occasion but not on another. We can explain this variation by considering the range of conversational situations where augmentative prosody might be used.

In general, when a speaker generates a message that is potentially ambiguous, it is often entirely obvious from the form of words used, the background knowledge and the situation what interpretation is intended, and in such cases no or little augmentative prosody is used. In other cases, however, we may find that the speaker suspects the listener may have difficulty in interpretation, so we find that here the speaker does indeed make use of prosody. In other words, prosody is usually used only when it is needed. Hence, if we consider a sentence in isolation, it is nearly impossible to predict what prosody a speaker will use. A consequence of this is that it is clearly possible to use more or use less prosody for any given sentence. We can think of this as a scale of prosodic intensity, and this applies to both affective and augmentative prosody. In many cases, speakers utter sentences at the lower end of this scale, in which case we say they are speaking with **null prosody** or **neutral prosody** to indicate that there is no true prosody being used.

2.3 Communication processes

In this section, we will build a general model of communication that will serve as a framework in which to analyse the various communicative processes involved in spoken and written language. For illustration, let us consider a dialogue between two participants, A and B. In a normal conversation, A and B take it in turns to speak, and these are known as **dialogue turns**. Suppose that A decides to say something to B. To do this, A firstly **generates** a linguistic form, which we shall call the **message**. He then **encodes** this as a

2.3 Communication processes

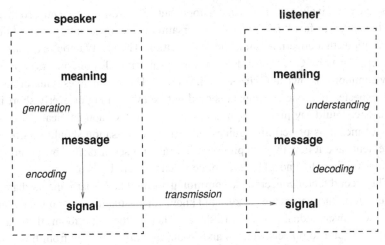

Figure 2.3 Processes involved in communication between two speakers.

speech signal, which is a form that can be **transmitted** to B (let's assume that they are in the same room). On receiving the signal, B first **decodes** the signal and creates her own message as linguistic form. Next she attempts to **understand** this message and so find the meaning.

In line with our model of language explained before, we have three types of representation: the meaning, the form and the signal. In communicating, A and B use four separate systems to move from one to the other; these are

> **generation:** meaning to form
> **encoding:** form to signal
> **decoding:** signal to form
> **understanding:** form to meaning

This model works equally well for written language: A converts his thought into a message, then encodes this as writing and sends this to B (in this case the "sending" can be quite general, for example an author writing a novel and a reader subsequently picking it up). Once received, B attempts to decode the writing into a message and then finally understand this message. A diagram of this is shown in Figure 2.3.

2.3.1 Communication factors

We start our discussion by simply asking, why do the participants say the things they do? While "free-will" plays a part in this, it is a mistake to think that "free-will" is the sole governing factor in how conversations happen. In fact there are guiding principles involved, which we shall now consider.

The first concerns the understood conventions of communication. In human conversation, participants don't just say everything on their mind; to do so would swamp the conversation. Rather, they say only what they need to in order to get their point across. We can state this as a case of **effectiveness** and **efficiency**; we wish to communicate our message effectively, so that it is understood, but also to do so efficiently, so that we don't waste energy. In fact it is clear that conversation works by adopting a set of understood conventions that govern or guide the conversation. The most famous exposition of these principles was by Grice, who defined four main principles of conversation, which are now commonly known as **Grice's maxims** [184]. These principles state for instance that it is expected that one participant should not be overly long or overly short in what is said; one should say just the right amount to make the thought clear. Grice's maxims are not meant as prescriptive guides as to how conversation should progress, or even statements about how they do progress, but rather a set of tacit conventions, which we normally adhere to, and which we notice when they are broken.

The second factor concerns the **common ground** that the participants share. This is a general term that covers the background knowledge, situation and mutual understanding that the conversational participants share. When in the same room, they can point to objects or refer to objects, actions and so on, in a different way from how such things can be referred to if they are more distant. If one says "have a seat" and there is a chair in the room, this means something straightforward. The situation can significantly alter the intended meaning of an utterance. Take for instance the sentence "my car has broken down". Now this can be taken as a simple statement of fact, but if someone says this upon entering a mechanic's workshop, the implication is that the person's car has broken down and he'd like the mechanic to fix it. If, by contrast, this is the first thing someone says when arriving for dinner, the intention might be to apologise for being late because an unforeseen travel delay occurred. To take another example, if two experts in the same research field are talking, they can talk assuming that the other has a detailed knowledge of the subject matter. This is different from when an expert talks to a lay person.

Finally, the knowledge of the **channel** helps form the message. Participants don't generate messages and then decide whether to speak them or write them down; they know what medium is going to be used before they start. This allows the participants to exploit the strengths of each form of communication; in spoken language extensive use may be made of prosody; in written language the author may take the liberty of explaining something in a complicated fashion knowing that the reader can go back and read this several times.

2.3.2 Generation

In our model, the process of converting a meaning into a message is called **message generation** (elsewhere the term **language generation** is often used, but this is too potentially confusing for our purposes). To demonstrate how generation works, let us use the above-mentioned communication factors to consider a situation where a participant in a conversation wants to know the score in a particular football game. We should realise that the participant has considerable **choice** in converting this thought into a message. For example, the participant can make a statement such as "I want to know the score" or ask a question such as "what is the score?"; the participant has choices over mood, and may choose to say "can you tell me the score?", may choose between active and passive, "who has scored?" versus "have any goals been scored?". The participant has choices over vocabulary, may well qualify the message, may decide on a casual, formal, polite, demanding tone and so on. In addition the participant can choose to emphasise words, pick different intonational tunes, perhaps make a pointing gesture and so on. The important point is that the participant has a wide range of options in turning the thought into the message. The result is the message, and a perfectly transparent communication system would send this to the receiver with all this detail and with no chance of error.

It is sometimes assumed that messages themselves convey or represent the communicated meaning. While in certain cases this may be possible, we have just demonstrated that, in most conversations, to extract (understand) the correct meaning one must know the situation and there must be shared knowledge.

2.3.3 Encoding

Encoding is the process of creating a signal from a message. When dealing with speech, we talk of **speech encoding**; and when dealing with writing, we talk of **writing encoding**. Speech encoding by computer is more commonly known as **speech synthesis**, which is of course the topic of this book.

The most significant aspect of speech encoding is that the natures of the two representations, the message and the speech signal, are dramatically different. A word (e.g. HELLO) may be composed of four phonemes, /h eh l ow/, but the speech signal is a continuously varying acoustic waveform, with no discrete components or even four easily distinguishable parts. If one considers the message, we can store each phoneme in about 4–5 bits, so we would need say 2 bytes to store /h eh l ow/. By contrast, a reasonably high-fidelity speech waveform[2] would need about 20 000 bytes, assuming that the word took 0.6 seconds to speak. Where has all this extra content come from, what is it for, and how should we view it?

Of course, the speech signal is significantly more **redundant** than the message: by various compression techniques it is possible to reduce its size by a factor of 10 without noticeably affecting it. Even so, this still leaves about three orders of magnitude difference. The explanation for this is that the speech waveform contains a substantial amount of information that is not directly related to the message. First we have the fact that not all speakers sound the same. This may be a fairly obvious thing to say, but consider that all English speakers can say exactly the same message, but they all do so in subtly different ways. Some of these effects are not controllable by the speaker and are determined by purely physiological means. Some are due to prosody, where, for instance, the speaker can have different moods, can affect the signal by smiling when talking, can indicate different emotional states and so on. This fact is hugely significant for speech synthesis, since it demonstrates that encoding the message is really only part of the problem; the encoding system must do several other tasks too if the speech signal is to sound genuinely human-like to the receiver.

---

[2] Say 16 bits dynamic range and 16 000 Hz sampling rate. (See Chapter 10 for an explanation.)

2.3.4 Decoding

Decoding is the opposite of encoding; its job is to take a signal and create a message. **Speech decoding** is the task of creating a message from the signal; the technology that does this is most commonly known as **speech recognition**. The challenges for speech-decoding (speech-recognition) systems are just the converse of those for speech encoding; from a continuous signal, which contains enormous amounts of information, the system must extract the small amount of content that is the message. Speech recognition is difficult because of the problem of **ambiguity**. For each part of a message, a word or a phoneme, if one studies the speech patterns produced from that, one sees that there is an enormous amount of variability. This is partly caused by the factors we described above, namely speaker physiology, emotional state and so on, and partly arises from the encoding process itself, where a phoneme's characteristics are heavily influenced by the other phonemes around it. This variability wouldn't be a problem *per se*, but it turns out that the patterns for each phoneme overlap with one another, such that, given a section of speech, it is very hard to determine which phoneme it belonged to. These aspects make speech recognition a difficult problem.

Text decoding, also called **text analysis**, is a similar task in which we have to decode the written signal to create a message. If the writing is on paper, we face many of the same problems as with speech recognition; namely ambiguity that arises from variability, and this is particularly true if we are dealing with hand writing. Computerised writing is significantly easier since we are dealing with a much cleaner representation in which the representation is discrete and each character is uniquely identified.

For many types of message, the text-decoding problem is fairly trivial. For instance, for simple text such as `the man walked down the street`, it should be possible for a simple system to take this and generate the correct message all the time. Significant problems do arise from a number of sources. First, ambiguity exists even in computerised written signals. One type is **homograph** ambiguity, where two different words have the same sequence of characters (e.g. `polish` can mean "from Poland" or "to clean", `console` can mean "to comfort" or "a part of a machine").

One final point should be made when comparing text and speech decoding. In speech recognition, the problem is understood as being difficult, so error rates of, say, 10% are often seen as good system performance on difficult tasks. As we have seen, text decoding is fundamentally easier, but expectations are higher. So, while for some tasks the vast majority of words may be decoded successfully, the expectation is also high and so error rates of 0.1% may be seen as too high.

2.3.5 Understanding

Understanding is the process of creating a thought from the message. Understanding can be thought of as the reverse of generation and also bears similarity to decoding in the sense that we are trying to recover a thought from a representation (the message), which also includes information generated from other sources such as the background, situation and so on.

In understanding, we also have an ambiguity problem, but this is quite different from the ambiguity problem in decoding. For example, it is perfectly possible when hearing a speech signal to decode it into the correct message, but then fail to understand it because of ambiguity. For instance, if someone says "did you see the game last night?", it is quite possible to identify all the words correctly, but fail to determine which game is being referred to.

2.4 Discussion

This chapter has introduced a model of communication and language that will serve as the basis for the next chapter, which looks at text-to-speech conversion in detail, and the subsequent chapters, concerning the various problems that occur and the techniques that have been developed to solve them. We should stress that what we have presented here is very much a *model*; that is a useful engineering framework in which to formulate, test and design text-to-speech systems.

In building this model, we have drawn from a number of fields, including semiotics, information theory, psychology, linguistics and language evolution. These fields have overlapping spheres of subject matter, but unfortunately are often studied in isolation such that none really gives as complete and detailed a model of communication as we might wish for. Each of these separate disciplines is a vast field in itself, so we have only touched on each area in the briefest possible way.

Considering semiotics first, we should look at the works of Ferdinand de Saussure, who is widely acknowledged to have invented the field of semiotics and is considered by many as the first modern linguist. Saussure was the first to emphasise the significance of the arbitrary nature of form and meaning in communication and proposed that the study of the relationship between the two should be the central topic in linguistics [391], [392]. The philosopher Charles Sanders Pierce also pioneered early work in semiotics and semantics, and gave us the framework of iconic and symbolic semiotics which is used here [350]. The second main area we draw from is communication theory or information theory, and from this we have the principles of encoding, decoding, channels, messages and signals. The pioneering work in this field was produced by Claude Shannon [400], who was originally interested in the problem of accurate transmission of information along telephone wires (he worked for the telephone company AT&T). It soon became clear that the implications of this work were much broader (Weaver [401]) and it has been widely adopted as an approach to the analysis of signals arising from linguistic sources, such that this **information-theoretic** approach is now dominant in speech recognition and many other fields. Saussure did not in general deal much with signals, and Shannon and Weaver did not deal much with meaning, but their work clearly intersects regarding the issue of form and message (the term form is used in linguistics and semiotics and message in information theory, but they both refer to the same thing).

The field of linguistics is often described as the science of language, but in the main this field focuses very much on just the meaning/form verbal part of language, as opposed to the entire process which includes participants, conversations, signals and so on. Linguistics generally approaches language without reference to use; the idea is that a language is an entity and it can be studied independently of any purpose to which it is put.[3] This approach is part tradition and part pragmatic, since linguistics can trace its lineage back to the philosophical musings of ancient Greek philosophers. In addition there is the practical justification that by isolating effects it makes them easier to study. In modern times, the field of linguistics has been dominated by Noam Chomsky [88], [89], [90], who in addition to many other innovations has championed the idea that the primary function of language is not communication at all, but rather that it forms the basis of the human mind; it evolved for this reason and its communicative purpose was a useful by-product. Despite this mismatch between linguistics and our more practical needs, this field does provide the basics of a significant chunk of the overall picture of human communication, including the models of phonetics, phonology, words, syntax and so on that are central to our problem. Many linguistics text books tend to be heavily skewed towards one of the many incompatible theories within the discipline, so care needs to be taken when reading any book on the subject. A good introductory book that explains the "hard issues", is accessible to those with a scientific background and doesn't follow any particular theory is Lyons [290].

The study of human communication in terms of participants, conversations, background knowledge and so on is often performed outside the field of mainstream linguistics by psychologists and philosophers. Good books on human conversation and language use include Clark [95], Clark and Clark [96], Searle [395], Grice [184] and Power [361].

2.5 Summary

Communication
- A semiotic system has three levels of representation; meaning, form and signal.
- In affective communication, the meaning, form and signal are continuous and simply related.
- In iconic communication, the form resembles the meaning in some way. Iconic systems can be either discrete or continuous.
- In symbolic communication, the form does not resemble the meaning and the correspondences between these have to be learned.
- Symbolic systems are discrete, and this and the arbitrary meaning/form relationship mean that communication of complex meanings can occur more effectively than with the other types.
- Human language operates as a discrete symbolic system.

---

[3] In fact, in surveying more than 20 introductory text books in linguistics found in the Cambridge University book shop, I did not find even one that mentioned that language is used for communication!

Components of language
- Human communication can be divided into two main components.
  The verbal component describes the system of phonemes, words and sentences. It is a discrete symbolic system in which a finite number of units can be arranged to generate an enormous number of messages.
  The prosodic component is by contrast mostly continuous, and in general expresses more-basic meanings in a more-direct way.
- Affective prosody is when a speaker uses prosody to express some emotion, speech act or other information that is conveyed sentence by sentence.
- Augmentative prosody is used to disambiguate and reinforce the verbal component.
- Both types of prosody are optional, in that speakers can choose to use no prosody if they wish and just communicate via the verbal component. Writing in fact uses just this component.

Communication
- We can break communication into four processes in spoken communication:
  1. generation is the conversion of meaning into form by the speaker
  2. encoding is the conversion of form into signal by the speaker
  3. decoding is the conversion of signal into form by the listener
  4. understanding is the conversion of form into meaning by the listener
- The equivalent processes exist for written communication also, where we have a writer and reader instead of a speaker and listener.

3 The text-to-speech problem

We now turn to an examination of just what is involved in performing text-to-speech (TTS) synthesis. In the previous chapter, we described some of the basic properties of language, the nature of signal, form and meaning, and the four main processes of generation, encoding, decoding and understanding. We will now use this framework to explain how text-to-speech can be performed.

In TTS, the input is writing and the output speech. While it is somewhat unconventional to regard it as such, here we consider writing as a signal, in just the same way as speech. Normal reading then is a process of decoding the signal into the message, and then understanding the message to create meaning. We stress this, because too often no distinction at all is made between signal and form in writing; we hear about "the words on the page". More often an informal distinction is made in that it is admitted that real writing requires some "tidying up" to find the linguistic form, for instance by "text normalisation", which removes capital letters, spells out numbers or separates punctuation. Here we take a more structured view in that we see linguistic form as clean, abstract and unambiguous, and written form as a noisy signal that has been encoded from this form.

The process of reading aloud then is one of taking a signal of one type, writing, and converting it into a signal in another type, speech. The questions we will address in this chapter are the following.

- How do we decode the written signal?
- Does the message that we extract from the written signal contain enough information to generate the speech we require?
- When someone is reading aloud, what are the relationships among author, reader and listener?
- What do listeners want from a synthetic voice?

3.1 Speech and writing

It is clear that natural (human) language developed first using sound as its medium. While humans clearly also communicate with facial and hand gestures and so on, these systems are not able to express the same wide range of expressions as natural language. Quite *why* human language evolved using sound as its medium is not clear, perhaps it was down to an ability to make distinct patterns more easily with the mouth than, say, by waving hands. It is important to realise that this choice of sound may well have been a complete accident of evolution; today we communicate in many different ways with vision either by reading or by interpreting sign language. For most of human history, speech was the only (sophisticated) means of communication, but a second key invention, perhaps 10 000 years ago, was that of writing.

To begin with, we need to look at the nature of speech and writing in more depth. We will do this from a number of angles, looking first at the physical differences between these signals, before moving on to look at how these differences influence communication.

3.1.1 Physical nature

The speech signal is a continuous, acoustic waveform. It is created by the operation of the vocal organs in response to motor-control commands from the brain. We have only a partial understanding of how this process works, but it is clear that the coordinated movement of the vocal organs to produce speech is extremely complicated. One of the main reasons why the relationship between the speech signal and the message it encodes is so complex is that speech is continuous and the form is discrete. In addition to this, speech is highly variable with respect to the discrete message since many other factors, including prosody, speaker physiology, a speaker's regional accent and so on, all affect the signal. A single message can generate many potential signals. Finally, speech is not permanent; unless a recording device is present, once something has been spoken, it is lost for ever; and if no listener is present or a listener does not hear the speech then the message that the speech encodes will never be decoded.

Writing is fundamentally different from speech in a number of ways. Firstly writing is primarily seen as **visual**: traditionally, writing always took the form of visible marks on a page or other material, and today we are quite used to seeing writing on computer and other screens. This is not, however, a complete defining property; we consider braille a form of writing and its key benefit is of course that one does not need to be able to see to read it. It has been argued [199] that the main defining feature of writing is that it is **permanent** in a way that speech is not. Irrespective of whether the writing is created with a pen bearing ink or impressions on a page, the key point is that once written it does not disappear. We can of course destroy what we have written, but this is not the point; the idea is that with writing we can record information, either so that we can access it at a later date, or to send it over some distance (as a letter for instance).

What, then, does writing record? Writing can be used to record natural language, but in saying this we should make it clear that writing attempts to record the linguistic form of a message, rather than the actual signal. That is, if we write down what someone says, we do not attempt to record all the information in the speech, such that a reader could exactly mimic the original; rather we decode the speech into a linguistic message, and then encode that. Of course, quite often the message we wish to write has never been spoken, so when writing we generate a linguistic form from our thoughts, and then encode this as writing without ever considering speech.

We explained in Chapter 2 that spoken language had a dual structure of sentences, made of words, and words, comprised of phonemes. We have a similar dual structure in written communication, where sentences are made of words (as before), but where words are made of **graphemes**. Graphemes are in many ways analogous to phonemes, but differ in that they vary much more from language to language. In alphabetic languages like English and French, graphemes can be thought of as letters or characters (we will define these terms later); in syllabic writing like Japanese hiragana, each grapheme represents a syllable or mora, and in languages like Chinese each grapheme represents a full word or part of a word.

It should be clear that even apart from the visual and permanency aspects writing signals and speech signals are very different in nature. The most important difference is that writing is discrete and "close to the message", while speaking is continuous, more complex, and does not encode the message in any trivial way. The complexities of the speech encoding and decoding arise because we are using an apparatus (the vocal organs) for a purpose quite different from their original one. In fact, one can state that it is a marvel that speech can be used at all for communication messages. In contrast, writing is a cultural invention, and was designed explicitly and solely for the purposes of communicating form. If we consider it a "good" invention, it should not be surprising that the encoding mechanism between linguistic form and writing is much simpler. This is not to say that the reading and writing processes are trivial; it takes considerable skill for the brain to interpret the shapes of letters and decode messages from them; but this skill is one shared with other visual processing tasks, not something specific to writing. The difficulty of decoding a written signal depends very much on the specific message. Often the relationship between the writing on the page and the words is obvious and transparent; but often it is complex enough for the decoding and encoding processes to be non-trivial, such that we have to resort to sophisticated and principled techniques to perform these operations.

3.1.2 Spoken form and written form

We have just explained the differences between spoken and written signals. Turning now to form, we can ask how (if at all) do the spoken and written form differ? One way of looking at this is to consider the combined generation/encoding process of converting meaning into a signal. In general we can convert the same meaning into either a spoken signal or a written one, so we see that speech and writing (in general) share the same meaning representation but have different signal representations. The question now is how much of the overall language system do the two share? It is difficult to answer this conclusively, for to do so would require a much more thorough understanding of the processes than we currently have. It is, however, possible to consider a model where the meaning-to-form systems for both spoken and written language share a common system from meaning to the second level of form. In other words, the meaning-to-word system is the same for both. The divergence occurs when we consider the primary level of form: in spoken language words are comprised of phonemes; in written language, they are comprised of graphemes. From this point until the signal, the two systems are separate.

Depending on the language, the primary forms may be broadly similar (for instance in Spanish) such that one can deduce the pronunciation from the spelling and vice versa, or quite different (for example in Chinese).

While a few problems certainly occur, this model seems to fit the facts very well. Whether we write or say the sentence THE MAN WALKED DOWN THE STREET, there are six words and they occur in the same order in the spoken and written versions. This fact may be so obvious as to seem superfluous, but it is in fact vital to our discussion. One would find it difficult for instance to teach someone to read French text and speak English words: even if we learned that the pronunciation for chien was /d ao g/, we would find it difficult to understand why the written French sentence has different numbers of words in different orders from the spoken English version.

The main point in this realisation of the commonality between the first level of form in written and spoken language is that it shows that, if we can decode the form from the written signal, then that we have virtually all the information we require to generate a spoken signal. Importantly, it is *not* necessary to go all the way and uncover the meaning from the written signal; we have to perform just the job of text decoding, not also that of text understanding. There are two qualifications we need to add to this. Firstly, some understanding may be required, or may make the job easier when we are resolving ambiguity in the written signal. For example, when encountering the homograph polish it may help in our disambiguation to know that the topic of the sentence is people from Poland. Secondly, we may need to generate prosody when speaking, and this might not be obtainable from the writing. This issue is dealt with below (Section 3.5.5).

This conclusion is so important for our system that it is worth restating: by and large, the identity and order of the words to be spoken is all we require to synthesise speech; no higher-order analysis or understanding is necessary.

3.1.3 Use

Speech and writing are used for different purposes. When do we use one instead of the other? In certain cases we have no real choice: in face to face communication it is normal to speak; while we could write something down and pass it to our friend to read, this is in general cumbersome. When writing a book or an article we normally use writing; that is the standard way in which material is stored, copied and distributed. Before the computer era it was fairly clear which communication mode was used for what purpose.

Modern technology certainly gives us more choices; with telephones we can communicate over distance, whereas in the past this was possible only by letter. We can also now quickly send written messages electronically, by email, phone messaging, computer messaging and other means. Computer or "instant" messaging is particularly interesting in that it assumes many of the characteristics of speech, but exists as a written signal. When phone messaging ("texting" or "SMS") first arrived, many thought the idea absurd; why would you ever want to send a message in writing with a phone when you could simply call the person directly? In fact, many "functional" uses have been found; a common one is for someone to send an address as a text message since this is easier for the receiver than having to listen and write the address down in case it is forgotten. Uses go beyond this though, and today we see "texting" as particularly popular among teenagers. Anecdotally, we believe that, by using text, teenagers can think about, plan and edit their messages so as to seem more confident, funny and so on; this saves the embarrassment that may ensue from actually having to *talk to someone*.

One major difference is that speech is in general **spontaneous**. During a normal conversation, we usually speak with little planning or forethought; we can't simply suspend the conversation until we have thought of the best possible way to say something. This accounts for the **disfluencies** in speech, such as hesitations, pauses and restarts. Writing is *usually* free of these effects, but it is important to realise that this is not because of any property of writing itself, but more because the writer can spend time over difficult sections, or can go back and correct what has been written.

Another significant difference is that in spoken communication the speaker can often see or interact with the listeners. In face-to-face communication the speaker can see the listener and see how they react to what is being spoken; in addition to the listener responding directly ("Yes, I agree"), they may nod their head, give an acknowledgment or show confusion and so on. In response, the speaker can alter or modify their behaviour accordingly. This also happens over the phone, where verbal cues ("uh-huh", "yeah...") are given by the listener to help show the speaker their level of comprehension and agreement. This is less effective than with the visual cues; anyone who has taken part in a multi-person call can attest to the difficulty involved in sensing everyone's understanding and participation in the conversation.

If we compare this with a typical situation where written communication is involved, we see some important differences. For example, imagine someone is reading a newspaper on a train. Here the author of a particular column will not personally know the reader; in fact the author will probably know only a handful of people who read the column. The author and reader are separated in time and space and the communication is one way only (the reader may comment on the article to someone else, but this will not reach the author). Because of these factors, the situation with respect to common ground is very different. First, the author is writing for many readers, of whom all will have somewhat different degrees and types of background knowledge. Secondly, since no feedback is possible, it is necessary for the author to "get it right first time"; there is no possibility of starting an article and then checking to see whether the reader is following. These situational distinctions account for many of the differences in content between spoken and written language; for example, because the level of background information cannot be known, writing is often more explicit in defining references, times and places and so on. Informally, we can describe the writing as being "more exact" or "less ambiguous". Of course, these terms make sense only when considering the speech or writing in isolation; as explained above, most ambiguity is instantly resolved when the situation is taken into account.

3.1.4 Prosodic and verbal content

One of the most notable features of writing is that it nearly exclusively encodes the verbal component of the message alone: the prosody is ignored. This feature seems to be true of all the writing systems known around the world. Because of this it is sometimes said that written language is **impoverished** with respect to spoken language insofar as it can express only part of the message. It is of course possible to represent effects such as emotion in written language (one can simply write "I am sad"), but this is done by adding information in the verbal component and is different from the systematic use of prosody to indicate this emotion. Quite why prosody is not encoded in writing is not well understood. Possible explanations include the fact that it is often continuous, which creates difficulties if we are using a discrete coding system. Prosody is most powerful for emotive language; and while this is an important part of literature, the ability to express emotion directly is not needed so much for legal documents, newspapers, technical manuals and so on.

We do, however, have a limited prosodic component in written language. Punctuation can be used to show structure; and underlined, italicised or bold text is used to show emphasis. These are clearly encoding augmentative prosody in that the punctuation and italics are there to help the reader uncover the correct verbal form from the writing. Affective prosody, on the other hand, is nearly entirely absent; there is no conventional way to shown anger, pain or sarcasm. The only aspect that is present is the ability to express sentence types such as question, statement and imperatives with punctuation at the end of a sentence. It is interesting to note that in most early writing systems (e.g. Ancient Greek) no punctuation was used at all, but as writing evolved punctuation appeared and more recently italics and other font effects have become commonplace due to the wide availability of computer word processing. In the last few years, a system of "emoticons" has become common in email and other electronic communication, in which punctuation is used to indicate happy faces :-) and other effects, which attempt to encode affective prosody. Quite why this trend is present is again not clear; while we can ascribe the rise of smiley faces to writing being used for more informal purposes, this doesn't explain the trend from classical Greek to the English of 200 years ago with its commas, semi-colons and full stops.

3.1.5 Component balance

From the outset, a person knows whether a message should be spoken or written: it is not the case that the person first generates the message and then decides what channel it should be sent in. This principle gives rise to another set of differences between spoken and written signals. Because he knows that a certain message is to be written, the person will, from the outset, structure the message so as to take account of the component balance available in that signal. So if, for instance, a writer wishes to convey something happy, he is quite likely to express this happiness in the verbal component, since the ability to encode affective prosody is not available. So he might say "I'm thrilled that you got that place" instead of perhaps "well done that you got that place", which could give the impression of being somewhat unmoved. With appropriate expression, the "well done..." version may sound very enthusiastic if spoken. The different use of components in written and spoken language is called **component balance**.

Simply put, a speaker has an open choice of how much prosody or how much verbal content to put in a message; they will balance these against each other, to create a final signal that conveys the message they wish in a way that is balanced between efficiency and effectiveness. A writer, on the other hand, has only a very limited ability to use prosody; and so has to rely on the verbal component more. Because of this, he is likely to "beef-up" the verbal component, so that it is less ambiguous and more explicit than the spoken equivalent. In addition, the writer will "verbalise" any emotion if that is required (that is, explicitly explain the emotion by using words).

It is this awareness of the prosodic deficiencies of written language and the ability to get around them that enables good writers to make themselves understood. It should be noted that the majority of people don't really write anything more than the odd note or shopping list in their day-to-day lives. While we all learn to read and write at school, it is *reading* that is the essential skill to cope with modern life; only a few of us regularly write. This point is worth making because writing well, that is so that one communicates effectively, is actually quite difficult. One of the main aspects of this is that the abilities to encode emotions, make writing unambiguous and have a good picture of a potential reader in mind while writing are all skills that require some mastering. We have seen this with the advent of email – many new users of emails send messages that are far too brief (e.g. "yes" in reply to something that was asked days ago) or that come across as being quite rude – the writer has written the words that he would normally use if he were speaking face to face, but without the prosody, the emotion (which could be humour or politeness) is lost and the words themselves seem very abrupt.

3.1.6 Non-linguistic content

In our discussion so far we have described speech as being the signal used to communicate linguistic messages acoustically and writing as the signal used to communicate linguistic messages visually or over time. This in fact simplifies the picture too much because we can use our speaking/listening and reading/writing apparatus for types of communication that don't involve natural language. Just as we can use our vocal organs for whistling, burping, coughing, laughing and of course eating, breathing and kissing, we find a similar story in writing, in that we can use writing for many other purposes than just to encode natural language.

This is a significant concern for us, in that we have to analyse and decode text accurately, and to do so we need to identify any non-linguistic content that is present. When considering "non-linguistic" writing, it helps again to look at the evolution of writing. The key feature of writing is that it is more permanent than speech, so it can be used to record information. Even the earliest forms of Egyptian and Sumerian are known to have recorded numerical information relating to quantities of grain, livestock and other "accounting"-type information. Furthermore, there have been several cultures that developed methods for recording and communicating accounting information without developing a system for encoding their language: the most famous of these is the Inca quipu system, in which numbers could be stored and arithmetic performed. This ability for writing systems to encode non-linguistic information is therefore an ancient invention, and it continues today in printed mathematics, balance sheets, computer programs and so on [6], [19], [64].

Consider for a moment numbers and mathematics. Without getting too far into the debate about what numbers and mathematics actually are, it is worth discussing a few basic points about these areas. Firstly, in most primitive cultures, there is only a very limited number system present in the language; one might find words for one, two and three, then a word for a few and a word for lots.

More-sophisticated number-communication systems (as opposed to numbers themselves) are a cultural invention, such that, in Western society, we think of numbers quite differently, and have developed a complex system for talking about and describing them. Most scientifically and mathematically minded people believe that numbers and mathematics exist beyond our experience and that advancing mathematics is a process of discovery rather than invention. In other words, if we met an alien civilisation, we would find that they had calculus too, and it would work in exactly the same way as ours. What *is* a cultural invention is our system of numerical and mathematical notation. This is to some extent arbitrary; while we can all easily understand the mathematics of classical Greece, this is usually only after their notational systems have been translated into the modern system: Pythagoras certainly never wrote $x^2 + y^2 = z^2$. Today, we have a single commonly used mathematical framework, such that everyone knows that one number raised above another (e.g. x^2) indicates the power. The important point for our discussion is to realise that when we write $x^2 + y^2$ this is *not* encoding natural language; it is encoding mathematics. Now, it so happens that we can *translate* $x^2 + y^2$ into a language such as English and we would read this as something like "x to the power of two plus y to the power of two". It is vital to realise that $x^2 + y^2$ is not an abbreviated, or shorthand, form of the English sentence that describes it. Furthermore, we can argue that, while mathematics can be described as a (non-natural) language, in that it has symbols, a grammar, a semantics and so on, it is primarily expressed as a meaning and a *written* form only; the spoken form of mathematical expressions is derivative. We can see this from the ease with which we can write mathematics and the relative difficulties we can get into when trying to speak more complex expressions; we can also see this from the simple fact that every mathematically literate person in the Western world knows what $x^2 + y^2$ means, can write this, but when speaking has to translate it into words in their own language first (e.g. EKS SQUARED PLUS Y SQUARED in English; IX QUADRAT PLUS YPSILON QUADRAT in German).

3.1.7 Semiotic systems

A **semiotic system** is a means of relating meaning to form. Natural language is a semiotic system, and probably has by far the most complicated form-to-meaning relationship, and as we have just seen mathematics is another. Most semiotic systems are much simpler than natural language since they are cultural inventions rather than biological processes (recall that, while mathematics itself may be a property of the Universe, the mechanisms we use for its form and encoding are cultural inventions). In addition to these examples, we also have computer languages, email addresses, dates, times, telephone numbers, postal addresses and so on. Just as with mathematics, a telephone number is not part of natural language, it is a complete meaning/form/signal system in its own right.

If it were just a case of each semiotic system having its own writing system, we could design a separate system to process each. The problem arises because frequently we mix these systems in the same signal and using the same characters to do so. For example, in the above text, we saw the sentence

(2) Pythagoras certainly never wrote $x^2 + y^2 = z^2$.

in which we clearly mixed natural-language text with mathematical text. There are many other cases too; while the mathematics example uses different fonts and so on, many systems share the same characters as natural-language writing, so that we get

(3) a. I was born on 10/12/67
    b. It weighed 23 Kg
    c. I'll send it to paul@yahoo.com

and so on. The point is that these numbers, dates and email addresses are encoded using the same signals as natural-language writing. To be able to process this type of writing, we have to identify which parts of the text relate to which semiotic systems.

3.1.8 Writing systems

To the best of our knowledge,[1] the first writing system was that of Sumer, which developed into Cuneiform. Perhaps the best-known early writing system is the hieroglyphic writing of the Ancient Egyptians, although it is not entirely clear whether these two systems are related.

Initially these systems were **logographic** in nature, meaning that one symbol usually represented one word or one morpheme. Furthermore, the scripts were **pictographic**, meaning that the symbols were often iconic, i.e. pictorial representations of the entity the word stood for. As these systems developed, the iconic nature became less obvious, evolving into a more abstract logographic script. Additionally, **alphabetic** elements that allowed writers to "spell out" words were introduced into the language. These developments should come as no surprise; in Section 2.1 we saw that it is quite natural for a communication system to start highly iconic and then progress to a state such that the relation between meaning and form is more arbitrary, since this generally allows experienced users to communicate more quickly and more accurately.

Most writing systems can be described as belonging to one of the following groups, but in some cases, e.g. Japanese, the complete writing system uses more than one of these types. Logographic systems, as we have just seen, use one symbol to represent a word or morpheme. **Syllabic** systems use one symbol per syllable, such as in Japanese hiragana. **Alphabetic** systems use a small list of symbols, which roughly correspond to the phonemes of a language. The Latin alphabet is of course the most widely used, but its relation to phonemes varies greatly from language to language. **Abjad** systems are ones primarily based on consonants; vowels are usually not written. It is a basic fact of human language that most information that discriminates words is carried by the consonants, so

---

[1] If only someone had taken the trouble to write down this key discovery!

ignoring these in written language is possible (this is why the abbreviation for DOCTOR is dr and not oo). The semitic languages, including Hebrew and Arabic, are abjads.

Writing systems are an important consideration in TTS since in general we have to be able to design a system for any language, and this means being able to handle all writing systems. Problems that have to be solved include identifying the language, the writing system (for example Japanese uses a logographic and a syllabic system), the computer encoding used, the writing direction and many other issues specific to each language.

3.2 Reading aloud

We now turn to the problem of **reading aloud**, that is, how to read text and then speak what we have read.

3.2.1 Reading silently and reading aloud

The process of reading, as we normally understand it, is one of taking some text, decoding it into a message, and then understanding it. For clarity, we call this process **reading silently**. By and large, the process of decoding the text is relatively straightforward for us, and in, say, a newspaper, we can often successfully decode all the text without error. Understanding is more difficult, in that we might not completely understand an article, and we can perhaps think of various degrees of understanding that we might achieve with respect to a text.

The other participant in this process is the author, and exactly the same factors govern the way she generates a message as with our conversation example of Section 2; she assumes some background knowledge on behalf of the reader, makes an assumption about the reader's situation, and is guided by the conversational principles outlined before.

In addition to all these though, she generally assumes that the reader will be reading in the way just described; that is, decoding the writing into the message and then understanding this. This last point is highly significant because this carries with it an assumption that the spoken part of the communication process is not relevant to the task of normal, silent, reading. Examples of the effect of this include a situation where a journalist may include the name of a foreign person or place, but not have to explain how this should be pronounced. The word structure of the sentences can be different, because, unlike when speaking, there is no need to limit the length of phrase or sentences to chunks that are possible to say with one breath. Far more significantly though, there is little or no specification of prosodic information; it is assumed that the verbal content is enough to create the correct understanding, and emotional or other thoughts are communicated via the verbal component. These issues cause no problem for normal reading, because speech isn't involved in this process.

Now consider the act of **reading aloud**, that is, reading the text and simultaneously speaking it. This is considerably more complicated in that it involves three agents, an author, a reader and a listener, and two types of signal, text and speech. A diagram of this process is shown in Figure 3.1. This type of reading of course *does* involve

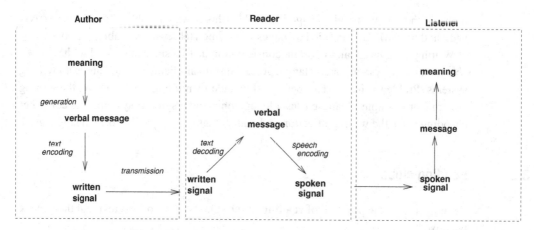

Figure 3.1 A basic model of reading aloud, which shows the information flow from the author, via a reader, to a listener.

speech; and herein lies the problem, because, in the majority of cases, the text that has been written has been written only to be read silently; it was never intended to be read aloud. The problem is that, as we explained in Section 3.1.5, the component balance is now wrong; a message that was intended to be communicated by writing is now being communicated by speech. While sometimes this is unproblematic, often difficulties can arise from this mismatch, and we can class these difficulties into those to do with prosody and those to do with the verbal content and style. We now investigate these in detail.

3.2.2 Prosody in reading aloud

We have seen already that the written signal contains little or no prosodic information, and, when this is decoded, it is decoded into a message that has verbal content only; the prosodic content is generally missing. This is not a problem for reading silently, since the reader can understand the thought without needing the prosodic information, and the assumption is that the author has made strong use of the verbal component to compensate for the lack of prosody. If the message is to be spoken aloud we face the issue of what and how much prosody to use. For messages that convey little emotion and/or are highly propositional in content, there should not be much of a problem. We decide to use neutral prosody, and the suprasegmental effects that need to be synthesised can be found from the verbal part of the message. For more emotive messages, or in cases where we think we need significant use of augmentative prosody, we have a fairly serious problem. This is because we have no means of knowing what prosody the speech should be encoded with; the message that we found from the text has no explicit clues as to what the prosody should be.

The only solution is for the reader to generate their own prosody. This is very different from what they are doing with the verbal content; there they are simply decoding from one medium and encoding in another; no imagination, creativity or understanding is required. With prosody generation, however, the message does not contain the necessary content, so the speaker must generate this themselves. We don't really know exactly how people perform this task, but it is plausible that some understanding of the message is undertaken, and from this an attempt is made to generate prosody that the speaker believes is appropriate for the verbal content. Another way of looking at this, is that they in some sense attempt to reproduce the prosody that the author would use if they had decided to speak the message rather than write it.

It is crucial to realise the difference between the process of finding the message by decoding the writing and the process of prosody generation. In the first, there is a clear right and wrong; the source generated this partial representation, it was clear and specific, and it is our job to find it. Any difficulties that may arise are solely because some distinctions in the structured representation are lost when the message is encoded in the signal (they are ambiguous). In the case of prosody generation, there can be no right or wrong; this information never existed in the first place and so we cannot recover it.

3.2.3 Verbal content and style in reading aloud

Another issue concerns the situation where the reading process is taking place. In our canonical example, the author is writing with the intent that the reader silently reads the text. If the text is then read aloud, we face a potential problem in that the communication-convention assumptions used to generate the message have been violated, and we are now converting the message into a signal that it wasn't intended for. A simple example will demonstrate this. Many written documents are quite long; simplisticly we can say that this is necessary because the author has a lot to say and it might not be easy to get this across except in a long document (say a book). The author of course assumes that the reader will read this at a time that is convenient and at his own rate. Furthermore, while the author might not explicitly desire this, it is understood that the reader can skip ahead, glance through sections, or reread sections.

If we now consider the situation of a human or computer reading this document out loud, we see that many of these assumptions no longer apply; the speech progresses linearly from the start of the book, the rate of reading is more or less constant, and the reader doesn't directly decide when the reading should start and stop. Potentially, this can cause severe problems and we may find that, although the system is reading and speaking the text perfectly adequately, the user is not satisfied with the experience because of these violations of communicative norms.

3.3 Text-to-speech system organisation

The above discussion shed light on some of the issues of spoken and written communication, and the particular process of reading aloud. From our understanding of these processes, we can now define a model of how text-to-speech conversion by computer

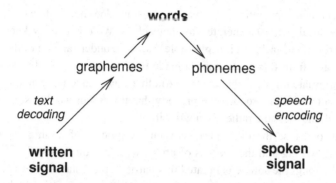

Figure 3.2 The common-form model showing the two processes of text decoding, which finds the words, and speech encoding from those words. For illustration we have shown the different primary levels of form for speech and writing as graphemes and phonemes to show that these are not directly connected.

can be carried out. The next section describes the model we adopt throughout the book, and the sections after that describe alternative models and how these compare.

3.3.1 The common-form model

In the **common-form** model, there are essentially two components; a text-analysis system, which decodes the text signal and uncovers the form, and a speech-synthesis system, which encodes this form as speech. The first system is one of resolving ambiguity from a noisy signal so as to find a clean, unambiguous message; the second system is one where we take this message and encode it as a different, noisy, ambiguous and redundant signal. The core idea is that writing and speech share a "common form" and that that, and that alone, is the single and correct intermediate representation between the two types of signals. A diagram of this is shown in Figure 3.2.

For purposes of basic explanation, we should take the word "form" here to mean "the words". Hence, in the analysis phase, we are trying to find the unambiguous, underlying words from the text; and in the synthesis phase we are attempting to create a speech signal from these words. In the basic common-form model, we always read the words as they are encoded in the text; every word is read in the same order as that in which we encounter it. No understanding or other higher-level analysis is performed; hence everything is read with a neutral prosody.

The key features of this model are the following.

- There are two fundamental processes; text analysis and speech synthesis.
- The task of analysis is to find the form, i.e. the words, from the text.
- The task of synthesis is to generate the signal from this form.
- No understanding is required; an accurate decoding of the text into a linguistic form is sufficient.
- Prosodic information cannot in general be decoded from the text, so everything is spoken with neutral prosody.

3.3.2 Other models

A number of different models are found in TTS systems and we will briefly examine these now. These models are not mutually exclusive and many real-life systems are combinations of two or more models.

Signal-to-signal models In this type of model, the process is seen as one of converting the written signal into a spoken one directly. The process is not seen as one of uncovering a linguistic message from a written signal, and then synthesising from this, but as a process whereby we try to convert the text directly into speech. In particular, the system is not divided into explicit analysis and synthesis stages.

Pipelined models Quite often the signal-to-signal model is implemented as a pipelined model in which the process is seen as one of passing representations from one module to the next. Each module performs one specific task such as part-of-speech tagging, pause insertion and so on. No explicit distinction is made between analysis and synthesis tasks. These systems are often highly modular, such that each module's job is defined as reading one type of information and producing another. Often the modules are not explicitly linked so that different theories and techniques can co-exist in the same overall system.

Text-as-language models In this type of model, the process is seen as basically one of synthesis alone. The text itself is taken as the linguistic message, and synthesis is performed from this. Since the text is rarely clean or unambiguous enough for this to happen directly, a **text-normalisation** process is normally added, as a sort of pre-processor, to the synthesis process itself. The idea here is that the text requires "tidying up" before it can be used as the input to the synthesiser.

Grapheme and phoneme form models This approach is in many ways similar to the common-form model in that first a grapheme form of the text input is found, and then this is converted into a phoneme form for synthesis. Words are not central to the representation as is the case in the common-form model. This approach is particularly attractive for languages in which the grapheme–phoneme correspondence is relatively direct; in such languages finding the graphemes often means that the phonemes and hence pronunciation can accurately be found. For other languages, such as English, this is more difficult; and for languages such as Chinese this approach is probably impossible. If it can be performed, a significant advantage of the grapheme/phoneme form model is that an exhaustive knowledge of the words in a language is not necessary; little or no use of a lexicon is needed. This fact makes this approach still attractive for small-footprint systems (i.e. systems with low memory usage).

Full linguistic-analysis models The common-form model is based on the idea that all we need to uncover from the text is primarily the identity of the words; we don't make heavy use of any other linguistic form or meaning representations. Some systems go much further in terms of linguistic analysis and perform morphological analysis, part-of-speech tagging and syntactic parsing. To some extent, all these are useful for finding the words; the issue is really whether these should be performed as separate stand-alone tasks whose output feeds into the word-finding system, or whether we can find the words with a single, integrated approach. This issue is addressed in detail in Chapter 5. In addition to word-identity detection, parsing and other types of linguistic analysis are often seen as being useful for helping with prosody.

Complete prosody generation The common-form model allows us to describe the three types of prosody independently. Suprasegmental "prosody" is modelled by the verbal component, and for utterances that have little affective prosody. This means that only augmentative prosody has to be generated explicitly. This contrasts with the many systems which state that the F0, phrasing, stress and so on in an utterance are all directly determined by prosody, and hence that, to generate speech, we have to generate all these quantities with an explicit prosodic model. If this is the case, then the prosodic part of the system plays a much bigger role.

Prosody from the text Following from the assumptions of the complete-prosody model, we find that, if every utterance requires a detailed prosodic specification, then this must somehow be generated. A common assumption is that the text does in fact contain enough information to determine prosody; hence many TTS systems have modules that try to predict prosodic representations directly from the text, often with the assumption that this is an analysis process with a right and wrong answer.

3.3.3 Comparison

While some of the above systems may seem quite different from the common-form model that we adopt, it should be realised that in many cases the differences can be somewhat superficial and that "under-the-hood" many of the same underlying operations are being performed. It formulates text analysis as a decoding problem, and this allows us to bring many of the powerful statistical text-processing techniques to bear on the problem. This is a more powerful approach than the alternatives of either treating text as the synthesis input, which in some way must be normalised, or performing a full traditional linguistic analysis, which can often be errorful: the reason why we have chosen the common-form model is that it simplifies the overall design of the system. The text-decoding approach allows us to focus on a single well-defined problem such that we are often able to perform this in a single, optimised system. Secondly, by avoiding the "complete-prosody" assumption, we bypass many of the difficulties (and, we believe, impossibilities) that arise from trying to determine the prosodic content from the text.

The common-form approach is what one can also term an *integrated* approach. In many TTS systems, one often finds a huge list of modules arranged in a pipeline; a typical system of this kind might include modules for pre-processing, tokenisation, text normalisation, part-of-speech tagging, parsing, morphological analysis, lexicon, post-lexical rules, intonation, phrasing, duration, F0 generation, unit selection and signal processing. The problem with this approach is that, if each module operates independently, it may be contributing marginally only to the overall process. Each module will have to make "hard decisions" to generate its output, and this can lead to a propagation of errors through the system. Furthermore, from an analysis of errors and performance, it may in fact be clear that many of these modules (e.g. a parser) would be of only marginal benefit even if their output were always correct. An integrated approach, on the other hand, aims to avoid these problems by optimising the system performance on a single metric. In the case of text decoding, the idea is to go from raw text to a word list in a single step. While some of the tasks performed by modules in a pipelined system may still be performed, they will be done implicitly, with few hard decisions until the final output is required.

3.4 Systems

3.4.1 A simple text-to-speech system

To demonstrate the common-form model, let us now sketch how a text-to-speech system actually works. The input text arrives as a sequence of ascii characters, which can be of any length. To make the processing more manageable, we break input text into separate sentences using a sentence-splitting algorithm. The input could have only one sentence, but we don't know this, so we always attempt to find sentence boundaries. For each sentence, we then further divide the input into a sequence of tokens, on the basis of the presence of white space, punctuation and so on. Often tokens are just the written encodings of individual words, but they can also be encodings of numbers, dates and other types of information. Next we find the semiotic class of each token. For non-natural-language tokens, we use a separate system for each type to decode the text into the underlying form, and then, using rules, we translate this form into a natural-language form with words. For natural-language tokens, we attempt to resolve any ambiguity and so find the words. We attempt a basic prosodic analysis of the text. Although much of the information we might ideally like is missing from the text, we do the best we can and use algorithms for determining the phrasing, prominence patterns and intonational tune of the utterance. This concludes the text- and prosodic-analysis phase.

The first stage in the synthesis phase is to take the words we have just found and encode them as phonemes. We do this because it provides a more compact representation for further synthesis processes to work on. The words, phonemes and phrasing form an input specification to the unit-selection module. Actual synthesis is performed by accessing a database of pre-recorded speech so as to find units contained there that match the input specification as closely as possible. The pre-recorded speech can take the form of a database of waveform fragments and, when a particular sequence of these is chosen, signal processing is used to stitch them together to form a single continuous-output speech waveform.

This is essentially how (one type) of modern TTS works. One may well ask why it takes an entire book to explain this then, but, as we shall see, each stage of this process can be quite complicated, so we give extensive background and justification for the approaches taken. Additionally, while it is certainly possible to produce a system that speaks *something* with the above recipe, it is considerably more difficult to create a system that consistently produces high-quality speech no matter what the input is.

3.4.2 Concept to speech

It is not strictly speaking necessary to start with text as input in creating synthetic speech. As we have clearly stated, our TTS model is really two processes, an analysis one followed by a synthesis one, and these are quite different in nature. Since the analysis system will never be perfect, it is quite reasonable to ask whether there are situations where we can do away with this component and generate speech "directly". Often this way of doing things is called **concept-to-speech** in contrast to text-to-speech. We shall stick with the generally used term "concept-to-speech", but we should point out that this can, however, mean a number of different things, which we will now explain.

Clearly, in our model, if we do away with text analysis, then the input to the system must be the input to the synthesiser itself, that is, the word-based form or message. We could call this "message-to-speech" or "form-to-speech" synthesis. The idea is that a message-to-speech system would be a component in a larger computer natural-language system, and that the message would be generated by a natural-language-generation system. Since this system knows whether it said PROJECT-NOUN or PROJECT-VERB, it can pass this directly into the synthesiser and so bypass any errors the text-analysis system may make here. While this would certainly bypass any inadequacies of text analysis, it does suffer from a significant problem in that we don't have a standard definition of what form actually is. Even with systems that conform to the common-form model, each may have a slightly different idea of what this form should take; one system may just use words, another may use syntax as well. We therefore face a situation of each system potentially exposing a quite different interface to an external system developer. A standard interface could be designed, but standardisations of systems often progress along paths determined by a lowest-common-denominator strategy and may end up being somewhat ineffective.

An alternative strategy is to propose a "meaning-to-speech" system, in which semantic representations are the input. Here the message-generation and speech-encoding parts would be combined into a single system. While this bypasses the question of how the message should look, it opens up an even thornier problem in that we have even less agreement about how semantic representations should look. Such a system would have a significant advantage in that it could solve the prosody-generation problem. Because we are synthesising directly from meaning, it would in principle be possible to generate prosody in the signal directly, and so bypass problems of prosodic impoverishment in the text, or problems with how to represent prosodic form.

A final solution is that we use a TTS system as before, but augment the input so as to reduce ambiguity and explicitly show the system where to generate prosodic effects. This can be done by a number of means, including the use of XML or other markup (explained in Chapter 17). While this will inevitably lack the power and fine control of the two above methods, it has the advantage in that the input is in a more standard form, such that a system developer should be able to get this working and switch from one system to another. It also has the advantage in that one can easily vary how close the system is to "raw" text or "clean" text.

The message-to-form and augmented-text approaches are well covered in this book in the sense that they can be adapted from our main common-form TTS system.

Meaning-to-speech is considerably more difficult and has been researched only in a very tentative manner to date. Mainly for this reason, we do not cover this approach in any detail here.

3.4.3 Canned speech and limited-domain synthesis

Today, most applications that use speech output do not use text-to-speech. Rather they make use of a set of recordings, which are simply played back when required. These are called **canned speech** or **pre-recorded prompts** depending on the application. A typical way these are deployed is that a designer creates an exhaustive list of the utterances that are needed for an application. Quite often these utterances are specific to that application and a new set will be created and recorded for a different application. Once the utterances have been designed, a speaker is asked to read these, and they are recorded and then stored as waveform files. The application then plays back the required file at the required time. A typical system of this kind might be deployed in a train-station announcement system or an automated credit-card booking system, where the users interact with the system by speech recognition and automated dialogue control.

Such systems initially compare unfavourably with TTS in that they require a new set of utterances for each application, compared with a TTS system that would just be deployed once and can be used for any application. Furthermore, the canned-speech approach can only say a very fixed number of things, which can limit the scope of the application (for instance it would be very difficult to speak a user's name). Finally, if the application has to be updated in some way and new utterances added, this requires additional recordings, which may incur considerable difficulty if say the original speaker is unobtainable. Despite these apparent disadvantages, canned speech is nearly always deployed in commercial systems in place of TTS. Part of the reason behind this is technical, part cultural. Technically, canned speech is perfectly natural and, because users show extreme sensitivity to the naturalness of all speech output, this factor can outweigh all others. In recent years, TTS systems have improved considerably in terms of naturalness, so it is becoming more common to find TTS systems in these applications. There are other non-technical reasons; canned speech is seen as a simple, low-tech solution whereas TTS is seen as complex and high-tech. The upshot is that most system designers feel that they know where they stand with canned speech, whereas TTS requires some leap of faith. There may be purely business reasons also; while the canned-speech approach incurs up-front cost in an application, it is a one-off cost, and does not increase with the size of deployment. Text-to-speech systems, by contrast, are often sold like normal software in the sense that extra licences (and hence cost) are required when the size of the deployment increases.

The main technical drawback of canned speech is that it can say only a fixed number of things. This can be a severe drawback in even simple applications where, for example, a telephone number is to be read in an answering-machine application. A common solution is to attempt to splice together recordings of individual words or phrases so as to create new utterances. The result of such operations varies greatly from acceptable (but clearly spliced) to comically awful. That said, even the resultant poor naturalness of this is often chosen over TTS in commercial situations.

Faced with the choice between fully natural but inflexible canned speech and somewhat unnatural but fully flexible TTS, some researchers have proposed **limited-domain synthesis** systems that aim to combine their benefits. There are many different approaches to this. Some systems attempt to mix canned speech and TTS. Black and Lenzo [52] proposed a system for cleverly joining words and carrier phrases, for use in applications such as a talking clock. The **phrase-splicing** approach of Donovan *et al.* [139] used recorded carrier phrases cleverly spliced with unit-selection synthesis from recordings of the same speaker. A somewhat different approach is to use what is basically the same system as for normal unit-selection synthesis but to load the database of recordings with words and phrases from the required domain [9], [394], [436], [440].

3.5 Key problems in text-to-speech

In this section we identify the main challenges in high-quality text-to-speech. First we describe four areas which we believe delimit the key problems that have been recognised as the main goals of TTS system building (but note that the problems are not always described in this way). Then we go on to describe two additional problems, which are starting to attract more attention and which we believe will be central to TTS research in the future.

3.5.1 Text classification with respect to semiotic systems

As we explained, it is a mistake to think that text is simply or always an encoding of natural language. Rather, we should see text as a common physical signal that can be used to encode many different semiotic systems, of which natural language is just one (rather special) case.

There are two main ways to deal with this problem. The first is the **text-normalisation** approach, which sees the text as the input to the synthesiser and tries to rewrite any "non-standard" text as proper "linguistic" text. The second is to classify each section of text according to one of the known semiotic classes. From there, a parser specific to each class is used to analyse that section of text and uncover the underlying form. For natural language the text-analysis job is now done; but for the other systems an additional stage is needed, where the underlying form is translated into words.

Let us consider the semiotic-class approach. Assume for a moment that we can divide an input sentence into a sequence of text tokens, such that the input sentence

(4) Tuesday September 27, 10:33 am ET

would be tokenised as

(5) Tuesday
     September
     27,
     10:33
     am
     ET

Semiotic classification is therefore a question of assigning the correct class to each of these tokens. This can be done on the basis of the patterns within the tokens themselves (e.g. three numbers divided by slashes (e.g. 10/12/67) are indicative of a date) and optionally the tokens surrounding the one in question (so that if we find "1967" preceded by "in" there is a good chance that this is a year).

Once the class has been identified, it is normally quite straightforward for classes other than natural language to analyse this token to find its underlying form. In dealing with times for instance, the text token might say 10:45 am and from this we would find a form something like (hours=10, minutes=45, time_of_day=morning). Finally, we can **translate** this into words. In one sense this is often straightforward, since there are often simple well-accepted rules that describe this. In our time example, the word version could be read as TEN FORTY FIVE A M, and it is fairly easy to see how the form-to-words rules would be written for this example. Two issues complicate the matter, however. Firstly, often there are many different legitimate ways to translate the same form; with our time example we could have

(6) a. A QUARTER TO ELEVEN
    b. A QUARTER OF ELEVEN
    c. A QUARTER TO ELEVEN IN THE MORNING
    d. TEN FORTY FIVE IN THE MORNING
    e. TEN FORTY FIVE
    f. TEN FORTY FIVE A M

and so on. All are legitimate, but since the system can read only one, we have to be sensitive to which one this should be. Sometimes preferences depend on the user; in particular a British English speaker may prefer one, an American speaker another. In addition, the context and genre of the text may have an effect; in an airline reservation system it would be useful to make absolutely sure the user knows the time is in the morning, whereas in a business appointment system this may be obvious. The important point to realise is that there is often no single correct answer to the translation issue, so systems need to be sensitive to the user's expectations.

A further difficulty with translation is that for some content there is no commonly agreed rendering in words. While times, dates and currencies normally present no problem, classes such as email addresses, computer programs and other technical information have no agreed reading. For example, taking the email address:

(7) pat40@yahoo.com

We can easily parse this into

```
user              = pat40
top level domain  = com
local domain      = yahoo
```

In effect all email programs do just this. The problem is, how should we read this? Should pat40 be read P A T FOUR ZERO, P A T FORTY, or PAT FORTY?

In summary then, there are three tasks involved here.

1. Identify which semiotic class a token belongs to. This is often the most difficult task.
2. Analyse each token to find its underlying form. Apart from the case of natural language, this is often quite easy because the semiotic systems are cultural inventions and have been designed to be easy to understand.
3. Translate into natural language. It is often quite easy to generate at least *one* legitimate translation but sometimes difficult to generate the particular version that a user expects.

3.5.2 Decoding natural-language text

While most semiotic systems have a fairly simple and direct signal-to-form mapping, this in general is not so for natural language. Once the text has been classified properly in terms of its semiotic class, the general level of ambiguity with respect to form is greatly reduced; it is only in the area of natural language that significant ambiguity still remains.

Ambiguity occurs in many different guises. The most basic is **homograph** ambiguity, where two words share the same form. In cases such as

(8) let's now project the image

and

(9) the project was running behind schedule

we see that two separate words PROJECT-VERB and PROJECT-NOUN share the same letter sequence, but have different meanings and different pronunciations. In addition to homograph ambiguity, **syntactic ambiguity** is also reasonably common. In the example (from [363])

(10) Police help dog bite victim

we have (at least) two different possible syntactic patterns that could give rise to these tokens:

(11) a. (POLICE HELP DOG) BITE VICTIM
     b. POLICE HELP (DOG BITE VICTIM)

The question when building a TTS system is just how much of this ambiguity we actually need to resolve. A comprehensive approach would attempt to resolve all the ambiguity in the text and create a rich structure for the form. In recognition that the more we attempt the more likely we are to make an error, and also in recognition of the basic engineering principle of efficiency, we could also take a minimalist approach, which would adopt the principle that we should resolve only those ambiguities that affect pronunciation. The word SET is used in many different ways in English, but it is always spelt the same and always pronounced the same, so we can ignore the multitude of meanings and uses. If we know that some text could be two or more underlying forms, but realise that all these forms are spoken the same way, then we shouldn't bother resolving the ambiguity.

3.5.3 Naturalness

We take the view that a major goal of TTS research is to make the system as **natural** sounding as possible. By natural we mean that the system should sound just like a human; so eventually we would like our TTS system to be able to create the full range of human speech.

Sometimes the question is raised as to whether we really want a TTS system to sound like a human at all. The concern is raised that if the system sounds too much like a human listeners will get confused and mistake it for a person. We think that in reality this is highly unlikely. All the time we hear "perfect" recordings of people's voices, view photographs, watch films and so on, and are rarely confused. In fact, most of us are used to hearing recorded voices on the telephone in the form of people's answering machines, introductory messages or recorded prompts in IVR systems. Rarely do these confuse us. So, no matter how good a system is, it will rarely be mistaken for a real person, and we believe this concern can be ignored.

We still have to ask, however, whether people really *want* natural-sounding voices: wouldn't they be happy with a "robotic"-sounding one instead? (which would after all be much easier for us to produce). This is really an empirical question that needs to be resolved from user trials. From what we know, however, the answer is overwhelming in that people give much higher satisfaction and acceptance ratings to natural-sounding systems. In fact we can go so far as to say that most listeners are extremely intolerant of unnaturalness, to the extent that they will refuse to use non-natural-sounding systems regardless of what other benefits are provided. Quite why this should be so is something of a mystery; after all in many situations we are quite happy with visual caricatures of people as evidenced by cartoons and animation. Somehow we don't consider a drawing of Homer Simpson a pathetic rendition of reality, but we are insistent on him having a natural voice, such that in *The Simpsons* (and all other cartoons) a real actor is used for his voice.

What then makes one voice sound more natural than another? Again, this is a question that we are not in a position to answer fully, but we can provide a sketch of the factors involved. Firstly, any system that produces obviously non-human artefacts in the speech will readily be judged as unnatural. It is all too easy to generate speech with pops, clicks, buzzes and an endless variety of other mechanical sounds. Even speech that is devoid of any of these "error" types of sounds can readily sound unnatural. While speech exhibits considerable variability in the realisation of a particular phoneme, this is controlled by a number of factors, namely phonemic context, position in sentence, suprasegmental influence and so on. Getting this variation just right is a key task. Beyond this, we can consider a number of issues that are related to individuality. In normal speech, any given instance of speech is obviously speech *from someone*; there is no such thing as speaker-neutral speech. So, in a sense, we are completely conditioned to the fact that speech comes from people, all of whom sound slightly different. So, when listening to a synthesiser, one naturally thinks that it is from someone, and, to sound natural, this someone must have a believable-sounding voice. This aspect of naturalness has virtually nothing to do with the message being spoken, it is entirely a property of the system that encoded the speech. Hence we must make the synthesiser sound like someone, either a copy of a real person's voice, or the creation of a new voice that could pass for a real person unproblematically.

3.5.4 Intelligibility: encoding the message in signal

The final central task is **intelligibility**. For our purposes, we define intelligibility as the ability of a listener to decode the message from the speech. As such, this does not include any measure of comprehension directly, but the idea is that if the listener can decode the message properly they should be able to understand it with the same ability as they could the same message spoken by a human. Because of that we can sometimes measure intelligibility with comprehension tests; this is usually done with content that may be hard to decode, but is often not difficult to understand (e.g. a simple sentence address or telephone number).

Of all the TTS problems, intelligibility is probably the easiest to solve. In fact, it is possible to argue that this problem was solved some time ago, since it has been possible to generate fairly intelligible-sounding speech in TTS systems from the late 1970s onwards. It can be shown in tests that the intelligibility of a modern TTS system is often only marginally better than with much older systems such as MITalk [10], once any text errors have been dealt with.[2] From this it is wrong to conclude that little progress has been made on this issue. Crudely speaking, the story is that, while the intelligibility of older systems was reasonably good, research focused on improving naturalness, which was uniformly rated as being very poor with these systems. Hence the idea in the research community (though rarely stated explicitly) was to improve naturalness without making intelligibility any worse.

The relationship is by no means deterministic or simple, but it is possible to see that in many synthesis paradigms there is an inverse correlation between naturalness and intelligibility. To achieve high intelligibility, formant systems and early diphone systems used very "safe" speech, which was well articulated and close to the average realisation for the message in question. In a sense, this speech was too safe, containing none of the fluency, slurring, rhythm and heavy coarticulation found in real speech. As these

---

[2] Today's systems are massively more accurate with respect to text analysis, so any text-processing errors in systems such as MITalk have to be eliminated if a comparison of synthesis encoding with respect to intelligibility is to be performed.

effects are added, there is a chance that they will be added inappropriately, and will actually cause a loss of intelligibility. The trick therefore is to add in extra naturalness by synthesising variation more appropriately, but to do so in a way that does not cause unnatural variation to occur.

3.5.5 Auxiliary generation for prosody

We now consider the first of the more-advanced problems; specifically how to create speech that encodes the full range of prosodic effects, not just the neutral prosody of most of today's systems. Recall that this is difficult because, while the text encodes the verbal message that the author generated, it does not do the equivalent with prosody. The author of course (subconsciously) knows this. Thus it is not the case that he generates a message with verbal and prosodic content as he would do if speaking; rather he generates a message that contains only verbal content. The prosody was neither generated nor encoded, so we can never recover this from the writing. How then are we to generate prosodic content if we can't uncover it from the text?

Firstly, let us separate the issue of generating a signal that contains a particular prosodic effect (for example surprise, or accentuation of a particular word) from the issue of deciding how to do this automatically from the text. The issue of the realisation of a particular prosodic form is by no means easy, but it is certainly more tractable than the second issue. In essence it is no different from other parts of the synthesis problem in that we can collect data with this effect, study how suprasegmental features vary with respect to this effect and so on.

It is really the second issue, that of how to determine automatically which prosodic effects to use from the text, that is the more difficult problem. We term this process **auxiliary generation** to show that it is a generation (rather than text-decoding) issue, and that this is done as a separate, auxiliary task in addition to the process of verbal decoding and encoding. How then are we to approach this problem of auxiliary generation? First let us state that the fact that there is no definite right or wrong doesn't imply an "anything goes" policy. One would not think highly of a system that read a message like "we are regret to inform you of the sad demise of Mr..." with a humorous or flippant paralinguistic tone. Furthermore, we probably wouldn't think too highly of a system that placed emphasis as follows: "THE share price is AT an all TIME high because of uncertainties caused BY the forecast of PROFESSOR Merton..." and so on. Now, there is nearly always some context we can construct that makes the emphasis of any particular word seem reasonable. However, using the same knowledge and context as we used to decode the writing, we can see that some emphasis patterns seem more appropriate than others. This leads us to the proposal that the approach to prosody generation should be one of **assumed intent**, which means that we have to make a *guess* as to what the author *might* have specified in the structured representation if he had intended to speak rather than write the message. This is by no means a certain or easy task but it does give us a reasonable basis on which to solve this problem. We should make two final points about this. Firstly, as we are making an assumption as to what the author might have done, there is no clear right or wrong about this task. Secondly, we must from the outset build the notion of generation choice into our model of auxiliary completion: just as the author can choose to represent a thought in a number of different ways, we have to assume that he could have exercised choice over the prosodic and paralinguistic components. Therefore, as well as being uncertain because we are making an assumption, it is also perfectly valid to assume that there is a number of equally valid choices.

3.5.6 Adapting the system to the situation

The issue here is that, unlike the "normal" communicative situations of reading silently or having a conversation, in the case of reading aloud we have three agents: the author, the listener and the reader (i.e. the TTS system). Most text was never written with the intention that it be read aloud, and, because of this, problems can occur in that faithful readings of the text can lead to situations where the reader is speaking something that the listener cannot understand, has no knowledge of or is not interested in hearing. When people take on the role of the reader, they often stray from the literal text, using explanatory asides, paraphrasing, synonyms and other devices to make the author's message understood. Just how faithful the reader is to the text depends very much on the situation; we sometimes describe a good reader as someone who has a good awareness of both the listener and the intent of the author and as such can steer a good compromise path between faithfulness and acceptability.

Very few TTS systems themselves make any serious attempt at solving this issue; it is normally seen as a problem for the application to resolve such issues. Even so, it is possible for the TTS to facilitate better control; this can be done by providing natural and easy-to-control ways to change the speed of the speaking, to have a phrase repeated, a phrase skipped and so on.

3.6 Summary

Speech and writing
- We can consider both speech and writing/text as signals that encode a linguistic, or other, message.
- The primary functional difference is that writing is an invention used to record messages; as such it is significantly "closer" to the message than is the case with speech.
- Most writing is written with the intention that the reader will read it silently. This can present difficulties if this writing is then read aloud.
- Not all writing encodes linguistic messages; it is used to encode messages in many other semiotic systems also.

Reading aloud
- To a large extent, we can perform the task of reading aloud, by decoding a written signal into a message, and then re-encode this into a speech signal.
- In many cases, no prosody is required, but in others a more-sophisticated approach is to attempt to generate prosody that is appropriate for the message. This cannot, however,

be decoded from the text since that information was never encoded there in the first place.
- These observations lead us to build our system within the framework of the common-form model. This has two fundamental processes.
    1. Text analysis; a decoding process that finds the message from the text
    2. Speech synthesis; an encoding processes that creates a signal from the message.

Text-to-speech key challenges
- We can identify four main challenges for any builder of a TTS system.
    1. Semiotic classification of text
    2. Decoding natural-language text
    3. Creating natural, human-sounding speech
    4. Creating intelligible speech
- We can also identify two current and future main challenges.
    1. Generating affective and augmentative prosody
    2. Speaking in a way that takes the listener's situation and needs into account.

4 Text segmentation and organisation

The next three chapters of the book deal with how to extract linguistic information from the text input. This chapter covers various pre-processing issues, such as how to find whole sentences in running text and how to handle various markup or control information in the text. Chapter 5 describes the main processes of text analysis itself, such as how to resolve homograph ambiguity. Finally, Chapter 6 describes how to predict prosody information from an often impoverished text input. In many ways, this subject shares similarities with text analysis. There is an important difference, however, in that, while we can view text analysis as a decoding problem with a clear right and wrong, prosody prediction has no strict right and wrong since we are attempting to determine prosody from an underspecified input.

4.1 Overview of the problem

The job of the text-analysis system is to take arbitrary text as input and convert this into a form more suitable to subsequent linguistic processing. This can be thought of as an operation whereby we try to bring a sense of order to the often quite daunting range of effects present in raw text. If we consider

(12) Write a cheque from acc 3949293 (code 84-15-56), for $114.34, sign it and take it down to 1134 St Andrews Dr, or else!!!!

we can see that this is full of characters, symbols and numbers, all of which have to be interpreted and spoken correctly.

In view of our communication model (Chapter 2), we can more precisely define the job of the text-analysis system as being to take arbitrary text and perform the task of classifying the written signal with respect to its semiotic type (natural language or other, e.g. a date), **decoding** the written signal into an unambiguous, structured, representation, and, in the case of non-natural language, **verbalising** this representation to generate words. Importantly, the text-analysis process only involves finding an underlying word form from the input; it does not attempt any understanding or further analysis. A simple overview of the processes involved in text analysis and prosody prediction is given below.

1. **Pre-processing**: possible identification of text genre, character-encoding issues, possible multi-lingual issues.

2. **Sentence splitting**: segmentation of the document into a list of sentences.
3. **Tokenisation**: segmentation of each sentence into a number of tokens, possible processing of XML.
4. **Text analysis**
   (a) **Semiotic classification**: classification of each token as one of the semiotic classes of natural language, abbreviation, quantity, date, time etc.
   (b) **Decoding/parsing**: finding the underlying identities of tokens using a decoder or parser that is specific to the semiotic class.
   (c) **Verbalisation**: Conversion of non-natural-language semiotic classes into words.
5. **Homograph resolution**: determination of the correct underlying word for any ambiguous natural-language token.
6. **Parsing**: assigning a syntactic structure to the sentence.
7. **Prosody prediction**: attempting to predict a prosodic form for each utterance from the text. This includes
   (a) **Prosodic phrase-break prediction**
   (b) **Prominence prediction**
   (c) **Intonational-tune prediction**

In this chapter, we concentrate on a number of preliminary, document-processing and architectural issues. Firstly we attempt to define the precise goal of most of these operations, which is to uncover the underlying linguistic form. Mostly, this involves defining a precise notion of what we mean by a word and a sentence and how it relates to its spoken and written form. Secondly, we examine many of the initial tasks mentioned above, specifically those of classifying documents, handling markup, splitting documents into sentences and splitting sentences into tokens. The tokens which are an end result of this are then used as the input to the further text-analysis modules. Finally, we discuss how to store and communicate information within a TTS system.

4.2 Words and sentences

This section examines the notion of just what we mean by a **word** and a **sentence**. We may think we are so familiar with these entities that we don't have to consider them further, but, as we saw in Chapter 2, we nearly all make one basic confusion, which is to confuse words with their written form. The next sections therefore attempt to define the notions of words and sentences with some rigour.

4.2.1 What is a word?

In the model of text analysis used in this book, we always maintain a clear distinction between a word and its written form. The basic idea is that each word is a unique linguistic entity, which can be expressed in either speech or writing. To do so, it must be expressed as a **form**, so we talk of the spoken and written form of a word. People rarely confuse a word with its pronunciation, but often the identity of a word and its written form are confused, so we will take extra care to show the difference explicitly. This is highly beneficial because we can then see text analysis as a process of determining words from text, rather than a process of, say, assuming that there are words actually *in* the text, such that all text analysis essentially becomes a process of tidying up these words (e.g. expanding abbreviations). Making text analysis a process of uncovering words from superficial text is a cleaner and more powerful formulation.

The idea behind our use of underlying words is to try to define some notion of word that captures the essential essence of the linguistic unit of word, but is free of distinctions that are not directly linked to word identity itself. Hence, while the capitalisation difference between `Captain` and `captain` may provide some useful information, in essence, these are written variants of the same word. We use the concept of word to reflect this, so we state that CAPTAIN is the underlying word of `Captain` and `captain`.

Next, we use this concept to help us with the ambiguity problem, namely that sometimes different words have the same written form. If we consider the text form `Dr` we see that this is highly ambiguous as it can mean DOCTOR or DRIVE, or less commonly DOOR (`4Dr car`). `Dr` is an **abbreviation**, i.e. an *orthographically* shortened form of a written word. Abbreviations are often created by leaving characters out (usually vowels) of the original word, and the omission of these characters can create the ambiguity. So the full written forms of the words DOCTOR and DRIVE are not confusable, but the abbreviated forms are and so we term these **abbreviation homographs**. Note that the abbreviation happens only in the written form of the language: when people speak they say /d aa k t er/ and /d r ay iv/ in full, not /d r/.

We also have words such as `polish`, which is ambiguous (someone from Poland and something to shine your shoes with), but not because of a short cut: there is no other way to write this word. Such words are called **true homographs** as the ambiguity is not caused by writing style or abbreviation, but rather due to the defined spelling of the word. There are two ways in which true homographs come into a language. First, there are many words such as `polish` in which it is a pure coincidence that the two words have the same written form. These are called **accidental homographs**. Secondly, we have words that are related, such as `record` in `I want to record this and just for the record`. These words are clearly related and form a verb/noun pair. Such homographs are called **part-of-speech homographs**.

We finish this section with a note on the definitions and terminology we will use for the remainder of the book. The term **word** will always be used for the unambiguous, underlying identity of a word, and never used to indicate its written or spoken form. Words will always be marked in a SMALL CAPS FONT. The terms **writing**, **orthography** and **text** are used more or less interchangeably, and all text will be displayed in a `courier font`. A **token** is a term used for the written form of a word (for reasons explained below). The written form of a word is often described in terms of **graphemes**, **letters** or **characters**. Here we will use the term **character** throughout, since it represents the normal computer-science view as a single unit of writing. Characters are always defined in terms of a standardised computer character set such as ASCII, so when one is working with a character set each character is simply a member of that set. The term **grapheme** is often taken as the minimal unit of written language, but is quite confusing in some respects because often common character sequences such as `ch` and `th` in English are considered graphemes. Since we will deal only with computerised sequences of characters as input, we will use the term **character** throughout, and avoid grapheme and letter except to name the field known as **grapheme-to-phoneme** conversion. (This is the standard name and we will stick to it; in our scheme, strictly speaking, this would be termed **character-to-phoneme** conversion.) We will reserve the term **letter** to mean the word which we used to speak a character, so that we speak of "the letters ABC" when we are talking about three individual words called LETTER-A, LETTER-B and LETTER-C.

4.2.2 Defining words in text-to-speech

A word is a linguistic device that relates **form** and **meaning**. In spoken language, words are composed of sequences of phonemes, and we call this the form of the word. The key concept of words is that the relation between form and meaning is arbitrary: we have /d ao g/ being the form of DOG and /k ae t/ being the form of CAT but there is nothing canine about the particular phone sequence /d ao g/, virtually anything would suffice. The association of forms and meanings has arisen over the years during the development of the language, and it is important to note that people need no knowledge of how a word developed or where it came from to use it properly. With the advent of written language, words were given an additional form to the spoken one. In alphabetic languages, there is often a clear relationship between the spoken and written forms, as we see in the two forms `look` and /l uh k/. Often, though, this relationship isn't so clear and both the spoken and the written form of a word have to be learned. In logographic writing systems (such as Chinese) there is no connection whatsoever between the spoken and written forms. We can represent the relationship of spoken form, written form and meaning as follows:

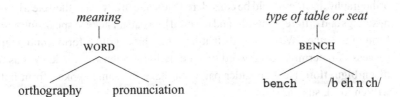

In a "perfect" language,[1] every word would have a unique meaning, written form and spoken form; upon encountering a written form, there would be no doubt as to what it meant, upon wishing to express a meaning, only one spoken or written form could be chosen. In such a language, enumerating the words in a language would be trivial. Of course, this is not the case, and *mainly by historical accident* many words have arisen that share one or more of the written form, spoken form and meaning. Because of this, identifying the set of words is not trivial, and we have to make decisions about whether `bank`, `polish` and `record` are one word or two. We will now study this issue in some

---

[1] Meaning, of course, one that would be easy for engineers to deal with.

detail with the aim of determining an exact definition of word that is best suited for text-to-speech. In doing so, our basic method is to use form and meaning *distinctions* between pairs of potential words. Idealistically, we can think of the following exercise as one where we have all the form–meaning pairs in a language and one by one compare them and make a decision as to whether they are the same word or not.

Let us begin by saying uncontroversially that we can assume that words that are different in orthography, phonetics and meaning are in fact different words. So if we have

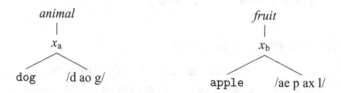

it is safe to assume that these are different words and so $x_a \neq x_b$. Let us also say, trivially, that if two pairs have (exactly) the same meaning, written form and spoken form then they are in fact the same word. Now let us consider words in which at least one form is the same, such as

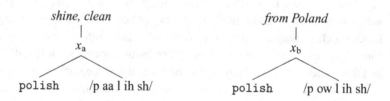

It is highly desirable that these be considered different words; it is only a quirk of orthography that they could be considered the same in any case; they clearly have different meanings and different sounds, and it would be a bad mistake to speak one with the spoken form of the other. We call words that have the same written form **homographs**, and the process of determining the word given only the written form is known as **homograph disambiguation**. Now consider pairs that have the same spoken form but a different written form, such as

Again it is intuitive to think of these as two different words; they mean different things, are spelt differently and are used differently. Pairs such as these are termed **homophones** and, as with the homograph example, it is only a historical quirk that these have the same spoken form; they are not related in any other way. Consider now the following:

4.2 Words and sentences

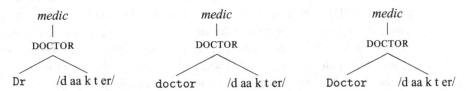

As with bear and bare these have different written forms but the same spoken form. But should we really regard them as different words? In general this is inadvisable because in doing so we wouldn't be able to make use of the fact that, apart from written form, these act identically in all other ways. The essential difference between the doctor cases and the case of bear and bare is in the *meaning*: in one case the meaning is the same while in the other it is different. Rather than count these as different words we say that the different text forms are written **variants** of a single word, DOCTOR (we can list more such as Dr., dr. and so on). A word such as DRIVE, which shares some written variants with DOCTOR, is considered a different word because it differs in meaning and spoken form:

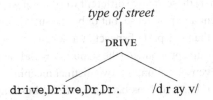

Thus words can be considered homographs if only one of their written forms is the same. A single word can also have **phonetic variants**, and this is found in words such as EITHER and ECONOMIC:

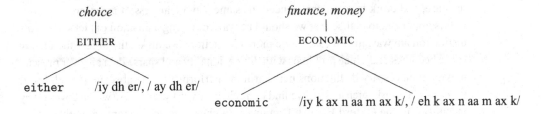

In general, written variants are more common than pronunciation variants because they are usually the result of creativity on the part of writing practices rather than aspects of the language itself.

In traditional linguistics, pairs of words that have the same meaning but different surface forms are called **synonyms** so that we have HIGH and TALL or QUICK and SPEEDY. Our intuition tells us that these are indeed different words that just happen to have the same meaning. In modern linguistics, the existence of synonyms is disputed in that there is always some context where a significant difference between the words seems to arise, so that `she is very tall` seems to mean something different from `she is very high`. There are a number of words, though, which do seem to have exactly the same meaning in that they refer to the same entity, but have both different written and different spoken forms. Consider `hercales`, pronounced /h eh r k ae l iy z/ and `hercules` pronounced /h eh r k y ax l iy z/. These clearly relate to the same (mythical) person, but they have different surface forms. Should we consider these as the same word? To answer this, we look at the issue of the author's intention and state that, while it is hardly a disaster if one reads `hercules` but says /h eh r c aa l iy z/, the author clearly intended one form over the other, and we should stick with his wishes. Most examples of this type relate to different surface forms of a name (e.g. `Peking` and `Bejing`) and while they are in some regards the same, they may carry subtle intended differences, which may be important with respect to the author's intentions. Hence we add to our definition of word the distinction that units in which both surface forms are different are different words, regardless of meaning.

Finally we consider cases where the written and spoken form are the same, but where we can argue that there is more than one underlying word. If we look at `sing` and `sang` it is clear according to the definitions given above that although related (one is the past tense of the other) these are two different words. But what about words such as `cut`? This too has a present and past tense, but only one surface form. Other words have the same form but more than one part of speech: `walk` can be a noun or a verb, but in both cases it is spelt the same and pronounced the same. As a final case, we can consider words such as `bank`, which, even as a noun, has two distinct meanings (*bank of a river* and *somewhere to keep money*). Pairs such as bank are termed **homonyms**, but traditionally this term covers both homographs and homophones; words that have difference in meaning and for which both surface forms are the same we shall call **pure homonyms**. Distinctions such as this pose the most difficultly for our definition of word. The problem is basically this: while it is clear that in general use these words are different, for text-to-speech purposes the distinction is largely unnecessary. `cut` is pronounced /k uh t/ regardless of its tense, and `bank` is again pronounced the same way regardless of meaning.

It is important to realise that we should be wary of taking a position of "let's make the distinction anyway, just in case". The more distinctions we make, the bigger the feature space becomes and, since problems with dimensionality and sparse data are ever present, making unnecessary distinctions may harm the performance of subsequent modules. If we consider words such as SET, we find more than fifty different uses (which cover many meanings) in the Oxford English Dictionary (a chess set, set in stone, a mathematical set, a film set and so on) and we run a severe risk of unnecessarily exploding the feature space for little gain. Because of this, we will complete our definition criteria by saying that words that have identical written and spoken forms will be considered the same word, regardless of meaning. We can justify this on the grounds that collapsing this distinction is acceptable for text-to-speech purposes, but it should be noted that for other purposes, say, machine translation between one language and another, distinguishing between these words may be vital. But in such systems we could probably make other assumptions, for instance completely ignoring differences in spoken form in a text-based machine-translation system.

We can summarise our definition in the following table, which describes the effect of similarities or differences between pairs of potential words. The table compares the meaning, written form and spoken form, gives a name to each, and shows whether we consider such pairs as the same or different words.

Meaning	Written form	Spoken form	Name	Distinction	
different	different	different	different	different	CAT, DOG
different	different	same	homophones	different	BEAR, BARE
different	same	different	homographs	different	BASS-FISH, BASS-MUSIC
same	different	different	synonyms	different	HIGH, TALL
different	same	same	pure homonyms	same	BANK
same	different	same	phonetic variants	same	EITHER, /iy dh er/ or /ay dh er/
same	same	different	orthographic variants	same	LABOUR: labor, labour,
same	same	same	identical	same	

4.2.3 Scope and morphology

Our discussion so far has assumed that we already have our units delimited in some way. That is, we have assumed that we know where one word stops and the next word starts. When first studying phonetics and speech, most people are surprised to learn that there are no gaps between the words in speech; we naturally think that language is writing, so that we think of words as well-delimited units (at least for speakers of European languages). In normal speech, however, the flow of speech is continuous with no acoustic boundaries marked. Furthermore, while some phonemes are more likely to occur near word beginnings and ends, there really is no bottom-up way to find word boundaries from acoustic evidence. The sceptical observer might well then ask, do words, as distinct units in time, really exist, or are they simply an invention of linguists (both traditional and modern)?

These problems of **scope** (that is, where one word stops and the next starts) can be seen as problems of **morphology**, the study of how words are composed. In practice, the main issue concerns whether we should explicitly regard all the **inflected** forms of a **lexeme** as words in their own right, or whether we should only consider the lexeme itself as a true word. A lexeme (e.g. BIG) is defined as the most basic form of a word, and its **inflections** are composed from this so that we have BIG, BIGGER and BIGGEST. In addition we have **derivational** morphology, which explains how words such as CREATION are formed from CREATE. In earlier text-to-speech systems it was often considered wasteful to list all the forms, rather the lexeme was stored and the other forms were derived. This approach of course is more expensive computationally since processing has to be used to derive the forms from the lexeme. Today it really is a matter of engineering expediency which approach is chosen, since it is quite possible to store all the forms of a lexeme.

4.2.4 Contractions and clitics

We will now turn to the issue of whether longer units form words in their own right. First let us consider **contractions**, which are written forms created from particular combinations of two words. Common examples are

(13) I've, can't, won't, they'll, he'd

Contractions are *not* abbreviations. Abbreviations are words that have a shortened written form only, such as (Dr ↦ DOCTOR). By contrast, contractions have shortened forms both in the written and in the spoken form, so they'll, formed from the contraction of THEY and WILL, is pronounced /dh ey l/, not /dh ey w ih l/. We can state more properly that contractions are shortened spoken combinations of words, whose properties are reflected in writing conventions. The question with contractions is whether we should regard the contraction as one word or two. First note that the contraction isn't always simply a case of concatenating a first written form to a shortened second written form. While they'll is created this way, won't, meaning WILL NOT, is not formed this way, and similarly with ain't. While don't can be formed from the concatenation of do and not followed by the replacement of o with ', phonetically the process is different in that the vowel of DO, /uw/, changes to /ow/ in don't. Furthermore, sometimes the apostrophe occurs at the place where we would break the forms (e.g. they'll) but sometimes not, e.g. don't would break into do and n't, not don and 't).

As with previous problems, we settle this issue by deciding what we think is best for TTS. While it is common in natural-language parsing to expand contractions and regard them as two words, in TTS, they are *spoken* exactly as a single word would be, and for this reason we regard all these forms as single words. In further support of this we find that contractions can be very ambiguous with respect to the second word, for example he'd can be either HE HAD or HE WOULD. If we chose to expand he'd we would have to decide which word the 'd was a form of. Luckily, in text-to-speech, both forms are written the same and are pronounced the same (/h iy d/), so we don't have to resolve the ambiguity. Contractions come close to forming a closed class, in that the vast majority of contractions are formed from pronouns plus auxiliaries (I'll, he'd) or with negations can't, won't, but it is in fact possible to form contractions from normal nouns as in

(14) The roads'll be busy tonight

(15) The boy's got to learn

(16) The team've really shown their strengths

which means that the set is not truly closed.

Clitics have similar written forms to contractions in that they use an apostrophe, and in that the written pattern follows through to the spoken form (unlike with abbreviations). Clitics differ, though, in that they are not contractions of two words, but rather function more like an affix joined to a normal word. There is only one use of a clitic in English, which is the genitive 's, as used in Paul's book. Other languages have much richer use for clitics. In this sense, clitics can be thought of as affixes; indeed some argue that it is only a historical quirk of English printing that the genitive apostrophe is used at all; there is no structural-linguistic reason why we shouldn't write `Pauls book`. As with contractions, we view a form with a clitic as a single word.

A final type of modified word is the **shortened form** where a normal word has been shortened in some way. Examples include `lab` (from `laboratory`), `phone` from `telephone` and so on. These are distinct from abbreviations in that both the written form and the spoken form are shortened. This is an important distinction because we assume that when an author writes `lab` he does in fact mean just this and if he were to speak it aloud he would say /l ae b/ not /l ax b ao r ax t ao r ii/. Despite the commonality in meaning, we therefore regard LAB and LABORATORY as two separate words.

4.2.5 Slang forms

An additional type of contraction is what we term **slang forms**, which include forms such as `gonna, wanna, gotcha` and so on. Despite the lack of the apostrophe that we find with the other contractions, these are similar to contractions in that the written form clearly represents the combination of two words as a single word. The issue here is again whether to represent `gonna` as two words (GOING and TO) or one. At first glance, the answer to this may seem to be the same as with the other contractions we looked at, such as `I've` etc., namely that they should be regarded as a single word. Recall, however, that our definition of a contraction was one of two words becoming a single surface form *both* in the written *and* in the spoken form. The problem with `gonna` is that it is not clear exactly what the status of the spoken form is. In reality, the sequence <GOING, TO> is realised in a wide variety of fine spoken forms, and it is only an accident that two (`going to` and `gonna`) have become standard in the orthography. While `gonna` may be a genuine case, there are actually quite a number of less-standard forms such as `gotcha` (GOT YOU), `whadaya` (WHAT DO YOU) and so on. If it weren't for the fact that these forms span two words, we would describe them as orthographic variants of the same word, in the way we would with `nite` and `thru` and so on. Weighing up these competing arguments and the discussion on pronunciation variation that we will have in Chapter 8, we decide that `gonna` and other similar forms should be represented by two words and should not be regarded as *bona fide* contractions.

4.2.6 Hyphenated forms

So far our discussion has centred on whether forms that we normally delimit by white space should be considered as full words or further broken down, and we have seen that, apart from a few exceptions, white space in conventional writing is actually a quite reliable guide as to word boundaries. But there are certain cases where we can argue that forms separated by white space should in fact be treated as a single word. Most of this involves issues to do with hyphens and compounding. Consider the following examples:

(17) a. `last ditch, last-ditch`
    b. `spokes woman, spokes-woman, spokeswoman`
    c. `web site, web-site, website`
    d. `new line, new-line, newline`
    e. `steam boat, steam-boat, steamboat`
    f. `vice chancellor, vice-chancellor`
    g. `lock nut, lock-nut, locknut`

Here we have a different set of words in which we can have the same two words combined to form up to three written forms: as separate written forms separated by white space, as forms separated by a colon, or as a single written form. These patterns raise a number of issues. Firstly, there can be a huge number of forms such as SPOKESWOMAN, in which two relatively common words combine to form a single token. While we could again put every form of this type in the lexicon as is, there are likely to be several unknown forms. Secondly, words like `spokeswoman` act like separate words for the purposes of inflection; the plural of `spokeswoman` is `spokeswomen` and not `spokeswomans`, so the orthographic form uses the irregular plural form of one of its constituents. Thirdly, as shown in the list of examples, the written conventions actually used vary considerably, and one finds that the `web site`, `web-site` and `website` forms are all common, when it is clear that they are all just written variant forms of the same word. Finally, many of these words act like a single phonological word. Of course, in all cases, there is no silence between the words when spoken fluently, but there is also the problem of stress. In normal adjective–noun patterns (e.g. `large site`) the stress normally goes on the second word, but in cases such as `web site` it goes on the first, as would be the case in single words such as PARASITE. This helps show the reason for the variability and behaviour of the written form; in speakers' minds this acts like a single word. There is no real way to resolve this issue since we feel that we must allow words like `blackbird` to be represented by a single word, but it is also impossible to know where to draw the line.

In summary, we use the following guidelines and rules to delimit words.

- Words include free stems and all inflected and derived forms of a stem.
- Contractions and forms containing clitics are single words.
- Slang forms created from two words are not single words.
- Compounds, hyphenated words and multi-words are considered single words on a case-by-case basis.
- In conventional writing, white space is quite a good guide to word boundaries.

4.2.7 What is a sentence?

As with our discussion of words, we find that there are many possible definitions of a sentence, and we need to investigate this a little further before discussing any algorithms to process sentences. It is frequently asked whether people really "speak in sentences" at all. When listening to politicians being interviewed, they have an uncanny ability to avoid ever ending a sentence; if they did they could be interrupted. As they wish to hog the floor, they avoid ending a sentence at all costs (to the extent that *even more* rubbish is spoken). In spontaneous dialogue, it is again uncertain whether sentences are the main unit. Thankfully this is one area where our cross-over between written and spoken language does actually help. While it is unclear whether people engaged in interviews or dialogues use sentences, it is clear that the vast majority of conventional writing is indeed clearly written in sentences. Furthermore, when people read aloud, they effortlessly find the sentence boundaries and adopt the appropriate sentence-final prosody, such that the boundaries are preserved in the spoken signal. For our purposes, the situation is complicated somewhat by the use of non-natural-language writing. In an email header, there are no sentences explicitly marked, and it is part of the verbalisation process that we find suitable sentence boundaries. Again, this problem is one we share with a human reader – when they are given an email header and told to read it, they also suffer from not knowing exactly how to go about this. Nonetheless, our task in sentence splitting is therefore to find the written sentence boundaries in conventional and all other types of writing.

In the linguistic literature, many different ways of defining what a sentence is have been proposed. One approach is what we can term the discourse approach, which states something like "a sentence is the smallest possible answer to a question", or that it is the smallest unit that can bear functional meaning and so on. Another approach is to define sentences using more formal mechanisms; this can take the form of saying that a sentence is a linguistic unit with a subject–predicate structure, or, even more formally, a sentence is the full expansion of any formal grammar (this final point of course doesn't attempt to define sentences "in nature" but just says that they are a logical by-product of a formal grammar). A third definition is based on writing, which would define a sentence as something like "the words between two full stops" or "it starts with a capital letter and ends with a full stop". With certain caveats (see below) we will adopt this definition since we can't rely on using any deeper linguistic analysis to assess the internal coherence of a sentence.

4.2.8 The lexicon

In all modern TTS systems we make extensive use of a **lexicon**. The issue of lexicons is in fact quite complicated, and we discuss this in full in Chapter 8 when we start to talk about pronunciation in detail. For our current purpose, its main use is that it lists the words that are known to the system, and that it defines their written form. A word may have more than one written form (`labour` and `labor`) and two words may share the same written form, `polish` etc. It is by using a lexicon as a place to define what is a possible word and what its written forms may be that we can use the decoding model we are adopting in this book.

4.3 Text segmentation

Now that we know what we are looking for (underlying words) we can turn our attention to the problem of how to extract these from text. While in principle this could be achieved in a single process, it is common in TTS to perform this in a number of steps. In this section and the next we deal with the initial steps of **tokenisation** and **sentence splitting** which aim to split the input sequence of characters into units that are more easily processed by other processes that attempt to determine the word identity.

4.3.1 Tokenisation

Formally, we can think of the input as a sequence of characters that encode a sequence of words:

$$\langle w_1, w_2, w_3, \ldots, w_N \rangle \mapsto \langle c_1, c_2, c_3, \ldots, c_M \rangle,$$

and so our problem in text analysis is simply one of reversing this process so that we can uncover the word sequence from the character sequence, as in the following:

(18) The old lady pulled her spectacles down $\rightarrow$ < THE, OLD, LADY, PULLED, HER, SPECTACLES, DOWN >

In most cases of conventional writing, the words don't overlap with respect to the characters, so we can also represent this relationship as:

(19) The old lady pulled her spectacles down
    THE OLD LADY PULLED HER SPECTACLES DOWN

We can clearly see that the written sentence is easily segmented by using white space as a delimiter. This is possible only because the incidence of white space in the input always indicates the presence of a word boundary. There are numerous cases, however, where this simplistic approach fails. In a sentence such as

(20) In 1997, IBM opened a Polish lab

problems occur because a section of the text may be ambiguous as to its word:

(21) polish $\rightarrow$ POLISH_COUNTRY or POLISH_SHINE

problems where white space is insufficient to act as the only guide to segmentation:

(22) IBM $\rightarrow$ < LETTER_I, LETTER_B, LETTER_M >

cases where the token does not map to a word:

(23) , $\rightarrow$ NULL

and cases where both ambiguity and segmentation problems exist:

(24)
    a. 1997 $\rightarrow$ < NINETEEN, NINETY, SEVEN >
    b. or < ONE, THOUSAND, NINE, HUNDRED, AND, NINETY, SEVEN >
    c. or < ONE, NINE, NINE, SEVEN >

so we see in this example that the process of converting sentences into words is one of both segmentation and classification of the written form of the sentence. Because the classification and segmentation interact, it is not possible to do one and then the other.

Rather, our approach is to perform first a *provisional* segmentation into potential written forms called **tokens**, followed by a second step whereby we examine each token in turn and resolve any ambiguity. This first process is called **tokenisation**; the step which then generates the words from the tokens is called **text analysis** and is the subject of the next chapter.

So, in our first example, the tokenisation process would perform the following mapping:

(25) `The old lady pulled her spectacles down` → < the, old, lady, pulled, her, spectacles, down >

The advantage of dividing the problem this way is that it makes it much easier to perform the text-analysis step. By tokenising (segmenting) the text, we are making it possible for the text-analysis algorithms to focus on one token at a time: while the analysis algorithms may examine the tokens on either side of the one in question, only a few are usually consulted, and this makes the writing of rules and the training of algorithms quite a bit easier.

4.3.2 Tokenisation and punctuation

In languages like English, it is obvious that the presence of white space in the input is a very good guide to tokenisation. In most conventional writing, splitting on white space will often leave a one-to-one mapping between tokens and words. The difficulties soon arise, however: when we encounter sequences of capital characters such as `IBM` we could guess that this will be a letter sequence and therefore go someway towards the classification and rewriting process by splitting this into three tokens. But there are several forms such as `NATO` and `UNICEF` that don't follow this convention; furthermore, sometimes normal words are put into upper case for effect: `lets GO!`.

Dealing with punctuation properly is one of the main issues in tokenisation. In sentences such as

(26) `So, he went all that way (at considerable cost) and found no-one there.`

we see the classic uses of punctuation. Firstly it creates a boundary between two tokens that might otherwise be assumed to be linked, or it creates a link between two tokens that might otherwise be assumed to be separate. So a full stop ends a sentence, a comma ends a clause and both create a boundary between the tokens that precede and follow them. Within this use, we have **status markers**, so that while ., ? and ! all indicate the end of a sentence, they each assign a different status to that sentence. Hyphens and apostrophes link tokens that might otherwise be assumed to be more separate (assuming that white space would be used if there were no hyphen or apostrophe).

More difficult problems arise with technical language. First the rules of using white space in conjunction with punctuation are often different; so that in a time we find

10:34am in which there is no white space after the colon. Furthermore, some of the verbalisations of technical language speak the punctuation as words, so that the words for yahoo.com are <YAHOO, DOT, COM>, where the . is spoken as DOT. It helps in cases like these to make some distinctions. First we need to distinguish **underlying punctuation** from **text punctuation**. Underlying punctuation is the abstract notion of what we normally think of as punctuation in conventional writing, and text punctuation is its written form. These are clearly the punctuation equivalent of words and tokens. So, while we may have two types of underlying punctuation DASH and HYPHEN, there may in fact be only one written form in our character set. This use of one character for two underlying punctuation marks causes ambiguity and this must be resolved. In addition, as we have just mentioned, punctuation is sometimes *spoken* in technical language. We deal with this phenomenon by using our model of verbalisation, so that in yahoo.com the sequence is analysed firstly as an email address, then verbalised into natural language and then spoken. In such cases, we will represent the word that the punctuation is spoken as WORD_COMMA, WORD_COLON and so on, to distinguish from the silent forms of COMMA and COLON.

4.3.3 Tokenisation algorithms

We will now describe a basic tokenisation algorithm.

1. Create an empty list for the tokens to be placed in.
2. Define a start point before the first character of the sentence.
3. Move forwards through the sentence, examine each character and the character following it, and decide whether there should be a token boundary at this point.
4. If yes, append an entry to the token list and copy (without modification) the characters from the start point to the boundary point.
5. Reset the start point to be at the boundary point
6. If at the end of the sentence stop, otherwise go to step 3.

The heart of the algorithm is therefore the part which decides whether a boundary should be placed between two characters of the input. This works on the following criteria.

1. Split on white space, but don't include the white space itself in any token.
2. Split on punctuation, create a separate token for each punctuation character.
3. Split when a contiguous non-white-space sequence changes from one character grouping to another (e.g. 10cm → 10 cm).
4. Don't split in any other cases.

Sometimes more-complex (or more-accurate) tokenisation algorithms are used. The amount of resource put into tokenisation to some extent depends on the sophistication of the process that follows it. If this can handle errors or noise in the tokenisation, then a fairly simple algorithm should suffice. If, however, the subsequent algorithms require more-accurate tokenisation a number of approaches can be applied. The deterministic approach can be extended by considering greater numbers of cases and writing rules by hand to deal with each case in turn (e.g. email addresses, times, dates and so on). Alternatively, a data-driven approach can be taken, whereby we label a corpus with token boundaries and train an algorithm to learn the correspondence. Such an approach is a classic use of a data-driven technique in speech and language processing, and discussion of this will be deferred until the general discussion of this in Section 5.2.1.

4.3.4 Sentence splitting

Many of the algorithms in TTS work a sentence at a time. This is because most linguistic units smaller than this (words, syllables) etc. are heavily influenced by their neighbours, which makes autonomous processing difficult. Sentences, on the other hand, don't interact with each other much, and, apart from some specific phenomena, we can by and large process each sentence independently without problem. The input to the TTS system is not necessarily in sentence form, however, and in many cases we are presented with a document that contains several sentences. The task of **sentence splitting** then is to take the raw document and segment it into a list of sentences.

While sentence splitting is not the most complex of tasks in TTS, it is important to get it right and this is mainly due to the fact that **sentence-final prosody** is one of the phenomena that listeners are most sensitive to. Generating the high-quality sentence-final prosody is hard enough in its own right, but if the sentence boundary is in the wrong place, then the system has no chance.

For conventional writing we make use of the fact that in most cases the writer has clearly marked the sentence boundaries; and so our job is simply to recover these from the text. Perhaps ironically then, our algorithm is based on the lay notion of finding upper-case characters at beginnings of sentences and full-stop characters at the end, and marking sentences as what lies between. The situation is not quite that simple; as we know the full-stop character can be used for a variety of purposes, as can upper-case characters. The task is therefore one of finding instances of full stops and related characters, and classifying them as to whether they indicate sentence boundaries or not.

A basic sentence-splitting algorithm for conventional writing can be defined as follows.

1. Start a new sentence at the beginning of the input.
2. Search forwards through the input to find instances of the possible end-of-sentence characters ., ! and ?.
3. Using the immediate context of these characters, determine whether a sentence boundary exists; if so complete the current sentence and start a new one.
4. End the current sentence at the end of the document (regardless of whether the last characters are ., ! or ?.)

The heart of the algorithm is the decision as to whether a instance of a text form of ., ! or ? constitutes a sentence or not. The most difficult case is . since this can be used as part of an abbreviation, a number, an email address and so on. In conventional writing, a space always follows the full stops at a sentence end, and the next sentence always starts with an upper-case letter. In fact the only *conventional* other use of . is in abbreviations, so that we might find

(27) Dr. Jones lives in St. James Road

In fact, in most modern writing, the use of . in this sort of abbreviation is not common. Professional publishers make use of **style manuals** to help authors with various issues in writing and layout. Conventions differ, but instructions from one academic journal include the following.

- Do not use periods in titles, hence Dr Jones, Mr Mark Johnson Jr
- Do not use periods in letter sequences, hence USA, IBM
- Do use periods for abbreviations in personal names, hence J. R. R. Tolkein (note that a space follows the period)
- Do use periods for Latin letter sequences, hence e.g. and et al.

These guidelines show that the incidence of full stops in abbreviations is probably less common than expected, but there are still potentially confusing situations. Remarkably, many style guides are aware of this and even advise writers "do not start a sentence with an abbreviation". Of course, real life is never so simple, and, even in conventional writing, one sees frequent variations on the above guidelines, such that we must in principle allow for a full stop after any abbreviation. (For example, one sees JRR Tolkein, J.R.R. Tolkein, J R R Tolkein and J. R. R. Tolkein all in writing from the Tolkein Society alone.) Usage patterns are potentially significant, in that, while we may allow a full stop after any abbreviation, a probabilistic algorithm could make use of the fact that they are rarely found in some cases. Of course, if these really were universal conventions they wouldn't have to be spelled out in a style guide; authors would simply *know*.

For casual conventional writing, the situation can be more difficult. In email messages, one finds that some writers do not use a upper-case character at the start of a sentence, or that new lines are used as sentence boundaries and no ending punctuation is found. We also come across the problem in email that the form of the headers is intended for email programs to parse and process, not for reading aloud. So the notion of "sentence" is blurred here. To get around this, we simply have to decide in what form we will read this information and so what form the sentences will take in these genres. One approach is to have each line as a sentence, another is to process the header somehow into a spoken list or other structure.

4.4 Processing documents

4.4.1 Markup languages

It is useful to be able to control the behavior of the TTS system at run time. While global parameters can be set in a number of ways, often it is useful to have very precise instructions that affect specified words. This is most effectively done by interleaving instructions to the TTS system with the actual text to be spoken. In older TTS systems, it was common for each to have its own type of markup. So for example in the Bell Labs system [411] there are special sequences of escape characters that help the user specify what behaviour the system is to follow.

The disadvantage of this is that these special characters would mean nothing to a different TTS system, so a document marked up for one TTS system would not work with another. To get around this, Taylor and Isard [435], [443] proposed the **speech synthesis markup language** (**SSML**) which aimed to provide a general set of markup tags, which could in principle be used by any TTS system. This was extended by a number of further projects, including the **spoken text markup language** by Sproat *et al.* [411] and the **java speech markup language** developed by Andrew Hunt at Sun Microsystems. These served as the basis of the synthesis part of the W3C standard **VoiceXML** (also called SSML, but not to be confused with the original).

All these schemes are based on the idea of **generalised markup**, which was first developed in the typesetting industry. There, the idea of **markup** had been around for a long time, where a typesetter would mark some text to be printed in 12 point, in italics or so on. The problem with such markup instructions is that they are **physical**, meaning that they are very specific to the type of printing being used. If someone wished to print the document in a bigger font for instance, the insistence on using 12 point (which may have seemed large in a 10pt document) means that the intended effect is not achieved. Generalised markup attempts to get round this by splitting the type setting into a two-stage process. The author, or editor, adds **logical** markup to the text, for example by indicating that something is a heading, or requires emphasis. In the second stage, this is then rendered by a process much more specific to the actual printing mechanism being employed. As the document itself contains only logical instructions, it is much more portable, and can be printed on a number of devices and in a number of styles.

The speech-synthesis markup languages followed a similar philosophy, so instead of specifying that a word should be spoken at 150 Hz (which would have different effects for different speakers), instead more abstract or "logical" instructions are used, which for instance state that a particular word is to be emphasised. A few simple examples will now be considered.

Tag	Example	Explanation
<s>	This tag can be used <s> to override the decisions of the <s> sentence splitting algorithm.	This tag indicates that a sentence break should be placed at this point. It is a good way for the author to override any possible shortcomings in the sentence-splitting algorithm.
<emphasis>	Sometimes we <emphasis> need </emphasis> to give a word special emphasis.	This is used to give particular emphasis to a word or group of words. Note how abstract or "logical" this tag is; no reference is made to the phonetics of how to render the emphasis as heightened pitch, longer duration and so on.
<voice>	<voice gender="female" age="20"> This is a female voice </voice> <voice name="mike"> And this is a male one <break/> Hope you're OK. </voice>	Many TTS systems have the capability of speaking in more than one voice. This tag can switch between voices by specifying general attributes or specific voice names.

In an SSML document, the tags are interleaved with the text to be spoken. It is of course important that the two are distinguished, so the first stage in processing an SSML document is to perform an XML parsing step. There are several standard parsers that do this and subsequently produce a structure in which the text and tags are clearly separated. From there, it is up to the TTS system itself to decide how to interpret each tag. The tags vary considerably in their purpose, and hence they require processing by different modules in the TTS system. The standard way of dealing with this is to do a partial processing of the XML initially, and then leave hooks in the tokens to tell subsequent modules what to do. As can be seen with the <s> tag, XML, tokenisation and sentence splitting can interfere, so it is often a good idea to perform all the required processing in a single step.

4.4.2 Interpreting characters

Ever since computers have been around they have been used to store text, but as computers operate only in ones and zeros, text cannot be represented directly. Rather a **character-encoding** scheme is used, which maps internal computer numbers onto characters. So long as the same convention is maintained by the readers and writers, chunks of memory can be used to represent encoded text. The issue for a TTS system is to identify the character encoding being used and process it appropriately. Partly due to the sheer diversity of the world's writing systems and partly due to historical issues in the development of character-encoding schemes, there are several ways in which characters can be encoded.

One of the earliest and most common encodings was **ASCII** (the American Standard Code for Information Interchange), which gave a character value to the first 127 values in an 8-bit byte of memory (the final bit was left unused). Ascii is of course a standard that was developed in the USA, which could represent only the 26 characters of the standard English alphabet. Most of the world's languages of course require different characters and so ascii alone will not suffice to encode these. Extensions were developed primarily for use with other European characters, often by making use of the undefined eighth bit in ascii.

The **ISO 8859** standard defines a number of encodings. For example **ISO 8859-1** encodes enough extra characters for most western-European languages such as French and German. **ISO 8859-2** encodes eastern-European languages, while **ISO 8859-7** encodes Greek characters. These are all 8-bit encodings that can be used without trouble so long as it is known which is in use. All contain ascii as a subset and so have the advantage of backwards compatibility. Many languages, in particular southeast-Asian languages, have far more characters than can be included in 8 bits, and so cannot be accommodated in single-byte encoding schemes. Various incompatible **wide-character** encodings were developed for these languages before a universal standard **unicode** was developed in the early 1990s. This provides an encoding for virtually all languages, although some controversies (largely political rather than technical) still prevent its completely universal adoption.

Unicode simply defines a number for each character; it is not an encoding scheme in itself. For this, various schemes have been proposed. **UTF-8** is popular on Unix machines and on the internet. It is a variable width encoding, meaning that normal ascii remains unchanged but wider character formats are used when necessary. By contrast **UTF-16**, which is popular in Microsoft products, uses a fixed-size 2-byte format. More recent extensions to unicode mean that the original 16-bit limitation has been surpassed, but this in itself is not a problem (specifically for encodings such as UTF-8 that are extensible).

4.5 Text-to-speech architectures

The various algorithms described here all produce output, and this has to be passed to the next module in the TTS system. Since the next module will produce output too, it makes sense to have a general mechanism for storing and passing information in the system. This brings us to the issue of how to design our text-to-speech **architecture**.

Most TTS systems have adopted a solution whereby a single data structure is passed between modules. Usually, this data structure represents a single sentence, so the TTS system works by first splitting the input into sentences by use of the sentence splitter, forming a data structure containing the raw text for each sentence, and then passing this data structure through a series of modules until a speech waveform is generated. We will term this the **utterance structure**.

In many early systems, the utterance structure was no more than a string that was passed between modules. For doing this systems could adopt either an **overwrite** paradigm or an **addition** paradigm. For instance, in an overwrite system, if the input text was

```
The man lived in Oak St.
```

the text-normalisation module might produce output like

```
the man lived in oak street
```

The subsequent pronunciation module might then produce output like

```
dh ax m a n l i v ax d ax n ou k s t r ii t
```

The point is that, in each case, the module takes its input and outputs only the particular results from that module. In an addition paradigm, each module adds its output to the output from the previous modules. So in the above example, the output from the text-normalisation module might produce something like

```
The|the man|man lived|lived in|in Oak|oak St|street .|null
```

The subsequent pronunciation module might then produce output like

```
The|the|/dh ax/ man|man|/m ae n/ lived|lived|/l ih v ax d/
in|in|/ax n/ Oak|oak|/ow k/ St|street|/s t r iy t/ .|null
```

The advantage of addition systems is that it is often hard to know in advance of development what each module requires as input, and if we delete information we run the risk of supplying a module with impoverished input. For example, we might find that the phrasing module (which is called after the text-normalisation module) requires punctuation, which can clearly not be used if the information has been deleted. The trouble with string-based addition systems is that they can become unwieldy rather quickly. Even in the example above, with a short sentence and the operation of only two modules, we see that the string is rather difficult to read. Furthermore, each module will have to parse this string, which becomes increasingly more complicated as the string becomes richer.

It is common therefore for modern systems to adopt an approach that follows the addition paradigm, but does so in a way that allows easier and clearer access to the information. Two of the most widely used formalisms are the **heterogeneous relation graph** or **HRG** formalism [441], used in the Festival system, and the **delta** system [201], used in a number of commercial and academic systems.

Both these are based on the idea of building data structures based on linguistic **items**.[2] An item can be any single linguistic unit, including a word, phone, syllable, pitch accent or other entity. The item is a **feature structure**, which is a data structure common in computational linguistics that defines a collection of properties as an unordered list of **key–value pairs**, also known as **attribute–value pairs**. Attributes must be atomic and drawn from a finite set of defined values. Attributes must also be unique within the feature structure, so that, when given an attribute, at most one value will be found. From a computer-science perspective, feature structures are very similar to **associative arrays** or **maps**, and can also be though of as **lookup tables** or **dictionaries**. In mathematics, they can be thought of as **finite partial functions**; that is, functions that have a finite domain and for which not all domain values are defined. An example feature structure is given below:

$$word : \begin{bmatrix} \text{NAME} & abuse_1 \\ \text{POS} & noun \\ \text{TEXT} & abuse \\ \text{PRON} & /@buws/ \end{bmatrix} \quad (4.1)$$

In the above example, all the values are **atomic**. It is also possible to have feature structures themselves as values, which is useful for grouping particular types of information

---

[2] For clarity, we will adopt HRG terminology throughout our explanation.

together:

These items represent a single linguistic entity. Since there are usually several of these in each sentence we need an additional data structure to store these. In the HRG formalism, **relations** are used to group items of a similar type. In the standard HRG formalism, we have three types of relation: **list**, **tree** and **ladder**. Many types of linguistic data are best stored in lists, and this is how word, syllable and phone information is normally stored. Other types of linguistic data are most suited to trees, and this is how syntax, prosodic phrasing and syllable structure can be stored. Less common is the use of ladders, which are used to represent connections between linguistic data that are not organised hierarchically, as in a tree. The ladder structure can be used to implement the data structures of **autosegmental phonology** [179], which is used for instance to represent the relationship between syllables and intonation accents. The three types of structure are shown in Figure 4.1.

One way of configuring the data structures is to have a separate relation for the output of each module, so that overwriting the output of previous modules is avoided. While straightforward, this configuration can makes the utterance structure very dependent on the particular TTS set-up, so that a module that performs pronunciation in one step would generate one relation, whereas one that implements lexical lookup, letter-to-sound rules and post-lexical processing as three separate modules would generate three relations. An alternative is to separate relations from particular modules and instead use relations to store information for each linguistic type, so we can have a "phone" relation that comprises a list of phone items, a syllable relation, word relation, intonation relation and so on.

The relations are graphs of nodes arranged in the form of lists, trees and ladders. Each node can be linked to an item, and in this way the relation can show the contents of the items and the relationship between them. Named links such as **next()**, **previous()**,

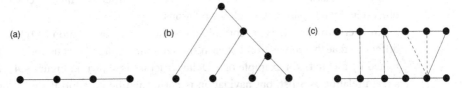

Figure 4.1 The three types of relation: (a) list relation, (b) tree relation and (c) ladder relation.

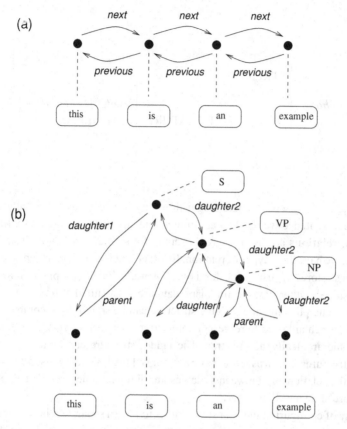

Figure 4.2 (a) A list relation containing word items. (b) A syntax relation containing syntax items.

daughter(), **parent()** and so on are used to navigate through the relation. Examples of a list and a tree relation are shown in Figure 4.2.

One of the key features of the HRG architecture is that items can be in more than one relation at once. For example, word items can be be in both the word relation and the syntax relation. This is useful because often we wish to process the words from left to right as a list: the word relation is used for this. Sometimes, however, we want to access the syntax relation, which is a tree in which the words form the terminal nodes. Having the same information in the two different places is unappealing because an update to one item requires an update to the other, which is often awkward to achieve. Figure 4.3 shows the syntax and word stream together. This ability to include items in multiple relations means that usually any information in the utterance structure can be accessed from any other item by navigating through the relations.

An alternative to the HRG formulation is the Delta formulation [201]. This mainly differs in that the connection between the relations (called streams in Delta) is determined by a **chart**. An example of a Delta structure is shown in Figure 4.4. Here, each stream/relation is a list, but navigation is performed by going to the edge of the item and ascending or descending through the chart. The Delta system is somewhat simpler

4.6 Discussion

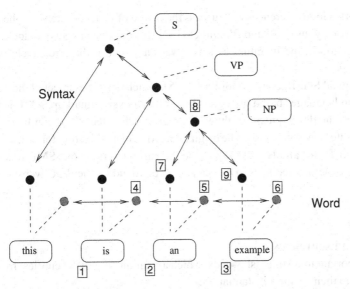

Figure 4.3 An example of a word and syntax relation in which the items in the word relation are also in the syntax relation.

word				structure					
syllable				syl				syl	
syllable structure 2		onset			rhyme		onset	rhyme	
syllable structure 1				nucleus	coda		nucleus	coda	
phoneme	s	t	r	u	k	ch	@	r	

Figure 4.4 Thing.

and often easier to use than the HRG formalism, but is somewhat less powerful in that it cannot represent general tree information.

4.6 Discussion

4.6.1 Further reading

The ontological status of linguistic entities was a prime concern of linguistics in the middle part of the twentieth century, but surprisingly has not been the subject of more recent work. Good references that discuss the nature of words, sentences, phonemes and so on include Lyons [290] and Bloomfield [57]. Tokenisation in general does not receive much attention in the literature as it is always assumed to be a relatively minor (and simple) task.

The original SGML book is Goldfarb [178], which describes in detail the justification for markup languages based on logical rather than physical markup. XML is inherently tied in with the development of the web, so good references for it can be found on the W3C website (www.w3c.org). The original SSML and justification for its use in TTS can be found in Taylor and Isard [443]. Since the current version of SSML was developed as a web standard, references to this can again be found on the W3C website.

4.6.2 Summary

Words and sentences
- It is important to distinguish **words**, which are unique linguistic entities, from **tokens**, which are their textural instantiation.
- **Homographs** are sets of words that share the same text form. There are three reasons why homographs arise.
  1. The same **lexeme** being used as a noun and as a verb, which have different pronunciations (e.g. project).
  2. From abbreviated forms of two words producing the same written form (e.g. DOCTOR and DRIVE producing Dr).
  3. By historical accident (e.g. BASS-MUSIC and BASS-FISH).
- We use a series of tests to define our set of words by looking at distinctions in written form, spoken form and meaning.
- Sentences have a number of possible valid definitions. We use one based on the writing since this is most helpful to TTS.

Text segmentation
- To make subsequent processing easier we require that the input be segmented into sentences and tokens.
- It is hard to divide the input into sequences of characters that align one-to-one with the words, hence tokenisation often produces a provisional segmentation of the text.
- Deterministic algorithms are often accurate enough to produce acceptable tokenisation and sentences.

Documents
- Various **markup languages** exist for speech synthesis. They are based on **XML**, the successor to **SSML**.
- These markup languages can be used to give extra instructions to the TTS system to override default behaviour.
- In modern systems, this markup takes the form of **logical** rather than **physical** instructions.

Text-to-speech architectures
- All TTS systems are composed of a series of modules and all systems have to pass data from one module to the next.
- Early systems typically used strings for this purpose.
- More-modern systems use sophisticated structures whereby each type of linguistic information (e.g. words, syllables, phonemes) can be accessed separately.
- The **HRG** system and **Delta** system are two popular systems for this purpose.

5 Text decoding: finding the words from the text

5.1 Overview of text decoding

The task of text decoding is to take a tokenised sentence and determine the best sequence of words. In many situations this is a classical disambiguation problem: there is one, and only one, correct sequence of words that gave rise to the text, and it is our job to determine this. In other situations, especially where we are dealing with non-natural-language text such as numbers and dates and so on, there may be a few different acceptable word sequences.

So, in general, text decoding in TTS is a process of resolving ambiguity. The ambiguity arises because two or more underlying forms share the same surface form, and, given the surface form (i.e. the writing), we need to find which of the underlying forms is the correct one. There are many types of linguistic ambiguity, including word identity, grammatical and semantic, but in TTS we need only concentrate on the type of ambiguity which affects the actual sound produced. So, while there are two words that share the orthographic form bank, they both sound the same, so we can ignore this type of ambiguity for TTS purposes. Tokens such as `record` can be pronounced in two different ways, so this *is* the type of ambiguity we need to resolve.

In this chapter, we concentrate on resolving ambiguity relating to the verbal component of language. At its core, this means identifying the correct sequence of words for every sentence, but we shall also consider in brief the issue of resolving ambiguity in the syntactic structure of a sentence also, since this is thought by some to have a lesser but still significant impact on speech. In the next chapter, we concentrate on the corresponding process for prosody, where we attempt to find a suitable underlying prosodic representation from the text.

This process boils down to the following tasks.

1. Determine the semiotic class (number, date, natural language etc.) for each token.
2. For each sequence of tokens in the same semiotic class, resolve any ambiguity and find their underlying forms.
3. For non-natural-language classes, perform a verbalisation process, which converts their underlying forms into a sequence of words that can be spoken.

Generally, non-natural-language tokens are quite easy to decode; the difficulty is not in knowing how to decode the token 1997 once we know it is a year, but rather in determining that it is a year in the first place. With natural-language tokens, even once we have established that they are indeed natural language, we still can face considerable difficulty in determining their underlying form because different words can share the same text form.

While there are many subtle issues involved in text decoding, in essence, then, it reduces to two main problems; **semiotic classification**, i.e. the assignment of a semiotic class to each token, and **homograph disambiguation**, whereby we determine the correct word for ambiguous natural-language tokens. Both these tasks involve assigning a category to a token, and, as we shall see, they can be solved with very similar algorithms. Because of this, we now turn our attention to a general overview of text disambiguation.

5.2 Text-classification algorithms

5.2.1 Features and algorithms

Both in semiotic classification and in homograph disambiguation, our goal is to assign a **label**, drawn from a pre-defined set, to each token. This process is one of **classification**, which, as we shall see, crops up all the time in TTS, to the extent that we use a number of basic approaches again and again for various TTS problems.

More formally, we say that for each unit (tokens in this case) we wish to assign a label l_i drawn from a set of labels $L = \{l_1, l_2, \ldots, l_N\}$. Our decision is based on analysing a number of **features**, each of which has set of legal **values**. These serve as information as to which label is best in each case. This process can be defined as a function C

$$C : F \mapsto L$$

which maps from a feature space F to a label space L.

Let us consider a simple example where we wish to disambiguate the token read which can encode the past-tense word READ-PAST, pronounced /r eh d/, or the present-tense/infinitive word READ-PRESENT, pronounced /r iy d/. In this case, our label set comprises two labels such that $L = \{l_1 = \text{READ-PAST}, l_2 = \text{READ-PRESENT}\}$. We now consider some example sentences:

(28) a. I was going to read it tomorrow.
 b. I read it yesterday.
 c. It is vital that young people learn to read at the correct stage.
 d. I read him the riot act.

We see that the other words/tokens in the sentence and the way they interact with the token read give us strong clues as to which word is correct in each case. These "clues" can be formulated as features, so in the above examples we might have features and values defined as

Name	Feature	Legal values
F_1	Time token	TOMORROW, NOW, YESTERDAY
F_2	Is the preceding word TO?	yes, no
F_3	Is the sentence part of a narrative?	yes, no

Hence a classification algorithm working on this is defined as

$$C(F_1, F_2, F_3) \mapsto l_1 \quad \text{or} \quad l_2$$

Below we describe a range of algorithms that operate in this way. Some are explicitly formulated by a human designer and by convention we call such algorithms **hand written** (although obviously it is the brain which really writes the rules). In these approaches, a human designer has considered the effect of the features and has by one means or another written a system that performs the above mapping. The other common approach is **data driven**, where the mapping is learned by analysis of a database of examples. Often this database is **labelled**, meaning that it comprises a set of pairs of feature values and labels, with the idea that an algorithm can be trained to reproduce this correspondence. Not all data-driven algorithms work in this way; some are **unsupervised**, meaning that they look at the inherent patterns in the data and attempt to come up with a mapping that way. An important sub-class of data-driven algorithms consists of algorithms that are **statistical** or **probabilistic**.

Regardless of which approach is taken, the goal is effectively the same, and all encounter the same problems to be overcome. The difference between hand written and data driven is not as large a difference as it is often portrayed in that, even if a human designs the rules, they presumably do this on the basis of their experience of the problem, in which they have encountered data with these features and labels. Conversely, using an arbitrary data-driven algorithm on data generally produces poor results; better results are obtained by pre-processing the features and constraining the model in some way, which of course requires knowledge of the problem.

The basic problem in all classification algorithms, no matter what type, comes from the issue of how features interact. In general, the features cannot be considered independent, such that we count the evidence from one feature, then the next and so on and then combine these to make a final decision. If we could do this, then the design of classification algorithms would be a simple matter. Rather the features tend to interact, so, in the above example, instead of examining a separate situation for each feature, we have to consider *all* combinations of feature values. If we had, say, three values for the first feature (TOMORROW, NOW, YESTERDAY), and yes/no values for the second and third, this would give use a total of $3 \times 2 \times 2 = 12$ possible feature-combinations.

The fundamental issue in classifier design is how to deal with this feature-combination issue. On the one hand, features hardly ever operate independently, so the effect of particular combinations needs to be taken into account. On the other, it is generally impossible to deal with every feature combination uniquely because there are simply far too many. If one uses hand-written rules there are too many cases to deal with; if one uses a statistical algorithm the amount of data required is too large. The worst case is where the feature combinations are completely unpredictable, in which case the only answer is to use a **lookup table** that lists every combination of feature values and specifies a label:

F_1	F_2	F_3	Label
YESTERDAY	yes	yes	READ-PRESENT
YESTERDAY	yes	no	READ-PRESENT
YESTERDAY	no	yes	READ-PAST
YESTERDAY	no	no	READ-PAST
NOW	yes	yes	READ-PRESENT
NOW	yes	no	READ-PRESENT
NOW	no	yes	READ-PAST
NOW	no	no	READ-PRESENT
TOMORROW	yes	yes	READ-PRESENT
TOMORROW	yes	no	READ-PRESENT
TOMORROW	no	yes	READ-PAST
TOMORROW	no	no	READ-PRESENT

Our above example is quite small in fact; more realistically we might have 10 or 20 features that we used and in such cases the number of feature combinations might be in the thousands or even millions. In general of course, the more features we use, the more accurately we should be able to make our prediction. However, as we add more features, the number of feature combinations increases and we become less likely to observe a full set. This problem is known in statistics as the **curse of dimensionality** and gives rise to the phenomenon known as **data sparsity**, meaning the inability to obtain sufficient examples of each feature combination. While using fewer features will result in less data sparsity, it also runs the risk that the classifier will be too crude and miss out on important distinctions. Thankfully, the **pathological case** where every feature combination operates differently is very rare; and, in general, while features rarely operate independently, they do have well-behaved effects, such that we can build models, or make assumptions about feature interactions, that allow us to avoid the lookup-table case.

As we shall see, all the classification approaches below work by using some knowledge or assumptions about how the features interact in order to reduce the number of parameters that need to be specified. One common approach is to **transform** the original features into a new feature space in which the new features do in fact operate independently. Another approach is to determine which features interact most, model the dependences among these and then assume that the other features do in fact operate independently. In our above example, we may decide for instance that, while F_1 and F_3 often interact, F_2 operates independently, which means that we can reduce the size of the lookup table to six entries, which can then just be combined with the value for F_2.

Γ_1	F_3	Label
YESTERDAY	yes	READ-PAST
YESTERDAY	no	READ-PRESENT
NOW	yes	READ-PRESENT
NOW	no	READ-PRESENT
TOMORROW	yes	READ-PAST
TOMORROW	no	READ-PRESENT

In general, hand-written rules can be quite successful when the number of features is low and the amount of training data is small. A knowledgeable designer can quite often use insight to design a classifier that gives reasonable performance. This becomes increasingly difficult as the number of features increases and the interactions among them get more complicated. Another way of looking at this is to say that, as we acquire more training data, it becomes possible to use more features and to model the interactions among them more easily. The complexity of this soon exceeds human capabilities, so a data-driven algorithm will tend to deliver better performance.

5.2.2 Tagging and word-sense disambiguation

Semiotic classification and homograph disambiguation are close enough in terms of formalism that by and large the same techniques can be used for either problem. We find that for these, and related problems, we have two main types of technique. The first is known as **tagging** and is the generalisation of the subject of **part-of-speech (POS) tagging**. In POS tagging, the goal is to assign a part-of-speech tag c, drawn from a pre-defined set C, to each token t. Hence for a sentence like

(29) The man took a walk in the park

we assign a POS tag to each word as follows:

(30) The/det man/noun took/verb a/det walk/noun in/prep the/det park/shelf

where the ambiguous token walk, which could be a noun or a verb, has been assigned a noun label. In general then, a **tagger** assigns a label to every token, and these are drawn from a single set C.

The second type of algorithm performs **word-sense disambiguation (WSD)**. The difference between this and tagging is that, in word-sense disambiguation, we define multiple sets of labels $L_1, L_2, \ldots, L_M$ and choose a label from one of these sets for only some of the tokens. In natural-language processing, WSD is often used for semantic disambiguation to resolve cases such as

(31) I took the money to the bank

where we are really concerned only with the disambiguation of the token bank. This has two senses, BANK_MONEY and BANK_RIVER, and these are particular to that token only (i.e. these labels are meaningless with regard to the token took). In TTS we are of course concerned with homonyms only if they affect pronunciation, so that instead we would look at examples such as

(32) He played bass guitar

with the aim of resolving it to

(33) He played bass/BASS-MUSIC guitar

Similar techniques can be used for both tagging and WSD; the reason why these are considered two different problems is that one is mainly concerned with syntactic ambiguity, the other with semantic ambiguity. In the following sections, we give as examples a number of algorithms. In these we use homograph disambiguation for all the examples; but this is purely for reasons of exposition. In general, all these techniques can be used for semiotic classification (and for other types of classification described in Chapter 6).

5.2.3 Ad-hoc approaches

While we like to imagine that beautiful well-constructed theories underlie our algorithms, we frequently find that when it comes to text classification many systems in practice simply use a hodge-podge of rules to perform the task. This is particularly common in approaches to semiotic classification in genres where the amount of non-natural-language text is very low, so that only a few special cases (say, dealing with numbers) are required.

As these systems are by their very definition ad-hoc we can't give a comprehensive guide to how they work, but often they are configured as a series of regular expressions that try to find a pattern (e.g. if the number is four digits long and has the form 19xx or 20xx it is a year). These rules are often ordered somewhat arbitrarily, in many cases with the most recently written rules last. Sometimes the output is directly rewritten over the input and sometimes systems keep running in a repeated fashion until no rules fire.

For text genres dominated by natural-language text, and in situations where very little training data is available, this approach may be adequate for semiotic classification. This is less likely to work for homographs, but again, depending on the language, the level of errors may be acceptable.

5.2.4 Deterministic rule approaches

A more principled approach is to define a general rule format, a rule processor or engine and a clearly defined set of rules. This at least has the advantage of **engine/rule separation**, meaning that the way the rules are processed and the rules themselves are separate, which helps with system maintenance, optimisation, modularity and expansion to other languages. One common type of rule is the **context-sensitive rewrite rule**, which takes the form

$$A \rightarrow B/d/C$$

meaning that token A is converted ("rewrites to") d when preceded by token(s) B and followed by token(s) C. In general, the contexts B and C can be of any length. This rule has the disadvantage that only the very immediate context can be used, so a second type of rule, called a **collocation rule**, is also popular. This takes the form

$$A \rightarrow d|T$$

which reads A converts to d if a **trigger token** T is present in the sentence. The main difference is that T does not have to be near to A. Often this approach is called a **bag-of-features** approach since it uses evidence based on other tokens without considering their order in the sentence. Token collocations are the most common type to consider to use, since we can observe the tokens directly, but it is also possible to use words as triggers. These are obviously not directly observable, so we have to perform some classification before we can determine the words, and of course, even once found, our algorithm might not have chosen the word correctly.

If we consider some sentences with an ambiguous token, we can see quite clearly how collocation information can help (from Yarowsky [506]):

```
... it monitors the lead levels in drinking...      LEAD-METAL
... median blood lead concentration was ...         LEAD-METAL
... found layers of lead telluride inside ...       LEAD-METAL
... conference on lead poisoning in ...             LEAD-METAL
... strontium and lead isotope zonation ...         LEAD-METAL

... maintained their lead Thursday over ...         LEAD-AHEAD
... to Boston and lead singer for Purple ...        LEAD-AHEAD
... Bush a 17-point lead in Texas, only 3 ...       LEAD-AHEAD
... his double-digit lead nationwide.               LEAD-AHEAD
... the fairly short lead time allowed on ...       LEAD-AHEAD
```

One simple rule-based system works as follows. Firstly, we require a lexicon, which gives the orthography for each word. With this, we can easily determine check whether two words in the lexicon have the same orthography, and from this compile a list of tokens that have two or more word forms. Next we write by hand a set of rules that **fire** (i.e. activate) on the presence of the trigger token, for example

```
concentration      LEAD-METAL
levels             LEAD-METAL

singer             LEAD-AHEAD
dog                LEAD-AHEAD
```

At run time, we then move through the sentence from left to right examining each token in turn. When we find one that is ambiguous (that is, it appears in our list), we look for trigger tokens at other positions in the sentence that will form a collocation with the current token.

In longer sentences we may, however, find several collocation matches, so exiting the list on finding the first may result in an incorrect decision. One extension then is to search the entire list and choose the word that gives the most matches. An alternative is to order the list consciously in some way, with the idea that the choices at the top of the list are more definite indicators than those elsewhere. For example, if we find bass followed immediately by guitar, we can be almost certain that this is BASS_1 (rhyming with "base") rather than the type of fish. Another way of using the rules is to run the most *specific* rules first, so if we have a rule that has several trigger tokens, or a very rare trigger token (e.g. strontium), these fire before more general rules. The idea here is that very specific rules are much more likely to be appropriate for that word only.

There are fundamental difficulties with regard to this and similar hand-written rule approaches. Firstly, it can be extremely laborious to write all these rules by hand. Secondly, we face situations where we have an ambiguous token but no trigger tokens are present in our collocation list. Both these problems amount to the fact that the mapping from the feature space to the label is only partially defined. Finally, we face situations where, say, two rules match for each word – which do we then pick? For these reasons, we now consider machine learning and statistical techniques.

5.2.5 Decision lists

One of the weaknesses of the simple collocation list is that it doesn't give any indication of how strong a collocation might be. While we can use ordering (as in the case of bass guitar), finding the correct ordering can be difficult and can still cause problems when contradictory rules fire. An alternative is then to use statistics to give us an indication of how strong or how likely a particular collocation is. One such method is to use **decision lists**, introduced by Rivest [373], and used in TTS by Yarowsky [505], [506], Gale *et al.* [125], [169] and Sproat [411].

The basic idea of decision lists is to determine the strength of the link between a token and its trigger, which is done by measuring the frequency of occurrence. In the following table (taken from [506]), we see the raw counts which have been found for the token bass:

Trigger	BASS-MUSIC	BASS-FISH
fish	0	142
guitar	136	0
violin	49	0
river	0	48
percussion	41	0
salmon	0	38

From this it is clear that the trigger words are very strong disambiguation indicators, but also that some triggers occur more frequently that others. From these counts, we can calculate the conditional probability of seeing a word given the collocation, e.g.

$$P(BASS-FISH|\text{guitar}, \text{bass})$$

$$P(BASS-MUSIC|\text{guitar}, \text{bass})$$

We then compute the log-likelihood ratios for each collocation:

$$\text{abs}\left(\log\left(\frac{P(BASS-FISH|\text{collocation}_i)}{P(BASS-MUSIC|\text{collocation}_i)}\right)\right)$$

which then give us a sorted decision list:

Log-likelihood	Collocation token	Word
10.98	fish	BASS-FISH
10.92	stripped	BASS-FISH
8.87	sea	BASS-FISH
9.70	guitar	BASS-MUSIC
9.20	player	BASS-MUSIC
9.10	piano	BASS-MUSIC

At run time, we simply start at the top of the list and progress downwards until we find the collocation which matches our input sentence. We then take the word that results from that as the answer. There are many subtle modifications, which for instance smooth the probability estimates, count all the evidence (not just the best case) and consider the conditional probability of a word being the answer when we have several triggers present. Yarowsky notes, however, that the simple approach is often good enough that many of these modifications don't add any extra accuracy.

5.2.6 Naive Bayes classifier

Use of the **naive Bayes classifier** is a simple and popular approach in word-sense disambiguation. In this, we attempt to estimate the probability that word w_i is the correct one for a token on the basis of a number of features:

$$P(W = w_i | f_1, f_2, \ldots, f_N) \tag{5.1}$$

This is difficult to estimate directly, so instead we make use of **Bayes' rule**, which gives us a more tractable formulation:

$$P(W = w_i | f_1, f_2, \ldots, f_N) = P(W = w_i) P(f_1, f_2, \ldots, f_N | w_i) \tag{5.2}$$

In this, the expression $P(f_1, f_2, \ldots, f_N | w_i)$ is easier to calculate than the expression of Equation (5.1) because all we need do now is find a word that gives rise to an ambiguous token and then count occurrences of the features $f_1, f_2, \ldots, f_N$ around this word. As we mentioned in Section 5.2.1, in general the interactions within a set of features $f_1, f_2, \ldots, f_N$ mean that calculating $P(f_1, f_2, \ldots, f_N)$ or $P(f_1, f_2, \ldots, f_N|w)$ can be quite complicated because we have to consider every possible combination of features. This approach is called the **naive** Bayes classifier because it makes the huge simplifying assumption that none of these features interact, such that they are statistically independent. This means that the expression $P(f_1, f_2, \ldots, f_N)$ can be simplified:

$$P(f_1, f_2, \ldots, f_N) = P(f_1)P(f_2), \ldots, P(f_N)$$
$$= \prod_{i=1}^{N} P(f_i)$$

which means that all we need find are the probabilities that a word is found with each feature individually:

$$P(W = w_i | f_1, f_2, \ldots, f_N) = P(W = w_i) \prod_{i=1}^{N} P(f_i)$$

When an ambiguous token is found, the classification is performed by calculating the above expression for each word and choosing the answer as the one with the highest probability. This algorithm has the advantage that the features can be quite heterogeneous; for instance, as well as simply the tokens themselves (as with the decision-list case), we can also use POS tags, position in sentence and any other features we think might help. While the algorithm is obviously crude in the sense that it ignores feature dependences, it does at least give a defined mapping from a point in the feature space.

5.2.7 Decision trees

Given the simplicity of the naive Bayes classifier, it performs surprisingly well. However, the assumption that the features are independent is in general too crude and better performance should be obtainable by a system that models interaction among features. To illustrate the problem of feature dependence we might see that the token Morgan is a strong indicator that we are dealing with BANK_MONEY, not BANK_RIVER. We might also find that the token Stanley is just as strong an indicator. But it is clearly false to assume that these tokens operate independently of one another and that it is just coincidence that they appear in the same sentence. Morgan Stanley is of course the name of a real bank and those two tokens very frequently co-occur. Once we have used one feature, seeing the second adds very little extra information that we are dealing with BANK_MONEY.

In general though, we have seen that it is intractable to assume that every combination of features is unique and needs to be calculated separately: we simply never have enough training data to estimate such models accurately. One popular solution to this is to use **decision trees**. These model interactions between features but do so in a particular way so as to concentrate on learning the influence of the most important features first, such that rare or undiscriminating feature combinations can be ignored with some safety. We will explore decision trees in full in Section 15.1.9; for now we will give a quick overview and show how they can be used for text classification.

At run time, a decision tree asks a series of questions in order, and, on the basis of the outcome of these, delivers a label. In essence then, this is no more than a series of nested "if then else" constructions in a computer program. For our cases, one set of questions might be "if the token `lead` and the following token contains the string `poison` and if the token `pollution` or `heavy` is also found in the sentence then mark this as LEAD_METAL".

Any series of questions can be formulated as a tree, and in fact decision trees can be written by hand. We do, however, have a number of algorithms that automatically build trees and it is because of these that the technique is popular. The decision-tree growing works as follows.

1. Create an initial group containing all the instances of one ambiguous token.
2. Design a set of questions based on our features, which we think will help in the word identification.
3. For each question
   (a) form two new groups based on the value of the feature,
   (b) measure the combined **impurity** of the new groups.
4. Find the question which gives the biggest reduction in impurity.
5. Delete this question from the list of features to be examined.
6. Form two new groups based on this question.
7. Repeat steps 3 to 6 on each new group until the stopping criteria have been satisfied.

The **impurity** of a cluster can be formulated in a number of ways, as we shall see in Section 15.1.9. Here even a simple measure will do, for example the ratio of word-1 over word-2. The stopping criteria usually involve specifying a minimum decrease in impurity. The decision tree gets round the problem of modelling all the feature combinations by effectively clustering certain feature combinations together. This is not always ideal, but it certainly does allow more accurate modelling than naive Bayes, since we can see that the same feature can appear in different parts of the tree and have a different effect on the outcome.

5.2.8 Part-of-speech tagging

A somewhat different approach to the problem is to use a **tagging** algorithm. This type of algorithm arose from the problem of **part-of-speech (POS) tagging**, which was concerned with finding the sequence of syntactic categories or parts of speech for each token in a sentence. These algorithms are of interest to us in two ways. Firstly, many homographs are **part-of-speech homographs**. In English, these are often noun/verb pairs where the same token can be used as a noun or verb from the same semantic root. In fact a huge number of words in English have noun/verb pairs, so that one can have GO FOR A WALK and I LIKE TO WALK. In most cases these are **syntactic homonyms**, meaning that, while they may operate differently with regard to syntax, they are spelt and pronounced the same way. A few, however, are spelt the same way but pronounced

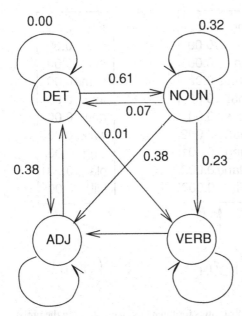

Figure 5.1 Example transition probabilities from a simplified four-state POS tagger. The probabilities exiting a state always add up to 1.

differently and these are the cases which concern us. The idea of POS tagging is that if we find the part of speech from the text we will know which word we are dealing with.

A second interest in POS tagging is that this technique can be used to assign any type of labels to the tokens, not just parts of speech. Hence we could use a tagging algorithm for general disambiguation purposes.

POS tagging has been a problem of concern in NLP for some time. There it is often used as a first required step before syntactic parsing is performed, but it is also used in other applications and seen as a problem in its own right. Many approaches have been tried but probably the most common approach is to use a **hidden Markov model**. These are dealt with fully in Chapter 15, but for now we can give a quick overview explaining their use for tagging purposes. An HMM tagger operates probabilistically to find the most likely sequence of tags (labels as we have previously called them) L for an input sequence of tokens T. We write this as

$$\hat{L} = \underset{l}{\operatorname{argmax}}\{P(L|T)\}$$

Rather than compute the probability that a token generates a tag $P(l|t)$, an HMM is a **generative model** that computes $P(t|l)$, that is, the probability of seeing a token given the tag. We can calculate one from the other by concatenating the HMM models to form sequences and then applying Bayes' rule:

$$P(L|T) = \frac{P(L)P(T|L)}{P(T)}$$

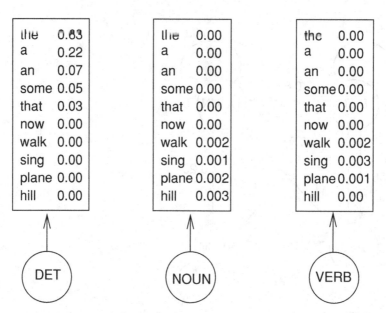

Figure 5.2 Example observation probabilities for three POS tags, showing the probability of observing a token given the state.

where $P(L)$ is called the **prior** and is the probability of a particular sequence of tags occurring. $P(T)$ is called the **evidence**, that is, what we actually see, and $P(T|L)$ is called the **likelihood**. An HMM can be viewed as a network of states, where each state represents one POS tag. In a tagger, all the states are connected, and the connection between each pair of states is given a **transition probability**. This represents the probability of seeing one POS tag follow another. A simple example of this is shown in Figure 5.1 and in this example we see that the probability of a noun following a determiner is high, whereas the probability of a determiner following another determiner is very low. In addition, each state has a set of **observation probabilities**, which give the probability of that state generating each token. An example of this is seen in Figure 5.2 for the determiner POS tag. As we would expect, tokens such as the have high probabilities, whereas content tokens such as walk have low probabilities, meaning that it is highly unlikely that this tag would generate that token.

We will consider the full issue of how to train an HMM in Section 15.1.8. For now, let us simply assume that we can calculate the transition probabilities and observation probabilities by simply counting occurrences in a labelled database. To see how the tagging operates, consider the issue of resolving the classic POS homograph record. This can be a noun or a verb, and a trained HMM would tell us, for instance,

$$P(t = \text{record}|l = NOUN) = 0.00065$$
$$P(t = \text{record}|l = VERB) = 0.0045$$

which basically states that in our training data one in every 1538 nouns that we encountered was the token record and one in every 220 verbs we encountered was record.

In addition to this we have the prior model, which gives us the probability that a sequence of tags will occur independently of any actual observations. In general this would take the form

$$P(L) = P(l_1, l_2, \ldots, l_N)$$

that is, the probability of seeing all the tags in the sentence. We then wish to find the probability of a particular tag given all its predecessors:

$$P(l_i | l_1, l_2, \ldots, l_{i-1})$$

We would never have enough data to count all these occurrences, so we make an assumption based on the chain rule of probability (see Appendix A). This states that a sequence as above can be approximated by considering only a fixed window of tags before the current one:

$$P(l_i | l_{i-2}, l_{i-1})$$

or

$$P(l_i | l_{i-1})$$

In this model we might find for instance that the probabilities of two particular tag sequences are

$$P(NOUN|DET) = 0.019$$
$$P(VERB|DET) = 0.00001$$

which indicates that the chance of a verb following a determiner is much much smaller than the chance of a noun following a determiner. Given an input token sequence that includes ... the record ... we would be certain that the token the would be classed as a determiner, so when we combine the two models and calculate the probability we see that the token record is a noun rather than a verb.

In a real HMM tagger system, we have to determine these probabilities from data. In general, the data have to be labelled with a POS tag for each word, but, in cases where labelled data are scarce, we can use a pre-existing tagger to label more data, or use a bootstrapping technique whereby we use one HMM tagger to help label the data for another iteration of training.

The prior probability is calculated by counting occurrences of sequences of tags. This is modelled by an *n*-gram model, which gives the probability of finding a tag t_i given a fixed-length sequence of previous tags $t_{i-1}, t_{i-2}, \ldots$. The longer the tag sequence the more accurate the prior model, but in general sparse-data problems prevent us from having an *n*-gram of length more that three or four tags. We also have to build a likelihood model, and we do this by defining a separate model for every tag. Each of these models then defines a probability distribution over the set of tokens, giving the probability of seeing any one token given the tag. At run time, a search is conducted through all possible tag sequences to find the single most likely one, and this is given as the answer. Another advantage of the HMM approach is that it is amenable to computing every possible sequence of tags and choosing the one with the highest probability. This is performed using the **Viterbi** algorithm, which is explained in Section 15.1.7. We will return to the entire subject of HMMs in Chapter 15 where we discuss the general nature of the likelihood models, the prior n-gram models, the search algorithms, the training algorithms and other practical matters.

5.3 Non-natural-language text

We now turn to the problem of dealing with text that does not directly encode natural language. In our model of communication we have a number of semiotic systems that are used to communicate information. Natural language is the richest and often the most common of these, but we also have numbers, dates, times, computer information, money, addresses and so on. These systems are separate from natural language, but in TTS we are required to speak them, and this can be done only if we convert data in this form into natural language. This is a two-stage problem where first we must analyse the text to determine its units and structure; this is done by a process of **semiotic decoding**. Then we have to convert this into natural language, and this process is called **verbalisation**. But we have an additional problem in that, since these systems share the same characters as natural language and hence as each other, the text forms of these systems often overlap, leading to considerable ambiguity. Hence, before we attempt decoding and translation, we first have to perform **semiotic classification** to resolve this ambiguity and determine which system each token of the input belongs to.

5.3.1 Semiotic classification

Semiotic classification can be performed by any of the techniques described in Section 5.2.1 above. First, we define which classes we believe our system will encounter. We can immediately identify some that crop up regardless of text type and genre, and these include

cardinal numbers	25, 1000
ordinal numbers	1st, Henry VIII
telephone numbers	0131 525 6800, 334-8895
years	1997, 2006, 1066
dates	10 July, 4th April 10/12/67
money	$56.87, 45p, Y3000, $34.9M
percentages	100%, 45%
measures	10 cm, 12 inches

In addition, we have a number of classes that are more genre-specific; for example we may have street addresses in a telephone-directory application, computer programs in a screen reader for the blind and any number of systems in specialist areas such as medicine, engineering or construction:

5.3 Non-natural-language text

emails	pat40@cam.com
urls	http://mi.eng.cam.ac.uk/ pat40
computer programs	for (i = 0; i < n; ++i)
addresses	4318 East 43rd Street
real estate	CGH, sng whit fem

So it is important to have a good knowledge of the application area in order to ensure a good and accurate coverage of the types of semiotic systems that will be encountered.

It should be clear that a simple classifier based on token observations will be sufficient only in the simplest of cases. If we take the case of cardinal numbers, we see that it would be absurd to list the possible tokens that a cardinal class can give rise to; the list would literally be infinite. To handle this, we make use of specialist **sub-classifiers** that use regular-expression generators to identify likely tokens from a given class. One way of doing this is to run regular-expression matches on tokens and use the result of this as a feature in the classifier itself. So we might have

(34) Match [0-9]+ ?

as a rule, which will return true if the token is composed entirely of an arbitrary number of digits. More specific regular expressions can also be used to handle numbers of a fixed length (e.g. years in the modern era being four digits long) or for real numbers, which contain a decimal point. In addition to these **open-class** rules, we also see that semiotic classes form a hierarchy. So the **money** class is composed of the **number** class and usually a currency symbol (e.g. $). A further issue is that often several tokens form a coherent semiotic group, such that if we have

(35) 8341 Belville Dr

we have three tokens that together form a single address. Hence we need an ability to group the tokens after or during classification.

The process of semiotic classification can therefore be summarised as follows.

1. Define the set $S = \{s_1, \ldots, s_N\}$ of semiotic classes.
2. For each decide whether this is a closed or open class.
3. Closed classes can be enumerated, open classes require regular-expression matches.
4. For each token
    (a) run the open-class expressions and record the answer for each as a feature,
    (b) using the open-class features and the normal features, run a classifier (e.g. the naive Bayes classifier), then record the most probable class s_i for each token.
5. Group consecutive tokens of the same class together.
6. Perform class-specific decoding and verbalisation.

To take a simple example, suppose that we have the sentence

(36) You should meet him at 3:30pm at 1994 Melville Drive

and we have the set of semiotic classes as just defined. For each open class we would have a regular expression as follows:

natural language	[A-Za-z]+
cardinal numbers	[0-9]+
telephone numbers	0131 525 6800, 334-8895
years	1[0-9][0-9][0-9] | 2[0-9][0-9][0-9]
time	hours : minutes part of day
hours	[0-9] | 1[0-9] | 2[0-4]
minutes	[0-6][0-9]
part of day	am | pm
addresses	cardinal number proper-name street-type
proper name	[A-Z][a-z]+
street type	Dr| St| Rd| | Drive| Street | Road

For each token we run each regular expression and record its answer as a feature:

text	Natural-language	Cardinal	Year	Hours	Minutes	Part of day	Proper name
You	1	0	0	0	0	0	1
should	1	0	0	0	0	0	0
meet	1	0	0	0	0	0	0
him	1	0	0	0	0	0	0
at	1	0	0	0	0	0	0
3	0	1	0	1	0	0	0
:	0	0	0	0	0	0	0
30	0	1	0	0	1	0	0
pm	1	0	0	0	0	1	0
at	1	0	0	0	0	0	0
1994	0	1	1	0	0	0	0
Melville	1	0	0	0	0	0	1
Dr	1	0	0	0	0	0	1

The results of these matches can now be used as features, along with general features such as "is this the first word in the sentence" (which is useful for telling a normal natural-language word from a proper name). All the features are then given to the classifier, which makes a final decision on semiotic class.

With regard to the issue of grouping, we can use a combination of the classifier and regular-expression matches to ensure that a sequence of tokens such as 3, :, 30, pm is seen as a single time. A more sophisticated approach is available if we make use of the HMM tagging algorithm since one of the features of an HMM is its ability to include multiple tokens in a single class, and also, via the Viterbi search algorithm, ensure that the most probable global sequence of classes is found, which helps resolves problems where two classes claim overlapping tokens.

5.3.2 Semiotic decoding

Once identified, each token or sequence of tokens must be decoded to determine its underlying form. Luckily, for nearly all classes except natural language, this is a fairly straightforward issue. In fact many classes (e.g. email addresses and computer programs) have been designed specifically to ensure that the underlying form can be found quickly and deterministically. We call this process **decoding** since it is seen as the process of reversing an encoding of the original form into text. This is often also called **parsing** in computer science, but we reserve this term for the specific problem of uncovering syntactic structure in natural language. We require a separate decoder for each semiotic class, but, due to the simplicity of each, this in general is not too difficult to design by hand. Let us take the example of dates to see how a decoder would work.

Let us take the **Gregorian calender** system (i.e. the date system we nearly always use) as an example. This is particularly illustrative as a non-natural-language semiotic class in that

1. It has a number of text forms, e.g. 10/12/67 or 10th December '67.
2. It is used nearly universally, regardless of which natural language is being used.
3. There are several ways to verbalise a date.

Gregorian dates have a day, a month and optionally a year and the task of the decoder is to determine the value of each of these from the text. The same date can give rise to a number of different text encodings, for example

(37) a. 10 December 1967
 b. December 10 1967
 c. 10th of December 1967
 d. 10th of December '67
 e. 10/12/1967
 f. 10/12/67
 g. 12/10/67

In each of these cases, the date is exactly the same, so here we see how a single date can give rise to multiple text encodings. Despite these possible differences, finding the day, month and year is normally possible with a little care (the only really difficult case being the handling of the convention often found in the USA, where the pattern is month/day/year rather than the more common day/month/year). A typical way of actually performing the decoding of a date is to use a list of rules, each of which decodes one particular date format. The rules are run in order, until a match is found.

5.3.3 Verbalisation

The final step in handling non-natural-language text is to convert it into words and this process is often called **verbalisation**. If we take our decoded-date example, we see that we have values for the three fields of day, month and year, and with these we can generate a sequence of words. For dates we have a number of conventions, such that the date day=10, month=12, year=December can be spoken in English as

(38) a. TEN TWELVE SIXTY SEVEN
 b. THE TENTH OF DECEMBER SIXTY SEVEN
 c. THE TENTH OF DECEMBER NINETEEN SIXTY SEVEN
 d. DECEMBER THE TENTH NINETEEN SIXTY SEVEN
 e. DECEMBER TENTH NINETEEN SIXTY SEVEN
 f. DECEMBER TENTH SIXTY SEVEN
 g. DECEMBER THE TENTH IN THE YEAR OF OUR LORD NINETEEN SIXTY SEVEN

and so on. In this case the issue of verbalisation is not difficult per se – a simple **template** approach can be used to create the word sequence. First we define two functions **ordinal()** and **cardinal()**, which convert a number into a word string representing the type of number. For example, for **cardinal()** we have

1	→	ONE
2	→	TWO
3	→	THREE
13	→	THIRTEEN
32	→	THIRTY TWO
101	→	ONE HUNDRED AND ONE
1997	→	ONE THOUSAND NINE HUNDRED AND NINETY SEVEN

and for **ordinal()**

1	→	FIRST
2	→	SECOND
3	→	THIRD
13	→	THIRTEENTH
32	→	THIRTY SECOND
101	→	ONE HUNDRED AND FIRST
1997	→	ONE THOUSAND NINE HUNDRED AND NINETY SEVENTH

Next we define a function **year()**. A simple version of this, which covers all years between 1010 and 1999, is

(39) **cardinal**($y_1 y_2$) **cardinal**($y_3 y_4$)

which when given 1387 for instance will produce the string THIRTEEN EIGHTY SEVEN. For the years immediately after the millennium the situation is more tricky; for these we require a function of the type

(40) **cardinal**($y_1$000) AND **cardinal**(y_4)

which when given 2006 would generate

(41) TWO THOUSAND AND SIX

Here we start to see the first case of why verbalisation can be tricky since some people prefer this rule not to have the AND so that the function should produce

(42) TWO THOUSAND SIX

and there is some debate over whether 2013 will be pronounced TWENTY THIRTEEN or TWO THOUSAND AND THIRTEEN. With the **year()** function we can now define a template for verbalising the full date. One template would then be

(43) THE **ordinal**(day) of **month-name**(month) **year**(year)

giving

(44) THE TENTH OF DECEMBER NINETEEN SIXTY SEVEN

In general, verbalisation functions for cases like these are not too difficult to construct; all that is required is a careful and thorough evaluation of the possibilities. The main difficulty in these cases is that there is no one correct expansion; from (38) we see at least seven ways of saying the date, so which should we pick? While sometimes the original text can be a guide (such that 10/12/67 would be verbalised as TEN TWELVE SIXTY SEVEN), we find that human readers often do not do this, but rather pick a verbalisation that they believe is appropriate. The problem for us as TTS system builders is that sometimes the verbalisation we use is not the one that the reader expects or prefers, and this can be considered an error.

For some such semiotic classes, we can specify a degree of **conventionality** with regard to verbalisation. We can state that verbalisations of ordinals and cardinals are very conventional in that nearly all readers verbalise a number the same way (although there may be slight differences between American and other types of English). With dates, we see that these are again quite conventional, but that there is quite a range of acceptable forms. We find that there is a significant problem with some other semiotic forms in that the degree of convention can be quite low. If we consider the email addresses

(45) pat40@cam.ac.uk

we might find that any of the following are acceptable verbalisations:

(46) LETTER-P LETTER-A LETTER-T FORTY AT CAM DOT LETTER-A LETTER-C DOT LETTER-U LETTER-K
LETTER-P LETTER-A LETTER-T FOUR ZERO AT CAM DOT LETTER-A LETTER-C DOT LETTER-U LETTER-K
PAT FORTY AT CAM DOT LETTER-A LETTER-C DOT LETTER-U LETTER-K
PAT FORTY AT CAM DOT AC DOT UK

the problem being that the particular owner of this address may consider only one of these as "correct" and will be dissatisfied if an alternative version is used.

5.4 Natural-language text

Having dealt with non-natural-language text, we now turn our attention to the issue of how to find the words from the text when that text *does* in fact encode natural language.

For many words, this process is relatively straightforward; if we take a word such as WALK, we will have its textual form walk listed in the lexicon, and upon finding the token walk in the text we can be fairly sure that this is in fact a written representation of WALK. At worst we might be presented with a capitalised version Walk or an upper-case version WALK but these are not too difficult to deal with either. If all words and all text behaved in this manner the problem would be solved, but of course not all cases are so simple, and we can identify the main reasons for this.

Alternative spellings Numerous words have more than one spelling. Often in English this is related to American versus non-American conventions so that we have tyre and tire and honour and honor. In many cases these differences are systematic, for instance with the suffix -ISE which can be encoded as organise or organize. Often it is possible to list alternative spellings in the lexicon.

Abbreviations Many words occur in abbreviated form, so for instance we have dr for DOCTOR and st for STREET. In many cases these are heavily conventionalised such that they can be listed as alternative spellings, but sometimes the writer has used a new abbreviation or one that we might not have listed in the lexicon.

Unknown words There will always be words that are not in the lexicon and when we find these we have to deal with them somehow.

Homographs are cases where two words share the same text form. In cases where the words differ in pronunciation, we must find the correct word. In natural-language homographs arise for three reasons.

1. **Accidental homographs** arise because completely words such as BASS-FISH and BASS-MUSIC just happen to share the same text form bass.
2. **Part-of-speech homographs** arise from words that have a similar semantic root, but where their use as a noun or verb gives them a different pronunciation. For instance PROJECT-NOUN and PROJECT-VERB both share the same text form project but are pronounced differently.
3. **Abbreviation homographs** arise because, since we use fewer characters in abbreviations, we are more likely to use the same text encoding as for another abbreviated word. Hence we find that DOOR, DOCTOR and DRIVE have the quite separate full text forms of door, doctor, DRIVE, but all share an abbreviated text form dr.

Mis-spellings Often if a text form of a word is mis-spelled a human reader is quite capable of guessing what word was intended, e.g. discovery, and ideally a TTS system should be able to cope with such instances.

This shows why we need a more-sophisticated model than a simple isomorphic correspondence between words and their text encodings.

5.4.1 Acronyms and letter sequences

At this stage it is worth mentioning two other types of word/token relationships that are often cited as problems for TTS. **Acronyms** are words formed from sequences of other words, usually (but not always) by taking the first letter of each. Examples include NATO, UNICEF SCUBA and AIDS. In our approach, these are treated as normal words so that we have a word NATO in our lexicon and this has a pronunciation /n ey t ow/; the fact that this was historically formed from other words or is normally found written in upper case is of no real concern. (In fact, with use many acronyms really do become indistinguishable from normal words such that RADAR is nearly always spelled in lower case and few people realise that it is an acronym, and fewer still what the original words were.)

In lay speak, the term acronym is also sometimes used for a different type of formation, which we shall call a **letter sequence**; examples of these include IBM, HTML and HIV. The difference between these and acronyms is that letter sequences don't form new pronunciations; they are simply spoken as a sequence of letters, hence we have IBM as LETTER-I LETTER-B LETTER-M, giving a combined pronunciation /ay b iy eh m/. It doesn't really matter whether we regard IBM as being a single word with pronunciation /ay b iy eh m/ or a token that represents three words (LETTER-I LETTER-B LETTER-M). Potentially, however, the productivity of letter sequences is very high, so regardless of whether we keep them in the lexicon or not, we still need a means of identifying new ones and pronouncing them as letter sequences rather than as normal words (i.e. we don't want to encounter a token RAF and think that this is just an upper-case version of an unknown word, and then pronounced it as /r ae f/).

A solid solution to the problem of upper-case tokens is to assume that in a well-developed system all genuine acronyms will appear in the lexicon. Upon encountering a capitalised token, we first check whether it is in the lexicon and, if so, just treat it as a normal word. If not, we then split the token into a series of single-character tokens and designate each as a letter. There is a possible alternative, where we examine upper-case tokens and attempt to classify them as either acronyms or letter sequences. Such a classification would be based on some notion of "pronounceability" such that if the token contains vowel characters in certain places then this might lead us to think that it is an acronym instead of a letter sequence. Experience has shown that this approach seldom works (at least in English) since there are plenty of potentially pronounceable tokens (e.g. USA and IRA) that sound absurd if we treat them as acronyms (e.g. /y uw s ax/ and /ay r ax/). If a mistake is to be made, it is generally better to pronounce an acronym as a letter sequence rather than the other way round.

5.4.2 Homograph disambiguation

The main problem in natural-language text decoding is dealing with homographs, whether they be accidental, part-of-speech or abbreviation homographs. Many systems choose to handle each of these separately, and in particular abbreviation homographs are often dealt with at the same time or in the same module as semiotic classification.

Typically this is done in systems that don't have an explicit semiotic-classification stage. An alternative approach is to handle these in a single module. Regardless, the general approach is to use one of the token-disambiguation techniques described in Section 5.2.1. Here we discuss some of the specifics of homograph disambiguation.

Turning to POS homographs first, we see that the classic way to resolve these is to use a POS tagger of the type described in Section 5.2.2. One problem with this approach is that unfortunately POS taggers have only a certain accuracy (figures in the range 96%–98% are commonly quoted [299]), and often it is in the hard cases of resolving noun/verb pairs that the errors are made. Recall that the POS tagger has two components, one that models sequences of tags and one that generates tokens given a tag. The tag-sequence n-gram is a very powerful constraint and nearly always gives the correct answer in cases such as

(47) The/det project/noun was/aux behind/prep schedule/noun

(48) The/det problem/noun is/aux how/wrb to/inf project/verb the/noun missile/noun accurately/adv

but in sentences such as

(49) The new images project well on to the wall

POS taggers often struggle because the n-gram sequence of tags alone is not sufficient to disambiguate in this case. Part of the problem with POS taggers is that they are often formulated as general-purpose NLP tools, and as such are not focused on the goals of how they might be used in a particular application. One simple improvement then is to adapt the tag set so that the discriminative power in the algorithm is geared towards the distinctions we really care about.

An alternative approach is to use one of the word-sense disambiguation algorithms since these can be more easily focused for each homograph case. A further advantage in using the WSD approach is that we can use this single approach for all types of homograph, not just the POS homographs that the tagger deals with. To give an example, consider the abbreviation homograph Dr, which can encode DOCTOR or DRIVE. Both of these are often collocated with a proper noun, but they tend to differ in their position relative to that proper noun. A simple feature that distinguishes many examples is then whether the Dr occurs before or after the proper noun:

(50) Dr + proper-noun → DOCTOR

(51) proper-noun + Dr → DRIVE

For example

(52) Dr Williams → DOCTOR WILLIAMS

(53) Williams Dr → WILLIAMS DRIVE

In a deterministic rule system this can be used as is, but since the rule is not 100% accurate an alternative is to use it as a feature in a WSD classifier.

We can summarise the procedure for building a homograph-disambiguation system as follows.

1. Build a lexicon that records all the text forms of each word.
2. Compare text forms and construct a list of homograph tokens.
3. For each form
    (a) search for cases of these in a large text corpus,
    (b) by hand mark the correct word,
    (c) find collocation trigger tokens by finding the tokens which most frequently co-occur with the word,
    (d) write feature rules that may describe any more-elaborate relationship between word and trigger tokens.
4. Train a WSD classifier.
5. Run over the test corpus, examine errors and repeat the training process using input based on the findings.

5.4.3 Non-homographs

Most tokens are not in fact homographs, and we will now outline how these non-homograph tokens are processed.

1. Check whether the token is present in the lexicon as the text form of a known word.
2. If it is found, take that word as the correct answer and conduct no further processing.
3. If it is not in the lexicon and is upper case, label it as a letter sequence.

Regarding this last rule, it is nearly impossible to be sure whether an upper-case token is an acronym or letter sequence, but analysis of data shows that unknown letter sequences are much more common than unknown acronyms and hence the most accurate strategy is to assign such an instance as a letter sequence. Furthermore, an acronym spoken as a letter sequence is deemed a lesser error than a letter sequence spoken as a single word.

If the token is not upper case and not in the lexicon we have a choice; it could potentially be a normal unknown word, an abbreviation for a known word, a different spelling for a known word or a mis-spelling of a known word. The probability of each of these cases mainly depends on the quality of our lexicon. If we have a very accurate comprehensive lexicon then it is less likely that we have missed a normal word or alternative spelling. This therefore increases the chance of the token being a mis-spelling. If the system has a spell-correction system, it should be run at this point. Assuming that the spell corrector has a confidence score, we can then decide whether the token is a mis-spelling of a known word or an unknown word. The final case occurs when we believe we have a token for a genuinely unknown word. This is assigned a new unique label (e.g. UNKNOWN-43), and the token and label are added to the run-time lexicon. The pronunciation is determined by the grapheme-to-phoneme convertor, which we will describe in Section 8.4. These operations can be found in the right-hand half of Figure 5.3.

5.5 Natural-language parsing

Our concern so far has been the task of identifying the word sequence from the token sequence, and to a large extent this is all the text decoding that we are required to do. Once we have unambiguously identified the word sequence, we can use the lexicon and grapheme-to-phoneme conversion to generate a sound representation, which can then be input to the synthesiser itself. It can be argued, however, that the word sequence alone is not enough, and that in addition it is advantageous to conduct further processing to determine the underlying structure of the sentence. Semantic analysis is generally thought too difficult and/or unnecessary, but some systems perform a **syntactic analysis**, which finds the correct or best **syntactic tree** for the input sentence. There are several justifications for wishing to perform this additional level of analysis.

1. Proper parsing can resolve part-of-speech ambiguity problems that finite-state (i.e. HMM) parsers fail on, such as Example (49) above.
2. The syntax can help determine sentential prominence patterns (explained in Section 6.8).
3. The syntax gives a basic grouping of words, and these groupings strongly influence the prosodic phrase structure of the utterance (explained in Section 6.7).

5.5.1 Context-free grammars

In Chomsky's early work [88], he proposed that the syntax of natural languages like English could be described at a basic level by a **context-free grammar (CFG)**. A CFG defines a number of rules of the form

(54) S → NP VP
NP → DET NP
VP → V NP
VP → PP
PP → P NP

The rules operate **generatively**, meaning that they are applied by starting with a basic top node (S standing for "sentence") and then applying rules in turn until a string of terminal nodes (i.e. words) has been generated. Chomsky's idea was that this finite set of these rules could generate the near-infinite set of sentences in natural language. In addition, the operation of the rules defined the **structure** of a sentence. The operation of context-free rules is easily represented by a syntactic tree:

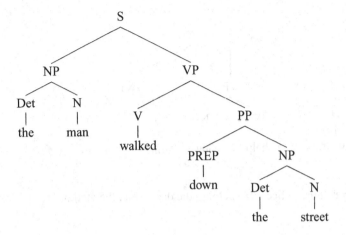

Formally, a grammar is a 4-tuple:

1. σ a set of **terminal** symbols
2. n a set of **non-terminal** symbols
3. P a set of rewrite rules
4. S an initial (non-terminal) symbol

Chomsky soon made many additions to this basic system and the field of modern linguistics has dispensed with the idea that simple context-free rules alone describe natural language. In language engineering, however, CFGs are still widely used since they are (reasonably) tractable models of syntax. The problem of finding the syntactic structure of a sentence is called **parsing** and a **parser** is an algorithm that assigns this structure to arbitrary text. All parsers work with respect to a grammar and effectively try to assign the best or correct structure to a sentence. Our problem is of course that we are starting from the other perspective in that we have the words already and wish to find which rules generated them. The job of the parser is therefore to find which rules can have generated the input, and from these rules we can find the tree and hence the syntactic structure of the sentence.

The first problem that we encounter in parsing is that any simplistic or exhaustive search for the correct set of rules is prohibitively expensive. Parsers can therefore be thought of as search algorithms that find a tree or trees for a sentence in a reasonable length of time. The second problem in parsing is that any non-trivial set of rules will generate more than one parse that fits the input, and in many cases a considerable number of perfectly legal parses will fit the input. A simple example is

(55) POLICE HELP DOG BITE VICTIM

which could have the parse

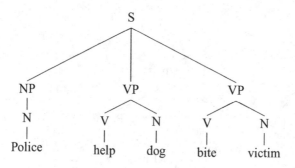

meaning that the police were helping the dog to bite the victim, or

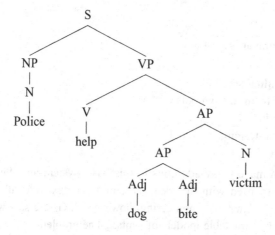

meaning that the police were helping a victim of a dog bite.

The **Cocke–Younger–Kasami (CYK)** algorithm is one of the most widely used parsing algorithms. It can (in general) efficiently parse text and is easily adaptable to probabilistic formulations. Its basic operation can be thought of as dynamic programming for context-free grammars.

This algorithm considers every possible group of words and sets entries in a table to be true if the sequence of words starting from i of length j can be generated from the grammar. Once it has considered sequences of length 1, it goes on to ones of length 2, and so on. For the sequences greater than length 1, it breaks the sequence into all possible subdivisions, and again checks to see whether a rule in the grammar can generate one of these subdivisions (this is why the grammar has to be in Chomsky normal form). If a match is found, this is stored in the table also. This process is continued until the full set of rules (i.e. parses) has been found. Deterministic parsers have no real means of choosing which parse is best, so **probabilistic parsers** (explained below) are usually preferred since these at least show that one parse is more probable than another.

5.5.2 Statistical parsing

The CYK algorithm as just explained will find the set of legal parses for a sentence, but the problem is in general that we wish to find just *one* parse, not several. We of course want the "true" parse and by this we mean we want what the author originally intended when the sentence was written. The major weakness with this type of parsing is that it considers only syntactic factors; semantics is completely ignored. That is why a parser often returns parses that are "correct" but seem silly. Some more-advanced models of syntax do incorporate semantic information [78], [177], [312], but in general most syntax models, especially in engineering, make use only of syntactic rules.

A sensible alternative then is, instead of trying to find the parse which the author intended, or which makes most sense, we can try to find the *most likely* parse. That is, we build a probabilistic model of the grammar and use the parser to find the most likely parse. There are several ways in which this can be done, but the most common is to assign probabilities to the rules themselves so that we say a non-terminal expands to each of its productions with a certain probability. When the parsing is complete, each set of rules has an associated probability and the highest can be chosen as the right answer. A number of specialist databases called **tree-banks** are available for training PCFGs. These have trees and nodes labelled by hand, and from these a grammar can be trained and probabilities assigned to each rule. For unlabelled data (that is, sentences for which we have no trees) the **inside–outside algorithm** can be used. This is the PCFG equivalent of the Baum–Welch algorithm used to train hidden Markov models, and is explained in Section 15.1.8.

We can therefore identity three common approaches to parsing.

1. Define a CFG by hand and use a deterministic search algorithm (i.e. the parser) to find the parse or parses which best fit the input.
2. Define a CFG by hand and use a probabilistic search algorithm to find the parse (or parses) which is most likely given the input.
3. Learn the CFG from data and use this with a probabilistic search algorithm to find the parse (or parses) which is most likely given the input.

5.6 Discussion

Figure 5.3 shows the flowchart for the text-decoding system described in this chapter.

At this stage, we will discuss a number of alternative formulations to the one presented here. In the (distant) past, text forms such as Dr were often seen as abberations insofar as text and language processing was concerned, to the extent that they were often called **text anomalies** [276]. The "real job" of text/language processing was to perform syntactic and morphological processing. For instance in the MITalk book 6 pages are devoted to text anomalies and 48 to syntactic, morphological and phonological processing. As can

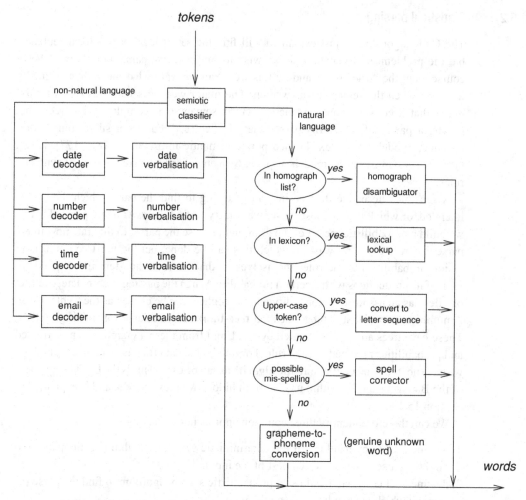

Figure 5.3 The flow of an entire text-analysis system (only four semiotic decoders/translators are shown for succinctness).

be seen from this chapter, in modern systems, the opposite balance is seen. Another view is that this problem is one of **text normalisation** in which a form like Dr should be *converted* into its true *text* form doctor. In such systems there is often no formal distinction between the concepts of words and tokens. A more-recent approach is to define such text forms as **non-standard words (NSWs)** [413]. This approach takes the notion of these text forms much more seriously, and for example notes that the text forms for a word are quite productive and can't simply be listed. The NSW approach uses decision trees to perform disambiguation and is in many ways similar to the approach taken here. Two notable differences are first of all that it again doesn't make an explicit distinction between the nature of words and that of tokens and secondly that the processes of non-natural-language and natural-language disambiguation are performed in a single step.

We argue that there are advantages to the text-decoding model used here. Taking the issue of natural-language decoding first, we see that this follows the information-theoretic source/channel paradigm where we have a clean, discrete underlying form that has generated the noisy surface form that we observe. This is a very general and powerful model and has been shown to be successful in many speech and language applications. It facilitates the ability to have abbreviations, mis-spellings and ambiguity all handled in a single unified powerful statistical framework. Probabilistic models can be built for the general process of text encoding that will facilitate successful decoding. In addition, once the words have been found in this model, all problems of ambiguity are resolved, which greatly simplifies the work of, say, the pronunciation module.

The model of semiotic systems described here makes explicit the notion that text encodes other information than natural-language information. This is well understood in the field of semiotics but is nearly universally ignored or brushed over in speech and language processing. There are two main advantages we see in adopting this model. First, it makes explicit the difference between analysis and verbalisation. In analysing text tokens (for example the time 3:30pm) there is one, and only one, correct answer, which is hour=3, minutes=30, part-of-day=afternoon. When it comes to verbalisation, however, there are many ways in which we can render this as words, e.g. THREE THIRTY, HALF PAST THREE, HALF THREE etc. and different users will have different preferences depending on their dialectal preferences and the system application. The upshot is that it is significantly easier to build a system where we know that only one part suffers from this subjective problem – the analysis/decoding part can be attempted with complete assuredness.

The second advantage comes when we consider language dependence. It might not be immediately apparent, but the text 3:30 is nearly universally recognisable no matter which natural language surrounds it, so we can have

(56) THREE THIRTY

in English and

(57) TROIS HEURES TRENTE

in French. If we consider 3:30 as simply a strange encoding of natural language, we would have to build a separate analysis system for every language. With the semiotic model, one classifier and decoder suffices for all languages; all that is required is a language-specific verbalisation component.[1]

We need to comment on the shift in emphasis in text analysis between older systems, which concentrated on syntactic, morphological and phonological analysis and paid little attention to text-decoding issues, and modern systems, which perform little linguistic analysis but devote considerable attention to ambiguous text forms. The are three separate factors influencing this shift. Firstly, with the commercialisation of TTS, we have found

---

[1] It is perhaps worth mentioning at this stage that books on teaching languages to non-native speakers are often particularly good at explaining the rules as to how numbers, times and dates are verbalised in particular languages.

that, in real use, non-canonical word encodings and non-natural-language text are rife and any TTS system must deal with this comprehensively and robustly if it is to be judged usable. Secondly, the simple length of time we have had to collect data and particularly lexicons over the years means that we have very comprehensive lists of words, their pronunciations and their text forms. This means that morphological analysis, for instance, is much less necessary than before and that data-driven methods are quite practical in terms of training (we have the data) and run time (we can perform searches or run algorithms with large numbers of parameters/rules). Thirdly, and perhaps most profoundly, pure linguistic analysis such as syntactic parsing has fallen out of favour for the simple reason that it has failed "to deliver". Few doubt that a full, accurate and true description of a sentence's linguistic structure would be very useful, but what form exactly should this take? Modern linguistics has more theories of syntax than even specialists can keep up with, and disagreements run to the very nature of what syntax is and what it represents (and there are now plenty who see it as an epiphenomenon). From a more-practical perspective, despite the use of tree-banks, there simply is no such thing as a theory-independent model of syntax that we can use for engineering purposes. All use a grammar and all such grammars make strong claims about the nature of language (such as whether it is context-free). Next, and maybe because of these problems, we find that even the best parsers are really quite inaccurate and relying on their output can cause quite noticeable errors in other parts of the system. Finally, we find that, even if we had a good model of syntax and an accurate parser, we still might not find syntactic analysis helpful. As we shall see in the next chapter, the main uses for syntactic structure are in determining prominence, intonation and phrasing; but even if we knew the syntax it is far from certain how to derive these other sorts of information. The relationship between syntactic and prosodic structure is a great void in linguistics.

Many regard these problems with syntax and other types of "deep" linguistic analysis as ones of expediency; with more work in these fields we will gradually improve the theories, grammars and parsers to the extent that accurate syntactic structures can be assigned to arbitrary text. An alternative viewpoint is that these difficulties are genuinely profound, and that, given the enormous amount of work that has been put into the area of theoretical and applied syntax with so little solid progress, perhaps the nature of the problem itself is incorrectly formulated and a new approach to the organisational properties of language is required.

5.6.1 Further reading

In commercial TTS, text analysis is seen as vitally important since any error can cause an enourmous negative impression with a listener. Despite this, the problem has received only sporadic attention amongst academic researchers. Notable exceptions include various works by Richard Sproat [410], [411], [412], [413], who tackles nearly all the problems in text analysis (and many in prosody, explained in the next chapter). The Bell Labs system that Sproat worked on is particularly well known for its very high accuracy with regard to text processing. In a separate line of work Sproat *et al.* investigated a number of machine-learning approaches to semiotic classification and verbalisation.

David Yarowsky [505], [506], introduced an approach that showed the use of the WSD approach to homograph resolution. A more-traditional view is given in the MITalk book [10].

Outside the field of TTS we find that a lot of the work in the field of natural-language processing (NLP) is directly applicable to these problems. While researchers in those fields are often attempting somewhat different tasks (e.g. homonym disambiguation rather than homograph disambiguation), the approaches are often essentially the same. A particularly good book in this area is by Manning and Schutze [300], who give a thorough overview of all aspects of statistical language processing, including WSD, POS tagging and parsing.

5.6.2 Summary

Classifiers

- Many tasks in text processing involve assigning a label to a unit; this is called **classification**.
- Usually this is done by making use of a set of **features** that are found elsewhere in the utterance.
- A classifier can be therefore be defined as a function that maps from a **feature space** to a **label space**.
- Treating evidence from features independently is often too crude and inaccurate. Examining every possible combination of features is nearly always impossible due to lack of data.
- All sophisticated classifiers therefore adopt an in-between solution, which gives an answer for every feature combination, but does not necessarily have to have seen all of these during training. The mapping for unobserved feature combinations is inferred from observed ones.
- A basic distinction can be made between **taggers**, which assign a label to every token drawn from a single set, and **word-sense disambiguators (WSD)**, which assign labels only to ambiguous tokens and do so from multiple label sets.
- Popular classifiers include context-sensitive rewrite rules, decision lists, decision trees, naive Bayes classifiers and HMM taggers.

Non-natural-language text

- Text is often full of tokens that encode non-natural-language systems such as dates, times and email addresses.
- The first stage in text analysis is therefore to perform **semiotic classification**, which assigns a class to each token. This is performed by one of the above-mentioned classification algorithms
- Once a token's class has been found, it is decoded into its underlying form and then translated into natural language by a process of **verbalisation**

Natural-language text
- The main problem here is to resolve **homographs**, which occur when two words share the same text form. These occur because of accident, part-of-speech changes or abbreviations.
- POS taggers can be used to resolve POS ambiguity.
- In general WSD approaches are more focused and produce better results.
- In addition some systems use syntactic parsers, which generate a syntactic tree for the utterance.

6   Prosody prediction from text

Informally we can describe prosody as the part of human communication which expresses emotion, emphasises words, reveals the speaker's attitude, breaks a sentence into phrases, governs sentence rhythm and controls the intonation, pitch or tune of the utterance. This chapter describes how to predict prosodic form from the text while Chapter 9 goes on to describe how to synthesize the acoustics of prosodic expression from these form representations. In this chapter we first introduce the various manifestations of prosody in terms of phrasing, prominence and intonation. Next we go on to describe how prosody is used in communication, and in particular explain why this has a much more direct affect on the final speech patterns than with verbal communication. Finally we describe techniques for predicting what prosody should be generated from a text input.

6.1   Prosodic form

In our discussion of the verbal component of language, we saw that, while there were many difficulties in pinning down the exact nature of words and phonemes, broadly speaking words and phonemes were fairly easy to find, identify and demarcate. Furthermore, people can do this readily without much specialist linguistic training – given a simple sentence, most people can say which words were spoken, and with some guidance people have little difficulty in identifying the basic sounds in that sentence.

The situation is nowhere near as clear for prosody, and it may amaze newcomers to this topic to discover that there are no widely agreed description or representation systems for *any* aspect of prosody, be it to do with emotion, intonation, phrasing or rhythm. This is not to say that description systems do not exist; the literature is full of them, but rather that none match the simplicity and robustness of the simple systems we use to describe the verbal component of language. There are many schools of prosodic theories, and, within one school, everything may have the illusion of simplicity, but the very fact that there are so many different theories, often with completely conflicting accounts, leads us to a position of extreme uncertainty with regard to how to describe, model, analyse and synthesise prosody. We mention this here, because firstly these difficulties prevent us from giving a single clean and simple explanation of prosodic phenomena, and secondly we wish to alert the reader to the very tenuous nature of the literature regarding prosody.

To some extent prosody can be seen as a parallel communication mechanism to verbal language. There we had meaning, form and signal, and broadly speaking we can use these terms when talking about prosody. Hence we can have a meaning (say anger), which is manifested in a form (say raised pitch levels), which is common to all speakers of a group, which is then encoded acoustically in a joint process with verbal encoding.

At the semantic (meaning) level, prosody is no more difficult to describe than the semantics of the verbal component (i.e. with great difficulty). It is when we get to prosodic form that we start to see more significant differences between prosody and verbal language. With verbal language, our basic units are words and phonemes, and, while these units have some difficulties, they are certainly coherent enough for the engineering purposes of text-to-speech. With prosody though, the situation is much more complex since it is very difficult to determine the basics of what prosodic form should be. We believe on the one hand that abstract prosodic form does exist (we all seem to use similar patterns when asking questions in particular ways), but on the other find it difficult to describe how this should operate in practice. In the next sections then we attempt to give a theory-neutral[1] account of some prosodic phenomena before describing some of the more well-known models.

6.2 Phrasing

We use the term **prosodic phrasing** or just **phrasing** to describe the phenomenon whereby speakers groups certain words within the utterances they speak. As with all aspects of prosody, there is acknowledgment that this phenomenon exists, but no agreement on how best to describe it.

6.2.1 Phrasing phenomena

Many cases are clear cut, so that when we say

(58) IN TERMS OF THE NECESSARY POLITICAL EXPEDIENCY | THE RESULT WAS CLEARLY A GOOD ONE FOR SHAW

there is a clear break in grouping between the words up to and including EXPEDIENCY and the words that follow (here phrasing is marked with a | symbol). The above case may be unproblematic but there are theories that state that many finer groupings exist, so that we would have

(59) IN TERMS | OF THE NECESSARY | POLITICAL EXPEDIENCY | | THE RESULT | WAS CLEARLY | A GOOD ONE | FOR SHAW

(where || indicates a strong phrase break) and furthermore, some theories express this as a bracketing to show that some groupings are stronger than others:

(60) ((IN TERMS) ((OF THE NECESSARY) (POLITICAL EXPEDIENCY))) ((THE RESULT) ((WAS CLEARLY) (A GOOD ONE)) (FOR SHAW))

---

[1] In reality, there is no such thing as theory-neutral; we always make some assumptions when we describe anything.

A question that has interested linguists for some time concerns the relationship between syntactic phrases (i.e. the syntax tree) and prosodic phrasing. While there is clearly some relationship (easily seen in the above examples), finding the general nature of this is extremely difficult. Factors that complicate the relationship include the following.

1. Prosodic phrasing seems "flatter" in that, while syntactic phrasing is inherently recursive and can exhibit a large degree of nesting, if levels in prosodic phrasing exist there are only a few, and there is no sense in which the phrases within an utterance seem to embed so strongly.
2. Prosodic phrasing is to some extent governed by purely phonological, phonetic or acoustic factors that can override the syntax. In the classic example, Chomsky [89] commented on the fact that in

    (61) THIS IS THE DOG THAT WORRIED THE CAT THAT KILLED THE RAT THAT ATE THE MALT THAT LAY IN THE HOUSE THAT JACK BUILT

    nearly every speaker says this with a flat prosodic phrasing like

    (62) THIS IS THE DOG | THAT WORRIED THE CAT | THAT KILLED THE RAT | THAT ATE THE MALT | THAT LAY IN THE HOUSE | THAT JACK BUILT

    whereas the syntactic phrasing is deeply recursive and embedded

    (63) (THIS IS (THE DOG (THAT WORRIED THE CAT (THAT KILLED THE RAT (THAT ATE THE MALT (THAT LAY IN THE HOUSE (THAT JACK BUILT))))))

    The speech patterns of the utterance seem to override the syntactic patterns.
3. Phrasing is particularly prone to speaker choice, and while "rules" exist for where phrase breaks should occur, quite often it appears optional whether a phrase break will be found in a particular utterance. Hence all of the following are possible:

    (64) JOHN AND SARAH WERE RUNNING VERY QUICKLY
    (65) JOHN AND SARAH | WERE RUNNING VERY QUICKLY
    (66) JOHN AND SARAH WERE RUNNING | VERY QUICKLY

So, as we can see, the relationship between syntactic and prosodic structure is complicated and a general theory that links syntax and prosody has yet to be developed. Some have even argued that the difficulty really lies with our models of syntax, so that if we developed syntactic models that took more account of the ways sentences were spoken some of these problems might be resolved [1].

6.2.2 Models of phrasing

While there is no single accepted theoretical model of phrasing, all models can be encompassed within the idea that a sentence has a **prosodic phrase structure**. The main point of disagreement revolves around the details of this phrase structure, and in particular how nested or deep it should be.

Some have proposed that phrasing should be represented by a tree and that this should act as a parallel tree to the syntax tree. Thus the prosodic tree can be generated by a CFG, and can have recursion and arbitrary depth. Perhaps the most-elegant model of this type is the model of **metrical phonology** from Liberman [284], [286]. The central idea is that any sentence, or part of a sentence, has a metrical tree that specifies its phrase and stress patterns. The lowest unit in the tree is the syllable, so metrical phonology provides a unified theory of syllable, word and phrasal patterns. Ignoring the syllable level for the moment, we can see the general form of a metrical tree in the example below:

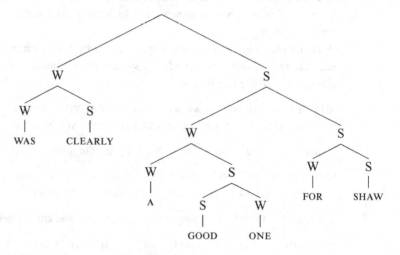

The S and W labels will be explained later, but for now simply note that the metrical tree groups words together. This model seems particularly good at explaining the patterns within compound-noun phrases, which we will re-examine in Section 6.3.1. A major criticism, though, is that the theory implies unlimited depth in trees, which implies that (a) there are many graded subtleties in the actual production or perception of phrase breaks, or (b) this representation maps to another surface representation that is flatter and has fewer levels. Many solutions have been proposed, both as amendments to the original theory[2] and as alternatives. The theory of **compound prosodic domains** proposed by Ladd [268] allows a more-limited recursion with arbitrary depth but not full recursion. Another variant, called the **strict layer hypothesis** [398] is much more constrained and states that every word has a fixed number of prosodic constituents above it. This and many other theories can be seen as forming a model called the **prosodic hierarchy**. In this, we have a tree as before, but with a fixed number of levels, each of which is its own type. A rich version of this might include the levels utterance, intonational phrase, phonological phrase, word, foot, syllable and segment. In this **intonational phrase** means a large phrase with a clear intonational tune (see below), a **phonological phrase** is a smaller phrase that spans only a few words, and a **foot** is a unit that starts with a stressed syllable and covers all the unstressed syllables until the next stressed syllable.

It is reasonable to say that nearly all other models are of this general type, and differ only in that they use differing numbers of levels or names for those levels. One of the

---

[2] The field of metrical phonology still survives today, but unfortunately bears little resemblance to the elegance and power of the original theory.

most popular, especially for engineering purposes, has two levels of sentence internal phrase, giving four levels overall:

1. Utterance
2. Intonational phrase/major phrase
3. Phonological phrase/minor phrase
4. Word

In this scheme, an utterance might look like

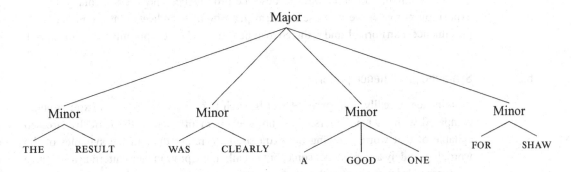

A slightly different view is to concentrate not on the constituent phrases, but rather on the **breaks** between the phrases. This view has been around for some time, but was formalised most clearly in the **ToBI** model in the concept of **break index** [406]. The break index is a number that exists between consecutive words, for example

(67) IN **1** TERMS **3** OF **1** THE **1** NECESSARY **3** POLITICAL **1** EXPEDIENCY **4** THE **1** RESULT **3** WAS **1** CLEARLY **3** A **1** GOOD **2** ONE **3** FOR **1** SHAW **6**

The numbers give the strength of the break such that between two normal words in running speech we have a **1**, for words that are deemed more separate we have **2** and so on, extending up to **6**. A simplified version of this is frequently used, where **3** is given to a minor phrase break and **4** to a major phrase break. Break-index systems and strict depth tress are isomorphic, so the distinction regarding which is used is often more of expediency rather than deep-rooted theory.

A major problem in resolving issues with prosodic phrasing is that there always seems to be a tension between what we can describe as a top-down approach, where we use the linguistic structure of the sentence to dictate the phrasing, and a bottom-up approach, where we listen to real speech and use clues from pausing, timing patterns and pitch to show where the phrases should be. No model yet proposed has come up with a satisfactory method of doing both of these and reconciling the results.

6.3 Prominence

Emphasis, **prominence**, **accent** and **stress** are all terms used to indicate the relative strength of a unit in speech. These terms are used with a variety of definitions in the literature, so to avoid confusion we will use these terms as follows. **Stress** indicates **lexical stress**, which is an inherent property of words, and indicates for instance that the first syllable in TABLE is stronger than the second, while the second syllable in MACHINE is stronger than the first. We discuss this type of stress more fully in the section on phonology in Chapter 7. **Prominence** is used to indicate the strength of a word, syllable or phrase when it is used in a sentence. We will use the term **accent** solely to indicate intonational phenomena associated with pitch, and **emphasis** to indicate a specific use of prominence in discourse.

There are many different ways to describe prominence and stress formally, but for exposition purposes we will use the term **prominent** to indicate that a unit has more prominence than normal, and **reduced** to indicate that it has less prominence than normal.

6.3.1 Syntactic prominence patterns

Prominence is really a property of words, such that one word is given extra strength compared with its neighbours. This, however, is manifested in the lexically stressed syllable of that word receiving the extra strength, rather than all the syllables of the word. It is widely accepted that some, but not all, of the prominence pattern in a sentence is governed by the syntax within that sentence.

A first type of effect is that of **nuclear prominence**,[3] which describes the particular prominence that often falls on the last content word in a sentence. In the following examples, we have marked this word in bold:

(68) THE MAN WALKED DOWN THE **STREET**.

(69) TELL ME WHERE YOU ARE **GOING** TO.

These examples are what are often termed **discourse-neutral** renderings, indicating the "normal" or canonical way these sentences are phrases are spoken, in contrast to some of the ways in which they may be spoken in particular discourse contexts.

Prominence patterns are particularly noticeable in sequences of adjectives and nouns that are collectively called **compound-noun phrases**. For example

(70) **PARKING** LOT

(71) CITY **HALL**

(72) THE UNION **STRIKE** COMMITTEE

The final main type of syntactic prominence concerns function words. A very simple rule is that content words (nouns, verbs etc.) can have normal or extra levels of prominence but function words are reduced in prominence. This certainly covers many simple cases, so that in

(73) THE MAN MADE THE OFFER ON THE BACK OF AN ENVELOPE.

---

[3] This is often termed a nuclear accent in the literature.

we feel that the three instances of THE, ON, OF and AN carry the least prominence and are reduced. Function words are not always reduced, however, and in cases like

(74) A BOOK LIKE THAT IS ALWAYS WORTH HOLDING ON TO

we see that THAT and TO are not reduced, and in some situations THAT would even be the most prominent word in the sentence. It can be argued that linguistic items such as THAT are really several distinct words that happen to carry the same form. This item can be used as a pronoun, an adjective, a conjunction, or an adverb as in the following examples

(75) THAT IS MY HOUSE (pronoun)

(76) I WANTED THAT BOOK (adjective)

(77) HE SAID THAT HE WAS AFRAID (conjunction)

(78) HE DIDN'T TAKE IT THAT SERIOUSLY (adverb)

where we see different degrees of prominence depending on use. Getting the prominence right in these situations may therefore not be so much an issue of prominence prediction per se, but rather of making the correct word definitions and finding these from the text.

A second type of prominence effect occurs when function words occur in sequences. When we have sentences such as

(79) I WAS WALKING THE DOG

it seems quite natural to have more prominence on WALKING than on WAS, which follows our simple content-word prominent/function-word unstressed rule. But when we have longer sequences such as

(80) WHAT WOULD YOU HAVE DONE IF IT HADN'T BEEN FOR ME?

we see a sentence made entirely of function words. If these are all spoken in reduced form the speech sounds terrible, so stresses must be placed on some words. A reasonable rendition of this with prominence might then be

(81) WHAT **WOULD** YOU HAVE **DONE** IF IT HADN'T BEEN FOR **ME**?

though many other patterns are possible. In fact the overall degree of prominence in sentences like this is no less than in "normal" sentences where there is a mixture of content and function words. This situation can be likened to the situation in phrasing where speakers feel the need to place phrase breaks at regular intervals somewhat independently of the top-down linguistic structure.

Each type of prominence can be "overridden" by the next type, so, while the normal prominence pattern for THIRTEEN is

(82) THIRTEEN

in many compound-noun phrases, the word undergoes **prominence shift** to give

(83) THIRTEEN BOYS

This is often explained by the notion that the main stress on BOYS "repels" the stress on -TEEN so that stressed syllables do not directly occur as neighbours.

6.3.2 Discourse prominence patterns

When sentences are spoken in discourses or dialogues, additional patterns can occur. One commonly cited affect is the difference in prominence levels between **given** and **new** information. This idea says that, when a word is first used in a discourse, it will often be quite strong, but on second and subsequent mentions it is actually weaker, so we might have

(84) FIRST MAKE SURE THAT PAUL ANSWERS THE PHONE.

(85) THEN TELL PAUL TO BRING THE BOOK OVER TONIGHT.

in which the first mention of PAUL is prominent and the second is not. While this given/new distinction has traditionally received considerable attention in the literature, many recent empirical studies on real speech have shown that this affect is not so common as had traditionally been thought [23], [328], [329], [330]. First, it is far more common that the second sentence would have a pronoun instead of the original name

(86) THEN TELL HIM TO BRING THE BOOK OVER TONIGHT.

and secondly, as Aylett [24] points out, even when words are repeated in full form in dialogues, there is very little evidence of prominence patterns being governed by the given/new distinction. What does occur, though, is that the degree of prominence is heavily influenced by the redundancy in the dialogue, such that if the speaker feels the listener has a high probability of understanding then he or she may well give a word less prominence. Crucially, the word receiving less prominence does not have to be the same as the original word, hence

(87) THEN TELL THE IRISH GIT TO BRING THE BOOK OVER TONIGHT.

here because since IRISH GIT is obviously related to PAUL it may receive less prominence. Prominence patterns of this sort are clearly augmentative and will be explained further in Section 6.5.3.

We can identify a second type of discourse prominence effect, which we shall term **emphasis**. This occurs when a speaker wants to draw attention to a particular word:

(88) IT WAS REALLY HUGE

(89) I DIDN'T BELIEVE A WORD HE SAID

(90) I'VE HEARD IT ALL BEFORE

(91) THAT'S JUST TOO BAD

(92) IT'S IMPORTANT TO DO THAT **BEFORE** YOU ENTER THE ROOM

(93) NOT **THAT** ONE

and so on. In general, emphatic prominence is used for augmentation purposes (see below in Section 6.5.3) when the speaker wants to make particularly sure that the listener understands what is being said. A particularly salient example is contrastive prominence of the type

(94) THE PROFIT WAS $10M **BEFORE** TAX, NOT AFTER TAX.

This, however, also extends into affective purposes, when for example the speaker knows that the listener will understand, but still uses extra emphasis. This could occur for example when the speaker wishes to express frustration or imply that the listener is stupid and therefore has to have everything spoken slowly or clearly.

6.3.3 Prominence systems, data and labelling

As with phrasing, we have a number of models and theories that attempt to explain prominence. Again as with phrasing, the theories range from simple models to ones that posit a rich and embedded prominence structure. Broadly speaking, the phenomenon of lexical stress is fairly uncontentious, with good agreement between labellers as to where stress values lie. Prominence and emphasis are more difficult to mark, since labellers often find it hard to distinguish among what is "normal", discourse-neutral prominence and marked emphasis. The distinction here is related to the issue of augmentative prosody, which is discussed in Section 6.5.3.

As with phrasing, the basic distinction is between systems that have a fixed number of prominence values or levels and ones in which nesting and recursion can occur. Yet again, we see that the issue is the tradeoff between a system that seems to match linguistic structure and one that is more bottom-up and perceptually verifiable. We have already examined Liberman's model of metrical phonology for phrasing, but actually this was primarily developed for explaining prominence patterns. In this, we have binary trees in which each node is marked either **strong (S)** or **weak (W)**. Using this, we can then describe prominence patterns as follows:

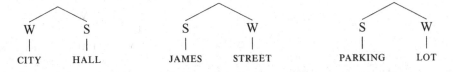

The "clever bit" is that this model allows us to combine smaller groups of words elegantly into bigger groups by further use of the metrical trees:

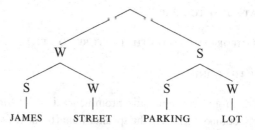

in which case there is one and only one node that is dominated only by S, and this is how we specify the most-prominent word or nuclear prominence of the sentence or phrase. Metrical phonology also neatly deals with the issue of prominence shift, which occurs in the following example for many speakers:

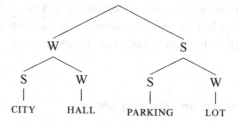

(the prominence has shifted from HALL to CITY). Just as with phrasing, however, it seems difficult to believe that when it comes to a whole sentence such as

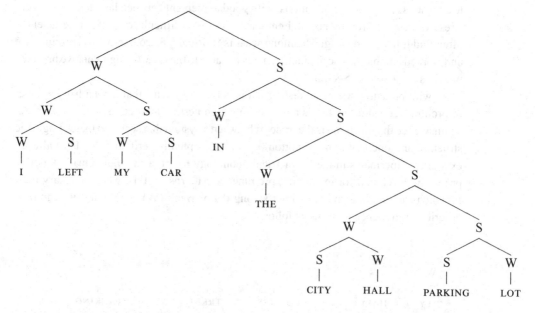

speakers can make and listeners discern all these different degrees of phrasing. This has led to researchers proposing a fixed number of types or levels of prominence, but, as with phrasing, there is virtually no agreement as to how many there should be. Probably the most-common models for practical engineering purposes are

1. a metrical tree in which the values on the high-level nodes are ignored
2. a system with three levels of prominence:
    (a) prominent/stressed
    (b) un-prominent/unstressed
    (c) reduced
3. as above, but with an extra level for emphatic prominence.

6.4 Intonation and tune

Intonation is sometimes used as a synonym for prosody, and hence includes prominence and phrasing as just described. Here we use a narrower definition in which intonation is the systematic use of pitch for communication. Often we describe intonation with reference to the **pitch contour** or the **F0 contour**, which is the pattern of pitch/F0 variation over an utterance.[4]

Most traditional accounts of intonation speak of intonational **tunes** that span the length of the sentence. In these accounts, we might have a question tune, an exclamation tune and so on. More-modern accounts advocate a more-compositional approach, where the global intonation tune is composed of a number of discrete elements. Most important of these is the **pitch accent**, which can be defined as a significant excursion (either positive or negative) in the F0 contour that lends emphasis to a syllable. Pitch accents occur only on prominent syllables, and hence prominence can be seen as defining the possible places where pitch accents can occur. Figure 6.1 shows a sentence with two clearly visible pitch accents. In addition to this, it is common to have **boundary accents** or **boundary tones**, which are pitch movements associated with the ends of phrases rather than syllables. A classic example of a boundary tone is where we having rising pitch at the end of a sentence to indicate a question, shown in Figure 6.1.

Beyond this, we again find that there is little agreement about how to describe pitch accents and boundary tones. Some theories state that there is a fixed inventory of these, whereas others describe them with continuous parameters. The nature of pitch accents and boundary tones is disputed, with some theories describing them as **tones** or **levels** whereas others say that their characteristic property is pitch **movement**. One prominent theory states that we have an **intonational phonology** that parallels normal phonology, and thus we have inventories of contrasting units (sometimes called **tonemes**) and grammars that state what can follow what.

As with phrasing, we find that there is a considerable degree of speaker choice with regard to intonation, such that, while it may be common to speak a question such as

(95) IS THIS THE WAY TO THE BANK?

---

[4] F0 is shorthand for fundamental frequency. Informally this can be thought of as being the same as pitch; the exact relationship is described in Section 9.1.1.

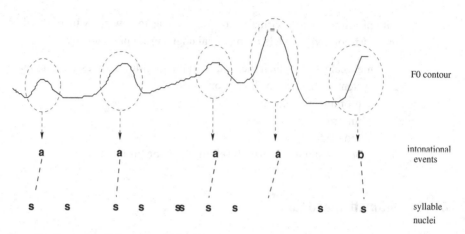

Figure 6.1 Example of F0 contour with pitch accents and boundary tones marked.

with the intonation appropriate to a question, quite often speakers speak this with the same intonation as for a normal statement. Just why speakers do this is addressed in Section 6.5.3.

It is difficult to discuss intonation further without reference to the phonetics and acoustics of this part of language, and for that reason we forego a more-thorough account of intonation until Chapter 9.

6.5 Prosodic meaning and function

One of the reasons why good models of prosody have proved so hard to develop is that researchers have often tried to study prosody without reference to its communicative function. In verbal linguistics it is quite common to study a particular sentence without reference to why a speaker said this, and by and large this separation provides a useful modularity for research purposes. So when we consider a sentence such as

(96) THE MAN WALKED DOWN THE STREET

We can say that THE MAN is the subject and WALKED is a verb and so on, and then perhaps go on to study the phonetics of the sentence, without having to know why the speaker actually said this sentence. It is common to follow a similar approach in prosody, but we believe this to be mistaken and that, in fact, it is more or less impossible to study the nature of prosody without taking into account the dialogue, the speaker's situation and other functional factors. Hence we now give an account of **prosodic function** that states the reasons *why* we use prosody. As we shall see, this significantly simplifies the practical issues concerning how to generate prosody in a text-to-speech system.

6.5.1 Affective prosody

Affective prosody is the most-basic type of prosody and in a sense can be regarded as **pure prosody**. Affective prosody includes the expression of meaning related to emotion, mental state and the speaker's attitude. Significantly, prosody is the *primary* channel for the communication of these meanings: if one is angry and wishes to express this, it is best done with a loud, shouting, growling voice, rather than a simple, natural statement I AM ANGRY. In fact simply saying that one is in an emotional state with words without any matching prosody is usually taken as a sign of insincerity (e.g. YES YOU LOOK NICE IN THAT DRESS), and if the words and prosody seem to contradict, the listener usually takes prosody to be the true expression of the speaker's meaning. (For example, if you ask a child whether they ate all the cookies, they find it easy to say NO but less easy to say this convincingly.)

Affective prosody has a certain degree of language universality. In all languages, raising one's voice and shouting are seen as expressions of angry or hostile states, while speaking in a softer voice is seen as conciliatory or calming. In fact the relationship is so direct that often we say "she spoke in a calm voice" etc. Affective prosody is not completely universal, however, and, in terms of primary emotion, there are significant differences across languages. Even within English it is common to find one group of speakers being regarded as brash and another meek, when in fact all that is different is the "settings" and "thresholds" with which the basic emotions are expressed.

In general, affective prosody has a fairly direct meaning-to-form relationship and hence the more extreme the form the more extreme the emotion. This implies that, unlike verbal language, the affective prosody system is not arbitrary with regard to meaning/form. Various emotion-categorisation systems have been proposed. One common theory holds that there are six basic emotions (anger, happiness, sadness, disgust, fear and surprise) from which follow a number of secondary emotions such as puzzlement and intrigue [110], [113], [153]. Some, however, argue that there is no fixed set of emotions and rather that emotion is best expressed in a multi-dimensional system. One such proposal advocates three independent axes of emotion called **activation**, **evaluation** and **power** [126], [393].

In addition to emotion, affective prosody includes the speaker's attitude. The difference between emotion and attitude is that only emotion can exist independently of communication. As an illustration, you can be happy on your own, but you need someone else to be sarcastic with. Apart from this distinction, attitude functions similarly to emotion in that it is fairly direct, has a significant degree of universality and does not have an arbitrary or dual nature.

One point of debate in the literature on both emotion and affective prosody concerns whether there is such a state as "no emotion" or "no affective prosody". The argument is that, if we are able to increase the level of intensity of anger, then we should be able to decrease this to the point where there is no anger. Similarly with the other emotions. This issue is important because it seems that the total affective content of an utterance can vary greatly and can be contrasted with the total propositional content of the verbal part of the utterance. These two effects seem to trade off such that we can have sentences that are nearly purely affective ("that's amazing") and others that are purely propositional ("the sum of the square of the opposite and adjacent sides...") The latter type of sentence is what we have previously termed discourse-neutral.

6.5.2 Suprasegmentality

If a sentence is spoken with little or no affective content, i.e. in a discourse-neutral manner, we still see characteristic patterns in the phrasing, rhythm, pitch, voice quality and timing. Typical effects include phones at the ends of sentences or phrases being lengthened, syntactically salient words (e.g. heads) having more emphasis, F0 levels being higher at the starts of sentences and so on.

These effects are often called "prosodic" but, according to the model we use here, these are not real prosodic phenomena but just another aspect of the verbal component. We can see this because speakers do not impart extra information by using these effects and they are as highly formulaic as the use of, say, nasalisation, voicing and tongue position. These effects are often called prosodic insofar as they use much the same acoustic means as expression – anger is expressed through pitch and voice quality, which is also used to distinguish the middle and ends of a phrase. In the model used here, we do not include these effects under the umbrella of prosody, but instead term these the **suprasegmental** part of verbal phonetics. The term suprasegmental here is used to show that, while these effects may be verbal in nature, they operate over the span of several phones (segments). It is best to think of these as a by-product in the production of the phonetic content.

While suprasegmental effects are rightly seen as the natural effect of phonetic production on F0, timing and voice quality, it is important to realise that the speaker still has conscious control over these. That is, while it is normal for a particular voice quality to be used at the end of a sentence, the speaker can override this if desired. However, while overriding default behaviour is possible, it is highly noticeable if this is done, so a speaker who deviates from default behaviour will be perceived as wanting to draw attention to the aspect of the utterance which has changed. Another way of looking at suprasegmentality is in terms of defaults and deviation from defaults. The key to this is to realise that the default for a sentence is not connected to mathematical simplicity where we might have a constant duration for each phone, a flat F0 contour and a constant state of vocal effort. Rather, the default behaviour exhibits considerable variation from these fixed values in ways dictated by verbal production. Deviations from the suprasegmental default are indications of real prosody, whether they be for affective or augmentative purposes (explained below).

In **tone languages** like Mandarin, pitch can be used to identify words. For example, liu spoken with one tone pattern means "flow", but when spoken with a different tone pattern means "six". In our model, this is a purely suprasegmental effect and treated in just the same way as nasalisation is used to distinguish words in English. Mandarin of course has intonation also, which is used for all the same affects as English. The fact that intonation and parts of word identity are expressed in the same acoustic variable does of course complicate analysis and synthesis somewhat, but with modern synthesis techniques we can model both without difficulty.

6.5.3 Augmentative prosody

Augmentative prosody is used as an aid to ensure successful communication of the verbal component of a message. Unlike affective prosody, augmentative prosody does not contain or convey any extra information; it is merely a means for the speaker to ensure that a message is decoded and understood more clearly. Augmentative prosody can be understood only with reference to a dialogue model of the type presented in Section 2.3. There we saw that participants in a dialogue have a mental image of each other and are constantly balancing the desire to be **efficient** in communication (whereby they would use the least effort in generating their utterances) with a desire to be **effective**, whereby they wish to ensure that the message is correctly decoded and understood.

In an ideal communicative setting, where the participants know each other, can see each other and are able to converse without background noise or other distractions, it is nearly always the case that the verbal message can be communicated without need for any augmentative prosody. Whether any prosody is used therefore is purely a matter of the affective nature of the utterance. Successful communication is possible because the verbal channel is highly redundant, to the extent that the speaker should always be able to decode and understand the message. In other conditions we face a different situation, so for example in noisy conditions we might see that the speaker speaks with greater vocal effort and more slowly. Likewise we may see the same effect if the speakers don't know each other so well, or in more polite situations in which the speaker wants to be seen to be making a clear effort to communicate effectively.

While it may be fairly obvious that increased articulation effort will lead to a higher probability of a phoneme string being correctly recognised, it is with other aspects of augmentative prosody that the situation becomes particularly interesting. For example, it is frequently asserted that **yes/no questions** such as

(97) IS THIS THE WAY TO THE BANK?

have rising intonation in English. In fact empirical studies of data show that the most-common intonation for such sentences is exactly the same pattern as that used for statements. While speakers *can* use rising intonation, they often don't. This behaviour is best explained with reference to the communicative model. In examples such as (97), it is usually *obvious* that a yes/no question is being asked, there really is little need for a speaker to use the additional cues that a rising intonation pattern would provide. However, if the speaker believes she may be misunderstood, or if she is trying to sound polite, the rising intonation pattern may appear. By comparison, it is common to ask "yes/no questions" that have the same syntax as statements, but in which the speaker does now in fact use a rising intonation pattern to make sure that these are interpreted correctly:

(98) THIS IS THE WAY TO THE BANK?

We see this effect in all aspects of prosody. The given/new distinction, where it is often claimed that subsequent mentions of an entity are less prominent than first mentions, is rarely observed in real speech, partly because it is quite clear from the verbal content which in fact is intended [24], [328].

In classic example sentences such as

(99) BILL DOESN'T DRINK BECAUSE HE'S UNHAPPY

which are spoken without phrase breaks may potentially be ambiguous but in general the listener will know enough of the dialogue context to decode this correctly. If, however, the speaker is less sure of a correct decoding, phrase breaks may be inserted to aid communication, and likewise with every other aspect of prosody.

In summary, augmentative prosody works by changing the default suprasegmental content of an utterance in particular ways. It is used by the speaker solely to increase the chances of the verbal message being correctly decoded and understood. Significantly, unlike affective prosody, it imparts no extra information into the message.

6.5.4 Symbolic communication and prosodic style

It is useful at this stage to compare and contrast the verbal and prosodic components of language. Verbal language was defined in Section 2.2.1, where we saw its defining properties as being **discrete** and **dual**, in the sense that every message is composed of a sequence of discrete word entities, which in turn are composed of a sequence of phonemes. Verbal language is **arbitrary** in the sense that one cannot infer the form of a word from its meaning or vice versa. Verbal language is **productive** in that completely new sentences can be created all the time. What, if any, of these properties does prosody have?

First of all, let us realise that, in verbal language, to construct a message we must always select a unit (e.g. word) and place such units in sequence. A word is either present or absent in an utterance and every utterance is composed of a sequence of words. With prosody the situation is different. Firstly, we need not always have a prosodic unit – it is completely acceptable and common to have an entire sentence with no particular emphasis, intonation or phrasing effects (recall that suprasegmental effects are considered part of verbal language, not prosody). Furthermore, if a prosodic effect is present, it can be there to a greater or lesser extent. So we can have a word that is somewhat emphasised relative to one that is heavily emphasised – this is unlike verbal language, where something is either present or not.

In verbal language, speakers really have very little choice in how they speak a particular word. So, while speakers when saying the word CAT may vary somewhat in their tongue positions or degree of aspiration, only so much variation is allowed, otherwise a different word may be perceived. Hence verbal language is inherently restricted and formulaic. With prosody, the fact that an utterance is not required to be "full" of prosody gives the speaker much more choice. Augmentative prosody may be used to aid communication but often it doesn't have to be; conversely, in certain polite or formal situations, a speaker may choose to use considerably more augmentative prosody than is strictly required for successful communication. The degree of affective prosody is again very variable – some conversations or messages (e.g. speech from a news reader) are nearly completely lacking in affective content while some are full (e.g. a chat between two friends).

An important consequence of the degree of choice in prosody is that it allows the speaker to deviate from the default patterns in certain ways without running the risk of changing the meaning of the sentence entirely. This then allows for the phenomenon termed **prosodic style**. This is most marked in situations where particular speakers, often actors or comedians, use prosodic style in a quite unconventional way. This use of prosody is not related to the communication of the message or even emotion, but is intended to draw attention to the speaker himself. Informally, we can talk about the speaker "sounding cool" by virtue of the way they can deviate from the normal patterns of prosody and be creative in their usage. Stylistic use of prosody is one significant factor in explaining why some speakers seem more interesting to listen to than others.

6.6 Determining prosody from the text

6.6.1 Prosody and human reading

In text-to-speech, our interest is of course in generating prosody from text. This is problematic in that text mostly encodes the verbal component of language; prosody is by and large ignored. Given that prosody is such a vital part of spoken language, how then is it possible that written communication can ever work? We examined this in some detail in Section 3.2, and can summarise the situation as follows.

- Quite often text encodes sentences that are rich in propositional content rather than affective content. Text genres like newspapers, non-fiction books, histories, company reports, scientific articles and so on simply do not have much affective content in the first place.
- The writer knows that prosody is absent from the text that is being written, and therefore *compensates* for this by using words in ways different from what would be the case when speaking. For example in literature we see phrases such as

    (100) 'that's right', he said angrily

    where the he said angrily is obviously written to help the reader understand how the 'that's right' section should have sounded.
- The awareness that prosody is lacking often leads to the writer writing in a style that is inherently less ambiguous; in such cases we say that the writer is writing "clearly" and so on.
- Quite often, people who are reading aloud add their own prosody into the speech. By this we mean that, when they read the words, they keep accurately to the word, but, because the prosody is not encoded in the text, the reader "makes up" what they think is appropriate for this situation. It is if the reader asks themselves "if I were saying this myself, what prosody would I use?" and then adds this into the speech.
- This requires considerable skill and is often based on an understanding of the text, rather than just a decoding as is the case for words. The reader is in effect adding in affective and augmentative prosody in places she believes are appropriate.

From this we can conclude that, in situations where the text genre is quite factual, it is usually sufficient to generate speech from the verbal message only, so all that is required is the generation of the suprasegmental part of the signal; the affective part of prosody is ignored. In other text genres the situation is significantly more difficult, and if, say, a dialogue from a play is read in this fashion (or, which is more likely, responses from a computer dialogue system), the generated speech can sound somewhat dull and removed. Finally, we see that mimicking a genuinely good human reader is very difficult indeed, since they will be performing an actual comprehension of the text and then generating their own prosody. In no way are they simply decoding the prosody from the text and then speaking it aloud, as they can do with the words. To date, no satisfactory solution to this problem has been found, so current text-to-speech systems are, unfortunately, lacking in this regard.

6.6.2 Controlling the degree of augmentative prosody

We face a different situation with regard to augmentative prosody. When a person speaks, she uses a mental image of the likelihood of a listener decoding and understanding the message to decide how much augmentative prosody to use and where to use it. We therefore face two problems in TTS: guessing the state of the listener and knowing which parts of the utterance may need disambiguating with prosody. Doing this properly is very difficult – even if we have a good idea about the likelihood of a message being misunderstood, knowing exactly where to place phrase breaks and additional emphasis is extremely hard since often this can require a full and accurate syntactic and sometimes semantic analysis of the sentence. But luckily, as with affective prosody, we have a "get-out" that enables us in many cases to bypass this problem.

Firstly, writers often use punctuation and fonts when they feel that the words in some way need disambiguation. So, when we see a comma or semi-colon, it is often a safe bet to place a phrase boundary there because we know the author felt that the words before and after the break should be separated. If a word is in italics or bold, it is again probably a safe bet to give this word extra emphasis. A second strategy arises from the fact that augmentative prosody is indeed only there to *help* in the decoding and understanding of an utterance. If we simply don't include any augmentative prosody, there is a good chance the utterance will still be decoded and understood, but perhaps just not with so much certainty. We can compensate for this specific lack of disambiguation information by ensuring that other parts of the signal do aid decoding; for instance if we synthesise the speech more clearly and slowly than normal the listener will have a better chance of understanding.

6.6.3 Prosody and synthesis techniques

In the remainder of this chapter we shall concentrate on **prosody prediction**, the term given to the task of generating the prosodic form from the text. In Chapter 9 we complete the task by considering the issue of **prosody synthesis**, the job of generating phonetic or acoustic information from the prosodic form. As we shall see, the degree to which prosody needs to be specified depends very much on the synthesis technique employed in waveform synthesis. Roughly speaking, in modern systems suprasegmental prosody can be generated at the same time as the verbal content, whereas in first- and second-generation synthesis systems the suprasegmental content had to be generated explicitly. Most of the following sections can be read without too much emphasis being placed on this distinction; it will, however, become more of an issue in Chapter 9.

6.7 Phrasing prediction

First we will examine the issue of **phrasing prediction**, which is how to generate a prosodic phrase structure for a sentence from the text. For purposes of illustration we will adopt the model described in Section 6.2.2, where we have a major phrase and a minor phrase. This is probably the most commonly used model in TTS.

6.7.1 Experimental formulation

To help in our exposition, we will formalise the problem in the manner suggested by Taylor and Black [442]. In this, our input is a list of tokens:

$$T = \langle t_1, t_2, t_3, \ldots, t_n \rangle$$

and between every token is a **juncture**, so that a full sentence is

$$T = \langle t_1, j_1, t_2, j_2, t_3, j_3, \ldots, j_{n-1}, t_n \rangle$$

The job of the phrasing algorithm is to assign values to every j_i drawn from an inventory of juncture types, which we will define to be {**non-break, minor-break, major-break, sentence**}. About one in five junctures are of **non-break** type (that is the gap between normal words), so we have to be careful in assessing the accuracy of any algorithm – if we use one that always assigns **non-break** we will have achieved 80% accuracy on a simple-junctures-correct measure. Because of this, Taylor and Black proposed a number of measures, later extended by Busser et al. [77]. These are calculated from the following basic quantities:

Breaks	B	Number of breaks in the corpus
Insertions	I	Number of non-breaks incorrectly marked as breaks
Deletions	D	Number of breaks incorrectly marked as non-breaks
Substitutions	S	Number of breaks of one type incorrectly marked as breaks of another type
Number of junctures	N	Total number of tokens in the corpus, minus one

These are then used to calculate the following scores:

$$\text{Breaks correct, BC} = \frac{B - D - S}{B} \times 100\% \qquad (6.1)$$

$$\text{Non-breaks correct, NC} = \frac{N - I - S}{N} \times 100\% \qquad (6.2)$$

$$\text{Junctures correct, JC} = \frac{N - D - I - S}{N} \times 100\% \qquad (6.3)$$

$$\text{Juncture insertions, JI} = \frac{I}{N} \times 100\% \qquad (6.4)$$

$$\text{Juncture deletions, BI} = \frac{I}{B} \times 100\% \qquad (6.5)$$

An alternative scheme more in line with that used in natural-language processing is based on **precision** and **recall**, which are specifically designed to give accuracy figures for problems where the number of occurrences is much less than the number of non-occurrences:

$$\text{Precision, } P = \frac{\text{number of breaks correct}}{\text{number of breaks predicted}} = \frac{BC - I}{BC} \qquad (6.6)$$

$$\text{Recall, } R = \frac{\text{number of breaks correct}}{\text{number of breaks in the test set}} = \frac{BC - I}{B} \qquad (6.7)$$

These are often combined in a **figure of merit**, F:

$$F_\beta = \frac{(\beta^2 + 1)PR}{\beta^2 P + R} \qquad (6.8)$$

where β is used to govern the relative importance of whether falsely inserting a break is a worse error than falsely missing one. If these are considered equal, then $\beta = 1$ and

$$F = \frac{2PR}{P + R} \qquad (6.9)$$

This is a commonly used measure of overall quality, but, as we shall see, it is almost certainly not true that human perception believes that an insertion and a deletion cause the same gravity of error.

6.7.2 Deterministic approaches

There are two popular and very simple phrasing algorithms.

1. **Deterministic punctuation (DP)**: place a phrase break at every punctuation mark.
2. **Deterministic content function (DCF)**: place a phrase break every time a function word follows a content word.

The first algorithm, DP, works very well in sentences of the type

(101) The message is clear: if you want to go, you have to go now

For the DP algorithm to work, we must have an accurate means of determining underlying punctuation. This is so that sentences like

(102) But really!? You can go -- but it won't do you any good

have the punctuation sequences !? and -- mapped to a single underlying form because we don't want to have an additional phrase break between the ! and the ?. The DCF algorithm works well in sentences such as

(103) It was taking a long time to gather all the information

where we place a phrase break after every function word that follows a content word:

(104) it was taking | a long time | to gather |all the information

These can be combined into a single algorithm, **deterministic content function punctuation (DCFP)**, which places a break after every content word that precedes a function word and also at punctuation. The algorithms were tested on a large corpus of read speech for which phrase breaks had been labelled by hand. The results for these algorithms given in Taylor and Black [442] are

Algorithm	Breaks correct	Junctures correct	Juncture insertions
Deterministic punctuation	54.274	90.758	0.052
Deterministic content function punctuation	84.40	70.288	31.728

These figures basically show that the DP algorithm massively underpredicts and the DCFP algorithm massively overpredicts. The DP algorithm result is worth examining further because it shows that placing a phrase break at a punctuation mark is in fact a very safe thing to do; this nearly always coincides with a real break. The only problem is that many other breaks are missed. The DPCF algorithm overpredicts far too much, which leads to the idea that, if finer distinctions than just content word and function word could be made, a more accurate algorithm would be possible.

Several more-sophisticated deterministic systems that make use of rules for specific cases have been proposed [10], [26], [186]. For example, the **verb-balancing rule** of Bachenko and Fitzpatrick [26] works through a sentence from left to right and compares the number of words in a potential phrase formed with the verb and the syntactic constituents to the left and the number of words in the constituent to the right. The potential phrase with the shortest number of words is chosen as the correct one.

6.7.3 Classifier approaches

The general historical trend in phrase-break prediction is the same as with most problems we examine: as more data have become available, researchers have moved away from rule-based systems towards trainable systems. This is partly because the new data showed how errorful rule-based systems were and partly because using trainable systems allowed algorithms to have their parameters and rules automatically inferred from the data. We now turn to data-driven techniques, where interestingly we see a similar pattern of algorithm type to that used in text classification. In this section we examine **local classifier** algorithms, which share many features in common with WSD algorithms. In the next section we examine HMM approaches, which are very similar to POS taggers.

Wang and Hirschberg [363] introduced the idea of using decision trees for phrase-break prediction. Decision trees allow a wide variety of heterogeneous features, examples of which are given below:

- total seconds in the utterance
- total words in the utterance
- speaking rate
- time from start of sentence to current word
- time to end of sentence from current word
- whether the word before the juncture is accented
- whether the word after the juncture is accented
- part-of-speech
- syntactic category of constituent immediately dominating the word before the juncture
- syntactic category of constituent immediately dominating the word after the juncture

These features (and others) were used as questions in the tree, which was grown in the normal way. Wang and Hirschberg trained and tested their system on 298 utterances from the ATIS corpus, giving the following results (the results aren't available in the formats we described above):

Boundaries correct	88.5%
Non-boundaries correct	93.5%

It should be pointed out that this approach can't strictly be used for TTS purposes since acoustic features (e.g. time in seconds) measured from the corpus waveforms were used in addition to features that would be available at run time. Following this initial work, authors of a number of studies have used decision trees [264], [418], and a wide variety of other machine-learning algorithms have been applied to the problem, including memory-based learning [77], [402], Bayesian classifiers [515], support-vector machines [87] and neural networks [157]. Similar results are reported in most cases, and it seems that the most important factors determining the success of a system are the features used and the quality and quantity of data rather than the particular machine-learning algorithm used.

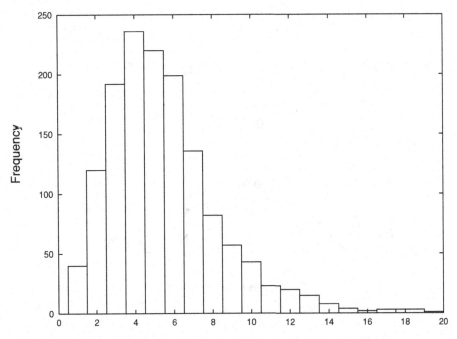

Figure 6.2 A histogram of phrase lengths.

6.7.4 HMM approaches

The HMM approach to phrase-break prediction was introduced by Taylor and Black [442] and has been extended by others [37], [369]. In their basic system, Taylor and Black proposed that the decision on whether a phrase break should or should not be placed was dependent on two main factors:

1. the parts-of-speech surrounding the juncture,
2. the distance between this proposed break and other breaks.

The first of these factors is simply a restatement of what we have already examined, i.e. that syntactic patterns have a strong influence on phrase-break position. The second factor is used in the classification approaches, but is given particular importance in the HMM model. The hypothesis is that, in addition to any high-level syntactic, semantic or dialogue factors, phrasing operates in a semi-autonomous fashion in that speakers simply like to place phrase breaks at regular intervals. This can be explained to some extent by the fact that speakers need to pause to get breath, but it seems that the phrase breaks occur much more frequently than is needed for breathing purposes, so this alone cannot explain the patterns. Regardless of why, however, it is clearly the case that speakers do exhibit strong tendencies to place phrase breaks at regular intervals, a fact noted by many researchers in this area [26], [156]. Figure 6.2, which shows a histogram of phrase lengths in words, clearly demonstrates this pattern.

The phrase-break-length histogram shows a preference for phrase lengths of about three to six words, and shows that very long and very short phrases are highly unlikely.

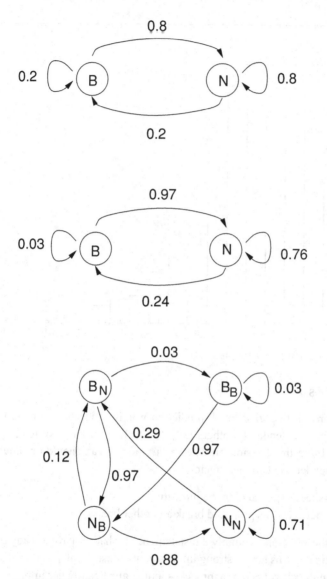

Figure 6.3 Finite-state-machine depictions of *n*-gram phrasing models.

Alternatively, we can view this distribution cumulatively, meaning that the longer we have been since the previous phrase break the more likely it is that the next juncture will have a phrase break. While it is easy to add this sort of information into a classifier, it is difficult to do so in a globally optimal way. The basic problem is, say, that we are at a juncture that is six words away from the previous juncture. Our distribution may say that it is now quite likely that we will observe a break at this point. If the evidence from the POS sequence seems reasonable, then a break will be placed here. The problem is that the next juncture, seven words away from the last break, may be an even better place to put a break, but we can't do this now because the break at word six has already been assigned. What is required is a way to search *all possible* sequences of phrase breaks, and this, via the Viterbi algorithm, is exactly what the HMM approach can deliver.

In the HMM approach, we have a state for each type of juncture, so, ignoring sentence breaks, we either have three states, if we distinguish major and minor breaks, or two states, if we don't. Taking the second case first, we now have a state that gives

$$P(L|j_i)$$

that is, the probability of a juncture after token i omitting a POS sequence L. Ideally the POS sequence L would be the tag sequence for the entire sentence, but, since these distributions are non-parametric models, which have to be learned by counting, the size of the window is quite limited in terms of what can be estimated robustly. Hence we approximate the POS sequence of the whole sentence as a window of POS tags around the juncture. This can be seen as a generalisation to the DCFP algorithm, where we used one tag before and after the juncture. Taylor and Black, however, reported that a window of two tags before and one tag after the juncture gave the best results, given limited training data. Informally, we can see that we would expect the following to give relatively high probabilities:

$$P(ADJ\ NOUN\ AUX|\text{break})$$

$$P(DET\ NOUN\ NOUN|\text{non-break})$$

The juncture-sequence model is given by an n-gram, which is calculated in the way explained in Section 5.2.2, that is by counting occurrences of sequences. An important difference from the n-gram in a POS tagger is that we now have only two or three types in the n-gram rather than 40 or so. This means that the number of unique sequences is much smaller, which means that for the same amount of data we can robustly compute longer n-grams. In these models we would see that

$$P(N, N, N, B, N)$$

would give a high probability and that

$$P(B, B, N, B)$$

would give a low probability. Taylor and Black report that the best results are found when one which uses a 6-gram language model, where there are two tags before the juncture and one after. This gave figures of 79.24% breaks correct, 91.597% junctures correct and 5.569% juncture insertions.

6.7.5 Hybrid approaches

More recently approaches that combine the advantages of decision-tree approaches (use of heterogeneous features, robustness to curse of dimensionality) and the HMM approach (statistical, global optimal search of sequences) have been proposed. In addition, there has been something of a re-awakening of use of syntactic features due to the provision of more robust parsers. Rather than attempting an explicit model of prosodic phrasing based on trying to map from the syntax tree, most of these approaches use the syntax information as additional features in a classifier [209], [237], [307].

While the pure HMM approach uses a discrete probability distribution over observations (in our case POS tags), in practice all we need is a means of finding the likelihood of seeing the observation given the juncture type $P(L|J)$. Many have therefore tried alternative ways of generating this. One of the most popular is actually to use decision trees, since these can be used to give probabilities as well as simple decisions on classification. As it stands, a decision tree gives a posterior probability, not a likelihood, but a likelihood value can be obtained either by ensuring that the tree is trained on an equal amount of break and non-break data or by normalizing the posterior by the basic frequency of breaks and non-breaks. The advantage in this combined approach is that we can use any features that we require, but also perform the global Viterbi search using n-grams to find the best overall sequence of breaks [369], [403], [423].

6.8 Prominence prediction

Prominence-prediction algorithms follow the same basic approaches as phrase-break prediction, where we have simple deterministic algorithms, sophisticated deterministic algorithms and data-driven algorithms.

6.8.1 Compound-noun phrases

The simplest prominence-prediction algorithm simply uses the concatenation of the lexical prominence patterns of the words. So for

(105) IN TERMS OF THE NECESSARY POLITICAL EXPEDIENCY TO ENSURE SURVIVAL | THE RESULT WAS CLEARLY A GOOD ONE FOR SHAW

we might generate a prominence pattern of

(106) IN **TERMS** OF THE **NECESSARY POLITICAL EXPEDIENCY** TO **ENSURE SURVIVAL** THE **RESULT** WAS **CLEARLY** A **GOOD** ONE FOR **SHAW**

The main ways in which this rule fails (again ignoring affective and augmentative effects) are in compound-noun phrases and certain uses of function words. The prominence patterns in compound-noun phrases in English are complex and can seem quite baffling. When we consider the prominence patterns in street names such as

(107) WINDSOR **AVENUE**

(108) BEAUFORT **DRIVE**

(109) PARK **LANE**

it seems quite obvious what the pattern is, until we consider examples with the word STREET

(110) **WINDSOR** STREET

(111) **BEAUFORT** STREET

(112) **PARK** STREET

in which the prominence pattern is reversed. Sproat [410] conducted a very thorough study into compound-noun phrasing patterns, and concluded that, in general, there is no simple syntactic pattern that governs prominence. Instead he proposed an extensive set of rules based partly on synactic and partly on semantic patterns. The most basic rule says that compound-noun phrases with two words assign their prominence according to whether the two words are taken as a single compound "word" or a phrase. In general in English, phrases receive their main prominence on the last word in the phrase. English words, and particularly nouns, by contrast tend to have their prominence on the first syllable. This can be seen in the difference between

(113) **BLACK**BIRD meaning the type of bird,

and

(114) BLACK **BIRD**, a bird that just happens to be black.

This principle can then be generalised to cope with all two-word cases, with the idea being that only conventionalised cases are treated as compound words that receive their prominence on the first word. In addition to this, there are some semantically driven rules, which range from the general to the more specific:

(115) Room + Furniture → RIGHT, e.g. KITCHEN **TABLE**,

(116) Proper name + street → LEFT, e.g. **PARK** STREET.

For more compound noun phrases involving three or more words, a CYK parser (see Section 5.5) is used to assign the internal structure to the noun phrase, after which prominence values can be assigned. To resolve ambiguity, a set of heuristics is used to find the best parse. These are the following.

1. Prefer parses determined by semantic rather than syntactic means. (This works partly because semantic parses are much more specific.)
2. Choose phrase-based parses over compound-word-based parses.
3. Choose parses with right-branching structure.

Additional rules are used to shift prominence in cases where the parses would produce two prominent syllables side by side. We also find that many types of non-natural-language entities have characteristic patterns. For example, it is commonplace to find telephone numbers read with a particular stress pattern (and indeed phrasing). Producing the correct stress patterns for these is often fairly straightforward and the creation of these patterns can be integrated into the verbalisation process.

6.8.2 Function-word prominence

Less work has been carried out on the second main problem of determining the prominence patterns in function-word sequences. The most we can say about this is that not all function words are equal when it comes to propensity for reduction. At one extreme, we have determiners (THE and A), which in their default form are always reduced, and at the other extreme closed-class words such as HOWEVER, WELL and SO can receive the most prominence in a sentence. Given a sequence of content words, a reasonable approach is then to give each a score according to where it comes in the strength list, and, starting with the strongest, add prominence at regular intervals. A possible hierarchy, from weakest to strongest, might be

articles	THE, A, AN
auxiliary verbs	WAS, HAS, IS
prepositions	TO, FROM, BY
particles	IF, THEN, WELL
conjunctions	THAT, AND, BUT
pronouns	HE, HIM, SHE, ME, THEM
modals	WOULD, COULD, SHOULD

Note that auxiliary verbs (e.g. HAVE) can function as main verbs also. We find a fairly clear pattern that when used as an auxiliary the word is often reduced, when used as a main verb it is often unreduced or prominent. With this table, we can then use a rule stating that prominent syllables should not occur side by side, so that in a sequence of function words we first assign prominence to the strongest word given in the table, ensure that its neighbours are reduced, and then look for the next-strongest word. This then can give the stress pattern for the sentence we saw in Section 6.3.1:

(117) WHAT **WOULD** YOU HAVE **DONE** IF IT HADN'T BEEN FOR **ME**?

6.8.3 Data-driven approaches

Prominence prediction by deterministic means is actually one of the most successful uses of non-statistical methods in speech synthesis. This can be attributed to a number of factors, for example the facts that the rules often don't interact and that many of the rules are based on semantic features (such that even if we did use a data-driven technique we would still have to come up with the semantic taxonomy by hand). Sproat notes [410] that statistical approaches have had only limited success because the issue (especially in compound-noun phrases) is really one of breadth, not modelling; regardless of how the prominence algorithm actually works, what it requires is a broad and exhaustive list of examples of compound nouns. Few complex generalisations are present (what machine-learning algorithms are good at) and, once presented with an example, the rules are not difficult to write by hand.

That said, with the provision of larger, labelled corpora, the natural progression has still been towards using machine-learning techniques to predict prominence. We should probably not be surprised to find that these techniques follow the basic paradigm used for phrase-break prediction, whereby we obtain a corpus of speech, label the prosodic content (in this case prominence levels) of this by hand and train an algorithm to predict these values from features available to the system at run time. Approaches tried have included decision trees [147], [207], [378], memory-based learning [303], transformation-based learning [200] and neural networks [72], [323], [389].

6.9 Intonational-tune prediction

Here we have explicitly separated prominence from the tune part of intonation, so for our purposes intonation prediction is specifically the prediction of intonational tune from text, rather than the broader definition of this problem that encompass predicting tune, accentuation and sometimes phrasing.

Of all the prosodic phenomena we have examined intonational tune is the most heavily related to augmentative and particularly affective content. In situations where these effects are absent, we can say to a first approximation that all utterances have in fact the same intonational tune; the only differences that occur concern where the pitch accents and boundary tones which make up this tune are positioned. Hence we can almost argue that, for discourse-neutral synthesis, there simply isn't any intonational-tune prediction to be done. In other words, the real task is to predict a suitable F0 contour that expresses the prominence and phrasing patterns and encodes the suprasegmental, rather than true prosodic, patterns of the utterance.

While we can describe prominence and phrasing in quite abstract high-level terms, this is significantly harder with intonation because all theories to a greater or lesser extent make explicit references to F0 patterns, levels and dynamics. Given this, and the fact that for the most part we are generating discourse-neutral suprasegmental intonation, we will leave the entire topic of intonation until Chapter 9.

6.10 Discussion

At first glance it may seem that predicting prosodic form from text is an impossible task. Two significant barriers stand in our way. First, the text is greatly underspecified for the type of information we require as textual encoding of prosody is more or less non-existent. Secondly, the uncertainty surrounding what the prosodic form should be makes it hard to label data, train algorithms and basically even know whether we are headed in the right direction.

6.10.1 Labelling schemes and labelling accuracy

Taking the second issue first, we see that in every area of prosody there is considerable disagreement as to how to represent prosodic form. The problem isn't so much that researchers disagree completely; there is widespread agreement about how to label many utterances; in many cases it is quite clear that a particular utterance has a strongly prominent syllable, has a clear phrase break and has a distinctive intonational tune. There are many "black and white" cases regarding which all theories and all labellers agree. The problem is rather that there is a significant grey area, where we can't tell whether this really sounds prominent, where we aren't sure that there is a phrase break between two words and where we can't really decide on the intonational tune for an utterance.

There are of course grey areas in other aspects of linguistics. As we saw in Section 4.2, pinning down the exact definition of a word can be tricky, and we will come across similar difficulties in Section 7.3.2 when we consider the definition of the phoneme. But in general the agreement among researchers is *much, much* higher with regard to these phenomena. Furthermore, so long as we are aware of the difficulties, the definitions of word and phoneme which we use in this book are *good enough*, meaning that, as engineers, using these models does not seem to result in significant loss of quality in synthesis. Studies that have examined **inter-labeller reliability** of prosodic schemes have shown the true extent of the problem; for instance in the original publication of the ToBI system, Silverman *et al.* reported that the agreement on phrase breaks was only 69% for four labellers, and 86% for whether a syllable is prominent (bears a pitch accent in their paper). These figures might not seem too bad, but are about an *order of magnitude* worse than the equivalent results for verbal, phonetic transcription and up to *two or three* orders of magnitude worse than for the transcription used for text analysis in Chapter 5. Furthermore, when we consider that the number of choices to be made in prosodic labelling is often small (say, choose one break-index value from five) we see just how difficult labellers find this task. These results for ToBI are not particularly unusual; consistently similar figures have been reported for many other schemes [426], [487], [486]. Furthermore, it is misleading to blame poor labelling agreement on the labellers not being "expert" enough – non-experts can readily label many of the types of data we require in TTS, so that in comparing the figures for other types of labelling we are in fact comparing like with like.

Why do these problems arise? Essentially we face two related issues: which labelling scheme, model or theory to use; and how to assign the labels arising from this to a corpus of speech. Most researchers involved both in the scientific investigation and in engineering implementation are aware that there is a huge number of differing and often incompatible theories and models of prosody. The temptation then is therefore to seek a "theory-neutral", "common-standard" or "engineering" system, which we can use to label the data and so avoid having to nail our colours to any one particular theoretical mast. We have to emphasise at this point that such an approach is folly; there is simply no such thing as a theory-neutral scheme. If we take the break-index scheme for instance, this explicitly states that there is a fixed number of types of breaks, that no recursion or limited recursion exists, that phrasing is not explicitly linked to prominence patterns and so on.

Furthermore, there is no sense in which we can just "label what we hear" in the data; all labelling schemes are based on a model, the fundamentals of that model are implied when we label speech, and, if our scheme is wrong, inaccurate or at fault in some other way, we run a high risk of enforcing inappropriate labels on our data. If we then use these labels as ground truth, we are running a severe risk of enforcing our training and run-time algorithms to make choices that are often meaningless.

6.10.2 Linguistic theories and prosody

Given these difficulties with labelling schemes and models, it is worth asking whether we need a model of prosodic form at all; perhaps it would be possible simply to define a mapping from the text or semantics to the acoustics. This viewpoint certainly has some merit, but we should be wary of dismissing the idea of prosodic form entirely. A brief digression into the theoretical history of prosody may help explain why.

It is probably fair to say that the field of linguistics has always had a troubled relationship with prosody. In fact many accounts of language completely ignore prosody (e.g. prosody is hardly ever mentioned in any work by Chomsky): whether this is because it is deemed to be quite separate from verbal language or simply too tricky to handle is hard to say. If *we*, however, want a full account of language and communication we have to include prosody.

Older linguistic accounts of prosody tended to treat it as quite a different system, and this to a degree was unproblematic. But with the rise of more formal theories of language, new theories of prosody were put forward that used the new techniques of formal language theory, feature structures and so on. The problem facing these researchers was that of how closely a formal theory of prosody should follow the formal theories of verbal language. Unfortunately (in my opinion) too much emphasis was placed on an insistence that prosodic phenomena could be explained with the same mechanisms as for verbal language, and hence we saw grammars of intonation, intonational phonology and schemes for pitch-accent description that were meant to be the direct equivalent of normal phonology. To be fair, external pressures pushed research in this direction; there was a strong desire to move from the very impressionistic and informal models of the past towards a more rigorous model and in addition there was the idea that if prosody could be explained with the same mechanisms as verbal language this would somehow be more parsimonious and would also ensure that prosody was taken seriously as a linguistic area of study (instead of just being ignored as mentioned above).

Ladd [269] points out that a major justification for the development of a phonology for prosody was to counter attempts in linguistics to correlate say F0 values with syntax patterns. While correlations may exist, building such a model ignores the fact that different speakers have different "parameters" in their F0 encoding model, for example that high-pitched voices will produce patterns different from those of low-pitched ones. A better model is to posit a system of abstract prosodic form, whereby all speakers of a language/accent would share the same representations for similar semantic and functional purposes. The actual acoustic encodings of these might vary with speaker physiology, but the prosodic form would be the same for all speakers. In formal terms, all we are saying is that there are levels, or interacting systems, of prosody. In summary the phonological school of prosody is nearly certainly right in advocating abstract description systems for prosodic form, but is perhaps less right in advocating that these should be formulated with the same mechanisms as verbal language. Prosody has an abstract structure, but it operates in a fundamentally different way from verbal language.

6.10.3 Synthesising suprasegmental and true prosody

Turning to the problem of predicting prosody from underspecified text, we find that the situation is less bleak that it might appear. Let us consider the issue of prosody not being encoded in the text. In many applications, we are not synthesising text with any particular emotive or attitudinal content, and are therefore going for a discourse-neutral reading. This simple fact saves us, and in such cases allows us to ignore the affective side of prosody. Many non-TTS synthesis systems are of course required to synthesize this type of text, and if emotive or attitude information is provided in the input there is no reason why these effects cannot be successfully synthesized.

When we consider augmentative prosody, we can again find ways around the problem. Firstly, as we discussed in Section 3.1.4, authors do often use punctuation and sometimes use italics and so on, and these are very good indicators that augmentative disambiguation is appropriate at this point. As we saw from Section 3.1.4, placing a phrase break at a punctuation mark is rarely wrong; human readers nearly always do this. However, recall that the whole point of augmentative prosody is that it acts as an *aid* to decoding and understanding and in many cases isn't strictly necessary. Hence if we *underpredict* phrase breaks, the only risk is that we may increase the chance of mis-comprehension by the listener. Significantly, speech generated with less augmentative prosody may sound completely natural – it's just the case that the listener may have to pay more attention. It is also important to realise that the most crucial phrase breaks are often just those marked by commas in the text, since it is here that the authors wants to be most sure of correct comprehension. While not ideal, underprediction is a safe strategy. Conversely, however, inserting a phrase break in a completely inappropriate position can sound terrible and often completely throws the listener, resulting in loss of naturalness and intelligibility. Hence, when developing algorithms for phrase-break prediction it is probably not sensible to balance insertions and deletions as equally bad errors; insertions are in general far worse than deletions. In practical terms, this means that the figure of merit scores which combine precision and recall should be weighted strongly in favour of precision rather than recall.

One area where augmentative prosody does, however, hurt us is in the data. While we can bypass the augmentative-prosody problem at run time by underpredicting, when we come to consider real data we effectively a priori have no idea what augmentative prosody the speaker will choose to use. While we can measure the acoustic encoding of this (pause lengths and positions and so on), this does not help us in determining the high-level features required to predict it. We can imagine the speaker "inventing" these features and using them when speaking; they are not available to it. This to a large extent explains the widely reported phenomenon of "speaker choice" or "speaker variability" in prosody. Basically, when speaking or reading, because of the redundancy in the verbal component, the speaker has considerable choice regarding where to place the phrase breaks; it all depends on their mental image of the listener. For practical purposes this has the effect of adding **noise** into the data. The features we have access to (the ones we can derive from the text) explain the presence of phrasing to a certain degree. The other features explain the rest of the behaviour, but, since we don't have access to these, they effectively appear as noise in our data and variance in our models.

The consequence of this, taken in conjunction with the issue of a very low level of transcribed agreement, may well mean that many prediction algorithms are as accurate as theoretically possible; no amount of extra refinement in the techniques or extraction of alternative features will make much difference. The reason why the algorithms cannot improve is due to noise in the data, from either or both of these sources.

6.10.4 Prosody in real dialogues

In Sections 6.8 and 6.7 we have seen how to predict prominence and phrasing patterns from text. The techniques described, though, in general operate only on single sentences, which are usually declarative in nature. Much of interest in prosody takes part at a different level, where many of the main phenomena occur in conversational, discourse, emotive or spontaneous communication. Hence most TTS systems model only a small range of the possible prosodic phenomena. We can, however, imagine systems that go beyond current capabilities. If speech synthesis were used in more-conversational applications, it would become important to model the discourse effects of prosody. In many current dialogue systems, the interaction feels very artificial because the system is capable of only simple questions and declarative sentences. The study of prosody in discourse is a rich topic of research in linguistics, but little has been done in terms of well-adopted practices in TTS systems. Hirschberg [208] gives a review of many discourse and functional aspects of intonation and describes much of the work that has been done in discourse prosody in TTS.

We need to reiterate some points about the augmentative function of prosody and how this affects studies of prosodic phenomena. While it is widely accepted that prosody does undergo more variation, or is subject to more speaker choice, than other parts of language, it is our belief that the *degree* to which this occurs is vastly underestimated in most studies. If we compare prosodic form with verbal form, we see that, while speakers can indeed vary the articulation of a word such as HAT, they are very constrained in so doing, and too much deviation from the canonical pronunciation runs the risk of the word being decoded incorrectly.

A key reason for the difference in degree of variation between verbal and prosodic language is to do with their functional use in communication. To a large extent, we can study the verbal part of language without reference to *why* a particular utterance is used in a discourse, but the same situation simply does not hold for prosody. The way in which prosody is used is primarily driven by the discourse, and taking sentences out of their discourse and studying them without reference to this leads to problematic analyses. In verbal phonetics a common practice is to elicit carefully spoken sentences under laboratory conditions. These conditions are artificial and can lead to discrepancies between the phonetic/acoustic patterns observed in "canonical" speech and those in real, spontaneous speech. In general, though, a considerable amount can be learned from controlled experiments in this way. The situation is radically different in prosody, however, and eliciting individual sentences under laboratory conditions is extremely risky in that these bear virtually no relation to how prosody is used in real, spontaneous speech. Authors of many studies unfortunately impose a strong discourse bias, so that when we examine utterances elicited for the purposes of studying, say, the given/new distinction, what we are doing is asking speakers to produce sentences in a very artificial discourse context. The effect of this is to exaggerate grossly the way the prosody is generated, such that effects such as given/new and contrastive prominence can seem quite clear, distinct and orderly. In real, spontaneous speech, by contrast, the discourse situation and the speaker's and listener's mental models of each other dictate how prosody is used. In these situations, the discourse context and information content are often so obvious and redundant that prosody is simply not required for further disambiguation. This helps explain why it is that authors of earlier laboratory-based studies of prosody claimed to find clear patterns and effects, but, when studies of real data are performed [24], [328], these effects are often far less clear.

6.10.5 Conclusion

The message to take home from this discussion is that the theoretical difficulties with prosodic models and the underspecification of prosody in text mean that this is an inherently difficult problem and should be approached with a degree of caution. While it is possible to continue to drive down the error rate with regard to semiotic classification, homograph disambiguation and so on, it is unrealistic to assume that the same can be done with prosody.

6.10.6 Summary

Prosodic form

- It is usual to consider three basic components of prosodic form: **phrasing, prominence** and **intonation**.
- Phrasing is used to group words together in the utterance, and is influenced by top-down factors, such as syntax, and bottom-up factors that place phrase breaks at roughly equal intervals.
- Prominence is used to give extra strength to certain words. It is partly determined by syntax, where in particular it governs stress patterns in compound-noun phrases and function-word sequences.
- Intonation is the use of pitch to convey prosodic information.
- Very little agreement exists on how to represent prosodic form, which makes the design of practical algorithms more difficult than in other areas of TTS.

Prosodic function

- **Affective prosody** conveys emotion and attitude in communication.
- **Augmentative prosody** is used to help disambiguate the verbal component of communication.

- **Suprasegmental prosody** arises from the natural patterns of how words are joined together in utterances.

Prosody prediction from text
- This is inherently difficult since prosody is largely absent from text.
- In practice though, the genres that we often use in TTS are not high in affective content.
- Phrasing and prominence prediction can be performed by a variety of algorithms, including simple rules, sophisticated rules, data-driven techniques and statistical techniques.
- Most popular machine-learning algorithms have been applied to prosodic prediction. The results are often similar, and difficult to improve upon because of the inherent difficulties in representing prosodic form and underspecification in the text.

Prosodic labels and data
- In general prosodic labelling by hand has low levels of inter-labeller agreement compared with other labelling tasks in TTS.
- This may stem from the fact that we have yet to develop a robust system for describing prosody.
- Speakers change their prosody depending on the discourse context and so it is difficult to associate one prosody with a particular sentence.
- Since speakers naturally vary their prosody and labels are often unreliable, the labelling of prosodic databases is often very noisy.

7 Phonetics and phonology

This chapter gives an outline of the related fields of phonetics and phonology. A good knowledge of these subjects is essential in speech synthesis because they help bridge the gap between the discrete, linguistic, word-based message and the continuous speech signal. More-traditional synthesis techniques relied heavily on phonetic and phonological knowledge, and often implemented theories and modules directly from these fields. Even in the more-modern heavily data-driven synthesis systems, we still find that phonetics and phonology have a vital role to play in determining how best to implement representations and algorithms.

7.1 Articulatory phonetics and speech production

The topic of **speech production** examines the processes by which humans convert linguistic messages into speech. The converse process, whereby humans determine the message from the speech, is called **speech perception**. Together these form the backbone of the field know as **phonetics**.

Regarding speech production, we have what we can describe as a *complete* but *approximate* model of this process. That is, in general we know how people use their articulators to produce the various sounds of speech. We emphasise, however, that our knowledge is very approximate; no model as yet can predict with any degree of accuracy how a speech waveform from a particular speaker would look like given some pronunciation input. The reason for this lack of precision is that the specifics of the production process are in fact incredibly complex; while we can make solid generalisations about how a [s] sound is pronounced for instance, it is another matter entirely to explain how this varies from speaker to speaker, how [s] is affected by speech rate, how surrounding sounds influence it and how all this is affected by prosodic interaction. The fact that this knowledge is incomplete should of course not be a surprise to readers of this book; if we fully understood this process we could simply implement it in a computer and thereby solve the speech-synthesis problem. In fact, as we shall see, in the field of text-to-speech, our lack of knowledge has led us to abandon (for the time being at least) the path of trying to mimic the human production process directly. However, what knowledge we have can be put to good use, and hence a solid grounding in what knowledge of speech production and perception we do have is an essential part of constructing a high-quality synthesiser.

Probably the main difficulty in discovering the mechanisms of human speech production is that of measurement and data collection. Of course, this is a problem common to all science; before telescopes only the simplest models of astronomy were possible; afterwards detailed and accurate observations could be made, which allowed the rapid development of theories. We see the same in the scientific study of phonetics; in the early days there were no instruments available for examining speech (phonetics even predates the invention of sound recording). With the invention first of sound recording, then spectrographic analysis, use of X-ray films and other techniques, speech scientists were able to examine the processes in more detail. Such instruments and more elementary techniques (such as looking directly into a speaker's mouth or simply trying to determine the position of one's own tongue) have led to the level of understanding we have today and shall briefly describe next.

Progress in speech perception has been significantly slower. The lack of progress can be explained simply by the difficulty in extracting information about the process. Although there exist some techniques for examining the cochlea and analysing brain patterns when listening, in general it is very difficult to collect concrete data on the problem, and as a result scientists have little to go on when developing theories. Whatever progress has been made is typically conducted within the framework of experimental psychology (e.g. [58]). This significant imbalance in our knowledge of production and perception helps explain why phonetics is in general approached from a speech-production point of view; that is simply the area of the speech process that we best understand.

7.1.1 The vocal organs

We generate speech by the coordinated use of various anatomical **articulators** known collectively as the **vocal organs**. Figure 7.1 shows a mid-sagittal section of the head showing the vocal organs. If we consider just the tongue in isolation for a moment, it should be clear that it can move in all three dimensions, and can create a complex variety of movements and trajectories. The movement and position of the tongue directly affect the speech produced and even a movement of a few millimetres can dramatically alter the speech sound produced. While the other organs are somewhat simpler, it should be clear that a substantial number of vocal-organ **configurations** is possible, and each gives rise to a different sound. This alone gives some insight into the complexities of speech.

7.1.2 Sound sources

Nearly all sounds in English are created by air moving from the lungs through the vocal organs to the lips and then outwards. This flow is called an **egressive pulmonic air stream**; egressive, because the air is flowing outwards, and pulmonic because the source is the lungs. During the passage of the air flow, one or more **constrictions** is applied, the effect of which is to generate a sound. We call constriction that causes the sound the **source** and the sound produced the **source sound**. We shall consider the types of source in turn.

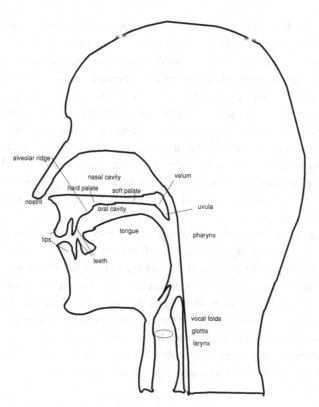

Figure 7.1 A diagram of the vocal organs or articulators.

The **vocal folds** are two folds of tissue that stretch across the **larynx**. A speaker can control the tension in his or her vocal folds (Figure 7.2) so that they can be fully closed, narrow or open. The gap between the vocal folds is called the **glottis**, and we usually refer to this type of sound production as a **glottal source**. When the vocal folds form a narrow opening, the air stream moving through them causes them to vibrate, giving rise to a **periodic** sound. We term the rate of vibration of the vocal folds the **fundamental frequency**, denoted **F0**. The term **pitch** is used for the rate of vibration that is perceived by the listener, and in general the pitch and fundamental frequency can be taken as the same thing. By varying the tension in the vocal folds, a speaker can change the fundamental frequency of the sound being produced. When the vocal folds operate in this way, they are said to be generating a **voiced** sound. A typical male speaker can vibrate his vocal folds between 80 and 250 times a second, so, using the standard terminology, we say that his fundamental frequency varies from 80 Hertz (Hz) to 250 Hertz. By comparison a female speaker might have a fundamental frequency range of 120 Hz to 400 Hz. All vowels are voiced sounds – you can experience the effect of the vocal folds by placing your finger on your larynx and speaking a constant "aaah". You should be able to feel the vocal folds vibrating. If you then whisper "aaah" the vibration will stop.

In addition to the fundamental frequency, a periodic signal usually has energy at other frequencies known as **harmonics**. These are found at multiples of the fundamental, so, if

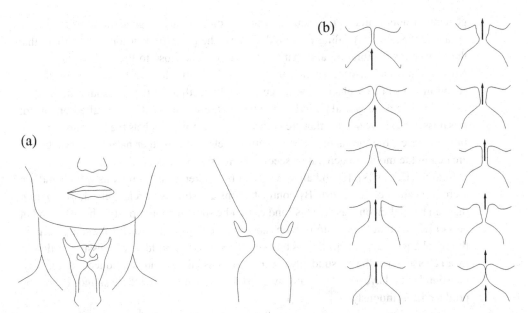

Figure 7.2 Vocal-fold operation. (a) The vocal folds. (b) Vocal-fold vibration. If the vocal folds are tensed in a certain way they will close. If air is pushed from the lungs, it will build up pressure under the closed folds until it forces an opening. The released air will cause the pressure to drop, and the tension in the vocal folds will cause them to close again. This process repeats itself, leading to a periodic noise being emitted from the folds. The amount of tension controls the rate at which this cycle of opening and closing will happen.

we have a glottal sound with a fundamental of 100 Hz, we will find harmonics at 200 Hz, 300 Hz, 400 Hz and so on. In general the harmonics are weaker than the fundamental, but nonetheless are heard and can be thought of as giving the sound its basic **timbre**. A musical analogy might help here. One can create a sound of the same fundamental frequency by either bowing or plucking a violin string. The fundamental frequency is the same in both cases, however, the different action creates a different pattern of harmonics and this is why the timbre of bowing and plucking sounds different.

If the glottis is opened slightly further, periodic vibration will cease. Instead a non-periodic turbulent air flow will be created, which generates a different type of sound, termed **noise**. Here the term "noise" is used in the technical sense (as opposed to the general definition which is a synonym for "sound") to mean the random sound created, somewhat like an untuned analogue radio. This is the natural way to generate the [h] sound in HEDGE, but is also the way in which whispering works. If the vocal folds are brought together completely, the air stream stops and no sound is produced. By carefully controlling the timing of this opening and closing, it is possible to produce a **glottal stop**, which is the normal way to realise the [t] in words such as BUTTER in some accents of English (e.g. Glaswegian and Cockney).

Sound can be created from sources other than the glottis. If the glottis is open, it will not generate any sound, but will allow the air stream to flow through. It is then possible to use combinations of the tongue, lips and teeth to form a constriction, and thereby generate a non-glottal sound source. The [s] sound in SIT is generated this way – on speaking this, and extending the length of time the [s] is spoken for, you can feel that the glottis is not vibrating and that the tongue is very close to the roof of the mouth. Noisy sounds can also be created by holding the teeth close to the lips (as in the [f] sound in FALL) or by placing the tongue near the teeth as in the [th] sound in THANK. Because there is no glottal periodic vibration in these sounds, they are called **unvoiced**. It is possible to have sounds that are both voiced and noisy, such as the [z] sound in ZOO. In these cases, the glottis operates as with vowels, but a further narrow constriction is created in the mouth to generate a secondary, noisy, sound source.

Sounds such as [s], [f] and all vowels can be spoken as a continuous sound, and are therefore called **continuants**. By contrast, **stops** are sounds of relatively short duration that can be spoken only as "events" and cannot be spoken continuously. The [p] sound in PEN or [t] sound in TIN are stops. Stops are produced by creating a **closure** somewhere in the vocal tract so that all air flow is blocked. This causes a build up of pressure, followed by a **release** when the air suddenly escapes. Because the sound is produced in this way, the sound must have a finite, relatively short duration and hence these sounds cannot be produced continuously.

7.1.3 Sound output

For most speech sounds, the acoustic signal leaves via the mouth – such sounds are termed **oral**. For a few, however, the sound leaves via the nose; such sounds are called **nasal**. In English the first sounds in NO and ME are nasal, as are the last sounds in THEM, THIN and THING. As evidence that these sounds leave via the nose rather than the mouth, say "mmmmm" and then pinch your nose so as to close it – the sound will stop. Sometimes sound exits through both the nose and the mouth. There are no canonical sounds in English that do this, but in French, for instance, the last sound in ONT exits through both the nose and the mouth. Such sounds are called **nasalised**.

The **velum** is a piece of skin at the back of the mouth. When it is raised, the nasal cavity is blocked off and the speech is completely oral. When the velum is lowered, the sound will be nasal or nasalised, depending on whether the mouth is closed or open. It is important to note that, in an oral sound, the velum blocks the entrance to the nasal cavity, so this plays no part in determining the sound. In nasal sounds, however, while the oral cavity is blocked, it is not blocked at the entrance; rather it is blocked at some other point (at the lips in an [m] sound and near the alveolar ridge in an [n] sound). During the production of a nasal, air enters the oral cavity, is reflected and eventual escapes from the nose. The shape of the oral cavity plays an important part in determining the sound of the nasal, even though no air escapes from it directly. Figure 7.3 shows vocal tract configurations for these output types.

7.1.4 The vocal-tract filter

The diversity of sound from the source and output are further enriched by the operation of the **vocal tract**. The vocal tract is collective term given to the pharynx, the oral cavity

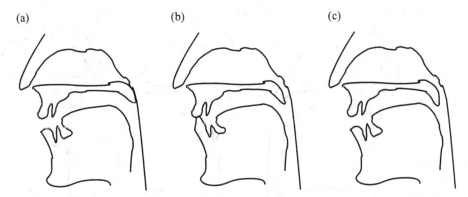

Figure 7.3 (a) Oral sounds occur when the velum blocks the nasal cavity, and the sound escapes from the open mouth. (b) Nasal sounds are caused first by blocking sound escape from the mouth and secondly by lowering the velum so that sound can escape from the nose. (c) Nasalised sounds can have sound escape from both the mouth and the nose.

and the nasal cavity. These articulators can be used to modify the basic sound source and in doing so create a wider variety of sounds than would be possible by using the source alone. Recall that all voiced sounds from the glottis comprise a fundamental frequency and its harmonics. The vocal tract functions by modifying these harmonics, which has the effect of changing the timbre of the sound. That is, it does not alter the fundamental frequency, or even the frequencies of the harmonics, but it does alter the relative strengths of the harmonics.

In general it is the oral cavity which is responsible for the variation in sound. The pharynx and nasal cavity are relatively fixed, but the tongue, lips and jaw can all be used to change the shape of the oral cavity and hence modify the sound. The vocal tract can modify sounds from other sources as well by operation of the same principle.

This model, whereby we see speech as being generated by a basic sound source and then further modified by the vocal tract, is known as the **source/filter** model of speech. The separation into source and filter not only adequately represents the mechanics of production but also corresponds to a reasonable model of perception in that it is known that listeners separate their perception of the source in terms of its fundamental frequency from the modified pattern of its harmonics. Furthermore, we know that the main acoustic dimension of prosody is the fundamental frequency, whereas the main dimensions of verbal distinction are made from a combination of the type of sound source (but not its frequency) and the modification by the vocal tract. The mathematics of both the source and the filter will be fully described in Chapter 10.

7.1.5 Vowels

All **vowel** sounds are voiced and hence have the same sound source (the glottis). What distinguishes one vowel sound from another is the shape of the oral cavity and to a lesser extent the shape of the lips and duration for which the sound is spoken. The oral cavity and lips operate to create differently shaped cavities through which the source sound

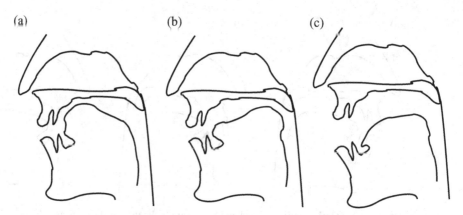

Figure 7.4 Vocal-tract configurations for three vowels: (a) HEAT, (b) HOOT and (c) HOT.

must pass. By moving the jaw, lips and tongue, different filter effects can be created, which serve to modify the harmonics of the glottal sound source and produce the wide variety of vowel sounds. It is important to note that the pitch of the speech is controlled entirely by the glottis, whereas the type of vowel is controlled entirely by the tongue, jaw and lips. As evidence of this, generate any vowel and keep the pitch constant. Now vary the shape of your mouth in any way you wish – it is clear that the pitch remains the same, but the mouth alone is determining the quality of the vowel. Conversely, say an "aaah" sound and convince yourself that you can change the pitch without changing the vowel itself.

It helps to describe vowels if we define a number of dimensions of oral cavity and lip configuration. First, the height of the tongue combined with the position of the jaw defines a dimension called **height** (sometimes called **open/close**). If one compares the positions of the mouth when speaking HEAT and HOT one should be able to tell that the jaw is raised and the tongue is high in HEAT, whereas for HOT the jaw and tongue are lower. Secondly, the position of the raised part of the tongue in relation to the front of the mouth defines another dimension, which we can call **front/back**. Consider the difference in tongue positions for the vowels in HEAT and HOOT – while this is slightly more difficult to feel or measure than in the case of height, it should at least be clear that the tongue is in a different position and that the jaw remains relatively fixed. Figure 7.4 shows the articulator positions for HEAT, HOOT and HOT. A third dimension is called **lip rounding** or simply **rounding**. This effect is caused when the lips are protruded from their normal position. Some vowels (e.g. HOOT) have noticeably rounded vowels, whereas others are noticeably unrounded (e.g. HEAT). A final basic dimension of vowel classification is **length**. This feature is very accent-specific, but in many accents the vowel in HEAT is considerably longer than the one in HIT.

In languages such as English, there is a particular vowel sound known as **schwa**. This is often called the **neutral vowel** in that its position is neither front nor back, high nor low, rounded nor unrounded. It is the first vowel in ABOUT, SEVILLE, COLLECT, and the second vowel in SARAH, CURRENCY, and DATA. Most other sounds are simply called by how they sound; schwa is given its special name because it is difficult to pronounce in isolation: if one attempts this, the sound that comes out is normally too strong sounding. Schwa is spoken with a tongue position close to that when the tongue is in its rest position, thus the tongue doesn't create any particular cavities or shapes in the vocal tract; rather the vocal tract is in the shape of a fairly uniform tube.

Some vowels are characterised by a movement from one vocal-tract position to another. In HEIGHT, for instance, the vowel starts with a low front-mouth shape and moves to a high position. Such vowels are termed **diphthongs**. All other vowels are called **monophthongs**. As with length, whether or not a vowel can properly be considered a diphthong is heavily accent-dependent. In particular the vowels in words such as HATE and SHOW vary considerably in their vowel quality from accent to accent.

7.1.6 Consonants

Consonant sounds are more heterogeneous than vowels in that they are produced by a wider variety of sound sources. While consonants share some inherent characteristics, it is not inappropriate to define consonants as the set of speech sounds that are not vowels.

Consonants can be classified in terms of **voicing, manner** and **place of articulation**. As we have mentioned, voicing describes whether the sound source is produced by the vibration of the glottis or by another means. Consonants such as [f] and [s] are unvoiced whereas [v] and [z] are voiced. This can be seen by speaking a [v] continuously and turning it into an [f] – it should be possible to do this without altering the shape of the vocal tract, simply by changing the vibration mode of the glottis.

The first important category of sounds is known as **oral stops**. In English, stops occur in voiced/unvoiced pairs, so that [b] and [p] use the same manner and place of articulation, and differ only as to whether the glottis is vibrating or not. The three nasals (the last sounds in THEM, THIN and THING) are often also described as **nasal stops**. While it is possible to utter these three phones continuously, they usually follow the pattern of the stop in normal speech, with a closure and release. During the production of nasals, the mouth is closed, so the sound escapes only from the nose. The position of the tongue in the mouth is important, however – the nasal tract is fixed and hence does not distinguish the three nasals. The production of nasals can be thought of as the glottis sending a periodic sound into the mouth, which is modified (filtered) by the oral cavity and then passed out through the nose, which performs further modification.

Fricatives are generated by creating a narrow constriction in the vocal tract, by placing (say) the tongue near to the roof of the oral cavity, thus creating a turbulent, noisy sound. Fricatives can be voiced (e.g. [v]) or unvoiced (e.g. [f]). In voiced fricatives, there are *two* sound sources, the glottis, which produces a periodic sound source, and the constriction, which adds turbulence to this. To demonstrate that this is indeed the case, try saying [v] and [f] with different pitches. This is possible with [v] but not with [f]. The final class of true consonants is the **affricates**, such as the first and last sound in CHURCH. These are spoken as a stop followed by further frication.

What then distinguishes the unvoiced fricatives such as [f] and [s]? The answer lies in *where* the constriction occurs. For [f] the teeth combine with the lips to form the constriction – this is termed **labiodental**. In [s] the sound is generated by the tongue

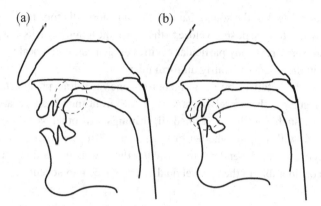

Figure 7.5 Places of articulation for alveolar and labio-dental fricatives: (a) alveolar constriction and (b) labio-dental constriction.

coming into close proximity with the roof of the mouth near the alveolar ridge and hence this sound is called **alveolar**. See Figure 7.5. There are many other possible positions, each of which will produce a slightly different sound. While in reality a continuum of constriction positions can occur, for convenience it is useful to identify and name a few key positions for constrictions. These include **bilabial**, in which both lips combine to form the restriction, and **dental**, in which the tongue combines with the teeth. The alveolar position is where the tongue approaches the roof of the mouth in a forward position. The tongue can create a constriction at positions further back, which are termed **post-alveolar**, **palatal**, **velar** and **uvular**. The place of articulation can be described on a scale starting at the lips and moving to near the back of the vocal tract. The important point is not so much which articulators (teeth, lips, tongue) are involved, but rather where the constriction occurs, since this governs the shape of the oral cavity.

One important point worth noting about fricative sounds is that the constriction *as a sound source* does not define the difference between an [f] sound and a [s] sound. In fact the sound source produced by all fricatives is more or less the same. Rather, it is the fact that the constriction gives rise to a vocal-tract configuration, and it is this acting as a *filter* that modifies the sound source and thereby produces the difference in sound. As we shall see in Chapter 11, it is not the case that sound always moves forwards in sound production. In dental fricatives, while some portion of the sound source is immediately propagated from the mouth, the remainder is propagated back into the vocal tract, where it is modified by the filter, after which it is then reflected and propagated forwards. It is by this mechanism that the vocal tract generates different sounds even if the sound source is near the mouth opening. The term **obstruent** is used as a collective term for fricatives, affricates and oral stops to indicate their noisy character.

The **approximants** are a class of sounds that are interesting in that they share many of the properties of both vowels and consonants. All approximants are voiced and all are produced in roughly the same manner as vowels, that is by varying tongue, lips and jaw position. [j] and [w] are know as **glides** and have similar sounds and means of articulation to the vowels in HEAT and HOOT. In some analyses, diphthongs are described not as compound vowels but as a combination of a vowel and a glide; since the glide comes after the vowel it is termed an **off-glide**. It is also possible to have **on-glides**, so CURE is described as a [y] sound followed by a vowel. Glides also function as proper consonants, for example as the first consonant in YELL, WIN and so on. If these functioned as vowels rather than consonants, the perception would be that these words have two syllables rather than one. The approximants [r] and [l] are known as **liquids** and act more like normal consonants, but still have many unusual properties (see Section 8.2.4).

7.1.7 Examining speech production

In traditional phonetics, the tools available to the phonetician were extremely limited and much of what we have so far described was developed with virtually no equipment. Progress was made by simply "feeling" the position of the articulators when speaking, or by looking down someone's throat as they spoke (which is not very helpful for examining nasal stops). The ability of these phoneticians to construct a model of speech production with such limited means is quite astonishing, and, by and large, they got the basics of the model correct (or correct in terms of our current understanding). One aspect that was not, however, realised at this time was the complexities of the syntagmatic dimension, meaning the influence that sounds had on one another when spoken as a sequence. Furthermore the role that "top-down" knowledge was playing in the phoneticians' examination was also largely unrecognised.

As recording and examination apparatus improved it became clear that what had seemed like clearly identifiable distinct sounds of speech were in fact quite hard to classify on the basis of articulatory evidence alone. We will address this question fully in Section 7.3.2, but for now let us simply note that it was found that what were thought of as single sounds (e.g. [p]) in fact had a huge range of possible articulatory realisations.

Over the years various analysis tools have been developed to examine the production process. The first was the use of X-ray still photography and then X-ray motion pictures. These techniques, especially the X-rays movies, clearly showed the various articulators in motion. Unfortunately, the availability of X-ray movies is extremely limited due to the now-understood harmful effects of X-ray exposure and hence this technique cannot be used much. More recently other techniques have been developed. These include the following.

Electropalatography This involves the use of a plate fitted to the roof of the mouth, which measures contact between the tongue and palate.

Electoglottography or laryngography This is the use of a device that fits around the neck and measures the impedance across the glottis. From this, a signal measuring glottal activity can be found.

Air-flow measurement A mask and measuring apparatus can be used for measuring air flow from the mouth and separately from the nose. This is particularly useful in that we can measure the outputs from the nose and mouth separately, and we can do this in terms of air-flow signals rather than the pressure signals which microphones record.

Magnetic resonance imaging (MRI) This technique, which was originally developed for medical use, can generate three-dimensional images of the head and hence measure the position of the articulators.

Electromagnetic articulography (EMA) This involves a device that works by recording signals from small electronic coils attached to the tongue and other articulators. During speech, the movement of the coils causes fluctuations in a surrounding magnetic field, which allows articulator movement to be tracked.

A description of the operation of many of these devices can be found in Ladefoged [272]. Examples of studies using EMA and MRI can be found in [13], [331], [332], [380], [497].

7.2 Acoustics, phonetics and speech perception

We term output of the speech-production process the **speech signal**. This acoustic signal travels through air to the listener, who can then decode this to uncover the message that the speaker uttered. Speech signals can be recorded by means of a microphone and other equipment, such that, with a little care, we can store a perfect representation of what the speaker said and the listener will hear. The field of **acoustic phonetics** studies speech by analysis of acoustic signals. The idea here is that, since this is what the speaker receives when listening, all the required information for perception should be present, so by suitable analysis we can discern the processes of human speech perception itself. That said, it must be emphasised that to date we have only a very limited understanding of speech perception. We know some of the initial processes (which for instance separate pitch from other aspects of the signal), but virtually nothing about how people actually distinguish one sound from another. Hence speech analysis should be thought of as an independent study of the signal, in which we are trying to find the dimensions of variability, contrast and distinctiveness from the signal itself, rather than as an actual study of perception.

7.2.1 Acoustic representations

The acoustic waveform itself is rarely studied directly. This is because **phase** differences, which significantly affect the shape of the waveform, are in fact not relevant for speech perception. We will deal with phase properly in Chapter 10, but for now let us take for granted that unless we "normalise" the speech with respect to phase we will find it very difficult to discern the necessary patterns. Luckily we have a well-established technique for removing phase, known as **spectral analysis** or **frequency-domain analysis**. We can use analysis software to transform the signal to the frequency domain and when this is done, it is a simple matter to remove the phase. Figure 7.6 shows what is known as a **log magnitude spectrum** of a short section (about 20 ms) of speech.

In this figure we can see the harmonics quite clearly; they are shown as the vertical spikes which occur at even intervals. In addition to this, we can discern a **spectral**

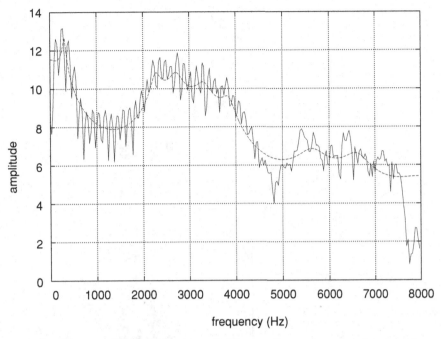

Figure 7.6 The log magnitude spectrum shows the pattern of harmonics as a series of evenly spaced spikes. The pattern of the amplitudes of the harmonics is called the spectral envelope, and an approximation of this is drawn on top of the spectrum.

envelope, which is the pattern of amplitude of the harmonics. From our previous sections, we know that the position of the harmonics is dependent on the fundamental frequency and the glottis, whereas the spectral envelope is controlled by the vocal tract and hence contains the information required for vowel and consonant identity. By various other techniques, it is possible to separate the harmonics from the envelope, so that we can determine the fundamental frequency (which is useful for prosodic analysis) and envelope shape.

While useful, these spectral representations show the speech characteristics only at one point in time; a further representation called the **spectrogram** is commonly used to show how the (phase-free) speech representation evolves over time. Figure 7.7 shows an example, in which the vertical axis of the spectrogram represents frequency, the horizontal axis time and the level of darkness amplitude, such that a dark portion states that there is significant energy at that frequency at that time. Some patterns can be seen immediately. Firstly the fricatives and vowels contrast strongly – the vowels have characteristic vertical streaks of the harmonics, and dark bands that move with time. To a certain extent the fricatives also have bands, but the important difference is that the vertical streaks are completely absent.

It is important to realise that the spectrogram is an artificial representation of the speech signal that has been produced by software so as to highlight the salient features that a

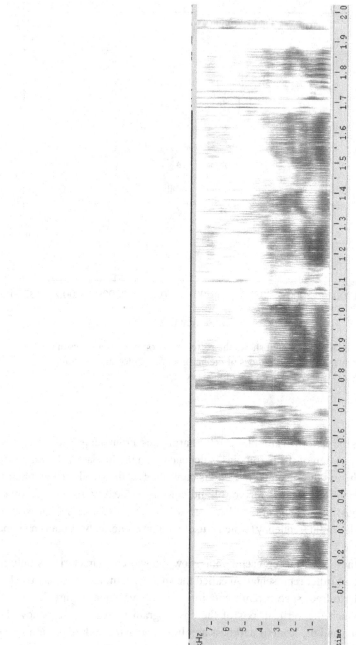

Figure 7.7 A wide-band spectrogram.

phonetician is interested in. The software for generating spectrograms can be configured in various ways, for example to vary the contrast between the levels of darkness. A second way to vary the spectrogram is to emphasise either the time or frequency resolution of the speech (it is generally not possible to do both at the same time). A spectrogram with high frequency resolution is known as a **narrow-band** spectrogram (shown later in Figure 12.9), one with high time resolution as a **wide-band** spectrogram. Wide-band spectrograms are actually more useful for examining the frequency patterns in speech because the lack of frequency resolution somewhat blurs the frequency information and makes the patterns within more visible. Finally, it is often useful to "zoom" with respect to the time domain, so that sometimes a full sentence is displayed, while at other times only a few phones are shown. These points illustrate some of the artefacts of spectrogram display so as to inform the reader that two spectrograms of the same utterance can look quite different because of the particular settings being used. In most modern spectrogram software, these settings can easily be varied. With these and other acoustic representations it is possible to study speech from an acoustic (and pseudo-perceptual) point of view.

7.2.2 Acoustic characteristics

At integer multiples of the fundamental frequency we have the harmonics. Speech with a low fundamental frequency (say 100 Hz) will have closely spaced harmonics (occurring at 200 Hz, 300 Hz, 400 Hz, ...), whereas speech with a higher fundamental frequency (e.g. 200 Hz) will have widely spaced harmonics (400 Hz, 600 Hz, 800 Hz etc.). The tongue, jaw and lip positions create differently shaped cavities, the effect of which is to amplify certain harmonics while attenuating others. This gives some clue as to why we call this a vocal-tract filter; here the vocal tract *filters* the harmonics by changing the amplitude of each harmonic.

An amplification caused by a filter is called a **resonance**, and in speech these resonances are known as **formants**. The frequencies at which resonances occur are determined solely by the position of the vocal tract: they are independent of the glottis. So, no matter how the harmonics are spaced, for a certain vocal-tract position the resonances will always occur at the same frequencies. Different mouth shapes give rise to different patterns of formants, and in this way the production mechanisms of height and loudness give rise to different characteristic acoustic patterns. Since each vowel has a different vocal-tract shape, it will have a different formant pattern, and it is these that the listener uses as the main cue to vowel identity. The relationship between mouth shapes and formant patterns is complicated, and is fully examined in Chapter 11.

By convention, formants are named **F1, F2, F3** and so on. Somewhat confusingly, the fundamental frequency is often called **F0**. Note that the fundamental frequency/F0 is *not* a formant, and has nothing to do with formants – it is determined by the glottis alone. Studies have shown that not all formants are of equal perceptual importance; in fact, the identity of a vowel is nearly completely governed by the frequency of the first two formants (F1 and F2). Figure 7.8 shows a **formant chart** in which the axes represent F1 and F2 values. Typical positions of each vowel (determined experimentally) are shown

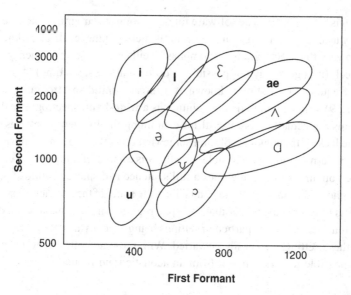

Figure 7.8 A chart of measured mean positions of vowels plotted in terms of first and second formant positions. The chart shows that, to a large extent, the two formant positions separate most vowels. This has led to the general assumption that F1 and F2 positions are used to discriminate vowels. This chart is shown in the form standard in phonetics; note that the axes have been specifically set up to make this plot easier to interpret.

on the graph, and from this we can see that each vowel occupies a different position on the graph, giving support to the idea that it is in fact the first two formants that distinguish vowels.

Other speech sounds have characteristic spectrogram patterns also. Nasals are generally weak and have a wide formant at about 300 Hz, which is caused by resonance in the nasal cavity. Since the nasal cavity is fixed, the resonance will always occur at the same position. Each nasal has its own oral-cavity shape, however, and the resonances in this are the main distinguishing feature between [m] and [n]. In stops, it is often possible to see the distinct phases of closure and the subsequent burst. Although the differences are subtle, it is possible to tell one stop from another from the resonance patterns in the burst and in the immediately neighbouring vowel. Approximants look like weak vowels, which is what we would expect.

7.3 The communicative use of speech

We have shown in the previous sections how the vocal organs can organise into a rich variety of configurations, which in turn can produce a rich variety of speech sounds. We have also seen that the differences among vocal-organ configurations can produce patterns that are discernible in the acoustic signal and representations derived from it. We now turn to the question of how this capability can be used to communicate.

7.3.1 Communicating discrete information with a continuous channel

Recall that natural language is a discrete symbolic semiotic system. In this system we combine words in various ways to produce a near-limitless number of sentences, each with a separate meaning. One might imagine a system of speech communication whereby each word was represented by a single unique speech sound, so that DOG would have a high central vowel, CAT an alveolar fricative and so on. The problem with such a system would be one of distinctiveness: if we had 10 000 separate words we would need 10 000 different speech sounds. It is unlikely that such a system would work simply because it would not be possible for listeners to distinguish all the separate sounds. Neither would it be likely that speakers would be able to utter each sound precisely enough. To get round this problem, we can make use of a much smaller set of sounds, but use multiple sounds in sequence for each word. Since the set of fundamental sounds is small, it should be easier to distinguish one from the other; and, so long as we can "segment" a word into a sequence of separate sounds, we should be able to create a large number of distinct words with a relatively short sequence for each.

This is of course how speech communication actually works; we have a small number of units that we can distinguish from one another and we use these in different sequences to create the forms of a large number of words. We call these units **phonemes**. The number of phonemes varies from language to language, but all languages use a roughly similar set of units ranging in size from about 15 to about 50. There is clearly a tradeoff between the number of phonemes we use and the length of words; languages with small numbers of phonemes will need on average to use longer sequences of these phonemes to produce distinct forms.

The question now is, out of all the possible vocal-organ configurations, which do we use for this process? First, let us say that the evidence shows that we do not simply pick an arbitrary set of 40 particular sounds from the entire set and make use of these alone. We know from our principle of semiotic contractiveness that it makes sense to pick a set of sounds that are readily distinguishable from one another; hence we don't find languages that have only fricatives or only front vowels. Secondly we want sounds, or more precisely a system of sound sequencing, that produce sequences that are relatively easy to speak and relatively easy to "segment" when listening. (For example, we would not expect a language to have sequences of velar stops followed by bilabial nasals followed by oral stops, because these sequences are difficult to enunciate). The choice of phonemes in a language is therefore a tradeoff between picking a large number of sounds (short words) and a short number (easier to identify each), sounds that are distinctive from one another, sounds that are easy to speak in sequence and sounds that are easy to segment when heard in sequence.

Even given a set of sounds obeying these criteria, it is not the case that each phoneme has a unique articulatory position or acoustic pattern. To take an example, when speaking the word SHOE, most speakers utter the first sound (represented by the symbol [ʃ]) with a mouth position in which the lips are pursed or **rounded**. Compare this with the lip position in SHINE and we see that the lips are in another position, which is more spread and in a position somewhat similar to a smile. However, despite the fact that the acoustic and articulatory differences between the initial sounds in these two words are quite big, they are perceived cognitively as a single sound. The reason why this variation occurs is quite obvious; SHOE has a rounded vowel, so all the speaker is doing is anticipating speaking the rounded vowel by already having the mouth in this position beforehand. This is therefore a consequence of wanting to produce a sequence that requires less effort to speak. This does not produce confusion because, although the rounded [ʃ] is different from the normal [ʃ], the rounded [ʃ] is not confusable with any other phoneme, hence the listener should not have difficulty in identifying the correct phoneme. Note that it is quite possible that the rounded and unrounded versions of [ʃ] could be different phonemes in another language; it is just a property of English that they are grouped together.

Hence the language system makes use of a set of contrasting sounds as its basic units, but within each unit a considerable degree of variability is allowed. These properties are entirely to be expected from the semiotic principles of contrast and sequencing, and the communicative principles of ease of production and ease of perception.

7.3.2 Phonemes, phones and allophones

We will now define four terms commonly used in phonetics. The term **phoneme** is used for the members of the relatively small (15–50) set of units which can be combined to produce distinct word forms. Phonemes can have a range of articulatory configurations, but in general the range for a single phoneme is relatively limited and occupies a contiguous region in the articulatory or acoustic space. The term **phone** is used to describe a single speech sound, spoken with a single articulatory configuration. The term **allophone** is used to link the phoneme and phone: different ways of realising a single phoneme are called allophones. In our example, we identified a single phoneme at the beginning of the two words SHOE and SHINE. Since each of these has a different articulatory position, each is a separate phone, but because they both "belong" to the same phoneme they are termed allophones of this phoneme. Finally, we will use the term **segment** as a general term to refer to phonemes, phones and allophones.

Phones can be thought of as units of basic speech production and perception, whereas phonemes are specific to a particular language. The way phonemes and phones are connected varies from language to language and is not necessarily a simple relationship. The easiest way to demonstrate this for speakers of English involves demonstrating a case where two sounds are phonemes in English but allophones in a different language. The classic case of this is the [r] versus [l] distinction in South-East-Asian languages; all English speakers readily tell these apart and usually have no idea that they are even related phonetically. In Japanese, there is no such distinction, and they are allophones of a single phoneme (by convention usually represented [r]), such that Japanese speakers speaking English often seem to get [l] and [r] "wrong". To many speakers of English, it seems incredible that such a "basic" distinction can be missed, and this unfortunately becomes the subject of many a lame comedy parody. In fact mastering the allophonic and phonemic distinctions is one of the most difficult tasks when learning a new language because our whole perceptual mechanism seems geared to treating all allophones in our native language as the same.

The above in fact shows the most accepted way of determining the phoneme inventory for a language. This is based on the principle of the **minimal pair**, and in effect says that, if one can find two distinct words (such as RUSH and LUSH) that differ only in one sound, then those sounds are cognitively distinct in the language, and are classed as phonemes. If, on the other hand, we take a word and change the manner of articulation of one of the sounds, but in doing so don't change its meaning, or the cognitive impression of that word, then the original and modified sound do not form a minimal pair and are not distinct phonemes. Theoretically (and only theoretically) one could randomly combine all the phones in various ways, perform tests on every change, and in so doing cluster all phones into sets of allophones and sets of sounds that form phoneme contrasts for the language in question.

In traditional phonetics, phones were seen as a set of distinct, discrete, separable units. This notion is no longer defensible because we now know that in reality the phonetic space is a multi-dimensional continuum (describable with either articulatory or acoustic dimensions), and there are no "bottom-up" divisions within it. The concept of using phones should now be seen as an idealised division of phonetic space and this can be justifiable as a handy notational device; it is cumbersome to describe every sound in terms of continuous dimensions (of, say, tongue height). It is important to realise, however, that phones really are a discrete abstraction of a continuous space; no procedure or process for objectively defining these units can be devised.

One of the advantages of using phones is that we can easily compare sounds from different speakers, accents and languages with some degree of standardisation. In doing this, it is common to make use of the standard International Phonetic Association symbol set (often call the IPA alphabet) to name each phone; having done so we can then transcribe a word to the extent that another person trained in this system can make a reasonable guess at determining the phonetics of the speech from this transcription alone. There are over 100 basic IPA phone symbols and this set is further enhanced by a system of diacritics that adds detail, such as nasalisation, rounding and so on. The standard IPA vowel chart and consonant chart are shown in Figures 7.9 and 7.10. In the IPA vowel chart, the vowels are arranged in a pattern that reflects the position of each vowel in terms of height and front/back.

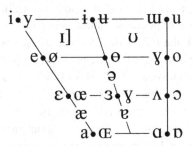

Figure 7.9 The IPA vowel chart. The position of each vowel indicates the position of the tongue used to produce that vowel. Front vowels are to the left, back vowels to the right. Where an unrounded/rounded pair occurs, the unrounded version is on the left.

For comparison, we also show in Figure 7.11 a revised version of the formant chart of Figure 7.8. We see that the two charts bear a striking resemblance. We have of course arranged the axes to help show this (e.g. F1 has its values going downwards), but even so the diagrams do exhibit a clear similarity. The usefulness of this correspondence is arguable – while on the one hand it demonstrates the distinctness of the vowels, it is important to realise that a vowel being high does not in itself cause a high F2. The relationship between mouth shape and formant patterns is very complicated – and is not a one-to-one relationship. In fact, as we show in Chapters 10 and 11, many different mouth positions can give rise to the same formant shapes.

Although it is rarely described this way, the IPA symbol set can be thought of as defining a set of possible sounds based on the union of phonemic definitions. If we imagine each language as being a set of phonemes, each described by boundaries in phonetic space, then the set of phones is the union of the sets of phonemes for all languages. This is of course how the IPA symbol set evolved historically; phoneticians started with a basic set of symbols, and, as new languages were discovered or analysed, new symbols so as to divide the phonetic space further. Hence the set of phones should be thought of as a set of named regions in phonetic space, which are defined from the union of all known phonemic boundaries.

Yet another way of thinking of phones and phonemes is in terms of cognitive and physical spaces. Cognitively, we represent all the instantiations of a particular phoneme as being the same; in our SHOE/SHINE example, the first sound is clearly the same in the sense that most people are completely unaware that there is even any difference until the differences are explicitly pointed out. By the same measure, although the physical difference between, say, [p] and [t] may be very slight, people readily agree that they are different sounds. It is important to realise that we don't group allophones together because we can't tell the difference, but rather because they group via the cognitive map of the phonetic space. To see this, we can again repeat the experiment in which we use an unrounded ʃ at the start of SHOE and a rounded ʃ at the start of SHINE. Most people can tell the difference here, and, despite maybe not being able to attribute this difference to the articulation itself, can nonetheless hear the difference. So it is not the case that somehow the ear or low-level perceptual hearing filters these changes, since we can hear them if we try.

It is now timely to comment on the practice of phonemic notation. Unlike with the IPA, there is no universally agreed method of phonemic notation, but a few conventions are common. The first is simply to pick the IPA symbol for one of the allophones of that phoneme. Hence we would represent the first phoneme in SHONE as ʃ. This has the advantage that we are drawing from an established symbol set, and also helps a newcomer to that language quickly grasp the basics of the phonemic system. This method does have its drawbacks, however. If a phoneme has two or more allophones only one symbol can be chosen, and this can be seen as unfairly biasing one allophone over another. Secondly, when using a phonemic transcription using IPA symbols, it is sometimes tempting to step outside the defined phoneme set and pick a symbol representing an allophone or other phone.

	Bilabial	Labiodental	Dental	Alveolar	Post-alveolar	Retroflex	Palatal	Velar	Uvular	Pharyngeal	Glottal
Plosive	p b			t d		ʈ ɖ	c ɟ	k g	q ɢ		ʔ
Nasal	m	ɱ		n		ɳ	ɲ	ŋ	ɴ		
Trill	ʙ			r					ʀ		
Tap/Flap				ɾ		ɽ					
Fricative	ɸ β	f v	θ ð	s z	ʃ ʒ	ʂ ʐ	ç ʝ	x ɣ	χ ʁ	ħ ʕ	h ɦ
Lateral fricative				ɬ ɮ							
Approximant		ʋ		ɹ		ɻ	j	ɰ			
Lateral approximant				l		ɭ	ʎ	ʟ			

Figure 7.10 The IPA consonant chart. Where there is an unvoiced/voiced contrast, the unvoiced symbol is shown on the left.

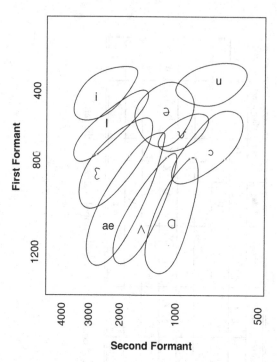

Figure 7.11 A rearranged version of the formant chart of Figure 7.9. The similarity between the F1/F2 positions as shown here and the height/front positions as shown on the IPA vowel chart can clearly be seen.

An alternative is to define a completely different set of symbols. A common approach in speech technology is to use one or more ascii characters for each phoneme, so that the consonant at the start of SHOE is /sh/, the vowel is /uw/ and so on. Note that the use of ascii characters is relatively easy when our task is to represent 50 or fewer phonemes; coming up with a unique ascii representation for each of hundreds of phones is much more difficult. For now we will use the **modified TIMIT** ascii character set, defined for general American English: a full description and justification of this will be given in Chapter 8. In addition to any practical, computer benefit, the use of a notation system for phonemes different from that for phones helps emphasise the difference between the two. This is helpful for our purposes because it makes it clear that phonemes are a set of contrastive linguistic units, rather than being entities expected to have well-defined or invariant physical manifestations. Notationally, it is conventional to use square brackets [ ] when describing phones and segments, and slash "brackets" / / when describing phonemes. Hence for the word HELLO we have a phone sequence [h ə l oʊ] and a phoneme sequence /h ax l ow/.

7.3.3 Allophonic variation and phonetic context

As has already been mentioned, phonemes do not represent a single point in phonetic space; if they did then each phoneme would have a unique acoustic pattern and the process of synthesis and recognition would largely be trivial. We have already touched on the subject of **allophonic variation**, which we define as the realisation of a phoneme as a range of possible phones. We have already discussed anticipatory rounding for ʃ, but other anticipatory affects are quite common. For example in the [k] of {COT, /k aa t/} and {KIT, /k ih t/} we see that in COT the tongue position for [k] is in a back position, whereas in KIT it is much further forwards. This can be explained by invoking the position that the tongue will be in during the following vowel.

The allophonic possibilities for a phoneme can depend on its position in relation to other phonemes. For example, the [n] in {NIT, /n ih t/} needs to sound different from the [m] in {MIT, /m ih t/}, but the [n] in {HAND, /h ae n d/} does not need to contrast with [m] since there is no word /h ae m d/ that it could be confused with. It does need to be distinct from say [r], [sh] and [v] otherwise it could be confused with HARD, HASHED and HALVED. This shows that the [n] still needs to be a nasal in that it still needs to be identified with respect to some other consonants; it doesn't need to be identified with respect to other nasals, so its distinctiveness with respect to [m] and [ng] need not be so precise as in other positions. Since there are constraints on what phoneme sequences occur; only distinctions between possible phonemes at a given point need to be maintained. The variability mainly arises from the fact that it is physically easier for a speaker to produce a given phoneme in one way in one position and in a slightly different way in a different position.

It turns out that a great degree of verbal variability can be explained once we take **phonetic context** into account. By this we mean that, if we consider the identity of the phonemes surrounding the one we are interested in, the identity of those surrounding phonemes helps explain why we see the variation in the speech. For example, we find in general that nasals before stops often share the place of articulation with the stop; so we naturally get words such as {HUNT, / h ah n t/} and {HUMP, / h ah m p/}, where the [n] and [t] are both alveolar, and the [m] and [p] both bilabial; but we don't get words /h ah m d/ and / h ah n p/, where the nasal has one place of articulation and the following stop has another. This phenomenon is called **assimilation**, reflecting the idea that the place of articulation of one phone assimilates to the other.

Another common effect is called **colouring**, whereby the identity of the phoneme in question isn't affected or limited, but the phoneme takes on some characteristics of a neighbouring phoneme. A common type of colouring is **nasalisation**, in which a vowel followed by a nasal itself becomes somewhat nasalised (that is, the velum is not completely closed during the production of the vowel). Colouring in vowels frequently happens when they are followed by nasals and also to some extent [l] and [r] (in rhotic accents). In English, this is again usually anticipatory. Other effects include **elision** or **deletion** in which a sound is "deleted" and **epenthesis** or **insertion**, where an extra sound is inserted. For these phenomena, consider the words MINCE and MINTS; they sound indistinguishable and clearly a process of elision or epenthesis has occurred. Interestingly, it is not clear which has happened; we could have either elision, in which case both would have the phoneme sequence /m ih n s/, or epenthesis, in which case the sequence for both would be /m ih n t s/.

Basic phonetic context on its own is not enough to explain all the variation. Consider the behaviour of [t] in the words TOP, LAST and STOP. In the first two, the [t] is normally

aspirated, that is, there is a clear burst of energy after the release, which is heard as a short section of noise. In STOP there is no burst; the vowel starts immediately after the release. This cannot be explained simply by context alone; after all the [t] is aspirated when preceded by a [s] in LAST and is also aspirated when followed by a vowel in TOP. These and other examples require reference to other structural factors such as the position in the syllable and word.

7.3.4 Coarticulation, targets and transients

If contextual allophonic variation were all we had to contend with, the phonemic-to-acoustic process would still be quite simple: all we would need is a model whereby each phoneme had a range of possible realisations; the process of conversion from one domain to the other would still be relatively trivial.

In fact, even if we take into account any allophonic variation that may occur, we find that the articulatory and acoustic patterns for a single phoneme are quite complex. This is because it is not the case that a single allophone is spoken with a single articulatory position; rather the articulators are constantly in motion. This is readily discernible from any of the acoustic or articulatory representations already encountered. The articulators require time to move from one position to the next; and, even though this can be done quite quickly, it is usually the case that by the time the articulators have reached the position required to speak one phoneme the speaker is already moving them to a new position to speak the next phoneme. This means that the articulators are constantly moving and that the pattern of movement for a particular phoneme is heavily dependent on the phonemes preceding and following. This phenomenon is known as **coarticulation**, indicating that, to a certain extent, two neighbouring phonemes have a joint articulation. Since the number of possible contexts in which a phoneme can occur is very large, this massively increases the possible variability in phonetic patterns.

An impressionistic model of the process can be thought of as follows. Each allophone has a set of canonical articulator positions. When each phoneme is "input" to the production apparatus, the articulators start to move to those canonical positions. Because of this "aiming" effect, these positions are often referred to as **targets**. At some point, the next phoneme will be input, and the articulators will start moving to the target for the required allophone of this new phoneme. Importantly, the target positions of the first phoneme need not have actually been reached; it is possible to start moving to the second phoneme before this happens. Sometimes this process can continue with no target positions ever being reached. Again our model of semiotic contrast helps explain why communication is still possible under such circumstances; so long as the positions reached are sufficiently near the canonical target positions that the phoneme can be distinguished from other possible phonemes, the listener should be able to decode the speech. Again, while it would be possible for the speaker to reach the target position and hold this for some time before moving on, to do so would substantially lengthen the time taken to say the message, which could be disadvantageous to both speaker and listener. The degree of "undershoot" with regard to targets is significantly affected by the rate of speaking; when speaking slowly the targets are often met; when speaking rapidly, the speech tends to "blur" and can become less distinct. To some extent, speakers can control this effect; we notice that some speakers "pronounce" their words distinctly; we say they "enunciate" clearly or are "articulate". It should also be clear that the degree of undershoot does affect the chance of being understood; slow, "careful" speech is in general easier to recognise.

It is important to realise that coarticulation and allophonic variation are distinct phenomena. Coarticulation is largely a physiological process outwith the speaker's control; while it may be possible to speak more slowly or with more effort, it is very difficult explicitly to change the trajectory of the articulators. By contrast, although allophonic variation may have originally arisen for reasons of minimal effort, it is a property of the particular language and we find that speakers of one language can easily control and differentiate an allophonic effect that seems automatic to speakers of another language.

7.3.5 The continuous nature of speech

One of the most startling facts about speech is that there are no specific acoustic cues in the signal to indicate boundaries. That is, the speech waveform is in general a completely continuously evolving signal, with no gaps or other markers that show us where the boundaries between phonemes, syllables or words lie. In general sentences and often long phrases are delimited by silence, but this is about all we get from the signal in terms of specific markers. This is of course in sharp contrast with writing systems such as that used for English, where the letters are distinct and words are usually delimited by spaces or punctuation. Note that this fact of speech should be less surprising to speakers of languages that which do not have gaps between the written word or that use little or no punctuation.

It is important not to confuse the notion of phoneme with the speech that is encoded by these units. Strictly speaking, when we look at a waveform or spectrogram, we are looking not at a sequence of phonemes, but rather at the output of the speech-production process that has encoded these phonemes. Hence, strictly speaking, we cannot identify phonemes in acoustic representations; rather we identify a section of speech that we believe is the output of the speech-production process when that phoneme was spoken. Making this distinction is not mere pedantry. Of course informally we often fail to distinguish a representation of an object from the object itself; if I show a photograph of a car and ask what it is, and someone answers "a Porsche", it would seem extremely pedantic if I then said, "no, it's a photograph of a Porsche": usually it is clear what is intended and no-one really mistakes a photograph for a real object. In speech analysis, however, this mistake is sometimes indeed made, which does lead to genuine confusions about the difference between the cognitive unit of contrast (the phoneme) and a particular example of speech that encodes that phoneme.

This has practical implications in speech analysis because the coarticulation effects can be so strong that the linearity of the phoneme sequence is lost in the speech. The targets generated by the phonemes may interact so strongly that they do not encode in sequence and therefore it is impossible to point to a particular section of speech and say "that is a representation of a /m/ phoneme". There is a limit to these interaction effects; in broad terms one can locate a section of speech in an utterance and say what word, syllable or phoneme this represents, but it is not always possible to do this with a high degree of precision.

This phenomenon questions the concept of whether we can in fact describe instances of speech using a linear sequence of symbols at all. While we can describe the canonical pronunciation of each word in terms of a sequence of phonemes, it does not follow that we can describe speech generated from these linear representations in a linear way. Since the effects of the phonemes overlap when they are uttered, there is no simple or fully legitimate way of dividing the speech into a sequence of discrete units. That said, contiguous-sequence descriptions are highly attractive from a practical point of view; people find them easy to use (after all, normal writing is done in this style) and, furthermore, sequence representations are highly amenable to computational analysis because this type of representation allows us to use many of the standard tools in signal-processing and finite-state-processing methods.

7.3.6 Transcription

We are often confronted with the problem of having to determine the correct sequence of words, phonemes or phones from a given waveform of speech. The general name given to this process is **transcription**. Transcription is involved in virtually every aspect of analysing real speech data, since we nearly always want to relate the data we have to the linguistic message that these data encode.

In considering this problem, let us first try to determine just *what* exactly we should transcribe. This depends on our purpose, and we find many different transcription systems, notations and devices that have been employed for various tasks. We can, however, describe most systems as lying on a scale between ones that attempt an "abstract" or "close-to-the-message" transcription and ones that attempt a "literal" or "close-to-the-signal" transcription. Along these lines, in traditional phonetics it is common to make reference to two main types of transcription, called **broad** and **narrow** transcription. For our purposes we find it useful to divide broad transcriptions further into **canonical** transcriptions and **phonemic** transcriptions. Each will now be explained.

Most words have a single **canonical pronunciation**; it is this that we think of as being stored in the lexicon; it is this which is the "definition" of how that word sounds. Hence a **canonical transcription** aims to transcribe the speech in terms of the canonical pronunciations which are given for each word in the lexicon. In such a system, the task can be described as one whereby we first identify the words, then transcribe the speech with the phonemes for each word and finally mark the boundaries between the phonemes. Complications lie in the fact that the speaker may say a filled pause (i.e. an "um" or "err"), or that they may "mispronounce" a word (i.e. they might say /n uw k y uw l er/ instead of /n uw k l iy er/ for NUCLEAR). In such cases, one must decide whether to take account of these effects and label them, or whether to stick to the canonical pronunciation. A second type of broad transcription, known as **phonemic transcription**, takes a slightly more "literal" approach, whereby the transcriber marks the sounds as he or she thinks they occur, but in doing so draws exclusively from the inventory of defined phonemes.

Thus, in such a system, the transcriber would be able to describe the differences between the two renditions of NUCLEAR.

Narrow transcriptions, also simply known as **phonetic transcriptions**, take this process further, and attempt to transcribe the speech in terms of a set of phone symbols (such as the IPA set) rather than phoneme symbols. This type of transcription is trickier, since of course the range of labels to choose from is far greater and it is usual in narrow transcriptions to find vowels marked with nasalisation diacritics, stops with aspiration and so on. Narrow transcriptions have a somewhat tenuous existence, which follows from our description of the problems of segmentation and classification described above. In traditional phonetics, narrow transcriptions were meant to be an indication of "phonetic reality", a sort of high-level recording of the speech uttered. The problem is of course that we now know that such a task is impossible because the speech is simply too greatly underspecified for a pure bottom-up determination of the phones to be found. What traditional phoneticians were implicitly doing was trying to listen and understand the speech, so that for each word the canonical or general phonemic sequence became known. From that, detail could then be added to describe the specifics of nasalisation, rounding, deletion of consonants and so on. While some phoneticians did in fact attempt to perform narrow transcriptions for languages they did not speak or understand, this task was considerably harder and many results fell short of the accuracy achieved with known languages.

One benefit of narrow transcription is that it is somewhat free of the problems of enforcing a fixed pronunciation (canonical transcription) or a fixed sound set (phonemic transcription) on speech that clearly is saying something different. For instance, if a speaker of English says a word in French with a good French accent, we would find it impossible to express this accurately with a broad system.

As a final example, consider the example of MINTS and MINCE that we mentioned before. While the canonical pronunciation of these words can be different (/m ih n t s/ and /m ih n s/), when transcribing speech, we are in the position of having two words that sound the same. We could use a single canonical pronunciation for both, but this would mean that in a process of word formation (described below in Section 7.4.2) we would need some process or rule that would delete the /t/ of MINT when the plural morpheme was added. Or we could use different transcriptions for these two possibilities, meaning that we would have to acknowledge that two transcriptions can sound the same. Or, two words could have the same narrow transcription but different broad transcriptions; but again we would have to decide what this was and whether the [t] should be deleted. The main point is that there is no single sure way to determine the transcription for real speech. While useful in various ways, all transcriptions are enforced symbolic descriptions of what was said and none reflect the "full reality" of the message and signal from every perspective.

7.3.7 The distinctiveness of speech in communication

Given these two problems of segmentation and classification, one can rightly ask how then is it possible that humans can perform these tasks and thereby decode the speech when listening? The answer again lies in our model of semiotic communication. We find that people make heavy use of top-down information when listening to speech; such that, using their internal grammar and knowledge of the conversation and context, they form strong hypotheses regarding the words that they expect to hear. This limits the choice of phoneme sequences that could agree with these words, and by effectively narrowing the search in this way they make the problem tractable. This is also how modern speech recognisers work. They are composed of two main parts; an acoustic model, which models the acoustic patterns for each phoneme, and a language model, which dictates the probability that a word (and hence phoneme) will occur in a given context.

We find that the degree of "clarity" or distinctiveness between phonemes in speech therefore varies depending on the conversational context. In situations where the listener has a strong idea of what will be said next, we find that speakers tend to be less distinct since the chance of confusion is low. Hence we find that speech in conversations between friends and family can exhibit quite indistinct acoustic patterns. Conversely, in situations where the listeners cannot easily guess what will be said next, we find that the level of distinctiveness increases. Lectures or news broadcasts are good examples of this type of speech.

7.4 Phonology: the linguistic organisation of speech

Now that we have described the production process and the means by which discrete information can be encoded in a continuous signal, we turn to the issue of how speech sounds are organised linguistically. So far we have simply said that words are made from sequences of phonemes, but a closer examination will show, however, that it is far from the case that any sequence of phonemes can make up a word and, in fact, we find that there is considerable structure to the relationship between words and phonemes. This area of linguistics is called **phonology**. We can describe phonology as having a dual character because it is related to both the syntactic part and the phonetic part of the communication process. Firstly, it shares many properties with the field of syntax or grammar, in that it studies the patterns of sounds that occur within a word, just as syntax studies the patterns of words that occur within a sentence. Phonology is seen as being distinct from phonetics in that it is part of the discrete symbolic language system, rather than being part of the encoding process. That said, it is clearly the case that the primitives in phonology are firmly grounded in the means of speech production and perception; as we shall see, the rules, grammars and processes described in phonology make use of the sort of phonetic information and descriptions we have previously introduced.

7.4.1 Phonotactics

It is generally accepted that, just as sentences have a grammar that defines what words can constitute a legal sentence, so words in turn have a grammar that defines what phonemes can constitute a legal word. The study of this is known as **phonotactics**. As with syntax, it is common to use the mechanisms of formal language theory to define a **phonotactic grammar**. In fact, the use of formal grammars for this purpose is less controversial than

7.4 Phonology: the linguistic organisation of speech

their use for syntax, since the phenomena in question are simpler, and many adequate grammars exist as proofs of concept.

Let us first consider some sequences of phonemes that occur at the starts of words. In English, a word can start with a vowel, or one, two or, at most, three consonants. Even a casual glance at some English words with three initial consonants shows consistent patterns:

STRING:	/s t r ih ng/
SPLENDOUR:	/s p l eh n d er/
SPRING:	/s p r ih ng/
SQUASH:	/s k w aa sh/

In fact, it turns out that, if a word has three initial consonants, the first must be an /s/. Also, the second consonant can only be a /p/, /t/ or /k/, and the third must be an approximant. There are some restrictions within this; the sequence /s p l/ occurs, but the sequence /s p w/ does not. A simple finite-state grammar that defines sequences with three consonants is given below:

```
0  →  s  1
1  →  p  2
1  →  t  2
1  →  k  2
2  →  r  5
1  →  p  3
3  →  l  5
1  →  k  4
4  →  w  5
```

From examination of phone sequences, we can gradually build a grammar that generates all of, and only, those sequences which occur. Figure 7.12 defines one such set of rules for a grammar of the legal phoneme sequences for most single-syllable English words. The set of terminal symbols of this grammar is the set of phonemes, and the set of non-terminals is defined as a set $V = \{0, 1, 2, 3, 4, 5, 6, 7, 8, 9, 10, 11, 12, 13, 14, 15, 16\}$. A state and transition diagram of this is shown in Figure 7.13. Legal words can be created by starting at state 0 and following any path until an end state is encountered. This grammar can be extended to polysyllabic words by the addition of arcs from the terminal states back to the start states. Some care is needed, however, to ensure that phenomena that occur only at the starts and ends of words are excluded from the grammar which defines consonant sequences in the middles of words.

The grammar in Figure 7.12 captures many facts that we have already mentioned, such as the fact that /ng/ cannot occur word-initially and so on. While our finite-state grammar may accurately describe what sequences can occur, its form is very "verbose" in that there are many regularities that the grammar does not explicitly express. For instance after a /s/, only the unvoiced stops occur, but we have no way of expressing this directly, so we have to list these stops (/p/, /t/ and /k/) explicitly.

0 → p 1	0 → n 7	5 → l 7	7 → oo 8		
0 → k 1	0 → f 7	0 → s 6	7 → ai 8	9 → dh 14	
0 → b 1	0 → th 7	6 → p 7	7 → au 8	9 → z 14	
0 → g 1	0 → s 7	6 → t 7	7 → oi 8	9 → zh 14	
0 → f 1	0 → sh 7	6 → k 7	7 → i 8@	9 → l 10	
1 → l 7	0 → v 7	6 → m 7	7 → e 8@	9 → s 10	
1 → r 7	0 → dh 7	6 → n 7	7 → u 8@	10 → p 15	
0 → t 2	0 → z 7	6 → l 7	7 → 7 @@	10 → t 15	
0 → th 2	0 → zh 7	6 → r 7	8 → 8 @	10 → k 15	
0 → d 2	0 → l 7	6 → w 7	8 → E 9	9 → l 11	
2 → w 7	0 → r 7	7 → a 9	9 → t 14	11 → m 15	
2 → r 7	0 → w 7	7 → e 9	9 → b 14	9 → n 12	
0 → p 7	0 → y 7	7 → i 9	9 → d 14	9 → l 12	
0 → t 7	0 → h 7	7 → o 9	9 → g 14	12 → d 15	
0 → k 7	0 → E 7	7 → u 9	9 → ch 14	9 → f 16	
0 → b 7	0 → s 3	7 → uh 9	9 → dz 14	9 → p 16	
0 → d 7	3 → p 4	7 → ii 8	9 → ng 14	9 → k 16	
0 → g 7	3 → t 4	7 → ei 8	9 → f 14	9 → m 13	
0 → ch 7	3 → k 4	7 → aa 8	9 → th 14	13 → p 15	
0 → dz 7	4 → r 7	7 → uu 8	9 → sh 14		
0 → m 7	3 → p 5	7 → ou 8	9 → v 14		

Figure 7.12 A complete grammar for syllable structure using classical finite-state-grammar rewrite rules.

We can look at this problem by taking a different approach to how we define segments (which in the following can be either phonemes or phones). In our account of segments and their properties, we made use of such terms as "nasal", "high" and "stop" because these helped us categorise the different ways in which sounds are generated. One of the central ideas in modern phonology is that these terms are not simply labels for groups of somewhat similar segments but act as a set of more-basic building blocks. Hence we define a set of phonological primitives known as **distinctive features**, and from arrangements of these we form segments. This system is sometimes viewed as an additional level in the sentences/words/phonemes hierarchy, and indeed it follows on from this in that there are fewer distinctive features than phonemes, and arrangements of these few distinctive features give rise to a greater range of phonemes. One significant difference, though, is that distinctive features do not arrange in *sequences* to form segments; rather their values are all realised at the same instant of time. Thus we can see that the process of phoneme formation is one of "mixing" as opposed to "sequencing". One analogy therefore is the way that the full spectrum of colour can be formed by mixing the primary colours. Another comes from atomic physics: once atoms were thought the primary and hence "indivisible" building blocks of nature; but subsequently electrons, protons and neutrons were discovered. Various arrangements of these three types of particle give rise to the more than 100 known elements. Likewise, phonemes were thought of as the "atoms of speech", but now we believe that a more parsimonious model can be built with reference to more basic units.

When using distinctive features, a common approach is to define each segment as a **feature structure**, of the type we introduced in Section 4.5. So we can define a segment

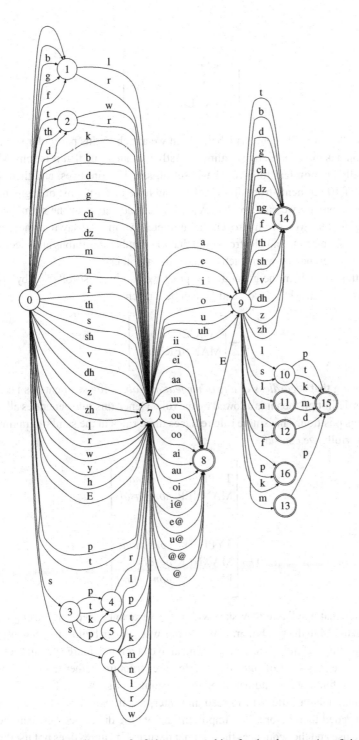

Figure 7.13 A diagram of a finite-state machine for the phonotactics of single-syllable words.

[t] as

$$\begin{bmatrix} \text{VOICED} & \textit{false} \\ \text{PLACE} & \textit{alveolar} \\ \text{MANNER} & \textit{stop} \end{bmatrix}$$

From this, "t" simply becomes a label, which we attach to this particular set of values. This notion has many attractive qualities. Firstly we can argue that it is somewhat more scientifically **economical** to have a basic set of, say, 12 primitives, and then to form a larger set of 40 segments from these, rather than have just a flat list of segments. Using our physics analogy we can say that, regarding classification of the elements, instead of just saying that we have hydrogen, helium, etc., we instead say that these are made from protons, neutrons and electrons, and that the number of protons and electrons fully defines the difference between each element and all others.

Using this formalism the generalisation that we can have /s/ followed by any one of /p/, /t/ and /k/ but not by /b/, /d/, and /g/ can be written

$$1 \to \begin{bmatrix} \text{VOICED} & \textit{false} \\ \text{MANNER} & \textit{stop} \end{bmatrix} 2$$

the idea being that *regardless of any other feature values* if a segment has features that match this definition then it is allowed by the rule. Hence this rule generates all unvoiced stops in this position. This single rule replaces three rules in the symbol grammar. Even more powerfully we can state

$$0 \to \begin{bmatrix} \text{TYPE} & \textit{consonant} \\ \text{MANNER} & \textit{non-nasal} \end{bmatrix} 7$$

$$0 \to \begin{bmatrix} \text{TYPE} & \textit{consonant} \\ \text{MANNER} & \textit{nasal} \\ \text{PLACE} & \textit{non-velar} \end{bmatrix} 7$$

which states that a syllable may start with any consonant except /ng/, using two rules, rather than the 14 in the symbol grammar. In this way, we can build a finite-state grammar based on feature definitions and in so doing arrive at a much more succinct set of rules. Figure 7.14 defines a set of rules that use feature structures rather than symbols as the terminals. A finite-state automaton of this grammar is shown in Figure 7.15. While the rules may take a little skill to read and interpret, it should be clear that far fewer rules are needed than before. It is important to note the difference between the formal devices we use and the sequences they define: the use of features does not itself give us a grammar that more accurately describes which phoneme sequences occur and which do not. Certain types of rules, feature systems and operators do allow us to express some

7.4 Phonology: the linguistic organisation of speech

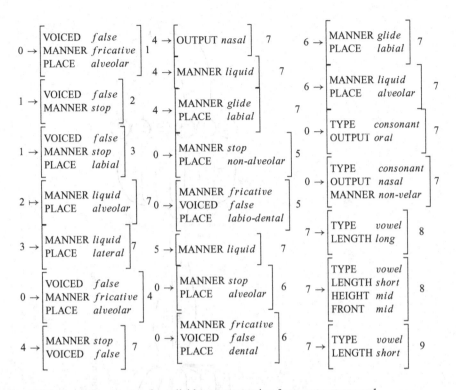

Figure 7.14 A partial grammar for syllable structure using feature structures and finite-state-grammar rewrite rules.

things more succinctly than do others, and it can be argued that systems that allows us to generate correct sequences succinctly with a small number of rules are more *natural*.

The grammar defined in Figure 7.14 is not without its problems and we shall now turn to these. Firstly, one sequence that this grammar does not generate is /s f V/, where V is a vowel. This sequence is found in only a very few words, such as SPHERE and SPHINX. Therefore it seems that we have to add a rule to account for these few words. Furthermore, this rule would be the only one to have two fricatives in the same consonant cluster. Another, somewhat more contrived, example is VROOM, which, while being an onomatopoeic word, is listed in most dictionaries as being a valid word of English. Again this requires its own rule, which would be the only rule which uses a voiced fricative in a consonant cluster. A further sequence this grammar does not allow is /y/ in a pre-vowel position, as in MULE /m y uw l/. The existing grammar is composed of three distinct parts; a set of initial consonant sequences (all the sequences leading from state 0 to state 7), a set of vowels (states 7 to 9) and a set of final consonant sequences (states 9 to 16). In the grammar, each is fairly independent, so that any initial consonant sequence can be attached to any vowel. The inclusion of /y/ in the set of initial consonant sequences would destroy this simplicity because /y/ can be followed only by a /uw/ vowel, thereby ending the separation. This phenomenon is even stranger in that it accounts for one of the main accent differences between American and British English, namely that in

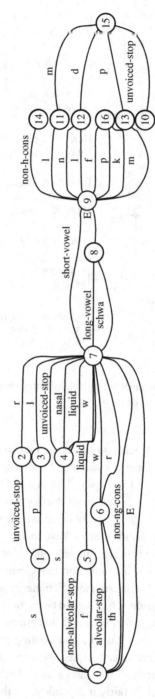

Figure 7.15 A diagram of phonotactics using features rather than atomic phoneme symbols.

American, the standard pronunciation of DUKE is /d uw k/, whereas in British English it is /d y uw k/. This generalisation covers all words with this pattern (with the possible exception of NEWS in which many Americans say the /y/). Hence there is some important generalisation going on here.

We will return to thorny problems such as these in the discussion, but the real point is that, within a rule and feature system, there will always be a tradeoff between accounting for all the facts and doing so in a neat and succinct way.

7.4.2 Word formation

One of the central topics in phonology is the study of what is called **word formation** or **lexical phonology**. Simply put, this concerns the processes by which individual morphemes combine into whole words. To demonstrate some aspects of word formation, let us look at the case of plural formation in English. In spelling, the "regular" plural in English is the character s, so that the plurals of DOG and CAT are represented as dogs and cats. In speech, however, the phonemic form differs for these words, so we get /d ao g z/ and /k ae t s/; that is, in one case the phoneme /z/ forms the plural, whereas in the other it is the phoneme /s/. What explains why one form is chosen rather than another? If we look at /d ao g z/ and /k ae t s/ we see that the last phoneme before the plural suffix is voiced in one case and unvoiced in the other; this in fact turns out to be the basis of the distinction, and we see that all words that end in an unvoiced consonant take a /s/ plural, whereas those that end with a voiced consonant take a /z/ plural.

As before, we could express this in purely phonemic terms as

$$g + PLURAL \rightarrow g + z$$
$$b + PLURAL \rightarrow b + z$$
$$v + PLURAL \rightarrow v + z$$
$$k + PLURAL \rightarrow k + s$$
$$t + PLURAL \rightarrow t + s$$
$$f + PLURAL \rightarrow f + s$$

and so on. As with the case of the phonotactic grammar, however, this entails a large number of rules, which are somehow missing the generalisation that this is simply a matter of voicing. If we use distinctive features, we can succinctly write the rule as

$$\left[\text{VOICING } true \right] + PLURAL \rightarrow \left[\text{VOICING } true \right] z$$

$$\left[\text{VOICING } false \right] + PLURAL \rightarrow \left[\text{VOICING } false \right] s$$

In fact, we can do better than this, by making use of some further devices common in phonological theory. One is the principle of **markedness**, which is a linguistic term for the concept of **default**. The idea is that, given a binary or other distinction, one form of behaviour is normal, unremarkable and therefore **unmarked**, whereas the other form is noticeable, stands out and is therefore **marked**. In our example, we could say that the unmarked plural morpheme is simple /s/, but some special behaviour is invoked if this is added to a word that ends in a voiced consonant. Hence we would not need a rule for joining /s/ to many words; this would be the default rule that applies when any morphemes are joined (i.e. no changes occur). Hence we need a rule only for voiced cases, which would be

$$\left[\text{VOICING } true\right] + s \rightarrow \left[\text{VOICING } true\right] z$$

We might well ask why it should be that /s/ is chosen as the default rather than /z/ (note that we shouldn't be misguided by the spelling convention here). This is in fact a prime concern in developing phonological theories, namely that we choose "natural" rules rather than simply ones that describe the data.

At this stage it is worth examining why we have chosen to describe the plural formation effect as a phonological one rather than a phonetic one. Wouldn't it be possible simply to say that this is a matter of allophonic variation or coarticulation? While the rule is clearly based on articulatory principles, it is included as part of the phonological because it is seen as a rule of *English* rather than a general process of speech. It is quite possible to overrule this effect and say /d ao g s/ and /k ae t z/. More importantly, however, we see that there are many non-plural English words that end in /s/ but don't follow this rule, for example {BASIS, /b ey s ax s/} and {SENSE, /s eh n s/}. It no more difficult or natural to say PENS as /p eh n z/ than PENCE as /p eh n s/. This demonstrates that this plural-formation rule is one of grammar and phonological organisation, not some speech-encoding effect.

English is rich with such word-formation rules. For example, one common process, known as **palatalisation**, describes the process whereby consonants move from their original place of articulation to a palatal place of articulation, usually resulting in a [sh] phoneme. So if we take examples such as

PRESS   PRESSURE
CREATE   CREATION
DIVIDE   DIVISION

we see that in the process of forming PRESSURE from PRESS the /s/ has palatalised to a /sh/, and this process has happened with CREATE and DIVIDE also. We can still observe the historical effects in the way the words are spelled; in pressure and creation the original consonant is preserved (even though the pronunciation has changed) whereas in division the spelling has changed to reflect the sound more accurately (but not completely accurately). Another common process is that of vowel tensing in derived forms that we see in words such as

DIVINE   DIVINITY
PROFANE   PROFANITY
SERENE   SERENITY

Phonological rules can extend beyond purely phonemic distinctions; stress shift (explained fully below) also happens in pairs such as

> PHOTOGRAPH   PHOTOGRAPHY
> TELESCOPE   TELESCOPIC
> ANGEL   ANGELIC

In studying word formation in this way, it becomes nearly inevitably necessary to adopt a paradigm whereby we have a basic representation of a word or morpheme, which when combined with other words or morphemes interacts in some way and then produces a final version. So, for example, we have two words {CAT, /k aa t/} and {DOG, /d ao g/} and a plural morpheme +S, /s/. We combine CAT and +s to form CATS, /k ae t s/, and this is simple enough. In the case of DOGS, however, the pronunciation is /d ao g z/, and we observe that the /s/ has "changed" into a /z/. In this paradigm, we start with a **deep** representation, and, by applications of rules, convert this into a **surface** representation. Sometimes the difference between the two is relatively simple, as in the plural example, but in others it is more complicated. If we consider the word SIGN, we see that it has derived forms SIGNAL, SIGNIFY and so on. What is noteworthy here is that these derived forms contain a /g/ whereas the root form does not. Rather than saying that a /g/ is inserted, it has been argued rather that the deep form of SIGN is really /s ay g n/ (or even /s ih g n/) and that, by process of rule, this is converted into the surface form /s ay n/. Thus the rules can operate even when there is no word formation per se. Such solutions have always attracted controversy in phonology. Firstly, it can be argued that the statement of such rules is heavily influenced by the orthography and that, since the exact nature of orthography often derives from some whim of a printer, it is a poor basis on which to form rules. Secondly, once such processes are allowed there seems no limit to what devices can be employed.

7.4.3 Distinctive features and phonological theories

In defining a feature system properly, one must formally define the attributes and the type of values they can take. So far we have described a somewhat-traditional set of features (manner, place etc.), each of which can take multiple values (e.g. dental, alveolar, or palatal), and have made use of descriptive devices, such as "assimilation", to describe how sounds change in particular environments. In phonology researchers attempt to improve on this, by defining features and rule mechanisms in an attempt to explain these sorts of effects in a more elegant and parsimonious manner.

The Sound Pattern of English [91] (known as the "SPE") was a seminal landmark in phonology, in that it was the first full account of how to describe English phonology *formally*, that is, with precise features and rules. The SPE uses a binary feature system, in which each phone is described by 13 features, each of which takes a +/− value. The SPE set of binary features is shown in Figure 7.16. While some of these have the familiar interpretations, a few are new. ANTERIOR is used for consonants and indicates

```
vocalic
consonantal
high
back
low
anterior
coronal
round
tense
voice
continuant
nasal
strident
```

Figure 7.16 Binary features in *The Sound Pattern of English*.

a front constriction. CORONAL, also used for consonants, indicates an arched tongue. Combinations of these are enough to distinguish most English consonants. TENSE is used for vowels in a very similar way to length and STRIDENT is used to indicate a "noisy" phone such as [s] or [f].

The SPE showed how many seemingly disparate phenomena could in fact be explained by rules that made direct access to the feature set. Nearly all rules were posited in terms of **context-sensitive rules** (these rules are used to define a third fundamental type of grammar in addition to the finite-state rules and context-free rules that we saw before.). For example, the tensing rule that describes how word patterns such as DIVINE, DIVINITY and PROFANE, PROFANITY arise is given as

$$V \rightarrow [-tense] \ /\_ \ CVCV$$

meaning that the vowel V in question becomes lax (the opposite of tense) when it occurs before a $CVCV$ sequence (such as the /n ih t iy/ of PROFANITY). Even for different feature sets, models and theories, rules of this type are very common. It is worth noting, however, that the use of context-sensitive rules here is merely a notational convenience; most phonological rules do not use recursion, so the above can readily be implemented with finite-state rules, if required for practical purposes.

Despite its success in many regards, several problems immediately appeared with the SPE. One of the main problems is that, if there are 13 binary features, this should give rise to $2^{13} = 8192$ possible phones. Not only is this considerably greater than most phonologists would agree exist, but also it is clear that many combinations (e.g. [+*tense*, +*anterior*, +*high*]) simply don't correspond to any known or even theoretically possible phone. For this and many other reasons, a huge flurry of activity started, with the aim of defining more succinct and more "natural" systems of features, and accompanying rules.

In the 1970s, two theories broke from the mould of the SPE by proposing that sound patterns could better be understood if more-elaborate rule and representation mechanisms were used. **Autosegmental phonology** [180] uses two (or more) **tiers**, in which we have a phoneme tier (roughly similar to the SPE) and a tonal tier. The idea is that rules can be written to operate on just one tier or the other, which is much simpler that would be the case if the phonemes and tones had to be represented on a single tier. The other theory is that of **metrical phonology** [284], which was designed to explain phonological phenomena of stress and intonation (described below). Together, these theories and some that followed were often called **non-linear phonology** (this is a different use of the term "non-linear" from that in mathematics), because their representations were not sequences of segments as in traditional phonetics, phonology and the SPE. Since then a bewildering variety of other theories and models has been produced. Among these are **natural phonology** [137], which argued against the sometimes very abstract positions taken in theories such as the SPE (for instance the example we gave where the underlying representation of SIGN would require a [g] segment).

Despite the large number of different theories, some trends have stood out in how phonology has developed. One strand of work has attempted to move away from the context-sensitive rule formalisms described above. Mainly this follows a general move in this direction in other fields of linguistics such as syntax, and is motivated by a desire to avoid the complications of determining what these rules really stand for cognitively. For instance, it is often difficult to believe that a speaker or hearer has to apply a whole sequence of rules in just the right order to generate the surface form of a word. Instead, a series of constraints are proposed, from which new words can be formed. Theories of this school include **dependency phonology** [145], [144] and **government phonology** [248]. Many theories use feature systems more elaborate than those expressible naturally in feature structures, the idea here being that features form hierarchies and that we can thus use these hierarchies as a natural way to limit the number of feature combinations and rules, thus avoiding the "segment-explosion" problem of the SPE. The theories of **articulatory phonology** [70] and **feature geometry** [100], [306], [386] are based on this idea.

In addition to the issues discussed in this chapter, it should be understood that the field of phonology tackles many issues that aren't directly relevant to speech-synthesis needs, such as, say, finding the set of language universals. **Optimality theory** is one such model that has attracted particular attention in recent years [362]. In addition to this, the emphasis on phonology even within directly relevant topics is often somewhat different from what we require from an engineering perspective. In particular the old issue of "economy" comes up again and again, with phonologists striving for ever more natural and economical representations, without usually much interest in assessing the processing requirements incurred. The general trends in phonology are clear, though: a move away from finite-state, linear models towards more-complex feature organisations, from ordered context-sensitive rules to systems that use constraints and dependences, and from procedural systems to declarative systems. Somewhat ironically, although most phonologists trace their work back to the SPE, the "formal" side of this approach has largely been forgotten, and few works in phonology since have anywhere near the level of explicitness and comprehensiveness of that work. A notable exception comes from the work of **computational phonology** [43].

7.4.4 Syllables

One of the most important aspects of phonology concerns the structural representation of sound patterns above the phoneme. We have already seen that morpheme boundaries are significant (e.g. in the difference between the realisation of PENS and PENCE), as are word boundaries. In addition to these, we find it very useful to make use of a third unit, the **syllable**, since this also helps explain many of the effects and patterns in speech.

As we have already seen, the syllable provides a solid basis for defining the phonotactic grammars of what constitutes a legal word. If we define a grammar for a syllable, it is an easy matter to define a grammar for any words by simply including more syllables and any possibly special sequences for word beginnings and ends. That said, it is interesting to ask what real evidence do we have for syllables as real entities? We seem to know instinctively that SET, FROM and BATH have one syllable, that TABLE, WATER and MACHINE have two, and that OPERATE, COMPUTER and ORGANISE have three. These judgments seem intuitive and are not dependent on any expert knowledge. Secondly, if we consider singing we see that in normal singing each note is sung with one syllable. If this principle is broken the singing can sound awkward if syllables are inappropriately squeezed in or elongated over several notes. Thirdly, syllables seem to act as the basic unit of many aspects of prosody. Syllables carry stress such that in THE PROJECT the first syllable of PROJECT is stressed, but in PROJECT THAT LIGHT the second gets the stress. There are no other possibilities – it is not possible to stress the /p r/ and not the vowel, or any other individual phone.

So what exactly is a syllable? We will take the view that the syllable is a *unit of organisation*. All words are composed of a whole number of syllables, and all syllables are composed of a whole number of phonemes. For an initial working definition, we will state that a syllable is a **vowel surrounded by consonants**. This definition seems to fits the facts fairly well – in all the above cases the numbers of vowels and numbers of syllables were the same, and, again from singing, we can see that the vowel seems to be the "centre" of the note. There is one commonly raised exception to this, namely the existence of so called **syllabic consonants**. Some phoneticians give two-syllable words such as BOTTLE and BUTTON the transcriptions /b aa t l/ and /b ah t n/, in which the second syllable is comprised only of /l/ and /n/ sounds. Conventions differ on whether to allow this, or use the alternative, which gives these words the transcriptions /b aa t ax l/ and /b ah t ax n/. There is probably no real way to decide from a phonetic point of view, but practical TTS experience has shown no deficiencies in using /ax/. Since this makes the definition of the syllable simpler, we shall use that instead of using syllabic consonants.

So, taking the vowel as the centre of the syllable, all we have to do is decide which consonants belong to which syllables. In words such as HOTEL, we can safely say that the /h/ belongs to the first syllable and the /l/ to the last, but what about the /t/? There are valid alternative positions to take on this. One position is the so-called **maximal onset principle**, whereby consonants that potentially could be in the onset or coda of a syllable are taken as being in the onset. Using the symbol /./ to denote a syllable boundary, HOTEL is then represented as /h ow . t eh l/.

This will provide a satisfactory account for many words, and it can be argued that this has some cognitive reality because, again from singing, we find that consonants tend to follow this pattern. There are some problems, however. Firstly, in non-word-initial syllables that have /s t r/ and other such sequences, e.g. {INSTRUCT, /ih n s t r ah k t/}, it can be argued that the /s/ attaches to the first syllable, not to the second. Secondly, consider such words as {BOOKEND, /b uh k eh n d/} – here it definitely seems that the /k/ attaches to the first syllable. In fact a syllable-final /k/ and a syllable-initial one sound quite different and so /b uh k . eh n d/ sounds different from /b uh . k eh n d/. There is an obvious reason for this, namely that BOOKEND is a word formed by compounding BOOK and END, and it seems that the word/morpheme boundary has been preserved as a syllable boundary.

One of the main reasons why we are interested in syllables is that the pattern of syllables within a word and the pattern of phonemes within a syllable help determine much of the phonemic variation due to allophones and coarticulation. Hence it is often useful to assign a basic constituent structure to describe the structure of phonemes internal to a syllable. We will explain one way of doing this, in which we use four basic constituents, called the **onset**, **rhyme**, **nucleus** and **coda**. The syllable structure for the single-syllable word STRENGTH in these terms is

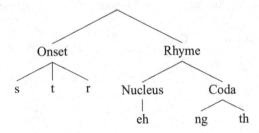

The first distinction is between onset and rhyme. Naturally enough, that which have the same rhyme do in fact rhyme, for example LEND and BEND:

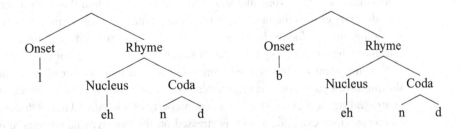

Within the rhyme, the nucleus is the vowel, and the coda is the post-vocalic consonants. If extra-metrical phones are included in the syllable, these will be included in the coda. So, for a word such as CONSISTS, the syllable structure would be one of

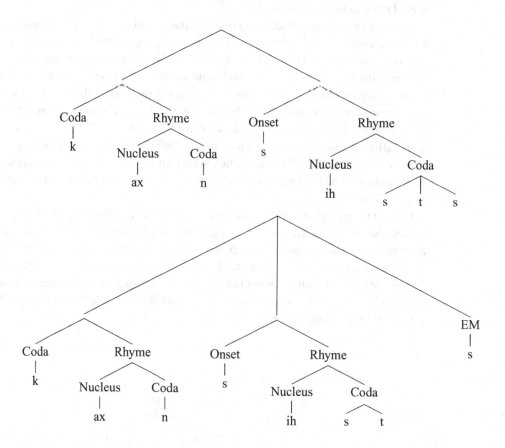

7.4.5 Lexical stress

Some syllables sound "stronger" or "heavier" than others. For instance, taking a word such as TABLE, we can say that the first syllable of this sounds stronger than the second. This contrasts with a word like MACHINE, for which we find that the second syllable sounds stronger than the first. Because of these types of distinctions, we view stress as being a property of syllables rather than of phonemes or words, and hence we always talk about the stress of a syllable, rather than the stress of a word or phoneme.

It is important to note that the term stress is used in a number of different ways in the linguistics literature. The most fundamental distinction is between **lexical stress** and **sentential stress**. Lexical stress is a core property of a word, and hence doesn't change. In our previous example, TABLE is stressed on the first syllable whereas MACHINE is stressed on the second; these are properties of the words themselves. Sentential stress, on the other hand, describes the effect whereby one syllable in the sentence receives more stress than the others; this is normally due to some pragmatic or communication reason and is not a fundamental property of the word itself (e.g. I SAID I WANTED THAT ONE). We shall ignore this second type of stress for now, since it is dealt with in detail in Chapter 8.

As usual, there are many different theories and models of lexical stress. Here we will adopt a relatively simple model, which describes the stress patterns without necessarily tackling the issue of how these patterns arise in the first place. In this model, each multi-syllable word has exactly one **primary stressed** syllable. Using the IPA symbol ' to indicate primary stress, we can write TABLE and MACHINE as /'t ey b. ax l/ and /m ax. 'sh iy n/. It is also common practice to use numbers to indicate stress, so that we have /t ey1 b. ax l/ and /m ax. sh iy 1 n/, where, again by convention, the number immediately follows the vowel. We can indicate stress in longer words such as INFORMATION in the same way, as /ih n . f ax . m ey1 . sh ax n/. What of the other syllables? It appears that the first syllable of INFORMATION is stronger than the second and fourth, but less strong than the third, so we call this **secondary stress**. Some linguists go further and have tertiary stress and so on, but primary and secondary stress patterns are usually taken as sufficient. Secondary stress is denoted by the IPA symbol `, or alternatively by a 2 following the vowel.

We can identify a third category called **reduced**. This is mainly used to indicate the stress level of syllables containing the schwa vowel. Reduced vowels occur in at least three subtly different cases. First, some words naturally, or always, have a schwa in a certain position, for example in the first syllable of MACHINE, /m ax sh iy n/. But consider the second syllable of INFORMATION. We have given this a schwa in our annotation and most speakers would agree with this pronunciation. Now consider the word {INFORM, /ih n f ao r m/ }; here the second syllable is a normal full vowel. We can say that INFORMATION is created from INFORM by the application of morphological rules, and in the process the second syllable has been *reduced* from a full vowel to a schwa. A third form of reduction occurs in function words, when for instance a word such as {FROM, /f r ah m/} is produced as /f r ax m/ in many instances. Finally, when people speak quickly, they often under-articulate vowels, which leads to a loss of distinction in their identity leaving them sounding like a schwa. This effect is also termed reduction. The number 0 is used to represent a reduced vowel, so the full transcription for INFORMATION is /ih2 n f ax0 r m ey2 sh ax0 n/.

In English, stress is a fundamental part of the pronunciation of a word, and thus weakens the notion we have been adopting until now that a word's pronunciation can be represented entirely by its phonemes. There are main related noun/verb pairs in English that are distinguished by stress, such that we have {RECORD_NOUN, /r eh1 k ax0 r d/} contrasting with {RECORD_VERB, /r ax0 k ao1 r d/} and likewise {PROJECT_NOUN, /p r aa 1 j eh2 k t/} contrasting with {PROJECT_VERB, /p r ax0 j eh1 k t/}. That said, there are very few genuine minimal pairs in English. Some have gone to considerable lengths to argue that stress can in fact be derived from just the phonemes (this is advocated in the SPE). Valiant though these attempts are, there are simply too many words with unpredictable stress patterns for us to be able to derive stress from phonemes for all words.

Stress can be graphically represented in the sort of tree used to describe syllable constituent structure. In our previous example the nodes at the syllable level were unlabelled; we can now show the same tree but with stress labels at the nodes:

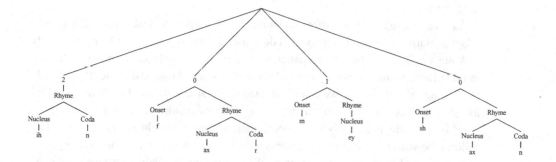

An alternative to the use of absolute levels of stress, such as primary and secondary, is the theory of **metrical stress** [284], [286] in which stress is purely relative. In this, a binary tree is used to represent each word and within this each syllable is defined as either **strong** or **weak**. So INFORMATION would be defined as

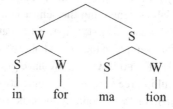

The strongest syllable is the one which only has S nodes above it; everything else is defined relative to this.

If lexical stress is a relative phenomenon, what of single-syllable words? Since there are no other syllables to contrast with, the stress status of such words is technically undefinable. However, by convention, most of these words are given a primary stress or strong stress in their syllable. The exception is content words, which are given a primary stress if in their full form (/f r ah m/) or no stress in their reduced form (/f r ax m/).

In order for us to perceive stress differences, stress patterns must of course manifest themselves in the acoustic signal. How is this achieved? Firstly, taking a contrast between "normal" syllables and their stressed version, we can say that, in the case of stress, the syllable is spoken with greater articulation effort, that is, the articulators are more likely to move to their canonical positions and, in so doing, produce a syllable with many of the idealised sounds which we would expect from our discussion of articulatory phonetics. As a by-product of this, we can see that the syllable seems more "distinct" in the spectrogram: the formants of the vowel are clearly observable, and it is often easier to ascertain the identity of the vowel from the formants of stressed syllables than from unstressed ones. Likewise with the consonants, stops often have more distinct closure and burst patterns in stressed syllables and so on. As a consequence of the great articulation effort, stressed syllables in general last longer than normal versions. The source acts differently in stressed syllables also. The voice quality is likely to be more clearly periodic (not creaky or breathy). Also, the pitch of the syllable is likely to be higher in a stressed syllables than in the syllables immediately surrounding it. This effect is difficult to determine in isolation however, because the intonational patterns of the utterance interact strongly here (this effect is described in more detail in Section 9.1).

By contrast, in reduced syllables, the opposite effects seem to occur. The articulation effort is diminished to the extent that any distinctiveness in the vowel quality is absent – all vowels are either /ax/ or /ih/. Consonants in reduced syllables tend to be less clear. The duration of reduced syllables can be very short indeed, even to the extent of their being undetectable in faster or more casual speech. Finally, it is unlikely that the pitch of an unstressed syllable will be higher than that of its surrounding syllables (note, however, that the presence of a reduced syllable does not cause a decrease in pitch).

7.5 Discussion

The study of phonetics has a long and rich history [97]. Just as the world's most famous detective is the fictional Sherlock Holmes, so indeed is the world's most famous phonetician a fictional character. Today the Henry Higgins of *Pygmalion* and *My Fair Lady* would probably appall any self-respecting linguist or phonetician with his dogmatic and prescriptive view of how English should be spoken. Nevertheless, Higgins probably does share some genuine features in common with the early British school of phonetics.

The study of modern phonology is widely regarded to have started with the "Prague School" of linguists, of which the main figures were Nikolay Trubetzkoy and Roman Jakobson. They established the principle of the formal (or scientific) study of phonology and the notion of the phoneme. Most importantly, they showed the semiotic basis of phonological contrast, with the idea being that the linguistic use of sounds could be studied only within a system and not in individual, purely phonetic terms. Jakobson collaborated with two other notable figures in speech and linguistics, Morris Halle and Gunnar Fant, to produce a coherent system of distinctive features. As already mentioned, *The Sound Pattern of English* (SPE) was a seminal work, both in its scope and because of its effect in proposing formal methods for phonological work.

7.5.1 Further reading

The most popular work in introductory phonetics is undoubtedly Peter Ladefoged's classic *A Course in Phonetics* [271], and is highly recommended. Another excellent book, which extends the subject and gives more detail on acoustic issues, is Clark and Yallop [97]. A more-advanced and comprehensive book is Laver's *Principles of Phonetics* [275].

There are many excellent introductory text books on phonology, including Giegerich [174], Katamba [245], Lass [274] and Anderson [17]. Clark and Yallop [97] is again particularly recommended because it gives a thorough introduction to phonology, does so with reference and comparison to phonetics, and discusses the issues involved from the type of traditional scientific perspective familiar to most engineers and scientists. Unfortunately, nearly all these works are quite introductory and few venture beyond SPE-type phonology. While there are plenty of books and articles covering more recent work, they typically focus on one particular theory or model, making it hard to perform the type of comparison that would be necessary for us to adopt any one of these for speech-technology purposes. Books and articles on the various schools of phonology are listed above.

7.5.2 Summary

Phonetics
- Speech sounds can be described with reference to either their articulatory or their acoustic features.
- Vowel sounds are described with the primary dimensions of high/low and front/back, and secondary dimensions of rounding and length.
- Consonants are described in terms of voicing, manner and place of articulation. The main values manner can take are stop, fricative, nasal, affricate and approximant.
- Computer analysis can be used to study the acoustic patterns in speech. The primary representation is that of the spectrogram.
- The resonances of the vocal tract are called formants, and these are thought to be the primary means by which listeners differentiate between different vowel sounds.

Communication with speech
- A phoneme is a linguistic unit of speech defined on the semiotic principle of contrast.
- We communicate via speech by using the speech apparatus to encode words as strings of phonemes, which are then fed through the speech-production apparatus.
- In traditional phonetics, phones are seen as concrete objective units, whereas phonemes are contrastive linguistic units. Phonemes are realised as phones and if a phoneme has more than one phone those phones are called allophones.
- Even for a single allophone, there is a large amount of variability. This is mostly a consequence of coarticulation effects from the continuous movement of the articulators.
- In the modern view, there are no objective phones; these are just descriptively useful points in a continuous space.
- There are no boundaries or markers in continuous speech; it is not possible to determine the phoneme or phone boundaries in speech in a bottom-up way.
- We can describe the linguistic content of speech using sequences of phonemes or phones; this is known as transcription. Transcription is inherently difficult because we are enforcing a discretisation of a continuous encoding.

Phonology
- Phonology is the study of the linguistic use of speech.

- A phonotactic grammar describes which sequences of phonemes are legal for a language.
- Lexical phonology describes how words are formed from constituent morphemes.
- Phonemes and phones can be described with distinctive features. More compact and "natural" rules can be created by making use of distinctive features rather than phonemes or phones.
- The syllable is a useful organisational unit that sits between the words and phonemes. All words have an integer number of syllables, each of which has an integer number of phonemes.
- Lexical stress is manifested at the syllable level. Exactly one syllable in a word is designated as the primary stressed syllable; all other syllables either have secondary stress or are reduced.

8 Pronunciation

We now turn to the problem of how to convert the discrete, linguistic, word-based representation generated by the text-analysis system into a continuous acoustic waveform. One of the primary difficulties in this task stems from the fact that the two representations are so different in nature. The linguistic description is discrete, the same for each speaker for a given accent, compact and minimal. By contrast, the acoustic waveform is continuous, is massively redundant, and varies considerably even between utterances with the same pronunciation from the same speaker. To help with the complexity of this transformation, we break the problem down into a number of components. The first of these components, **pronunciation**, is the subject of this chapter. While specifics vary, this can be thought of as a system that takes the word-based linguistic representation and generates a phonemic or phonetic description of what is to be spoken by the subsequent waveform-synthesis component. In generating this representation, we make use of a **lexicon**, to find the pronunciations of words we know and can store, and a **grapheme-to-phoneme**[1] **(G2P)** algorithm, to guess the pronunciations of words we don't know or can't store. After doing this we may find that simply concatenating the pronunciations for the words in the lexicon is not enough; words interact in a number of ways and so a certain amount of **post-lexical processing** is required. Finally, there is considerable choice in terms of how exactly we should specify the pronunciations for words, hence rigorously defining a pronunciation representation is in itself a key topic.

8.1 Pronunciation representations

In this section we shall define just what exactly the pronunciation component should do, what its input should be and what its output should be.

8.1.1 Why bother?

First of all, we should ask ourselves – why bother at all with a pronunciation component? Why not just synthesise directly from the words? To answer this, consider an approach in which we don't have any pronunciation component per se, but attempt synthesis directly from the words. In doing so, let us imagine a concatenative synthesiser in which we have a database of recorded samples for each word. If we take even a

---

[1] Also known as letter-to-sound (LTS) rules.

modestly sized vocabulary, we might have more than 50 000 words, so this would require 50 000 basic types in our synthesiser. It is not valid for the most part to join words directly; the nature of continuous speech means that the effect of one word influences its neighbours and this must be dealt with in some way. There are two possible solutions to this: either we could adopt an "isolated-word" style of synthesiser, whereby we record each word in isolation and then join the words with a short pause, or we could record one example of each word in the context of each other word. The first approach is taken by many simple concatenated prompt systems, and, while quite reliable, will never perform well in terms of naturalness. The problem with the second approach is that, for our vocabulary of 50 000 words, recording each word in the left and right context of each other word would mean $50\,000^3 = 1.25 \times 10^{14}$ types in all. Recording this number of words from a single speaker is clearly impossible, but even if some way of reducing this were found, it is clear that the number of required recordings could still be vast. Even if we avoid a concatenative type of approach, we would still have to build a model for every word in the database, which would still be an enormous undertaking.

This is clearly a problem of *data sparsity* in that we can't collect or design models for such a large number of types. When we consider that the required vocabulary for a real system may in fact be even larger and that unknown words will have to be handled somehow, we see that in reality the situation may be even worse.

8.1.2 Phonemic and phonetic input

Now consider an approach whereby we transform each word into its phonemes and use those as the input to the synthesiser. In doing so, we have drastically reduced the number of base types from 50 000 words to the 40 or so phonemes. We still have to deal with the influence of continuous speech, but, if we assume that the scope of this is limited to the immediately neighbouring phonemes, then we will need a left and right context for each type. This will give us a total of $40^3 = 64\,000$ types, which is clearly more manageable. The total number is, however, still quite large, so it is natural to ask whether we can reduce it further. The source of the large number is of course having to take the left and right contexts into account, rather than the number of phonemes themselves. Hence it is reasonable to ask whether we can transform the phonemic representation further into a space that has a further reduction in the number of base types.

We can do this by attempting to predict the effect of the context of a phoneme's neighbours. As we saw in Chapter 7, the traditional model of phonetics states that phonemes are realised as phones and that in general this realisation process is controlled by context. For example, an /l/ has two allophones, described as **clear** and **dark**, such that the clear phone is found at the start of a syllable and the dark one at the end; a /t/ is aspirated at the start of a word, but unaspirated when following an /s/ in a consonant cluster. So, if we could apply this process to every phoneme, we could generate a sequence of phones for the input, and in doing so render the base types free of many of the effects of context. With a phoneme representation, we might have a type /s-t-r/, representing a /t/ in the context of being after an /s/ and before an /r/. With a phonetic representation we replace this with just [t] (the IPA symbol for an unaspirated [t]), using [tʰ] for cases with aspirating. Context will still play a part, because coarticulation (as opposed to allophonic variation) still operates, but the degree of the contextual effect will be lessened. Since many phonemes have more than one allophone, the number of base types may well increase (to, say, 60), but, because we can do without a left and right context for each, the overall number will be considerably lower.

8.1.3 Difficulties in deriving phonetic input

There is, however, a drawback with this approach. The problem with phone representations was discussed previously in Section 7.3.2 where we stated that, although phone descriptions could be very useful in certain circumstances, they should be treated with caution since they are a discrete representation enforced on a continuous space; and as such reflect only crudely the actual processes going on. To give an example, let us consider again an example of place-of-articulation assimilation that occurs with words such as

$$\text{TEN} + \text{BOYS} \rightarrow \text{t e n} + \text{b oi z} \rightarrow \text{t e m b oi z}$$

where the alveolar place of articulation of the /n/ of the first word assimilates to the bilabial place of articulation of the /b/ of the second word (bilabial). There is no doubt that something is going on here; the [n] certainly seems somewhat different from a "normal" [n], and this can be verified from both acoustic and articulatory measurements. This is not the issue; rather, it is whether the [n] has *turned into* an [m]. Such a statement says that there are only two choices, an [n] or an [m], and that in certain situations the [n] keeps its identity and in others it changes. Consider the situation of saying TEN BOYS with a pause between the two words; it should be clear that the [n] sounds like a "normal" [n] and that if we say the (made-up word) TEM (/t e m/) as TEM BOYS with a short pause, we can tell the difference between the two. Now, if we say a sequence of utterances TEN BOYS starting with a long pause and gradually shortening it, until there is no pause, we see that there is a *gradual* shift in the nature of the /n/, until, when we talk quickly, it seems impossible to make the distinction between TEN and TEM. This shows that assimilation is certainly taking place, but that it is not the case that an underlying [n] is at one point realised by the phone [n] but at a certain point suddenly becomes a different phone, [m]. The process is one of continuous change, and it is quite inappropriate to model it discretely. We see exactly the same effects with /r/ colouring, nasalisations and reduction.

In a way, we are forcing a discrete level of representation onto a range of effects that have a substantial continuous component; and this is damaging in all sorts of ways. As an example, consider again the phenomena of reduction; whereby we have AND represented underlyingly as [a n d], but where this undergoes reduction, resulting in a range of possible phonetic forms including [a n], [ən d], [ən], [n] and so on. It may be clear that some level of reduction has occurred, but at what stage do we delete the [d]? Where do we draw the boundary between a full [a] and its being reduced to a schwa? Studies of hand human labelling have shown that even trained phoneticians often give different labels to the same utterance, and much of this disagreement stems from where the draw the boundaries as to whether something is reduced or deleted.

8.1.4 A structured approach to pronunciation

In essence, our situation is this. If we stick with high-level representations such as words, we can have a high degree of confidence about the accuracy of these, but the number of base types is far too large. As we generate representations nearer the signal, we see that the number of base types reduces considerably, but at a cost, in that we are more likely to create a representation that is inaccurate in some way. There is no single solution to this problem: the answer lies in picking a point where the compromise between number of types and accuracy of representation best suits a particular system's needs.

Perhaps unsurprisingly, the consensus as to where this point should be has shifted over the years. When more-traditional systems were developed, memory was very tight and hence the number of base types had to be kept low regardless of any errors. In recent years technological developments have eased the pressure on memory, making more-abstract representations possible. Given this, there is more choice over where exactly the ideal representation should lie. In fact, as we shall see in Chapter 16, the most-successful systems adopt a quite-phonemic representation and avoid any rewriting to a phonetic space if at all possible. Because of this, the pronunciation component in modern systems is in fact much simpler than was perhaps the case in older systems, and quite often the input to the synthesiser is simply canonical forms themselves, direct from the lexicon.

Hence the solution we advocate for a state-of-the-art unit-selection synthesiser can be described as a **structured phonemic representation**, whereby we use the phonemic information more or less as it is specified in the lexicon, but do so in such a way that the synthesiser also has direct access to the context, syllable structure and stress information that govern the allophonic and coarticulation processes. In doing so we are relying on the subsequent synthesis component to model the allophonic and coarticulation effects *implicitly*. This can be achieved very effectively by representing pronunciation as a phonological tree of the type introduced in Sections 7.4.4 and 7.4.5, since this allows us to describe the phonemes, syllables and stress information in a single representation. In such a scheme, the words STRENGTH, TOP, LAST and HAND would for example be represented as follows:

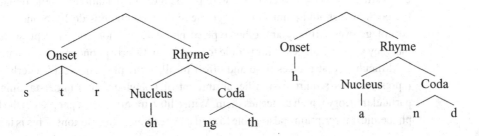

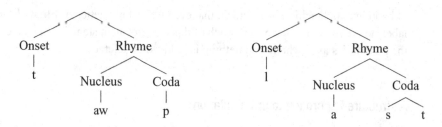

From these trees, we can access the information we require; the phonemic context can be read directly by accessing a phoneme's neighbours; the syllable structure is presented as a tree, and from this we can distinguish between /s t/ in a syllable onset in STRENGTH and offset in LAST.

8.1.5 Abstract phonological representations

We have just seen that, despite potential issues with data sparsity, in general, higher-level phonemic representations are preferred over lower-level phonetic ones. In pursuing this approach, however, we should ask ourselves whether the phonemic representations we have been using are in fact optimal; might there not be an even more abstract and less linear representation that would help us to overcome the dimensionality problem? Progress in modern phonology has in fact moved away from using linear sequences of phonemes or phoneme-like units as the "deep" or canonical representation; freed from the constraint that representations must be linear, the more structured "non-linear" representations reviewed in Section 7.4.3 could be considered instead.

Ogden *et al.* [334] discuss just this issue and point out that the problem with specifying phonemic-style representations is that this concentrates overly on the paradigmatic dimension; the phonemic system is based on what contrasts occur at a particular point, and the issue of what phonemes follow what other phonemes is left to a different, phonotactic part of the phonology. In fact, we see that it is indeed possible to construct representations in which both the normal notion of phonemic contrast and a mechanism for governing legal sequences are handled in a unified way. While phonemic representations are less susceptible to borderline decisions than phonetic representations, they are not entirely free of problems.

At present, we know of no unit-selection techniques that make use of such abstract phonological representations, but this possibility is sure to be explored eventually. While there is every chance the field will indeed move in this direction, because of the lack of practical proof we will continue with phonemic representations for the remainder of the book. We should point out that phonemic representations do have one significant advantage over virtually any other type of phonetic or phonological representation in that they are generally easy for people to work with. Overly phonetic representations can be difficult in that it takes time and effort to fill in the phonetic detail; overly abstract representations require a significant amount of knowledge and understanding of the particular theory which underlies them. With a little training, most people get the hang of phonemic transcription and are able to read and generate transcriptions. This is important for text-to-speech because at some stage someone must of course actually enter all the lexicon entries by hand. In doing so, we wish to have a system that is both quick (there might be hundreds of thousands of entries) and accurate, and the phonemic system seems to fulfil these criteria reasonably well. That said, we should make the point that it is perfectly possible to *enter* the pronunciation for a word in the lexicon in terms of phonemes and then by rule *convert* this to either a deeper or a more-superficial representation.

Finally, using a linear phonemic representation has the benefit that it makes automatic data-base labelling considerably easier. In Chapter 17 we shall consider the issue of labelling a speech database with phoneme units, both by hand and by computer. As we shall see, it is easier to perform this labelling if we assume a linear pronunciation model, because this works well with automatic techniques such as hidden Markov models.

8.2 Formulating a phonological representation system

Having in principle chosen a system that uses phonemes and syllable structure, we now turn to the issue of how to define this for a given speaker, accent or language. It is essential that this is done with care; if we decide, say, to use a single phoneme /ch/ instead of /t/ + /sh/, we have to use this consistently, such that we represent CHURCH as /ch u r ch/, not as /t sh u r ch/ or /ch u r t sh/ and so on.

The following sections explain how a phonological system is developed. The first thing that we must decide is which accent we are dealing with from the language in question. In general it is not possible to define a phoneme inventory that can cover multiple accents, so whatever we develop will be particular to that accent. One solution to this is to use the actual speaker the TTS system will be based on, and from analysis of his or her phonemic contrasts ensure that the correct pronunciation system is found. This solution can require too much effort, so more-general accents often have to be defined. In this chapter, we will conduct an exercise on developing a representation for just two accents of English, namely General American and British English received pronunciation (RP). The principles involved, however, are the same for most other accents and languages.

It does seem somewhat wasteful that separate pronunciation systems and therefore lexicons are required for British English and American English; while differences of course exist, pronunciations in the two accents aren't *completely* different. Fitt [161], [162], [163] proposed a solution to this whereby a more-abstract pronunciation drawn from a large set of phonemes was used as a base lexicon. From this, filters could be used to generate accents for any accent of English.

8.2.1 Simple consonants and vowels

Here we will adopt the standard approach of using the principles of **minimal pairs** and **phonetic similarity** to develop a phoneme inventory (see Section 7.3.2). This is not an exact procedure and it is possible to come up with two equally valid phoneme inventories for a given accent. There is therefore some discretion in determining a phoneme inventory; the key point is not to find the "absolute truth" of what constitutes the set of sound contrasts, but rather to define an inventory that suits our purposes, and then *standardise* on this, so that exactly the same conventions are used in all parts of the system. Experience has shown that "drift" in phoneme conventions, whereby, say, we transcribe BOTTLE as /b aa t el/ in one place and as /b aa t ax l/ in another, can have a disastrous effect since the data, rules, synthesiser and lexicon can become misaligned. The crucial point is to choose a standard as best we can and stick to just that standard everywhere in the system.

Recall that the principle of minimal pairs states that it is not the similarity or dissimilarity of two sounds per se which is the issue, but rather whether they distinguish two separate words. By using this test, it is straightforward to determine a set of contrasts in a particular position. So, if we take the form [X i p], we can find a set of contrasts as follows:

p	PIP	m	MIP	l	LIP
t	TIP	n	NIP	r	RIP
k	KIP	s	SIP	w	WHIP
b	BIP	sh	SHIP	y	YIP
d	DIP	z	ZIP	h	HIP

The /X i p/ pattern doesn't give the full list of possible contrast because not every possible legal phoneme sequence actually ends up as a real existent word. But if we take some further patterns, we can elicit further contrasts:

X	/X eh t/	X	/X ih n/	X	/X ay n/
p	PET	sh	SHIN	v	VINE
b	BET	f	FIN	dh	THINE
d	DEBT	th	THIN	f	FINE
g	GET	s	SIN	s	SIGN
m	MET			sh	SHINE
n	NET				

When we consider one particular pattern, the analysis seems clear: PET and BET are different words, so /p/ and /b/ must be different phonemes. It is reasonable to ask, though, how we know that the sound at the start of PET is the same sound as at the start of, say, PIT. There is no known deterministic or objective way to prove that this is the case, all we can do is rely on judgement and state that, in terms of *articulation*, the two sounds are both produced with the lips, are unvoiced stops and so on, and that if we examine them *acoustically* we see they are similar. But they are rarely so similar as to be taken as identical; their articulation and acoustic patterns are affected by the following vowel and this does have some effect. To a certain extent, there is a leap of faith required in order to say that PET and PIT start with the same phoneme. This problem is even more acute when we consider sound contrasts in different positions:

p	RIP	m	RIM	n	RAN
t	WRIT	n	RING	m	RAM
b	RIB	f	RIFF	n	RANG
d	RID	s	REECE	b	BRIT
g	RIG	d	WRITHE	t	STRING

So, while we have no problem is saying that the last sounds in RIP, WRITE and RIB are different phonemes, it is a harder matter to say that the last sound of RIP should be the same phoneme as the first of PIT. We can again only appeal to some notion of acoustic and articulatory similarity. We should bear in mind, though, that we are attempting a compromise between an accurate representation and a minimal one, and that various possibilities can be tried and tested.

If we consider the words SPAN, STAN and SCAN, we see that we have a three-way contrast, in which the second sound differs in terms of place of articulation. There is, however, no voicing contrast possible at these points, so there are only three possible phonemes, unlike in word-initial position where we have three place-of-articulation contrasts in addition to a voicing contrast, giving us six phonemes in all. The question is, then, should SPAN be transcribed as /s p ae n/ or /s b ae n/? The phonetic evidence tends to favour /p/, but this is not conclusive, and will vary from speaker to speaker and accent to accent. Clearly, though, there is a systematic effect going on. There is probably no objective answer to this, so we obey the normal convention (which may have been influenced by the spelling) and represent these words with the unvoiced phonemes /p/, /t/ and /k/.

A similar minimal-pair test can be used for vowels, in which pairs (or n-tuples) of words can be compared and phoneme identities determined. In

iy	HEAT	uw	HOOT
ih	HIT	ah	HUT
eh	HET	ae	HAT
ao	HAUGHT	aa	HOT

we can clearly identify a number of distinct vowel phonemes.

8.2.2 Difficult consonants

Now we turn to some more problematic cases. In English, the consonants /dh/, /zh/, /ng/ and /h/ are a little more difficult to deal with. There are clearly exactly four unvoiced fricatives; /th/, /f/, /s/ and /sh/ in English and, of these, /f/ and /s/ have the voiced equivalents /v/ and /z/, which are unproblematic. The voiced equivalent of /th/ is /dh/ and, while /th/ is a common enough phoneme, /dh/ occurs only in very particular patterns. For a start, only a few minimal pairs occur between /dh/ and /th/; real examples include TEETH_NOUN, / t iy th/, and TEETH_VERB, /t iy dh/, while most are "near" minimal pairs such as BATH /b ae th/ and BATHE / b ey dh/, where the quality of the vowel differs. Even if we except /dh/ as a phoneme on this basis, it occurs in strange lexicon patterns. In an analysis of one dictionary, it occurred in only about 100 words out of a total of 25 000. But if we take *token* count into consideration, we find in fact that it is one of the commonest phonemes of all, in that it occurs in some of the commonest function words such as THE, THEIR, THEM, THEN, THERE, THESE, THEY and THIS. When we take the remainder of the words in which /dh/ is found we see that it is found in many "old" or traditional-sounding English words such as FATHOM, HEATHEN, HITHER, LATHE, LITHE, LOATHE, NETHER, SCATHE, SCYTHE, SEETHE, SHEATHE and TETHER. In fact, it occurs in only relatively few "normal" English words such as MOTHER, LEATHER and HEATHER. Clearly /dh/ is a phoneme, and, for the purposes of defining our phoneme inventory, it gets a place alongside the other fricatives. There are consequences that arise from its use, though, in that, because of the dominance of its use in function words, it is difficult to obtain "typical" characteristics of the phoneme (say in terms of its average duration) that are not swamped by the fact that it is used as a function word (function words are often very short, regardless of what phonemes make them up).

Now consider /zh/, the voiced equivalent of /sh/. This occurs in words such as BEIGE, /b ey zh/. Again, while there are few direct minimal pairs, some exist (BEIGE versus BAIT or BANE), so we classify /zh/ as a phoneme. Problems again arise from patterns of use. It has been argued (e.g. [97]) that /zh/ is a relative newcomer to the English language since in the past words such as TREASURE and MEASURE were pronounced /t r eh z y uh r/ and /m e z y uh r/ and that it is only relatively recently that these have been spoken as /t r eh zh er/ and /m eh zh er/. /zh/ presents the same problems as arose with /dh/, but in this case they are even more difficult because of the very few words naturally occurring with /zh/. These problems with /zh/ and /dh/ can lead to problems in database collection and analysis (see Section 14.1).

The sounds [ng] and [h] are noteworthy because they are severely restricted as to the positions they can occur in. Primarily, /h/ cannot occur at the end of a syllable or a word, and /ng/ cannot occur at the beginning of a syllable or word. There are further issues with /ng/ in that, since it occurs only in a post-vocalic position, it suffers from the loss of contrast that affects all nasals in this position. So, while it is clear that SIN, /s ih n/, SIM, /s ih m/, and SING, /s ih ng/, all end in different sounds, it is less clear what the correct representation for words like THINK should be. In words such as this the last stop is in a velar position, and the nasal takes on a velar quality also (this is similar to the TEN BOYS effect we discussed earlier). The issue is then whether we should use a velar nasal /ng/ or a dental nasal /n/; there is no contrast possible in this position and so it is a question of convention which we should pick. As with other such issues, the main goal is to aim for consistency; if /ng/ is used here then it should be used in all other post-vocalic pre-velar stop positions.

8.2.3 Diphthongs and affricates

So far we have concentrated on the paradigmatic dimension, by contrasting (say) HAT, HEAT and HOT and showing that in a frame of /h X t/ we can distinguish a number of phonemes. We were able to do this because a word such as HOT clearly divides into three segments. We assign provisional phonemic status to each, swap in other sounds and thereby determine the system of phonemic contrasts. In many words, however, it is not so easy to segment the word into distinct parts, and in such words we face a question of not only what the contrasts are, but also how many positions there are to have contrasts within.

In words such as CHAT, the sound before the vowel can be represented as either a combination of /t/ and /sh/ or a distinct single phoneme /ch/. A similar case arises with words such as JUICE, which can be represented as a combination of /d/ and /zh/ or as a distinct single phoneme /dz/. The argument for using two phonemes stems from the principle of keeping the number of phonemes as low as possible; the argument for a single phoneme arises from the idea that these act as a single coherent unit and are more than the sum of the two parts. There is no real way to settle this argument, so we will somewhat arbitrarily decide to use single phonemes /ch/ and /dz/.

A similar situation arises with the diphthong vowels in words such as BOY, TOE, TIE and BAY. In each of these, the vowel does not hold a single steady-state position, either acoustically or in terms of articulation. Rather, say, TIE has an /ae/ sound followed by an /iy/ sound. We face the same situation as with affricates in that we could use a more-minimal paradigmatic representation by saying that two phonemes exist here, or invent new single phonemes to reflect the fact that these sounds seem to act as single coherent units. Again, there is no real objective way to decide, so we will follow our previous example and use a single symbol for each. One benefit from this is that, as we shall see below, this allows us to state a rule that every syllable is formed with exactly one vowel.

8.2.4 Approximant–vowel combinations

The approximants /r/, /w/, /j/ and /l/ in many ways operate like normal consonants, such that we have contrasts in words such as ROT, WATT, YACHT and LOT. Of these /l/ is probably the most straightforward and the main point to note is that the acoustic pattern of dark /l/ in words such as PULL, and SMALL is very different from that with clear /l/, such as in LOOK and LAMP. This does not pose much of a problem phonemically, but is an important difference to bear in mind when trying to decide whether two /l/ units will join well together.

The glides /w/ and /y/ are interesting in the ways they interact with vowels. Firstly, there is considerable phonetic similarity between /w/ and /u/ and /y/ and /i/, and, in many phonetic analyses, words such as SHOW are transcribed as /sh o w/. If we follow our diphthong principle above, we rule out these cases and transcribe this as /sh ow/. The process of /w/ or /y/ following a vowel is sometimes called an **off-glide**. As mentioned in Section 7.4.1, one of the difficulties in determining a phonotactic grammar for English is that /y/ often acts as what we can call an **on-glide**. So, in British English, speakers say

TUNE as /t y uw n/ and DUKE as /d y uw k/. There is evidence that the /y/ should be attached to the vowel, however, just like a diphthong, to give /t yuw n/ and /d yuw k/. Further evidence for this comes from other English accents; in the USA, most speakers say /t uw n/ for TUNE and /d uw k/ for DUKE. The effect is systematic, and would seem easier to explain if only one factor (the realisation of a phoneme /yuu/) were changing between accents. Despite this argument, it is most common to regard VIEW as being composed of three segments, and we shall adopt this convention and say that it is represented as /v y uw/.

Finally, we consider /r/. This is perhaps the strangest and most-interesting sound in human speech. Setting aside the phonemic question, consider for a moment how this differs phonetically across languages. In English, an alveolar approximant /r/ ([ɹ] in the IPA) is most common and it is quite common for this to have a retroflex (curling of the tongue) quality as well. In contemporary Scottish English, a tapped alveolar (ɾ) is quite common, which is separate again from the "classical" Scottish trill [r] (now usually seen only on stage or in film). In French and many accents of German an uvular [ʁ] sound is the norm, with other accents of German having a more alveolar quality. In Spanish, words such as PERRO are formed with a trilled [r], and this forms a minimal pair with the tapped [ɾ] of words such as PERO. It seems incredible, given all this variety, that somehow all these sounds are grouped together across languages and represented in writing as an r. Do they even have anything in common? While we will not go into cross-language issues here, it is worth noting that words are quite commonly shared across languages, for example the words EMAIL, APRIL and METRE. In the case of METRE, we see that it contains an r in all languages and is pronounced in a wide variety of ways. If we study the acoustics, however, we see that these different "types" of /r/ do share somewhat more in common in that they have characteristic energy bands at certain parts of the spectrum.

When we consider the phonemics of /r/, we find that it has an interesting effect. First we distinguish between **rhotic** accents, which have /r/ in post-vocalic position (in English, these are Scottish, Irish, General American), and **non-rhotic** accents, which have no realised /r/ (most England English, Australian and some New York accents). For example, in Irish English, there is no "long a" sound [aa], which is found in RP BATH /b aa th/; this word is pronounced /b a th/. However, in words that have an /a/ and a /r/ following it, the /a/ gets lengthened, and *phonetically* we have something like [h aa r d]. Being a phonetic-realisation effect, this doesn't concern us unduly, but it does show how significant (and unexpected to the unwary) the presence of an /r/ can be. More significant are the non-rhotic accents. In these, it is not the case that the /r/ is simply not present; rather (historically or by having some deep presence) it has the affect of altering the quality of the vowels before it. So, in a word like HER, which might be represented as /h uh r/ in a rhotic accent, we do not get /h uh/ but rather /h @@/, which has an extended vowel. It is almost as if the /r/ has turned into a /@/ and is still present. Because of this, we are faced with a choice between transcribing non-rhotic FIRE as /f ai @/ and inventing a new symbol /ai@/, which would be regarded as a triphthong.

8.2.5 Defining the full inventory

From the above we should have sufficient knowledge to construct a phoneme inventory for any accent or language we are interested in. In practice we find that we don't always have to design a phoneme inventory from scratch; for many languages or accents the existence of a significant computer-speech-research community means that phoneme inventories already exist. Since these communities often develop and distribute resources using these phoneme inventories, it can be highly beneficial to adopt one of these rather than starting from scratch. In the past, databases of speech that contained hand-labelled phone or phoneme transcriptions were highly valued, but, since phonetic/phonemic transcription can now be done automatically, today by far the most useful resource is in fact a comprehensive lexicon. Since the development of a large lexicon (>100 000 words) can take years of work, any existing resources greatly help. Lexicons are in general tied to a specific accent of the language, and hence come with a phoneme inventory predefined.

So in many cases a phoneme inventory may exist for the accent we are interested in. By and large, most phoneme inventories developed for computational work roughly follow the guidelines just described. As just pointed out, there are, however, several ways to resolve some of the issues, so it is worth examining any predefined phoneme inventories in the light of the above discussion to assess their suitability. Most predefined phoneme inventories have not been developed with TTS in mind; rather they were developed for purposes of speech recognition or are just "general" phoneme inventories developed for a variety of uses. Experience has shown that minor modifications in predefined phoneme inventories can sometimes be beneficial for TTS purposes. The commonest problem is that sometimes the phoneme inventory has some "phonetic" features (such as markers for aspirated or unaspirated, or a different label for the stop closure from that for the stop release), but these can normally be removed with some automatic processing.

Table 8.1 gives a definition of a complete phoneme inventory based on a modification of the TIMIT phoneme inventory. This was originally designed for use with the acoustic–phonetic database of the same name and is now one of the best-known phoneme inventories for General American [173]. Most of the modifications we have made have been to eliminate some of the purely phonetic distinctions in the original TIMIT inventory.[2] The main peculiarity of the TIMIT inventory relates to the fact that sequences of schwa followed by /r/ are represented by a single phoneme /axr/, and the vowel in words such as BIRD is treated similarly and represented as /b er d/. All other combinations of

---

[2] For the record, the differences between this and the standard phone inventory are the following.

1. The /axr/ unstressed schwa + /r/ phoneme has been replaced by /er/. /er/ and the old /axr/ are now distinguished by stress alone.
2. The closed phones bcl, tcl etc. have been removed.
3. The nasal tap /nx/ has been merged with the /n/ nasal.
4. /hv/ has been merged with /hh/.
5. The glottal-stop phone /q/ has been removed.
6. The syllabic consonants /en/, /em/, /eng/ and /el/ have been replaced by /ax/ + consonant sequences, e.g. /en/ → /ax n/.
7. /ix/ has been replaced by either /ax/ or /ih/, /ux/ by /uw/ and /ax-h/ by /ax/.

Pronunciation

Table 8.1 The modified TIMIT phoneme inventory for General American, with example words

Symbol	Example word	Transcription	Symbol	Example word	Transcription
b	BEE	B iy	l	LAY	L ey
d	DAY	D ey	r	RAY	R ey
g	GAY	G ey	w	WAY	W ey
p	PEA	P iy	y	YACHT	Y aa t
t	TEA	T iy	iy	BEET	b IY t
k	KEY	K iy	ih	BIT	b IH t
dx	MUDDY	m ah DX iy	eh	BET	b EH t
jh	JOKE	JH ow k	ey	BAIT	b EY t
ch	CHOKE	CH ow k	ae	BAT	b AE t
s	SEA	S iy	aa	BOT	b AA t
sh	SHE	SH iy	aw	BOUT	b AW t
z	ZONE	Z ow n	ay	BITE	b AY t
zh	AZURE	ae ZH er	ah	BUTT	b AH t
f	FIN	F ih n	ao	BOUGHT	b AO t
th	THIN	TH ih n	oy	BOY	b OY
v	VAN	V ae n	ow	BOAT	b OW t
dh	THEN	DH e n	uh	BOOK	b UH k
hh	HAY	HH ey	uw	BOOT	b UW t
m	MOM	M aa M	er	BIRD	b ER d
n	NOON	N uw N	ax	ABOUT	AX b aw t
ng	SING	s ih NG	axr	BUTTER	b ah d axr

vowel and /r/ are treated as two separate phonemes. Table 8.2 gives a phoneme inventory for the British English RP accent that is, based on the University of Edinburgh's MRPA (machine-readable phonetic alphabet) system [276]. The consonant system is more or less the same; this is what we would expect from the facts of English accent variation. The vowel system is different, and in particular the treatment of "r" is different because RP is a non-rhotic accent. The pseudo-diphthong vowels /i@/, /u@/ and /e@/ are used in cases where there "would be" an /r/ if the accent were non-rhotic. For the remainder of the book, we will use the modified TIMIT and MRPA phoneme inventories for General American and RP English accents, respectively.

8.2.6 Phoneme names

The above has focused on minimal-pair analysis, which is the major task in defining a phoneme inventory. In defining the phoneme inventories shown in Tables 8.1 and 8.2 we have also performed an additional task in that we have explicitly given a *name* to each phoneme. The processes of minimal-pair analysis and naming are not connected; the names in theory could be anything, such that we could just name the discovered phonemes 1, 2, 3 or a, b, c or anything else. In choosing a naming system, we have first made the decision to use plain ascii characters for all phonemes. The TIMIT phoneme inventory is exclusively alphabetic; the MRPA phoneme inventory is alphabetic apart from the use of "@" for sounds with schwa. Since there are more phonemes than alphabetic characters,

Table 8.2 A definition of the British English MRPA phoneme inventory, with example words

Symbol	Word-initial example	Word-final example	Symbol	Word-initial example	Word-final example
p	PIP	RIP	j	YIP	–
t	TIP	WRIT	ii	HEAT	PEEL
k	KIP	KICK	ih	HIT	PILL
b	BIP	RIB	ei	HATE	PATE
d	DIP	RID	e	HET	PET
g	GAP	RIG	aa	HEART	PART
ch	CHIP	RICH	a	HAT	PAT
dz	GYP	RIDGE	uu	HOOT	POOT
m	MIP	RIM	u	FULL	PUT
n	NIP	RUN	ou	MOAT	BOAT
ng	–	RING	uh	HUT	PUTT
f	FOP	RIFF	oo	HAUGHT	CAUGHT
th	THOUGHT	KITH	o	HOT	POT
s	SIP	REECE	ai	HEIGHT	BITE
sh	SHIP	WISH	au	LOUT	BOUT
v	VOTE	SIV	oi	NOISE	BOIL
dh	THOUGH	WRITHE	i@	HEAR	PIER
z	ZIP	FIZZ	e@	HAIR	PEAR
zh	–	VISION	u@	SURE	PURE
h	HIP	–	@@	HER	PURR
l	LIP	PULL	@	CHINA	ABOUT
r	RIP	CAR	w	WIP	–

we use two and sometimes three characters for a single phoneme. Often this is when we have a sound with two phonetic components such as /ch/ or /oy/, but it is also done for simple phonemes such as /th/ and /sh/.

In naming these phoneme inventories we have also not worried about whether the "same" phoneme in one accent has the same name in another accent. Since we have no need to mix phonemic transcriptions across accents in normal synthesis, this is unproblematic. Strictly speaking, the principles of phonemic analysis imply that this is a meaningless comparison anyway: phonemes have status only as units of contrast for a single accent, so it is of no consequence whether we use the same symbol for the vowel in BUY in British and American English. In practice, it is clear from the TIMIT and MRPA definitions that the vowel sets are different but the consonant sets are nearly identical (this of course partly reflects the fact that vowel contrasts are the main source of phonemic differences between accents in English). We have also avoided the possibility of naming the phonemes after one of their allophones in the IPA alphabet. This is because it is generally easier to use a separate and self-contained symbol set for phonemes rather than use IPA symbols since this can lead to confusion about the phoneme/phone divide.

In both the TIMIT and the MRPA phoneme inventories, we have used exclusively lower-case letters. To some extent this is a purely aesthetic issue since we wish to avoid upper case (which has a somewhat "aggressive" quality) and mixed case (which can look ugly). There is an additional benefit in that upper-case equivalents of the symbols can be used in cases where some emphasis is required, such as stress or other highlighting. In the above, we have chosen, though, to give simple mnemonic names to each phoneme, such that the first phoneme of TIP is named "t", the first of SHIP is named "sh" and so on. Mnemonic names are usually helpful because we often have to enter words in the lexicon ourselves or perform transcriptions, and this is easier if we can easily recall the symbols to use. These names can sometimes trip us up, especially with vowels, so a little care should be taken.

8.2.7 Syllabic issues

We have shown previously that, in addition to simple or "linear" phonemic context, a major factor in determining the realisation of phonemes is their place in the syllabic structure. Phonemes at the starts of syllables or words can be realised differently from ones at the ends; the position of a phoneme in a consonant cluster is important, and stress effects, which are expressed at syllable level, also have an important bearing on phonetic realisation. We therefore have to give the waveform-generation component access to this syllabic information.

If we actually know where the syllable boundaries lie within a word, it is a simple matter to assign the type of tree structures described in Section 7.4.4. Quite simply, the vowel serves as the nucleus, the consonants after the nucleus up until the end of the syllable are assigned to the coda, and the consonants between the vowel and the syllable start are assigned to the onset. In this way, the syllable structure is found deterministically from the syllable boundaries and phonemes. As we saw in Section 7.4.4, syllable boundaries can for the most part be determined by using the maximal-onset principle. In the few cases where this is not so (recall our BOOKEND example) the boundary needs to be marked explicitly. As system designers we therefore face some choices. On the one hand, we can choose to mark every syllable boundary explicitly in the lexicon. This is costly, however; many existing lexicons do not have syllable boundaries marked and the additional effort involved in marking these may be prohibitive. Alternatively we could determine the syllable boundaries automatically by application of the syllable-onset rule. In doing so we will of course get the syllable boundaries wrong for the difficult cases just mentioned. A third possibility is a compromise between the two, whereby the difficult cases to transcribe are marked by hand, whereas the rest are transcribed by algorithm.

For languages such as English, three levels of stress are necessary. While some phonological systems use more (e.g. SPE), our experience has shown that, at the lexical level at least, three is clearly sufficient. Following from a convention adopted for TIMIT, we use **1** for primary stressed syllables, **2** for secondary stressed syllables and **0** for reduced syllables. The stress numbers are placed adjacent to and after the vowel, so, for example, the entry for INFORMATION would be /ih2 n . f ax0 . m ey1 . sh ax0 n/.

All syllables receive a stress mark; hence most vowels will be marked 2. All words except function words must have a single primary stressed vowel. All normal schwa vowels (/ax/) receive a 0. There are several possibilities regarding whether to transcribe normal vowels with a 0 or a 2. One option is to regard 0 and 2 as weak and strong versions of one another, such that any vowel could receive either 0 or 2. Another option is to regard 0 as being only for schwa vowels; this means that in effect every other vowel is a 2 regardless. This option has the benefit of simplicity, but some feel that it disregards phonetic facts. In traditional phonetics, the last vowel of HAPPY is quite often transcribed as /ih0/ or /iy0/. In addition, many think that it is most appropriate to transcribe the last vowel of HABIT as /ih0/, giving /h ae1 b ih0 t/. Here we again face the issue of phonological organisation versus phonetic explicitness. Either option can be taken, but it is worth noting that direct experience showed that it was possible to take the 0 for schwa option everywhere, and label all other vowels as 2. This would then give /h a1 p iy2/ and /h a b ax0 t/.

Finally, we come to the issue of phonotactics. This is an important part of defining a phonology; there is little point in spending time defining a phoneme inventory to define the paradigmatic relations if we then ignore the phonotactics which define the syntagmatic relations. Defining a phonotactic grammar is necessary for a number of practical reasons. Firstly, we need to define what we think the synthesiser should be able to say in terms of phoneme sequences. Even if we create a very simple grammar, which states for instance that /ng/ can't occur at the start of a word, we can find any case where this is passed to the synthesiser by mistake and take corrective action. Secondly, we need to keep a check of the entries in the lexicon; since these entries are made by hand there is a chance that errors will be made. Hand in hand with the phonotactic grammar comes a set of conventions that we must establish regarding how words should be transcribed. Finally, having an explicit phonotactic grammar helps in database collection since we can easily tell how complete a particular database is with regard to phonemic-sequence coverage.

8.3 The lexicon

In our introduction to the nature of language, we explained that one of its fundamental properties is that there is an arbitrary relation between the meaning and form of a word. A word's pronunciation cannot be inferred from its meaning, nor vice versa. Obviously, for a language to work, these arbitrary relations need to be stored somehow, and the device used to do this is called a **lexicon**. We can talk about a **speaker's lexicon**, meaning the lexicon that a particular individual uses to store his or her words; we can also talk about the **language lexicon**, which is a store of all the meaning/form pairs (words) in that language (a traditional dictionary is this type of lexicon). For language-engineering purposes, however, we talk about the **computer lexicon**, which is a physical computer-data object containing known descriptions of words.

In machine-understanding and generation systems, or any systems that deal with semantics and meaning, the lexicon is a vital component in a system, because it is here that the connection between meaning and spoken or written form is stored. In speech-synthesis and -recognition tasks, however, the meaning element is largely ignored, so in principle the lexicon is not strictly speaking a *necessary* component. However, lexicons are in fact widely used in synthesis and recognition to store the correspondences between the two main forms of the word; that is, the spelling and pronunciation of a word. These are called **orthography–pronunciation** lexicons.

8.3.1 Lexicon and rules

While the relation between meaning and form is completely arbitrary for all languages, in general the relation between the two main forms, i.e. the spoken and written forms, is not. The two main forms do not have an equal status; speech is the primary form of the language and writing is a secondary, invented form designed to record the spoken form. The degree to which these forms are related is dependent on many factors, and it is clear that in some languages the written form closely follows the spoken form (e.g. Spanish), in others the relationship is discernible but complex (e.g. English and French), while in others it is nearly as arbitrary as the meaning/form relationship (e.g. Chinese). Hence for native words in some languages it is possible to infer one form from the other, and this mapping can be performed by a **grapheme-to-phoneme** algorithm. We will leave a discussion of such algorithms until Section 8.4, but for now let us assume that we can build such a system that always generates a pronunciation form from an orthographic form, even if the accuracy of doing this can be quite low for some languages. Owing to the low accuracy of these rules, for some languages we have to store the correspondences in a lexicon. It is important to realise that even for the more-direct languages, the widespread use of English and other loan-words, technical words and acronyms means that at least some use of a lexicon is required.

Most text-to-speech systems use a combination of a lexicon and rules to perform pronunciation. The justification for this two-pronged strategy has changed somewhat over the years. In traditional systems, the explanation usually went something like this: "grapheme-to-phoneme rules are used as the main pronunciation mechanisms, but in addition we use an **exceptions dictionary** that deals with those cases which don't follow the normal patterns of the language". Thus it was rules that were primary, and the lexicon was used for cases where the rules failed. The reasons for the primacy of the rule system were that, firstly, this was still in the days when using "knowledge" was a respectable way of building all types of speech, language and AI systems, so using rules was a natural way in which anyone would approach this problem; secondly, memory space was limited, so storing tens of thousands of words was prohibitive. Note that using rules to save memory typically incurs a processing cost, but at the time speed was not generally seen as a critical issue (i.e., it was simply impossible to have enough memory to store the words, but it was possible to use slow rules, even if it resulted in a non-real-time system). Finally, it takes time to develop a high-quality lexicon, and in the past there simply weren't the resources and experience to have built a comprehensive lexicon. More common today is the justification "We use a lexicon to store pronunciations, and resort to rules for missing entries in the lexicon". Hence the situation is reversed; the lexicon is seen as primary and the rules are there as back-up for when the lexicon fails.

The reader should be aware that there is still a lingering hangover in the literature from the earlier period regarding the rules versus lexicon issue. Rules are often inherently favoured over the lexicon and one often comes across statements stating that there is not enough memory to store all the words in a language. Such reasoning is rarely solid, since no decision about this can be taken until the actual size of the lexicon and the processing cost of the rules have been quantified. For relatively small-footprint systems (say <10 megabytes), lexicon size will definitely be an issue, but for larger systems the size of the unit-selection database will far outweigh the size of the lexicon. One also finds that alongside this we also see justifications for the claim that using rules is a more-sophisticated approach and that simply storing words in the lexicon is a mere brute-force approach. As we shall see below, this is a false argument; applications of rules and lexicons are simply different techniques that trade off memory and processing cost; both have their place and one is not a priori better, more sophisticated or even more accurate than the other. One occasionally encounters the opposite view, in which lexicons are seen as the only true way to handle pronunciation, and hence the use of very large lexicons is advocated. This can be dangerous too; while most people are aware that rules often make mistakes, there is sometimes a failure to realise that significant errors can exist within the lexicon also; too often this possibility is ignored and the lexicon is taken as the absolute truth.

Most accounts of the lexicon and rule set-up in a TTS system would give a reader the idea that these were two completely separate mechanisms, which just happen to have been brought to bear on the same problem. In fact, the distinction between lexicon and rule approaches is not as clear cut as it seems, and, with modern TTS systems, it is best to regard what we normally think of as "lexicons" and "rules" as being at certain points on a scale. From the outside, the distinction seems clear; a lexicon is a declarative data structure (e.g. a list) that contains entries, whereas a rule set is a series of processes that examine the grapheme sequence and generate a phoneme sequence. The distinction is, however, not always clear cut. Most rule systems (see below) use a context window on the orthography to help disambiguate which phoneme should be generated. If we take a word such as IRAQ for instance, it is quite possible that no other words with anything like this grapheme pattern will be found; so if the rules correctly predict this word, can we not say that the rules have learned the pronunciation for this and only this word, and that the rules have in fact implicitly stored the pronunciation?

In machine learning, this phenomenon is called **memorising the data**: this is often considered harmful if the system learns exactly those examples it was given but not any generalities; if this is the case we say that the model has **over-fitted** the data. For a trainable model with enough parameters, it is entirely possible that it will end up memorising all the training data rather than learning the generalisations behind it. This is normally seen as an undesirable property in a rule system, but, in our case, it can be argued that this is not strictly so. In most machine-learning problems, the *population* of data is effectively infinite, and the training data are a small *sample* of the population. Since the sample is not identical to the population, care must be taken not to learn it too exactly, otherwise the trained algorithm will not perform well when presented with different samples from the population. As we shall see below, learning grapheme-to-phoneme rules is somewhat different, in that the population (all existent words) is large but finite, and there is a good chance of encountering exactly the same words in different samples. Hence memorising common examples is not in itself harmful [76].

8.3.2 Lexicon formats

In speech technology, it is quite common to find existing lexicons in what we can term **simple dictionary format**. Such lexicons can be stored as simple ascii files, in which each entry starts with the orthography, followed by the pronunciation and possibly other information. The entries are ordered by the orthography such that a few typical entries would look like this:

```
deep        d iy1 p
deer        d ih1 r
defend      d ax0 f eh1 n d
define      d ax0 f ay1 n
degree      d ax0 g r iy1
delicate    d eh1 l ax0 k ax0 t
delight     d ax0 l ay1 t
demand      d ax0 m ae1 n d
density     d eh1 n s ax0 t iy2
```

It is clear that this format explicitly gives priority to the written form, since this is used to order and index the entries. Dictionary formats can be problematic, though, for the reason that the orthographic form of a word is not unique; homographs are fairly frequent and these must somehow be handled. Two common amendments to this style are to have multiple lines with the same orthographic form,

```
project     p r aa1 jh eh2 k t
project     p r ax0 jh eh1 k t
```

or to use more than one entry for the pronunciation on each line,

```
project p r aa1 jh eh2 k t, p r ax0 jh eh1 k t
```

Even if this is done, we still somehow need to know that /p r aa1 jh eh2 k t/ is the noun form of `project`; it is not sufficient to say that `project` simply has two different pronunciations. While POS markers could be added, we soon find that when we also consider the issue of pronunciation variants (which are arbitrarily different ways of pronouncing the same word) and orthographic variants (different ways of spelling the same word) we see that the simple dictionary style comes under considerable pressure and loses its main positive feature, namely that it is easy to read and process.

A solution to this is to build the lexicon as a **relational database**. In this, we have exactly one entry for each word. Each entry contains a number of uniquely identified fields, each of which has a single value. For a simple word, the entry may just contain two fields, ORTHOGRAPHY and PRONUNCIATION. It is a simple matter to add more

8.3 The lexicon

fields such as POS and SYNCAT. Each entry in a relational database can also be seen as a feature structure of the type we are now familiar with. Because of this similarity, we will use the feature-structure terminology for ease of explanation. Hence some simple words would look like

$$\begin{bmatrix} \text{ORTHOGRAPHY} & \text{defend} \\ \text{PRONUNCIATION} & \text{d ax0 f eh1 n d} \end{bmatrix}$$

$$\begin{bmatrix} \text{ORTHOGRAPHY} & \text{degree} \\ \text{PRONUNCIATION} & \text{d ax0 g r iy1} \end{bmatrix}$$

$$\begin{bmatrix} \text{ORTHOGRAPHY} & \text{define} \\ \text{PRONUNCIATION} & \text{d ax0 f ay1 n} \end{bmatrix}$$

In this the orthography has no preferred status, and the entries can be indexed by the orthography or pronunciation field. The main advantage of using this style is that it is straightforward to add more fields to the entries while still maintaining a well-defined structure. Hence we can add POS information to distinguish between homographs:

$$\begin{bmatrix} \text{ORTHOGRAPHY} & \text{project} \\ \text{PRONUNCIATION} & \text{p r ax0 jh eh1 k t} \\ \text{POS} & \textit{verb} \end{bmatrix}$$

$$\begin{bmatrix} \text{ORTHOGRAPHY} & \text{project} \\ \text{PRONUNCIATION} & \text{p r aa1 jh eh2 k t} \\ \text{POS} & \textit{noun} \end{bmatrix}$$

Throughout the book we have been using a convention whereby we give a name to each word, for instance PROJECT-NOUN for the last entry above. We should state that this name has no objective status; it is simple a "handle" that allows us to identify the word uniquely. We normally use the spelling as the handle and add some additional identifier if this isn't sufficiently unique – again, this is just a convention, the handles could be $1, 2, 3, \ldots$ or $a, b, c, \ldots$ It is up to the system designer whether to include this in the entry or not: including it means that there is a unique value for each word in the lexicon, and this can have useful housekeeping functions. On the other hand, the developer needs to create these handles and ensure their uniqueness. For demonstration purposes we will include them in the following examples, so the above would look like

$$\begin{bmatrix} \text{NAME} & \text{PROJECT-VERB} \\ \text{ORTHOGRAPHY} & \text{project} \\ \text{PRONUNCIATION} & \text{p r ax0 jh eh1 k t} \\ \text{POS} & \textit{verb} \end{bmatrix}$$

$$\begin{bmatrix} \text{NAME} & \text{PROJECT-NOUN} \\ \text{ORTHOGRAPHY} & \text{project} \\ \text{PRONUNCIATION} & \text{p r aa1 jh eh2 k t} \\ \text{POS} & \textit{noun} \end{bmatrix}$$

$$\begin{bmatrix} \text{NAME} & \text{DEFEND} \\ \text{ORTHOGRAPHY} & \text{defend} \\ \text{PRONUNCIATION} & \text{d ax0 f eh1 n d} \\ \text{POS} & \textit{verb} \end{bmatrix}$$

It is often useful to make use of lists as values because we want to describe the full range of possibilities without necessarily ascribing any causes to them. For example, many noun/verb pairs don't differ in orthography or pronunciation, so for our purposes we treat these as a single word (see Section 4.2). We still wish to record the fact that the word can occur in several parts of speech; this may for instance be done by using the POS tagger:

$$\begin{bmatrix} \text{NAME} & \text{WALK} \\ \text{ORTHOGRAPHY} & \text{walk} \\ \text{PRONUNCIATION} & \text{w ao1 k} \\ \text{POS} & \{\textit{verb}, \textit{noun}\} \end{bmatrix}$$

Orthographic and pronunciation variants can be handled this way also:

$$\begin{bmatrix} \text{NAME} & \text{EITHER} \\ \text{ORTHOGRAPHY} & \text{either} \\ \text{PRONUNCIATION} & \{\text{iy1 dh er0, ay1 dh er0}\} \end{bmatrix}$$

$$\begin{bmatrix} \text{NAME} & \text{COLOUR} \\ \text{ORTHOGRAPHY} & \{\text{colour, color}\} \\ \text{PRONUNCIATION} & \text{k ah1 l axr} \end{bmatrix}$$

8.3.3 The offline lexicon

Now let us consider the issue of *certainty* or *knowledge* with regard to pronunciation. To help in this, let us make a distinction between two types of lexicon. The first is the **system lexicon**: this is the database of word pronunciations actually used by the system at run time. The other is the **offline lexicon**; this is a database of everything we know about the words in our language: it is the sum of our recorded knowledge. The offline lexicon has all sorts of uses; we use it to label speech databases, to train the POS tagger, for help with text analysis and so on. The system lexicon is created from the offline lexicon, and there are choices about what entries and fields to include. We should stress that, regardless of the needs and requirements of the system lexicon, it is highly beneficial to have as large and comprehensive an offline lexicon as possible.

When we consider the offline lexicon, it is not the case that a given word is strictly "in" or "not in" the lexicon. For instance, we may have entries that are incomplete, in that we know the values for some fields but not others. Most probably there will be missing words also, that is words that we know of but haven't entered. Not every value in every field will be correct; errors may occur or there may be entries of which we are not entirely certain. More formally, we can evaluate our offline lexicon in terms of **coverage**, **completeness** and **quality**:

Coverage This is a measure of how many words in the language are in the lexicon. For every word in the language, we can assign it to one of three categories:
known, words that are present in the lexicon;
missing, words that exist and are known to the developers, but are not present in the lexicon; and
unknown, words that exist, are not known to the developers and are not in the lexicon.
Completeness Every lexicon entry has a number of fields, but some of these may be unfilled.
Quality A measure of how correct the values in an entry are.

What is perhaps less clear is that these measures can be traded off against one another. Recall that we said (Section 4.2) that for our purposes the derived forms of a word are also considered words in their own right, so WALK, WALKING and WALKED are all distinct words. It is quite common for dictionary writers to define only the main, base form, so if we come across an existing lexicon it is quite likely that we would find WALK but not WALKING or WALKED. We can easily create the empty entries for WALKED and WALKING, and in this way we have improved our coverage, but have left these entries incomplete since we have specified only the orthography, not the pronunciation. To complete these entries, we can use rules to add the pronunciations, for example by simply concatenating /ih0 ng/ to the end of the stem to create WALKING. In doing so, we don't, however, know whether the rule generates the correct form; so at this stage the quality of the new entry is unknown. Finally, by performing a manual check, we can tell whether the rule has operated correctly.

In this example we have seen how our knowledge of missing words (that is words known to us, but not in the lexicon) can be used to improve coverage. We have also seen that we can use rules to improve the completeness of the lexicon, but when doing so we don't necessarily know whether the rule has generated the correct answer. This is one example of how the rules and lexicon can interact. We can generalise this approach to providing more extensive coverage and completeness. With regard to the unknown versus missing distinction, with a well-developed lexicon the most likely missing entries are of the type just mentioned, that is absent derived forms of known words: we can improve our coverage of missing words by the process just outlined. By contrast we can be sure that the number of unknown words is vast; this will include a huge number of technical or medical terms and of course many names. For unknown words, the task is really to find these words from sources outside our personal knowledge. One way of doing this is to conduct a search for unique words in any text corpus, or any collection of text material (e.g. the Internet). Given such a list, we have effectively improved our coverage but done so in such a way that the completeness for all these new entries is limited to an orthographic definition. We can improve completeness as before, but this time we have to generate the base form of the pronunciation as well as any derived forms. We can find the base form by a grapheme-to-phoneme algorithm, and use the morphological rules to find the derived forms, and in that way generate a complete lexicon. Our only issue now is with quality; since we have used a grapheme-to-phoneme algorithm there is a good chance that many entries will be completely wrong. All the automatically generated entries can be labelled as such, so that we know not to have the same degree of confidence in these as in the other, hand-checked, entries. In due course all these can of course be checked and edited to improve the certainty and quality of the lexicon.

These processes show the value of the offline lexicon; it is in effect the full depository of our knowledge about words in the language. Creating this is highly beneficial in that it gives us a full database of words from which to generate the system lexicon, to train or develop a grapheme-to-phoneme algorithm, and for other uses mentioned above.

8.3.4 The system lexicon

If we generate the offline lexicon by the above process we will have a lexicon that is very large, but will have a considerable number of entries whose quality is unknown. In traditional TTS, the normal approach was to have only carefully checked entries in the lexicon, the idea being that the rules would be used for all other cases. However, it should now be clear that whether we use the rules at run time to handle words not in the lexicon or use the rules offline to expand the lexicon will not have any effect on the overall quality: all we are doing is changing the place where the rules are used.

Given this large offline lexicon, we see then that the real debate about rules versus lexicons is not one of quality, but rather one of balance between run time and offline resources. If we take the case of including the entire offline lexicon in the system lexicon, we will have a system that uses a considerable amount of memory, but the processing is minimal (simply the small amount of time taken to look up a word). If, on the other hand, we create a system lexicon that is only a small subset of the offline lexicon, this will result in a smaller footprint, but, since the pronunciations of absent words will have to be generated at run time, the processing costs could be considerable. One very useful technique for lexicon reduction is to take a set of rules, run them over all the words in the database and record which entries agree with the pronunciations generated by the rules. If we then delete all these entries, we have saved considerable space but will generate output of exactly the same quality. In fact this technique is just the opposite of what we did before when we used the rules to create pronunciations for incomplete entries. There we used the rules to create the largest possible lexicon, whereas now we have used the rules to create the smallest possible lexicon. The quality of the output is the same in these two cases.

In fact, there are other possibilities beyond this. Considering the last point again, what we did was to use the rules to reduce the size of the lexicon; in effect the rules summarise the redundancy in the lexicon. This technique can be thought of as **lossless compression**; that is, the output is *exactly* the same as when the full lexicon is used, it's just that the generalities of the lexicon have been captured exactly in a set of rules. If we desire a smaller system still, we can opt for **lossy compression**, whereby the output is not the same as before and can be expected to contain some errors. After we have deleted the entries which the rules correctly predict, we further delete some less frequently occurring entries, resulting in a smaller system, but one in which a few more words will be wrong. We can tune this exactly to our space requirements right down to the point where we have no run-time lexicon at all and simply rely on the rules every time. We can move in the other direction too; if we have some additional memory and we find that the "decompression time" needed to use the rules is too slow, then from a lossless-compression system we can add some correctly predicted common words back into the lexicon. This will have the effect of increasing memory but speeding the system up.

8.3.5 Lexicon quality

Since the lexicon is the repository of all knowledge regarding the orthography and pronunciation of words, it is vital that the quality of the entries that it contains is as high as possible. All too often this issue is overlooked, and unfortunately many lexicons that are used today have significant numbers of errors. It should be obvious enough that, if the pronunciation of a word is entered incorrectly in the lexicon, then, when the TTS system comes to say that word, it will sound wrong. Often listeners are completely thrown by this; they don't perceive that the correct word has been spoken wrongly, rather they suffer a complete loss of comprehension. Other consequences arise from errors in the lexicon; if this is used to label the speech database, we might find a case where an instance of an /uh/ vowel is wrongly labelled as /uw/. When this unit is used in synthesis to say a word such as /b uh k/ the result will be /b uw k/; again throwing the listener. This is even more serious than the first case because this will affect words other than the one where the mistake occurs. Personal experience has shown that, after the recording of a high-quality speech database, the quality of the lexicon is the single biggest factor in determining overall TTS quality.

A certain amount of skill is required to add entries by hand; the primary requirements are that one should have a familiarity with the pronunciation scheme being used and a knowledge of the accent in question. In many cases the person adding or checking entries has the same accent as the lexicon itself, but this isn't always the case, so particular care needs to be taken when adding entries of a different accent; it is all too easy mistakenly to think that we know how someone else says a word. That said, it is certainly not the case that one needs to be an expert phonetician; personal experience has shown that most developers can perform this task if given adequate training.

Many errors result from a small number of basic causes. A primary source of error is that people make the pronunciation too close to the orthography. Inexperienced developers could for instance add /e/ phonemes at the ends of words where the orthography has an e, such that we would get /h ey t e/ instead of /h ey t/ for HATE. Other simple errors include using /j/ for /jh/, using /c/ for /k/ and so on. One simple trick is to find an existing checked word that we know rhymes with the word we are about to enter; so that when entering HATE we look up LATE and simply change the first phoneme. This is particularly useful when working with an accent one isn't sure of. Another is to use an existing version of a synthesiser to play back the word synthesised with the new transcription; this technique very quickly exposes errors such as adding an /e/ to the ends of words. Finally, we can build an automatic pronunciation-verification system that simply checks whether the pronunciation for an entry is legal within the phoneme inventory, stress system and phonotactic grammar definitions.

Those more experienced in phonetics often make the opposite mistake of creating lexicon entries that are too close to the phonetics. Here we see cases where many multiple variants are added because of a slight perceived difference or possibility of difference in a pronunciation. For the word MODERATE we might get /m o l d ax0 r ax0 t/, /m o l d r ax0 t/, /m o l d eh0 r ax0 t/ and so on all being added because of perceived differences. In general, the tendency to "split hairs" and add many pronunciation variants for each word should be avoided. Even if two speakers do seem to differ in how they pronounce a word, it should be remembered that the phonemic representation is only approximate in any case, and cannot be expected to cover all differences. Using too many variants can be harmful because it leads to a fragmentation of the data, and makes it difficult to learn and generalise.

8.3.6 Determining the pronunciations of unknown words

Even with experienced lexicon developers, severe difficulties can arise because of unknown words. Recall that these are legitimate words that exist in the language but with which the developer is not personally familiar. To see the extent of this problem, take any large (i.e. unabridged) dictionary, open it at random and have a look – it usually doesn't take long to find unfamiliar words. The profusion of words unknown to us is quite considerable because the dictionary is full of medical, chemical and nautical terms, old words that have fallen out of usage and so on. If we then add to these the issue of names we see that the problems mount up. For many an unknown word, even though the word's meaning and form are unfamiliar to us, the pattern of letters seems familiar, and because of this we have a reasonable chance of guessing the correct pronunciation. Many are in groups of words that may be personally unfamiliar to the developer, but are common enough for other people. Sailing terms are a good example of this; for example, most people can correctly read aloud bobstay, cringles and jibe even though they don't know what these words mean. Other terms are more difficult, however, such that most struggle with how to read aloud fo'c'sle, boatswain and gunwale if they are not familiar with the terms. For such groups of words, the only answer is to consult a knowledgeable source; either a book on sailing or a sailor in person.

In considering this issue, we come across an interesting philosophical issue regarding the pronunciation of words. Quite simply, how do we *know* what the correct pronunciation for *any* word is? Saying that word pronunciations are properly defined in traditional dictionaries avoids the point; this would imply that words not in a dictionary have no standard pronunciation, and misses the fact that dictionary writers try to listen to real speakers to elicit the actual pronunciation [483]. The best we can do is say that the "correct" pronunciation for a word is X because that is what the population of speakers believes. For most common words, this is unproblematic; and everyone of a particular accent will say the words WALK, DOG and HOUSE in the same way. Problems do arise, though, because people do not in fact always pronounce words in the same way, even within a single accent. The issue is whether we can establish a correct pronunciation, relegating the others to some "incorrect" status, or whether we have to accept multiple correct forms. Establishing the boundaries here is difficult; it seems equally absurd to insist on a "gold standard" of a single correct pronunciation and to adopt an "anything goes" policy whereby all pronunciations are equally legitimate. There is a personal element in this as well; it doesn't bother me whether someone says CONTROVERSY as /k o1 n t r ax0 v er2 s iy/ or /k ax0 n t r o1 v ax0 s iy2/, but somehow when someone says /n uw1 k uw2 l ax0 r/ for NUCLEAR this just seems wrong. To take another example, it is now quite common to hear American English speakers pronounce the "silent" /l/ in words like CALM and SALMON – while this may seem clearly "wrong", how can it be if millions of people now follow this habit?

Difficulties arise most with regard to rare names. In most European countries, the native pronunciation of names is well established, and even if a name is borrowed from one language to make a new name in another language, we can often find the correct pronunciation in the original language. The situation in the USA is significantly different, in that the sheer diversity of names outweighs that in any other country. In the USA, we see firstly that many names have moved considerably from their original pronunciation, but secondly many names are genuinely made up or so altered from the original that they have effectively become new names.[3] This situation creates many cases where there is only a handful of families with a particular name, making it difficult for the lexicon

---

[3] Note that there is a consistent myth that peoples' names were changed by officials during transit through the Ellis Island immigration service. What evidence we have suggests that this is not in fact the major source of new names, since some degree of consistency had to be kept between embarkation in the origin country and immigration procedure. Many minor changes in spelling and so on certainly happened, but it was not the case that immigration officials simply invented new names because they thought the real name was too awkward to handle.

developer to find the pronunciation. Even more problematic is that we find cases where families share the orthography for their name but have different pronunciations. It is often unknown even whether these names share a common origin. How then do we determine the correct pronunciation for such a name? As Spiegel [409] says,

> Complicating the issue is that everyone is a pronunciation expert for their own name, and (only slightly less so) for the people they know. Sometimes people are unaware of other (valid) pronunciations for their names, even when the pronunciation is more common.

Hence even if we decide that the correct pronunciation for smith is /s m ih1 th/ we will still find enough people who complain that the system got it wrong because their name is pronounced /s m ay1 th/. There is no objective real solution to this problem since it is one of system usage and user requirements. The real point is that developers who deploy TTS systems should be aware that there might not be any objective truth as to whether a system is saying a name correctly or not.

8.4 Grapheme-to-phoneme conversion

A **grapheme-to-phoneme (G2P)** algorithm generates a phonemic transcription of a word given its spelling. Thus it generates a sequence of phonemes from a sequence of characters.[4]

8.4.1 Rule-based techniques

Grapheme-to-phoneme conversion by rule is perhaps the classic application of traditional knowledge-based rules in TTS. Indeed, in the past it was thought that this was how humans performed the task; when presented with an unknown word they would apply a set of rules to find the pronunciation. The most common approach is to process the character sequence left-to-right and, for each character, apply one or more rules in order to generate a phoneme. It should be clear that these rules can't operate in isolation, otherwise a t character would always result in a /t/ phoneme. Rather the character context of the character is used and in this was we can generate /th/ when t is followed by h.

Most approaches use context-sensitive rewrite rules of the type introduced in Section 5.2.4. Recall that these are of the form

$$A \rightarrow B/d/C$$

---

[4] In this book, the use of the term **grapheme-to-phoneme** conversion relates specifically to algorithms that perform just this task; elsewhere this term is sometimes used for the process which uses algorithms and the lexicon, and indeed is sometimes used for the entire process of text analysis. The term **letter-to-sound rules** is also encountered, but this is increasingly misleading because most modern techniques don't use rules in the traditional way. A final complication regarding the terminology of this field is that strictly speaking a grapheme is a minimal unit of spelling rather than a single character, so in that the th is described as a single grapheme. Since our input is always a sequence of characters, we are really performing **character-to-phoneme** conversion.

which reads as "A is converted (rewrites to) d when preceded by B and followed by C". In any well-developed system, multiple rules will match a given context, so a means of choosing amongst rules is required. A common technique for this is to **order** the rules and apply each in turn, so that the first matching rule is the one chosen. Since this is a purely hand-written technique the developer has to order these rules him or herself. A common technique is to order the rules with the most specific first. Rule systems of this kind include [8], [136], [155], [226], [307].

These systems can often produce quite reasonable results for highly regular systems with a shallow orthography (e.g. Spanish). The results are generally very poor, though, for languages like English and French, which has increasingly led to the adoption of data-driven and statistical techniques [300].

8.4.2 Grapheme-to-phoneme alignment

All the data-driven techniques described below use algorithms to learn the mapping between characters and phonemes. They use an existing lexicon as training data, and often words in this are held out as test data. Thus these data-driven techniques are usable only when a comprehensive lexicon is available. In general words do not have a one-to-one correspondence between characters and phonemes: sometimes two or more characters produce a single phoneme (e.g. gh → /f/ in rough) and sometimes they don't produce a phoneme at all (e.g. the e in hate → /h ey t/). A prerequisite for training then is to **align** the characters with the phonemes for each entry in the lexicon.

Most techniques for performing this are based on a **dynamic time warping (DTW)** technique [343]. This works by assigning a cost to every phoneme that can align with a grapheme. "Good" correspondences are given low costs; "bad" correspondences are given high costs. The DTW technique is a dynamic-program algorithm that effectively searches all possible alignments and picks the one which has the lowest total cost. This is a significantly easier task than either writing the rules by hand or aligning every word in the lexicon by hand. For example, the character s maps to the phoneme /s/ with a low cost, but maps to /n/ with a very high cost. In many cases there is an obvious phoneme that maps from the character, but in other cases, e.g. h, this is used to represent many different phonemes (/sh/, /th/, /dh/ etc.) in writing and thus has a number of low costs.

A probabilistic version of this was used by Jelinek [239], who used the hidden-Markov-model Baum–Welch training technique (described in Section 15.1.8) to learn the probability that a character would generate a phoneme. This was extended by Taylor [433], who showed that it was possible to do this from a **flat start**, meaning that no initial seeding is necessary. In general, the alignment method is independent of the actual algorithm used, meaning that any alignment technique can be used with any of the algorithms described below.

8.4.3 Neural networks

The first data-driven G2P algorithm was the NetTalk algorithm developed by Sejnowski and Rosenberg [377], [396], [397]. This work became genuinely famous, not for its raw performance or use in practical text-to-speech, but because it was one of the first real demonstrations of a neural network in operation. A popular demonstration of the system speaking the pronunciations it had learned added to the fame of the system as it showed how the system performed better with each training iteration. NETtalk uses a feed-forward network consisting of an input layer, a hidden layer and an output layer. The input considers a character at a time, and uses a context window of three characters before and three after. This results in a total of seven groups of units in the input layer of the neural network. Each group has a series of nodes, which can be used in a binary-coding fashion to represent each character. (For example, if we have five nodes, then a would be 0001.1, b would be 00010, c would be 00011, and z would be 11010.) A single group of units is used in the output and a single phoneme is therefore generated.

Authors of a number of further studies have also used neural networks for G2P [191], [240] but this technique's importance is mostly due to its having been a powerful demonstration in the early days of machine learning.

8.4.4 Pronunciation by analogy

Pronunciation by analogy is a popular data-driven technique that was first introduced by Dedina and Nusbaum [127] and popularised by Damper and colleagues [422], [122], [123], [300], [27]. Its motivation comes from studies of human reading and draws from studies by Glushko [175] and Kay and Marcel [247]. The basic idea is that, given an unknown word, a human considers known words that are orthographically similar, and adapts a pronunciation of the nearest word to use for the unknown word. For example, if we are confronted with a new word zate, we consider that, although we don't know the pronunciation of this, we do know of many similar words, e.g. date, late, mate, all of which have the same ending pronunciation. In addition, we also know that most words that start with z generate a /z/ phoneme, so we apply both these analogies and decide that zate should be pronounced /z ey t/. Marchand and Damper [300] give a review of this technique and describe it as belonging to the family of machine-learning techniques known as **memory-based learning** [119].

The algorithm works by comparing the characters in the input word with each entry in the lexicon. For each matching substring, a lattice is built with the input characters and the output phonemes (which are found by the alignment process). The lattice comprises a path of the characters $\langle c_1, \ldots, c_N \rangle$ and a set of arcs that span character sequences in the path. Each arc represents one match between a sequence of characters and a sequence of output phonemes. Hence in through we would have an arc from c_0 (a dummy start character) to c_2, showing the match between the letters t and h and the phoneme /th/. In addition a frequency count is kept of how many times this match is found in the lexicon. This lattice then represents a summary of all the matches in the lexicon for this word. Finally, a decision function is used to find the single best sequence of phonemes.

The decision function can take a number of forms, and finding better methods of designing this function is the main goal of research in pronunciation by analogy. Marchand and Damper [300] give a review of some techniques. The original approach is to find the shortest paths through the lattice, meaning choosing the matches which have the longest individual spans. If more than one shortest path is found, then the one with the highest frequency count is used. Other approaches include calculating the product of arc frequencies, the standard deviation of the path structure, the frequency of the *same* pronunciation, the number of different pronunciations generated and the minimum of arc frequencies.

8.4.5 Other data-driven techniques

Several other data-driven techniques have also been used for G2P conversion. Pagel *et al.* [343] introduced the use of decision trees for this purpose, a choice that has been adopted as a standard technique by many researchers [249], [198]. The decision tree works in the obvious way, by considering each character in turn, asking questions based on the context of the character and then outputting a single phoneme. This can be seen as a way of automatically training the context-sensitive rules described above.

Other data-driven techniques include using support-vector machines [18], [111], [114], transformation-based learning [67], [518] and latent semantic analysis [38]. Boula de Mareüil *et al.* [62] describe a formal evaluation of grapheme-to-phoneme algorithms and explain the complexities involved in ensuring that accurate and meaningful comparisons can be performed.

8.4.6 Statistical techniques

The statistical approach has been something of a latecomer to G2P conversion, perhaps because of the success of other data-driven techniques such as pronunciation by analogy or owing to the impression that context-sensitive rewrite rules are adequate so long as they can be automatically trained, e.g. by a decision tree. In recent years, however, various approaches that give a properly statistical approach have been developed.

As mentioned previously, Jelinek [239] proposed a system to align graphemes and phonemes using the Baum–Welch algorithm originally developed for speech recognition. His intention was to expand the lexicon to include unknown words that would then be used in a speech recogniser. His use of the statistical approach was limited only to the alignment step, however; the actual G2P conversion was performed by other techniques. Taylor [433] took this work to its logical conclusion by using HMMs to perform the whole task. In this approach, states represent the phonemes, each of which has a discrete output probability distribution over the characters which form the observations. The system therefore mimics a "writing" process whereby the underlying form (phonemes) generates the surface representation (characters). G2P is then the opposite process of recovering the phonemes from the characters. If performed with a standard HMM, the results are quite poor, but with various refinements the technique can be made to perform as well as any other. These include using context in the models, using an n-gram over the phoneme sequence and performing pre-processing steps to make the alignment task easier. The main advantage of this technique is that no explicit alignment is required; the system implicitly generates a "soft" alignment during training. The use of the n-gram phoneme constraint also means that, even if the answer is wrong, the generated phoneme sequence should be a likely one, rather than simply a random sequence that is unpronounceable.

A more-common approach is to use a technique proposed by Galescu and Allen whereby a search is performed using a joint n-gram model over both the grapheme and the phoneme sequences [44], [86], [172], [223]. Typically, the alignment is performed in the normal way. This set of training data then gives a collection of short sequences of phonemes and graphemes, which are aligned. For example, the word PHONEME gives {ph, f}, {o, ow}, {n, n}, {e, iy}, {m e, m}. Once the entire training lexicon has been converted to this format an n-gram is calculated over all sequences. At run time, a simple search gives the most likely sequence given the characters. The technique is completely general and allows for a variety of refinements in terms of n-gram smoothing and estimation techniques. It avoids the pre-processing steps of the HMM technique described above but does, however, require a separate alignment step.

8.5 Further issues

8.5.1 Morphology

It is common in linguistics to say that words are not atomic entities with regard to syntax or semantics but rather are composed of **morphemes**. If we consider the words BIG, BIGGER and BIGGEST we see that there is a clear connection relating the three words. The issue that concerns us is whether we should consider all of these as separate words, and store them in the lexicon, or just store the root form BIG and then use rules to derive the other forms.

There is no right or wrong answer to this; the fundamental issue is one of size versus speed. If we have plenty of storage space and require quick processing, it is better to store all words. If storage space is limited it may be better to store only the root and generate other examples on the fly. In modern TTS systems, the first approach is generally taken, since memory is cheap, and even for large lexicons, the size of these is swamped by the size of the unit-selection database. It is traditional to define two types of morphology: **inflectional**, whereby (in English) we typically add well-defined suffixes to words to make other words in the same class (as above); and **derivational**, whereby we make new words that often have different parts of speech or different meanings, such that we form CREATION from CREATE. While in general it is possible to store all common derived and inflected forms of words, the set of possible words is very large indeed; for example we can attach the suffix -ESQUE to virtually any noun or proper noun (e.g. TAYLORESQUE).

We can deal with this in a number of ways, but the most common approach is to store all roots and common inflections and derivations in the lexicon. When other inflections or derivations are encountered, these are treated as unknown words and passed to the G2P convertor. This then can attempt a basic **morphological decomposition**, which attempts to find known morphemes within the word. Either this succeeds for all parts of the word, in which case the pronunciation can be formulated by using these parts, or only some of the morphemes are identified, in which case a mixture of a lookup table for the known morphemes and a G2P conversion for the unknown morphemes can be used. The difference may be slight for English, but this approach is certainly more attractive for languages with rich morphology such as German [370], [459], [482].

8.5.2 Language origin and names

A further clue to how to pronounce a word is given by the origin of the word. While we know that late is pronounced /l ey t/ and that the final e is not pronounced, we also know that latte is pronounced /l ah t ey/ with a pronounced final e. The reason is of course that LATTE is an Italian word and so receives an "Italian" pronunciation. We should note, however, that the word is *not* pronounced the way an Italian person would say it, but rather in an English pronunciation using an approximation of the Italian phoneme sequence. Quite how "natively" words such as this should be spoken is by and large a cultural issue; many native English speakers pronounce the final /t/ in RESTAURANT, many do not. Given this fact, there have been approaches that have used analysis of character sequences to find the language of origin, typically using n-grams trained on separate lexicons from each language [94], [289].

A final point concerns the types of words that are most likely to be used by G2P algorithms. The approach taken by most machine-learning research is to divide the lexicon arbitrarily into training and test data, for example by assigning every tenth word to the test set. When using G2P algorithms for lexicon-compression purposes this is a reasonable thing to do, but when using the G2P algorithm for rare words that are unknown to the system, it becomes less valid. In a system with a large, well-developed lexicon, the words most likely to be processed by a G2P convertor are very rare, new formations (e.g. email), loan words or proper names. The chances are that G2P algorithms will fare worse on these than on normal words, so care should be taken in assuming that results taken from a normal word test set apply to the words we are likely to find in real usage. Taking this into account, a better approach would be to design or train G2P systems on just those words which we are most likely to see. For instance Spiegel [409] describes a system that uses an extensive set of rules to model proper names. The system includes all the types of names found in the USA and is designed primarily for directory enquiry and other telephony applications.

8.5.3 Post-lexical processing

Some systems use the output from the lexical lookup process and/or G2P conversion directly as the input to the synthesizer. In some systems, however, an additional step often known as **post-lexical processing** is performed, which modifies the output in some way. The most significant factor in determining how much post-lexical processing is performed is the specific form of the input to the synthesizer. This was discussed in Section 8.1.3, where we saw that we could choose high-level phonemic representation or a low-level detailed phonetic representation. A low-level representation will require significant processing, for example to turn an /n/ into an /m/ in a labial context, but, because many modern systems use high-level representations, often no post lexical processing of this type is required.

Some processes operate at a purely phonological level, so, regardless of the type of input to the synthesizer, some processing may still be required. English doesn't really have any clear demonstrations of this, but when we consider French we see that the process of **liaison** must be modelled. In general, the word LES is pronounced /l ey/, but, when followed by a word starting in a vowel, it is pronounced /l ey z/ as in LES AMIS → /l ey z a m iy/.

8.6 Summary

Basis for pronunciation
- Using a phonemic or phonetic representation helps reduce the size of the specification for the synthesizer.
- A choice has to be made between a high-level abstract representation and a low-level detailed representation
- In general, modern systems use high-level representations.

Pronunciation systems
- Phoneme inventories can be designed using the principles of phonetic similarity and minimal pairs.
- There is no single correct phoneme system for a language, but if care is taken an adequate system is normally not too hard to define.
- In many cases, suitable phoneme sets already exist, especially for well-studied accents and languages

Lexicons and rules
- Pronunciations can be generated by lexicon lookup or by algorithm. These two techniques should be viewed as points on a scale.
- Lexical lookup is fast but expensive in memory.
- Using rules is expensive in processor time but cheap in memory.

Grapheme-to-phoneme conversion
- Grapheme-to-phoneme conversion is the process of guessing a word's pronunciation from its spelling.
- Traditional systems used context-sensitive rules. These are generally inaccurate for languages like English, but can still produce good results for languages with more regular spelling.
- Many data-driven techniques exist, including the use of neural networks, pronunciation by analogy and decision trees.
- Some statistical techniques have recently been proposed.

9 Synthesis of prosody

This chapter is concerned with the issue of synthesising acoustic representations of prosody. The input to the algorithms described here varies but in general takes the form of the phrasing, stress, prominence and discourse patterns which we introduced in Chapter 6. Hence the complete process of synthesis of prosody can be seen as one whereby we first extract a prosodic form representation from the text, as described in Chapter 6, and then synthesize an acoustic representation of this form, as described here.

The majority of this chapter focuses on the synthesis of **intonation**. The main acoustic representation of intonation is the fundamental frequency (F0), such that intonation is often defined as the manipulation of F0 for communicative or linguistic purposes. As we shall see, techniques for synthesizing F0 contours are inherently linked to the model of intonation used, so the whole topic of intonation, including theories, models and F0 synthesis, is dealt with here. In addition, we cover the topic of predicting intonation form from text, which was deferred from Chapter 6 since we first require an understanding of theories and models of intonational phenomena before explaining this.

Timing is considered the second important acoustic representation of prosody. Timing is used to indicate stress (phones are longer than normal), phrasing (phones become noticeably longer immediately prior to a phrase break) and rhythm.

9.1 Intonation overview

As a working definition, we will take **intonation synthesis** to be the generation of an F0 contour from higher-level linguistic information. Intonation is probably the most important aspect of prosody (the others being phrasing and prominence) in that it has a richer degree of expression. As we explained in Chapter 6, the primary semantic content of prosody is in affective representations such as emotion and attitude, and intonation is the primary way in which this is communicated. A sentence can be made to sound believing or incredulous simply by changing the intonation. Intonation is also used augmentatively in that many prominences are indicated with changes in local F0. Phrasing is also often indicated by a lowering of F0 at phrase ends and a subsequent raising of F0 at phrase beginnings. Finally, even if intonation is not used for any affective or augmentative purpose, all utterances have characteristic intonation patterns, which must be modelled correctly if the speech is to sound natural.

9.1.1 F0 and pitch

In this book we have used the terms F0 (fundamental frequency) and pitch interchangeably, since there are few cases where the difference matters. Strictly speaking, pitch is what is perceived, such that some errors or non-linearities of perception may lead this to be slightly different from fundamental frequency. Fundamental frequency is a little harder to define; in a true periodic signal this is simply defined as the reciprocal of the period, but since speech is never purely periodic this definition itself does not suffice. An alternative definition is that it is the input, or driving, frequency of the vocal folds.

In prosody, F0 is seen as the direct expression of intonation; and often intonation is defined as the linguistic use of F0. The relationship between the two is a little more subtle than this, though, since it is clear that listeners do not perceive the F0 contour directly, but rather a processed version of it. The exact mechanism is not known, but it is as if the listener interpolates the contour through the unvoiced regions so as to produce a continuous, unbroken contour.

9.1.2 Intonational form

There is probably no area of linguistics where there is less agreement than when it comes to the nature of intonational form. In the following sections we will cover many of the well-known models and theories, and it will be clear that there is enormous variation in how researchers have chosen to represent intonation. Disagreements occur in all other areas of linguistics; it is just the degree to which the theories differ in intonation that is so noticeable. For instance, when we consider the theories of verbal phonology described in Section 7.4, we do in fact find that there are quite a number of approaches to the problem, but all researchers at least agree that there *is* such a thing as phonology.

We can give an overview of the situation as follows. The function of intonation is to express emotion and attitude, to augment the verbal component firstly by emphasising or marking words and secondly by giving phrases and utterances characteristic patterns that show their structure and type. Although it is rarely expressed as such, this much is uncontentious.[1] At the other end of the scale, it is widely accepted that F0 is the main acoustic manifestation of intonation. The disagreement is about how any intermediate intonational form should be represented. In some early studies, researchers did actually try to map directly from intonational semantics to acoustics [107], [287], but these studies were criticised by many, summarised clearly by Ladd [269]. Basically, in measuring say the F0 of anger, we must of course do this with reference to particular speakers and if we did this the difference in basic physiological behaviour between speakers might well swamp any subtleties in the effects of anger itself. Put more plainly, regardless of how high or low a speaker's voice, a woman, child or man plainly exhibits clear regularities in how these effects are expressed, and we are missing something if we state that the mean value of anger is say 150 Hz. This leads us to the conclusion that there must be

---

[1] Any disagreement would be based around the notion of whether there was a pure linguistic intonational core and a separate paralinguistic or extra-linguistic component.

some sort of **intonational form** that speakers of a language/accent share, which is readily identifiable for those speakers. This is obviously somewhat similar to the idea of the form of words and phonemes in verbal language, and often just this parallel is drawn, such that we are trying to find the intonational equivalents of notions such as word, phoneme, phone and allophone.

The basic problem is that, while nearly everyone agrees that some representation of prosodic form must exist, we have no equivalent test to the minimal-pair test for phonemes. Recall from Section 7.3.2 that this said that, if two sounds could be found that distinguished two words, then we could call these phonemes. If, however, we had two sounds that clearly sounded different (say dark and light [l] in English), but which did not distinguish words, then these were not phonemes but allophones. Crucially, this test relies on the extremely convenient ability of listeners to tell whether two words are the same or not – this can be done without even knowing what the words mean. In intonation, and other aspects of prosody, we simply do not have such a test, and the "semantic space", while perhaps simpler overall, doesn't arrange itself into nice discrete categories the way the words in a lexicon do.

We should realise from the start, however, that not every model was designed with the same goals in mind; for instance, the **British school** (explained in Section 9.3.1) evolved from old-style prescriptive linguistics, and even quite recent publications were concerned with the teaching of "correct" intonation patterns to non-native learners of English. The aims of the **autosegmental–metrical (AM)** school (Section 9.3.3) were to provide a theory of intonation that worked cross linguistically such that phenomena from African and Asian **tone languages**, **pitch-accent languages**, such as Swedish, and **intonation languages**, such as English, could all be described with a unified model. An aim of the **Fujisaki model** (Section 9.3.5) was to create a model that followed known biological production mechanisms, whereas the aim of the **Tilt** model (Section 9.3.6) was to create a model solely for engineering purposes.

The models differ significantly in what they take as the primary form of intonation. In the AM model this is quite abstract, whereas in the Tilt model it is quite literal or "acoustic". These differences in primary form should not be taken to mean that the proponents of these models do not believe that there should be more-abstract or more-concrete representations, just that the "best" representation happens to lie where they describe it. In the many synthesis schemes based on the AM model there are other, more-phonetic or -acoustic levels, and in the Tilt model there is always the intention that it should serve as the phonetic description of some more-abstract higher-level representation.

9.1.3 Models of F0 contours

Imagine that we have an adequate representation of intonational form and that we have extracted such a representation from the input text. How then should we go about generating an F0 contour? We have seen a consistent pattern throughout the book, whereby mappings between one domain and another are performed by means of either explicit rules or, in more-recent systems, machine-learning algorithms such as decision trees or neural networks. The general formulation of this is as a mapping between a set of features and a label:

$$X(f_1, f_2, \ldots, f_N) \rightarrow L_i$$

This approach worked for semiotic, classification homograph disambiguation, POS-tagging phrase-break prediction, prominence prediction and many other problems described in Chapters 4 and 5. It would therefore seem reasonable to try such an approach for intonation synthesis. Strictly speaking, the algorithms used in text analysis are called **classifiers**; they attempt to classify an entity using features that describe it. This is a labelling process of assigning one label from a pre-defined set to the entity. In F0 and timing (discussed below) the problem is similar but now we have to assign a continuous value to our entity. Techniques used to perform this are called **regression algorithms**. In many cases the same algorithms can be used for both; for example decision trees are really classifiers but can be adapted to become regression algorithms by replacing the labels at the nodes with distributions. Neural networks are really regression algorithms but can be converted into classifiers by interpreting each output node as representing a binary part of a label.

In practice standard regression algorithms are difficult to use directly for F0 prediction, because of a serious mismatch between the natures of the input and output representations. In general the input representation is a representation of intonational form as just discussed (for example, pitch-accent types and positions), whereas the output is a continuous list of real-valued numbers. In particular, the feature combination needs to generate not one, but a sequence, of F0 values, which is further complicated because the number of values in this sequence can vary from case to case:

$$X(f_1, f_2, \ldots, f_N) \rightarrow F0_1, F0_2, \ldots, F0_M$$

Furthermore, when we come to performing a bottom-up analysis, where we wish to describe the continuous contour with a discrete set of linguistic units, we find the mapping even more complicated. Because of these problems, it is nearly universal to use some sort of **acoustic model**, which acts as an interim representation between the abstract intonational form and the F0 contour. This model has a fixed number of parameters per unit, so we can use one of the standard mapping algorithms (such as CART) to generate these parameters from the intonational form:

$$X(f_1, f_2, \ldots, f_N) \rightarrow P_1, P_2, \ldots, P_L,$$

As a second step we can use the specialised model to generate the F0 values themselves:

$$Y(P_1, P_2, \ldots, P_L) \rightarrow F0_1, F0_2, \ldots, F0_M,$$

These models usually work on a principle whereby they **encode** or **stylize** the real F0 contour. In other words, they give a tractable simplification of the F0 contour in which the salient linguistically important aspects of the contour are preserved, while less-necessary parts are discarded. At its simplest, the parameters would be a discrete list of **target points** in the contour. From these, a continuous F0 contour can be synthesized. If we keep the number of points constant for each linguistic unit, then we have a situation wherein the parameters of the model align nicely with the linguistic description, while, it is hoped, still allowing accurate synthesis of the contour. The values of these parameters can then be learned by a machine-learning algorithm or written by hand, since we now have a much simpler situation such that we can align input and output pairs. The models described below generally take a more-sophisticated approach, but in essence the principle is the same: to make the mapping between the discrete prosodic-form units and the F0 contour more tractable.

9.1.4 Micro-prosody

Micro-prosody is a term used to denote some of the lowest-level, more-detailed aspects of prosody. Often this includes the interaction between verbal effects and prosodic effects on the main acoustic components of prosody such as F0. It has been widely observed [270], [405] that the phones of an utterance have an affect on F0. Although these effects are less noticeable than the effect of prosody proper, they need to be handled if a complete model of F0 synthesis or analysis is to be generated. Typical effects include a sudden raising or lowering of F0 at a boundary between a voiced and an unvoiced section of speech. The degree to which micro-prosody needs to be synthesized is debatable. Most intonation algorithms synthesise smooth contours and simply ignore micro-prosody altogether and unit-selection systems have the fine micro-prosodic detail already included in the units they use. If, however, any automatic analysis of F0 is to be performed, the micro-prosody must be handled. One common approach is simply to use an algorithm to smooth the F0 contour, in an attempt to make it appear more like the stylised intonation contour that we believe represents the underlying intonation pattern. Modern statistical techniques can handle micro-prosody more directly, by considering it as noise in the data.

9.2 Intonational behaviour

In this section we describe some basic intonational phenomena and attempt to do so in a theory-neutral way.

9.2.1 Intonational tune

The intonation tune can be broadly described as the core pitch pattern of an utterance. Tunes differ from one another in **type** and in **association**. By using different *types* of tunes, the speaker can express for example an emotion such as surprise, disbelief or excitement. The **association** of the tune connects the prosodic part of intonation (the tune itself) to the verbal part, by associating parts of the tune with particular words. By shifting association from JOHN to MATCH in examples (118) and (119), one can convey different effects. By varying the type of tune, one can also express different effects as in example (120).

(118) JOHN WENT TO THE MATCH (as opposed to Harry)

(119) JOHN WENT TO THE MATCH NOT THE THEATRE

(120) JOHN WENT TO THE MATCH (disbelief: but he hates football!)

Describing tune type is perhaps the most difficult issue in prosody. Tune schemes can be broadly divided into those which classify tunes using dynamic features (rises and falls) and those which use static features (tones). Theories also vary in terms of the size of the units they use. **Global** descriptions make use of a few basic patterns that cover the entire phrase; **atomistic** theories make use of smaller units that combine together to form larger patterns. Jones [242] is at the global end of the scale, the British school [193], [333] uses sub-phrase units (see Section 9.3.1), while the AM school [351], [352] and the Dutch school [445] use units that are smaller still (Sections 9.3.2 and 9.3.3).

Much of the discussion on the subject of tune centres around how to describe **pitch accents**. A pitch accent is commonly manifested in the F0 contour as a (relatively) sudden excursion from the previous contour values. This is where association comes in, since pitch accents occur only in conjunction with prominent syllables, and in doing so attract attention to that syllable. Pitch accents can occur only in association with prominent syllables (see Section 6.3 on prominence), but need not occur on all prominent syllables.

Most agree that the prosodic phrase as described in Section 6.2 is the basic domain of intonational tune patterns. Hence the phrase not only groups words together but also serves as the start and end of each section of tune. In many models, we have a **nuclear accent**, which occurs once per phrase.

The other main area of interest in tune description concerns what happens at the ends of intonation phrases. Often F0 is low at a phrase boundary, but in many circumstances F0 is high. For instance, if another phrase directly follows the current one, a **continuation rise** may be present. If the tune is that of a yes/no question, the final pitch may also be high. The British school deals with these effects by using different nuclear accent and tail configurations. The AM model makes use of high and low **boundary tones**, which distinguish the different types of contour.

It would be impossible to show all the possible types of intonational tune for English, but six common tunes that vary because of their nuclear-accent types are shown in Figures 9.1–9.6. These examples are not comprehensive and other theories may classify these contours differently. The examples merely demonstrate some of the intonational effects that can be produced.

9.2.2 Downdrift

It has been observed by many researchers that there is often a gradual **downdrift** in the value of F0 across a phrase [107], [167], [285], [351], [445], [446]. How downdrift (often referred to as **declination**) is dealt with by different theories varies widely. Ladd [267] gives a review of some of the theories.

Many treat downdrift as an automatic physiological effect arising from changes in subglottal pressure during the course of an utterance [107], [287]. This account gives the speaker little conscious control over declination. The approach of the Dutch school [445] has been to use three parallel declination lines, which refer to a baseline, a

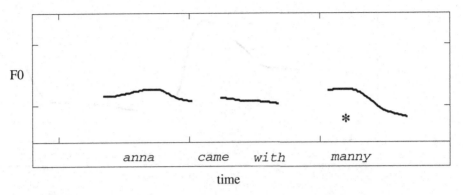

Figure 9.1 Typical fall accent. "Anna came with Manny." The nuclear fall is on the stressed syllable of "manny" denoted by an *. The fall is a commonly found intonation accent and is often used in neutral declarative situations.

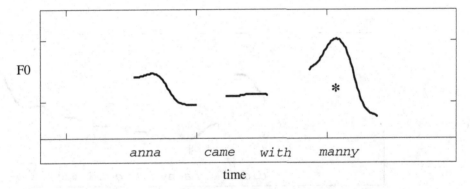

Figure 9.2 High fall, "Anna came with Manny!". This shape corresponds to a British "high fall", +raised or pitch level 4. In this particular utterance there is still a single intonation phrase, and the word "anna" also has an accent, but this accent is pre-nuclear. Some may argue that there is no phonological distinction between fall and high fall, and that the high fall is really just an extra-prominent fall.

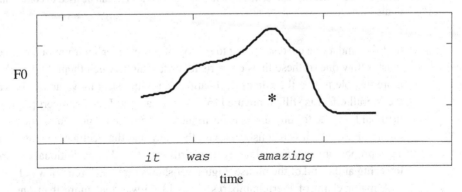

Figure 9.3 Rise-fall accent, "It was amazing!". Spoken with a sense of wonderment, this accent is similar to a fall, but with a much larger preceding rise. The peak value of the F0 contour is also later than with a simple fall accent.

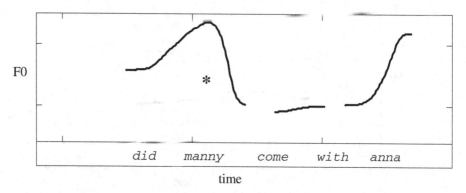

Figure 9.4 Fall–rise accent, "Did Manny come with Anna?" A peak in the F0 contour occurs in the stressed syllable of "manny" (*). After coming down from the peak, the contour rises slowly and finishes with a sharp rise at the end of the phrase. This type of accent is often used for simple yes/no questions.

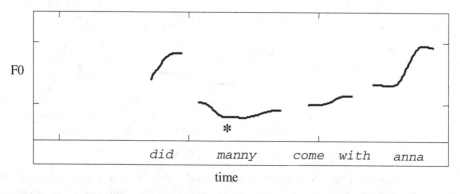

Figure 9.5 Low rise, "Did Manny come with Anna?!". This accent shape may at first glance look similar to the fall–rise, but differs in that the stressed syllable (*) of the word which carries the nuclear accent is not a peak but a valley. Thus the F0 contour rises from the nuclear accent. Quite often this accent is preceded by a falling F0. This accent can be used to convey incredulity or disbelief.

mid-line and a line corresponding to the top of the speaker's normal range. The contour must follow one of these lines or be rising or falling between them. Fujisaki's model is more flexible in that the rate of declination and initial starting value can be varied, but the overall effect is still automatic [167]. Liberman and Pierrehumbert [285] show that the final F0 value for utterances is invariant under a wide range of utterance lengths and pitch ranges, which is inconsistent with the view that the declination slope is constant. They propose an exponential decay downdrift effect, with the additional feature of **final lowering** at the end of the phrase. Figure 9.7 shows three different views of declination.

A major claim of Pierrehumbert's thesis [351] was that more than one factor was responsible for the downdrift of F0 contours. As with many other theories, she proposed that the phonetic declination effect exists, but also argued that the major contribution to the downdrift of utterances was **downstep**, which was a phonological effect and therefore

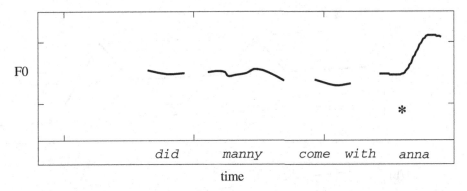

Figure 9.6 High rise, "Did Manny come with Anna?". Here the accent falls on the first syllable of Anna. There is no valley as with the low rise, and the F0 on the nuclear syllable is much higher. High rise accents are often used for yes/no questions where the speaker is looking for confirmation in a statement, as in "ok?" or " right?". It is similar in many ways to the low rise, with the F0 contour rising from the nuclear accent, the main difference being that the nuclear accent occurs considerably higher in the speaker's pitch range, and is often not preceded by a falling section of contour.

controllable by the speaker. Figure 9.8 shows a downstepping and non-downstepping version of the same sentence. These two sentences don't just differ in terms of F0 shape but also have subtly different meanings. (The first sounds more excited; the second sounds more relaxed and confident.)

9.2.3 Pitch range

In music, if a sequence of notes is repeated an octave higher than the original, the tune remains the same, even though the frequency values of the notes are different from the original. Rather it is the constancy of the *pattern* that gives the perception of the tunes being the same.

The same effect is observable in intonation: on increasing the overall pitch of an utterance while keeping the basic tune pattern constant, the perceived tune remains the same. The relationship between intonational tunes in different pitch ranges is not as simple as the musical equivalent. It has been shown that the increase or decrease in pitch range need not be constant throughout the phrase, insofar as utterances always tend towards a fairly constant final F0 value.

Pitch range varies for a number of reasons. In single isolated utterances it can be used for increasing the overall emphasis of the utterance. When one "raises one's voice" in anger one is using an increased pitch range. Pitch-range factors also have a role to play in longer utterances. If a speaker started at the same F0 level with every intonation phrase, the speech would sound very bland. Speakers use a variety of pitch ranges to be more expressive. The boundaries between pitch-range and prominence effects are not clearly defined. For example, in many systems it is difficult to say whether an unusually high accent is due to extra (local) prominence or a temporary shift in pitch range.

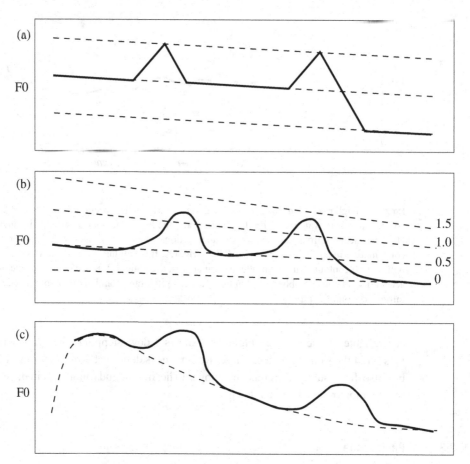

Figure 9.7 The Dutch model, (a), has three declination lines, which refer to a baseline, a mid-line and a line corresponding to the top of the speaker's normal range. The contour must follow one of these lines or be rising or falling between them. Pierrehumbert's system (b) scales the pitch range 0.0–1.0 for normal speech but allows higher levels. The contour is not required to follow any of these declination lines – they are merely the "graph paper" on which the F0 contour is produced. Note how the lines converge with respect to time. The Fujisaki model, (c) just specifies a baseline, which decays exponentially.

9.2.4 Pitch accents and boundary tones

Many theories of intonation define pitch accents as the fundamental unit of intonation. While we defer consideration of specific systems of pitch-accent descriptions to the section below on intonational models, here we attempt to describe some pitch-accent phenomena in theory-neutral terms. In most models that use pitch accents, we see a basic distinction between the pitch accent's inherent properties and its setting.

In general, by inherent properties, we mean the shape of the F0 contour of the pitch accent. To a degree, we can separate the F0 pattern of the pitch accent from the surrounding contour, but this is only for convenience; in reality the pitch accents rarely have distinct beginnings and ends.

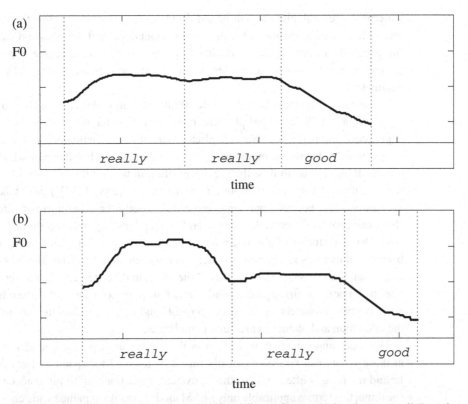

Figure 9.8 Two utterances of the phrase "really, really good". Part (a) has the two "really"s at the same pitch level, with a fall on "good". In part (b) each word is downstepped relative to the previous word.

The most common type of pitch accent takes the form of a peak in the F0 contour. The actual shape of this map be a rise followed by a fall, or a rise followed by a relatively level section and then a fall. The degree of the excursion is important, and generally speaking the higher the excursion the more prominent the pitch accent will sound. There are many other types of pitch accent also. One common type is the simple fall, where the contour is level, falls over a short region and then continues at a reasonably level rate. Another important type is the low-pitch accent, or rising-pitch accent. The local shape of this is a little harder to define, but basically this type of accent is marked by a negative excursion in the F0 contour. This can take the shape of a fall followed by a rise, or just a specific low point in the contour. This is often associated with question intonation, cases where the speaker is doubting the listener and so on. Beyond these, theories differ regarding the range of additional types. The important point about pitch-accent shapes is that too often these are thought of as simple peaks in the contour; falls and low-pitch accents are just as important.

The pitch accent's setting in the overall contour is also vital. This is governed by two factors: its **height**, meaning how high or low it occurs in the speaker's range, and its **alignment** with the phonetic content. In terms of height, we find a few characteristic patterns. In general, pitch-accent height decreases as a function of time throughout the utterance, as we just mentioned in the section about downdrift. We also find that, as with the size of the excursion, the accent height is correlated with the degree of prominence (in fact, in many models, accent height and excursion size are governed by the same parameters).

The issue of alignment is more subtle. While intonation does have a form of its own, its interaction with the verbal phonetic content of the utterance is important to how it is produced and perceived. We have already seen that association is a key link between the prosodic and verbal component since it determines which syllable or which word is accented. In addition to this, the precise position in time of the accent with respect to the accented syllable is very important. Some models [258], [259], [260], talk of **early** accents and **late** accents, and some argue that exactly the same intrinsic pitch-accent shape can give rise to quite different overall perceptions if merely the alignment differs. Quite detailed studies of alignment have been conducted [132], [266], [464] and it has been found that this is a complex issue; it is not even clear just what should be aligned with what. Some suggest that the start of the vowel in the accented syllable should act as one anchor point, with the peak or mid point of the pitch accent acting as the other. Many other alignment models are, however, possible and in general the interactions between the intonation and phonetics are quite complicated.

The other major intonational effect is the behaviour at phrase boundaries. The F0 at these points can be systematically raised or lowered for specific effect. The term **boundary tone** is often used as the equivalent intonation unit to pitch accent. Strictly speaking, this term is applicable only to AM models, but it has gained wide enough usage that we use it here to describe general behaviour at boundaries regardless of whether the particular model uses tones or not.

9.3 Intonation theories and models

9.3.1 Traditional models and the British school

The British School of intonation includes contributions made as far back as Palmer [344]. Other major contributions in this school have come from O'Connor and Arnold [333], Crystal [117], and Halliday [193]. All these variants on Palmer's original theme use dynamic features such as **rise** and **fall** to describe intonation.

In the account given by Crystal, the most important part of the contour is the nucleus, which is the only mandatory part of an intonation phrase. The nuclear accent can take one of several configurations, e.g. fall, fall–rise, low rise. Other parts of the contour are termed the **tail** (**T**), which follows the nucleus, the **head** (**H**), which starts at the first accented syllable of the intonation phrase and continues to the nucleus, and the **pre-head** (**P**), which precedes the head. The intonation phrase has a grammar of (**P**) (**H**) N (**T**), where the brackets denote optional elements.

The relationship between the form and acoustics of this school is the most loosely defined of all the models described here; this is hardly surprising, however, since none of the originators of this system had the technology to analyse F0 contours in detail. The description of form is related to actual contour shapes that are found, but the descriptions should not be interpreted too literally. Both "fall" and "rise–fall" accents have rises followed by falls, the difference being that the rise in the fall accent is much smaller and earlier in the syllable than the rise in the rise–fall accent. Halliday describes his tones using rise-and-fall terminology, but does away with the standard naming system, preferring simply to name his tones 1, 2, 3, 4 and 5.

Some more-formal descriptions have been proposed for use with the British-school phonology. In particular, two models designed for speech-synthesis purposes are those of Isard and Pearson [231], who use Crystal's phonology, and Vonwiller et al. [478], who use Halliday's. Both these synthesis models use the full range of the British-school tune descriptions, and Isard and Pearsons's scheme is capable of dealing with variations in prominence and pitch range.

9.3.2 The Dutch school

The Dutch School [445], [446] is based on the principle of **stylisation** of F0 contours. Stylisation in the Dutch system involves taking an F0 contour and attempting to fit a series of straight lines as closely as possible to the original contour. This stage is useful because it reduces the amount of data needed for further analysis: a small number of straight lines is easier to deal with than a continually varying F0 contour. From these stylisations, a series of basic patterns can be found – this process is called **standardisation**.

The version of the theory presented in 't Hart and Cohen [445] describes contours in terms of three declination lines – high, middle and low. Pitch accents are realised by rising and falling between these declination lines. An example of a stylised and standardized contour is shown in Figure 9.9 (from Willems [493]).

Because of the stylisation process, the continuously varying nature of the F0 contour is eliminated; and because of the standardization process, the contour description is further reduced to a small number of units (rises, falls etc.). The idea of F0 stylisation has proved popular in many models and techniques outside the Dutch school [121], [480].

9.3.3 Autosegmental–metrical and ToBI models

Ladd terms this group the **autosegmental–metrical (AM)** model of intonation. Its origins are in Liberman's [284] and Bruce's [286] work, but its first full exposition was in Pierrehumbert's seminal thesis [351]. Starting from that basic model, alternatives and amendments have been proposed, but the basics remained unchanged. Most significantly, the intonation part of the **ToBI** scheme comes directly from Pierrehumbert's work in this school. Before these early publications, it was rare to associate the word "phonology" with prosody, but these early works promoted the idea that intonation and prosody could be described and modelled with much the same technical apparatus as normal phonology. Hence this school is often referred to as **intonational phonology** [266], [269].

The AM model describes intonation as a series of high and low tones. By using a system of diacritics that distinguish tones located on accented syllables from those occurring at

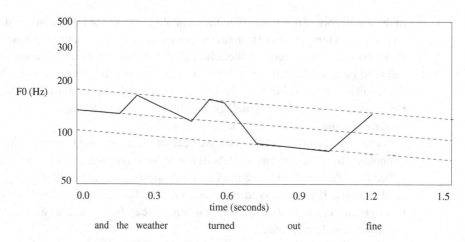

Figure 9.9 An example of a standardised contour in the Dutch system. The dashed lines denote the three declination lines and the thicker solid line shows the path of the F0 contour. The first excursion to the top declination line is a head accent (British school). The second accent, which rises to the top line and then falls to the baselines, is a fall accent. The rise at the end is a continuation rise.

boundaries and between accents, Pierrehumbert argued that intonation could be described in terms of patterns of two basic tones, which she called **H** (high) and **L** (low). Pitch accents can be represented as either a single or a double tone. Every pitch accent has a **starred** tone (*), which signals that it is that tone which is directly associated with the accented syllable. The possible pitch accents are **H*, L*, H* + L, H + L*, L + H*** and **L* + H**. At phrase boundaries, **boundary tones** can be found, which are marked with a **%**. **Phrase tones** are used to show the path of the contour from the last (nuclear) accent to the phrase boundary, marked with a -.

Unlike in the British-school analysis, there is no strict division of the contour into regions such as head and nucleus. Both nuclear and pre-nuclear accents can be any one of the six types described above. The nucleus accent is distinguished because the phrase and boundary tones that follow it allow a much larger inventory of intonational effects.

Each tone forms a target from which F0 contours can be realised by using interpolation rules. As with many other theories, the AM model retains the idea of a declination baseline, but says that the downdrift commonly observed in F0 contours is mainly due to the phonological effect of **downstep**, which again is controllable by the speaker. In her original work, Pierrehumbert proposed that the downstep effect is triggered by the speaker's use of a sequence of **H L H** tones, using evidence from African tone languages as justification (see Figure 9.8 for examples of downstepping and non-downstepping contours).

The history of the AM model is particularly interesting in that it has a dual character of being intended as a "pure" linguistic theory in the MIT, Chomsky/Halle phonology tradition and as a working model for the Bell Labs TTS system. It is important to realise that, from a theoretical linguistic point of view, the model as just described is not intended to be a "complete" model of intonation as used in human communication.

Rather, the AM school divides the general intonation system into a part that is properly linguistic, a part that is **paralinguistic** and a part that is **extra-linguistic**. The distinction is seen as important within this school because the properly linguistic part, i.e. that described above, is the main area of interest since it forms part of the general grammar of a speaker's language faculty. For a complete description, however, we require the paralinguistic effects too, which include for instance the degree of prominence of a pitch accent, and extra-linguistic effects such as global F0 settings. For our purposes, the distinctions between these subsystems are less important, but it is important to realise that the system of **H* + L** accents and so on is only a partial representation of an utterance's intonation.

9.3.4 The INTSINT model

The INTSINT model of Hirst *et al.* [211], [212], [213] was developed in an attempt to provide a comprehensive and multi-lingual transcription system for intonation. The model can be seen as "theory-neutral" in that it was designed to transcribe the intonation of utterances as a way of annotating databases and thereby providing the raw data upon which intonational theories could be developed. Hirst has described the development of INTSINT as an attempt to design an equivalent of the IPA for intonation. As stated in Section 6.10, there is no such thing as a completely theory-neutral model since all models make some assumptions. Nevertheless, INTSINT certainly fulfils its main goal of allowing a phonetic transcription of an utterance to be made without necessarily deciding which theory or model of intonation will subsequently be used.

INTSINT describes an utterance's intonation in terms of a sequence of labels, each of which represents a target point. These target points are defined either by reference to the speaker's pitch range, in which case they are marked Top (T), Mid (M) or Bottom (B), or by reference to the previous target point, in which case they are marked Higher (H), Same (S) or Lower (L). In addition, accents can be marked as Upstepped (S) or Downstepped (D).

Hirst [212] describes this system in detail and shows how it can be applied to all the major languages. Several algorithms have also been developed for extracting the labels automatically from the acoustics and for synthesizing F0 contours from the labels.[2] Applications of this model to synthesis include Veronis *et al.* [474].

9.3.5 The Fujisaki model and superimpositional models

Fujisaki's intonation model [166] takes a quite different approach from the models previously discussed in that it aims for an accurate description of the F0 contour that simulates the human production mechanism. Fujisaki's model was developed from the filter method first proposed by Öhman [335].

---

[2] In addition to being an expert on multi-lingual intonation, Hirst is also known for being able to do a particularly fine rendition of "Don't Cry For Me Argentina"

In the model, intonation is composed of two types of components, the **phrase** and the **accent**. The input to the model is in the form of impulses, used to produce phrase shapes, and step functions, which produce accent shapes.

This mechanism consists of two second-order critically damped FIR filters (these are introduced fully in Section 10.4). One filter is used for the phrase component, the other for the accent component. The F0 contour can be represented by Equations (9.1), (9.2) and (9.3),

$$\ln F_0(t) = \ln F_{\min} + \sum_{i=1}^{I} A_{p_i} G_{p_i}(t - T_{0_i}) + \sum_{j=1}^{J} A_{a_j}(G_{a_j}(t - T_{1_j}) - G_{a_j}(t - T_{2_j})) \quad (9.1)$$

where

$$G_{p_i}(t) = \begin{cases} \alpha_i^2 t e^{-\alpha_i t} & \text{for } t \geq 0 \\ 0 & \text{for } t < 0 \end{cases} \quad (9.2)$$

$$G_{a_j}(t) = \begin{cases} \min[1 - (1 + \beta_j t)e^{-\beta_j t}, \theta] & \text{for } t \geq 0 \\ 0 & \text{for } t < 0 \end{cases} \quad (9.3)$$

and the terms have the following meanings:

- $F_{\min}$ baseline
- I number of phrase components
- J number of accent components
- A_{p_i} magnitude of the ith phrase command
- A_{a_j} magnitude of the jth accent command
- T_{0_i} timing of the ith phrase command
- T_{1_j} onset of the jth accent command
- T_{2_j} end of the jth accent command
- α_i natural angular frequency of the phrase-control mechanism of the ith phrase command
- β_j natural angular frequency of the accent-control mechanism of the jth accent command
- θ a parameter to indicate the ceiling level of the accent component.

Although the mathematics may look a little complicated, the model is in fact very simple. Each phrase is initiated with an impulse, which, when passed through the filter, makes the F0 contour rise to a local maximum value and then slowly decay. Successive phrases are added to the tails of the previous ones, thus creating the type of pattern seen in Figure 9.10. The time constant, α, governs how quickly the phrase reaches its maximum value and how quickly it falls off after this.

Accents are initiated by using step functions. When these step functions are passed through the filter they produce the responses shown in Figure 9.11. The accent time constant, β, is usually much higher than α, which gives the filter a quicker response time. This means the shape produced from the accent component reaches its maximum value and falls off much more quickly than the phrase component. Phrases are initiated

9.3 Intonation theories and models

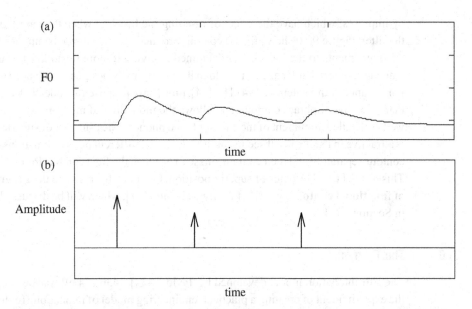

Figure 9.10 Three phrase components of differing amplitude. Graph (a) shows the path of the F0 contour, while graph (b) shows the input impulses.

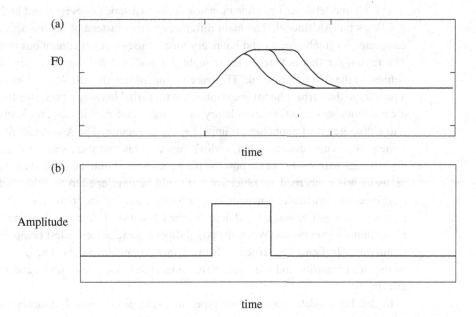

Figure 9.11 Three accents of differing durations. Graph (a) shows the path of the F0 contour; graph (b) shows the input step function for the second accent. As the accent duration increases, the accent becomes less like a peak and develops a flat top.

by impulses, which have the effect of resetting the baseline when they are fed through the filter. Figure 9.10 shows the F0 contour produced by the phrase component.

Amendments to the Fujisaki model aimed at giving it more flexibility for modelling languages other than Japanese, for which it was originally designed, have been proposed. For instance, Van Santen *et al.* [411], [464], [466] have proposed a model that allows more than one type of phrase contour, and allows the modelling of micro-prosodic effects as well as detailed alignment of the contour to the phones. The principle of superimposition is attractive in itself since it seems intuitive to model different parts or functions of the F0 contour separately and then combine these to produce the final contour for the utterance. This has led to other types of superpositional models, for instance the **superposition of functional contours** model of Bailly and Holm [31], which will be discussed in detail in Section 9.6.4.

9.3.6 The Tilt model

The **Tilt** intonation model developed by Taylor [434], [436], [439] was developed with the explicit intent of creating a practical, engineering model of intonation. To this extent, issues of purely linguistic concern (such as phonological rules in the AM model) or biological plausibility, as in the Fujisaki model, were ignored. While the Tilt model is no more plausible as a means of intonation production than say, concatenative synthesis, it *was* designed so that its parameters would have a clear linguistic interpretation.

The Tilt model describes intonation abstractly as a series of events, and in this sense it follows the AM model. The main difference is that instead of having a fixed set of categories for pitch accents and boundary tones it uses a set of continuous parameters. The reason for this is that Taylor thought the evidence for the particular categories defined in the AM model weak. The basic argument was that the AM model acted like a parallel to the verbal phonological model. With verbal language, phonetically we have a continuous space, either articulatory or acoustic, but cognitively this is divided up into a discrete set of phonological units, i.e. the phonemes. The AM model follows the same policy with intonation, but Taylor's concern was that there was no equivalent to the minimal-pair test to decide how the phonetic space should be divided up. In fact, as far as he was concerned, no objective test could be developed to find the pitch-accent categories of a particular language. The solution was to abandon the idea of a discrete number of categories and instead describe the intonational events with a small number of continuous parameters. While the possibility of some objective test being developed is still possible, being concerned with an *engineering* model purely, it made sense to err on the side of caution and use a model that at least described F0 contours accurately and effectively.

In the Tilt model, there are two types of event, **accent** and **boundary** (these are effectively the same as the pitch accents and boundary tones in the AM model). Each event comprises two parts, a **rise** and a **fall**, as shown in Figure 9.12. Between events straight lines called **connections** are used. All variation in accent and boundary shape is governed firstly by the relative sizes of the rise and fall components and secondly by how the event aligns with the verbal component (basically, whether it is higher or

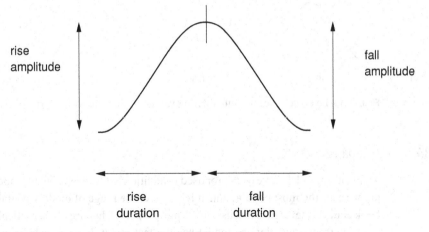

Figure 9.12 A pitch accent split into rise and fall components.

lower in the pitch range, or occurs earlier or later with respect to the accented syllable). This then gives a total of six parameters, four describing the shape of the event and two describing how it aligns. This basic model (called the rise/fall/connection model [439]) models contours accurately but is not particularly amenable to linguistic interpretation. The Tilt model itself is a transformation of the four parameters for each event (rise and fall amplitude, and rise and fall duration) into the three Tilt parameters. Amplitude and duration are given by

$$\text{Amplitude} = A = |A_{\text{rise}}| + |A_{\text{fall}}|$$
$$\text{Duration} = D = |D_{\text{rise}}| + |D_{\text{fall}}|$$

The Tilt parameter itself is used to define the general shape of the event, independently of its amplitude, duration and alignment. To find its value, we first calculate an interim value, the **Tilt amplitude**, which is found by taking the ratio of the difference of the rise and fall amplitudes and the sum of the rise and fall amplitudes:

$$\text{tilt}_{\text{amp}} = \frac{|A_{\text{rise}}| - |A_{\text{fall}}|}{|A_{\text{rise}}| + |A_{\text{fall}}|}$$

We do likewise for duration:

$$\text{tilt}_{\text{dur}} = \frac{|D_{\text{rise}}| - |D_{\text{fall}}|}{|D_{\text{rise}}| + |D_{\text{fall}}|}$$

These two quantities are highly correlated, which allows us to combine them into the single final Tilt parameter:

$$\text{tilt} = \frac{\text{tilt}_{\text{amp}} + \text{tilt}_{\text{dur}}}{2}$$

Figure 9.13 shows a number of event shapes for various values of the Tilt parameter.

Figure 9.13 Five pitch accents with differing values of tilt, from +1 (pure rise) to −1 (pure fall).

9.3.7 Comparison

Many other models have been proposed in addition to the above, but the models described are perhaps the most popular and at least cover the range of models available. An entire book could be devoted to discussing the philosophical, historical, theoretical and practical issues surrounding why one model has the features it does and why it might be better or worse than another model. For now, however, we will simply describe some of the differences among models in order to help the reader understand why these differences occur.

Purpose

Not all models were designed for the same purposes. The AM models are usually not described as models at all, but as theories of how intonation actually works in human communication. Issues such as cross-language studies and links with other components of human grammar and so on are particularly important in this group of models. The Fujisaki model is often said to be biologically justifiable, meaning that it mimics the actual articulation of the human production mechanisms. The Tilt model is described only in practical terms, as a tool to get the job done, and has no claim to model any aspect (cognitive or biological) of human behaviour. These different approaches are of course quite common in all areas of language study; what is of note is that intonation seems to be one of the few areas where engineering and scientific approaches still have enough commonality to constitute a single field.

Phonological versus phonetic versus acoustic

The AM model is phonological, the INTSINT model is phonetic and the Fujisaki and Tilt models are acoustic. While these differences are obviously important, one should be wary of attributing too much significance to them as being statements about what the model developers think is the "real" nature of intonation. For example, Hirst proposed INTSINT as an intonational equivalent to the IPA that would allow researchers to label what intonation information a contour contained independently of linguistic theory, in just the same way as a phonetician might make detailed transcriptions of verbal phenomena. Once this has been done, a phonological theory can be built using these transcriptions as data, so it is not correct to say that Hirst thinks intonation is "phonetic" in character. Likewise, in the AM model, there is the assumption that production includes phonetic and acoustic components, and in acoustic models the assumption again is that there are higher-level, more-abstract representations. In fact, there is nothing to stop one creating a single formulation wherein the AM model would serve as the phonological representation, the INTSINT model as a phonetic representation and the Tilt model as the acoustic representation.

Tones versus shapes
There is a basic distinction in most models as to whether intonation is an inherently tone or pitch level based system (as in AM models), or is based on pitch shapes or pitch dynamics (as in the Dutch model, Fujisaki model and Tilt model). In theoretical linguistics this debate is seen as particularly important since researchers are attempting to find the "true nature" of intonation, with "conclusive" evidence occasionally being produced for one view over the other. The issue is hard to resolve, and it may be the case that intonation is actually a mixture of both phenomena.

Superimpositional versus linear
Some models have a superimpositional nature (Dutch, Fujisaki) such that pitch accents are seen as being relatively short-term, "riding" on top of phrases, which are seen as being relatively long. By contrast, in linear models (AM/Tilt) contours are composed of linear sequences of intonational units. Often a grammar that states which units can follow each other is used, but, so long as the sequence lies within the grammar, any unit can follow any other.

In brief, the main reason for supporting the superimpositional model is that F0 contours do seem to exhibit patterns of global behaviour, whereby phrases define particular F0 patterns. The main reason for supporting linear models comes from speech-production concerns. Here, we see that if we support a superimpositional model then the speaker has to **pre-plan** the utterance some way ahead of time. This doesn't seem to correspond to known facts in speech production, where we know for example that a speaker can change the course of the F0 contour at any time.

Finally we should note that some papers have shown how labels in one model can be translated into labels in another [374], [436].

9.4 Intonation synthesis with AM models

We now turn our attention to the issue of intonation synthesis itself. In Chapter 6 we described a variety of techniques for generating the prosodic-form representations of phrasing and prominence from text. We made only passing reference to the issue of generating intonational form from text because this really required a thorough discussion of intonational form first. With this job now done, we can describe techniques both for generating intonational-form descriptions from text and for generating F0 controls from intonational form.

While it is possible to build synthesis algorithms with any of the models outlined in Section 9.3.3, we will focus our attention on the most-popular techniques, which divide into those based on AM models (described here) and those based on acoustic models (described next).

9.4.1 Prediction of AM labels from text

We approach the task of predicting intonation from text in much the same way as we predicted phrasing and prominence from text. With the AM model, our job is to assign an intonational label to each syllable, and, no matter what algorithm is used, this is formulated as a mapping from a set of features describing the syllable and its context to a label. A typical approach of this type is given by Lee [280], who trained a decision tree to generate ToBI labels using linguistic features such as position in phrase, part-of-speech tags and syntactic information. In general, though, any of the standard text-prediction machine-learning techniques introduced in Chapter 5 can be used on this task, and the approach is by and large the same as for prominence prediction and phrasing prediction. Many other approaches for the prediction of AM and other intonation representations from text have been proposed, including the use of HMMs [325], decision trees/CART [378], [512] and rules [365], [461].

9.4.2 Deterministic synthesis methods

Given a representation of intonational form predicted from text, we next attempt to synthesize an F0 contour from this. In this section we will examine two rule-based deterministic proposals for doing this, the first developed for use with the original Pierrehumbert model and the second more specifically for use with ToBI.

Anderson, Pierrehumbert and Liberman [16] proposed a synthesis scheme (hereafter called APL) that generated an F0 contour from an arbitrary AM tone description. Recall that a feature of the AM model is that it is linear in the sense that there is no global pattern to intonation; each intonational unit in the sequence operates largely independently. This is reflected in the synthesis algorithm, where an acoustic model made of target points is used. In the algorithm, each intonational event generates a set of target points, which define the basic shape of the event. Pitch range in the model is expressed by two lines, the **reference line**, which expresses the default or "at rest" value of F0, and the **baseline**, which is the minimum F0 value, the value that F0 falls to at the end of a phrase that is terminated by an **L - L%** boundary tone. Declination as a global trend is absent from the model; rather, downdrift is synthesized by imposing a trend whereby each successive pitch accent is in general placed lower in the pitch range. Between pitch accents, phrase accents and boundary tones, linear interpolation is used to "fill in" the F0 values. An important nuance in the algorithm is its treatment of phrase-final behaviour. Following conclusions drawn from extensive studies reported in previous work [285], Anderson *et al.* added to the algorithm a component that dropped the F0 contour to the baseline after the nuclear accent, regardless of whether the model itself would have predicted such a value. This greatly increases the naturalness of the contours produced in comparison with those obtained with other synthesis algorithms they investigated since they have a proper impression of finality.

A more-recent algorithm designed to work with the ToBI scheme was proposed by Jilka, Mohler and Dogil [241]. This in many ways extends the original work by Anderson *et al.* insofar as it too synthesises the ToBI events as isolated units, places them in context in the utterance, and then uses interpolation to cover the regions between. The algorithm improves on the original work in that it pays particular attention to where target points should be placed with respect to their syllables, and for instance models in some detail the fact that accents that occur early in phrases are often later with respect to their syllable position than accents that occur later in phrases (it's as if the phrase boundaries were repelling the accents in both directions). In this algorithm, a topline and baseline are used as the reference points. When downstepping occurs, it triggers a lowering of the topline, which is the general mechanism by which downdrift occurs. Example rules for the position of the target points are given below:

Intonational event	Context	Resultant time position	Resultant pitch range position
H*	First syllable in phrase	85%	Topline if first H in phrase, equal with most recent H otherwise
	Last syllable in phrase	25%	
	One-syllable phrase	50%	
	All other cases	60%	
L*	First syllable in phrase	85%	Baseline
	Last syllable in phrase	25%	
	One-syllable phrase	50%	
	All other cases	60%	
L + H* (H part)	First syllable in phrase	90%	Baseline
	Last syllable in phrase	25%	
	One-syllable phrase	70%	
	All other cases	750%	
L + H* (L part)	Normal case	0.2 s before H*	Baseline
	If voiceless region 0.2 s before H* (reference point)	90% of voiced region to left of reference point	

9.4.3 Data-driven synthesis methods

Black and Hunt [51] described a data-driven technique for synthesising F0 contours from ToBI labels. In fact, their work can be seen as a data-driven version of the APL algorithm, in which the target points are learned from the data rather than specified by hand. The approach was based on learning the mapping between ToBI labels, which had been marked on a database by hand, and F0 contours extracted automatically. The APL model defines target points at very specific locations, but these are difficult to determine bottom-up so the Black and Hunt approach instead uses three target points for each ToBI label. These are placed at the beginning, middle and end of the syllable associated with the label. This gives us a representation wherein the input is a ToBI label and a set of features describing the linguistic context of the label, and the output is three target points. In this form, the mapping is now amenable to use by a standard machine-learning algorithm.

Black and Hunt in fact used a linear-regression technique, with features such as lexical stress, numbers of syllables between the current syllable and the end of the phrase, and the identity of the previous labels. Once it has been learned, the system is capable of generating a basic set of target points for any input, which we then interpolate and smooth to produce the final F0 contour. Other data-driven techniques such as CART have proven suitable for synthesizing from AM representations [292], [340].

9.4.4 Analysis with autosegmental models

The biggest difficulty with the AM models is the difficulty in labelling corpora. Several studies have been conducted on labelling with ToBI, with the general conclusion that, while labellers can often identify which syllables bear pitch accents, they are very poor at agreeing on which particular accent is correct. As we explained in Section 6.10, the inter-labeller agreement with models such as ToBI is worryingly low, and in many practical database-labelling projects many of the pitch-accent distinctions have been dropped [486], [487].

Two major problems stem from this. Firstly, any database that has been labelled with ToBI will have a significant amount of noise associated with the pitch-accent-label classes. Secondly, for any large-scale machine-learning or data-driven approach, we need a considerable amount of labelled data, to the extent that it is impractical to label data by hand. As we shall see in Chapters 15 and 16, virtually all other aspects of a modern data-driven TTS system's data are labelled automatically, so it is a significant drawback if the intonation component cannot be labelled automatically as well. Because, however, the level of inter-labeller agreement is so low, it is very hard to train a system successfully on these labels; we can hardly expect an automatic algorithm to perform better than a human at such a task.

One solution that is increasingly adopted is to forgo the distinction between label types altogether; see for instance [487]. While the break-index and boundary-tone components are often kept, only a single type of pitch accent is used; in effect the labellers are marking whether a word is intonationally prominent or not. However, it should be clear that such an approach effectively reduces ToBI to a data-driven system of the type described below.

9.5 Intonation synthesis with deterministic acoustic models

Here we describe how to synthesise F0 contours using the Fujisaki and Tilt models. As we will see, the approach is similar in both cases. Synthesis from the Fujisaki or Tilt parameters is of course given by the models themselves, and so presents no problems. For TTS purposes, then, the issue is that of how to generate the Fujisaki or Tilt parameters from the text. Two basic approaches can be taken in this regard; either we try to generate the parameters directly from the text, or we take a two-stage approach whereby we first generate an abstract, discrete prosodic-form representation from the text and then proceed to generate the model parameters from that.

9.5.1 Synthesis with superimpositional models

The Fujisaki model is most commonly used with Japanese, but has been used or adapted to many other languages. In Japanese, we find that the range of pitch-accent phenomena is narrower than in languages such as English, which means that the model's single type of accent is particularly suited. In addition, the nature of intonation in Japanese means that accents are marked in the lexicon, which greatly simplifies the problem of prominence prediction. Hence a simple approach to this, which uses accent information from the lexicon alone, is often sufficient. A common approach therefore is to determine phrase breaks and prominent syllables from the text, and then phrase by phrase and syllable by syllable generate the input command parameters for the Fujisaki model using one of the standard machine-learning techniques.

Hirose et al. [205], [206] describe such an approach for Japanese. In doing so, they make use of the basic lexicon tone distinctions in Japanese to give where accents occur, thus simplifying the prominence- and accent-prediction process considerably. To generate the accent commands, they use parts of speech, syllable structure, relative positions of these, numbers of preceeding phrases and punctuation. Syntactic units are also optionally used. Neural-network algorithms are used, including recurrent networks. In addition, multiple-regression analysis in a vein similar to the ToBI Black and Hunt [51] technique was investigated. In most cases the results were very similar regardless of which algorithm was used (this of course follows the pattern for other prosody-prediction problems).

One significant advantage that Hirose makes note of is that, even when the prediction algorithm generates incorrect parameters for the Fujisaki model, the result is still a "natural" contour because the model itself is tightly constrained. In other words, the contour produced is *some* contour, just not the right one. By comparison, when prediction is poor with target models, the result may be a contour that bears no similarity to the type that a human could produce.

9.5.2 Synthesis with the Tilt model

Dusterhoff et al. [146] described a three-stage technique for synthesising intonation with the Tilt model. First, they predict where accents are placed; this is effectively a prominence-prediction algorithm of the type described in Section 6.8. Next, for each accent, the three intrinsic and two extrinsic Tilt parameters are generated. This is done with a decision tree, using features such as syllable position in phrase and length of phrase. Finally, the Tilt model itself generates the F0 contour. In general, though, there are many similarities between synthesis with Tilt and the Fujisaki model, so a technique that works with one can probably be adapted to the other.

9.5.3 Analysis with Fujisaki and Tilt models

An advantage that the Tilt model has over many others is that automatic analysis is significantly easier using this model. Taylor [436] describes a complete technique for first finding pitch accents and boundary tones and then determining the Tilt parameters from the local F0 contour in the surrounding region. Many algorithms have also been developed for automatically finding the Fujisaki model parameters from data [7], [71], [165], [314], [379].

9.6 Data-driven intonation models

A common criticism of the Fujisaki model [285], [436] is that there are many contours that the model cannot generate. These include gradual low rising contours, which occur when a low accent occurs early in a phrase, and the effect of phrase-final lowering, when the contour falls past the normal baseline to a final baseline in declarative utterances. Suggestions that fix the problems, for instance by including phrase-final negative phrase commands and so on, have been made, but such suggestions detract from the elegance of the original model. The Tilt model can be criticised as *over*generating in that there are contours it can generate that are never observed. When we consider simpler stylisation models we see that these models often massively overgenerate.

To solve these problems, we can of course refine the models, but this can result in endless "tinkering" in which the models lose their original elegance and simplicity. The basic problem is that, when we design a model, we wish it to be elegant and easy to interpret, but we always run the risk that some intonational phenomena that we haven't taken into account may appear and be outside the model's capabilities. We then face the choice of adding to the model, which often appears as a "hack", or designing a completely new model.

We can, however, consider a quite different approach whereby we completely abandon the idea of creating an explicit model, but instead infer the model from data. Bailly [29] has considered this issue and asked whether explicit intonation models are required at all; perhaps we should concentrate on databases and labelling, and leave the specifics of the model to be determined by modern machine-learning techniques. As we shall see in Chapter 13, it is convenient to describe synthesis techniques in terms of three generations, such that the first generation used explicit hand-written rules, the second generation uses basic and quite constrained data-driven approaches, and a third generation uses general data-driven and statistical approaches. The same pattern can be seen in intonation synthesis, where for example the APL algorithm is of the first generation and the Black and Hunt decision-tree ToBI algorithm is of the second generation. The techniques described here can be considered third-generation ones insofar as they match closely the approaches used in the HMM and unit-selection synthesis techniques described in Chapters 15 and 16.

In the following sections, we give an overview of the techniques used, but, since many of these follow the synthesis techniques described in Chapters 15 and 16, we wait until then to give a fully formal account.

9.6.1 Unit-selection-style approaches

The unit-selection synthesis technique described in Chapter 16 uses an entirely data-driven approach, whereby recorded speech waveforms are cut up, rearranged and concatenated to say new sentences. Given the success of this approach in normal synthesis, researchers have applied these algorithms to F0 synthesis [296], [310], [311].

F0 contours extracted from real utterances are by their very nature perfect. The basic idea therefore is to collect a database of naturally occurring F0 contours, and use these at synthesis time for new utterances. It is extremely unlikely that one of the complete contours in the database will be exactly what we want; for this to be the case we would have to have exactly the same text for the synthesis and database utterance. At a smaller scale, say at the phrase, word or syllable level, it is, however, possible to find exact or close matches and so the database is processed to create a set of **F0 units**. Each unit has a set of features describing it, and, at run time, these features are compared with features generated by the prosodic prediction. Sometimes exact matches are found, but often only close matches are found, in which case a distance, called the **target cost**, from the desired features to the ones for each unit in the database is calculated. The lower the distance, the closer a unit is to the desired features. Obviously we wish to avoid large discontinuities in F0 values when we move from one unit to the next, so a second distance, called the **join cost**, is used to give a measure of how similar two units are at their edges. For F0 contours, a simple metric such as absolute F0 difference is often sufficient.

The idea then is to find the best sequence of F0 units for our desired features, and this is calculated by summing the target and join costs for a complete sequence of units. Formally, this is given by

$$C(U, S) = \sum_{t=1}^{N} T(u_t, s_t) + \sum_{t=1}^{N-1} J(u_t, u_{t+1}) \qquad (9.4)$$

where U is a sequence of units from the database, S is the sequence of desired features, u_t is one unit, s_t is one set of desired features and $T(.)$ and $J(.)$ are the target and join costs, respectively. A search is then conducted to find the single sequence of units $\hat{U}$ which minimises this cost:

$$\hat{U} = \underset{u}{\operatorname{argmin}} \left\{ \sum_{t=1}^{N} T(u_t, s_t) + \sum_{t=1}^{N-1} J(u_t, u_{t+1}) \right\} \qquad (9.5)$$

The advantage of this approach is that we are *guaranteed* to synthesise a natural contour; the only question is whether it will be the "correct" natural contour for the linguistic context. This approach is appealing in that it is simple and should work for all languages and genres. Furthermore, as database sizes grow, the quality of this technique will continually improve as we will become more and more likely to find good matches in the database.

9.6.2 Dynamic-system models

The main drawback of the above unit-selection approach is that the algorithm suffers from a "hit or miss" problem with respect to finding matches in the database. In other words, we have a fixed number of units, and it is sometimes a matter of luck whether we find a good match or not. While the technique can produce excellent F0 contours, occasionally no good matches are found and the results can be poor. Unit-selection approaches can be seen as ones whereby certain F0 contours are memorised by the system. We can contrast this with statistical approaches, with which the aim is to learn the general nature of all F0 contours by learning model parameters (not memorising) from a limited amount of training data. The advantage is that the technique is more robust when we need to synthesize contours we have not previously come across. The downside is of course that in building our model we have made some assumptions, so the reproduction of F0 contours might not be as accurate for cases that we have in fact seen.

Ross and Ostendorf developed a model based on the **dynamic-system model**, a model well known in control engineering (also known as a **Kalman filter**). This model is described by a **state equation** and an **observation equation**, given respectively as follows:

$$x_{k+1} = F_j x_k + u_j + w_k \qquad (9.6)$$
$$F0_k = H_j x_k + b_j + v_k \qquad (9.7)$$

The idea here is that the state equation describes a trajectory through time. The above is a first-order IIR filter of the type we will describe in Section 10.4; for now all we need know is that by appropriate setting of the parameters F_j, u_j and w_k we can generate a wide variety of trajectory shapes. By inferring these parameters from data, we can train this equation to generate contours. The term u_j can be seen as an input, and can be thought of as the equivalent of the phrase or accent command in the Fujisaki model. w_k is a term for modelling noise and is what gives the model its statistical nature; for now we can think of this as an error term in the model, which allows us to learn the other parameters robustly and statistically.

Rather than use this equation as is, we also use the observation equation, which performs a mapping from each point in the state trajectory to an actual F0 value, again with trainable parameters determining the exact nature of this mapping. While we should be somewhat cautious about giving explicit interpretations to statistical models, we can think of the overall process as follows. The state equation generates a "true" intonation contour that properly expresses the actual linguistic manifestation of the prosodic form. These trajectories are smooth, clean and simple. We know, however, that real F0 values, as measured by pitch-detection algorithms (see Section 12.7) are noisy, in that the contours are rarely smooth or predictable at the lowest level. The observation equation models this part of the system. If we used the output of the state equation to model F0 values directly, our problem would be significantly harder because we would then have to model every perturbation in the contour and would miss the general and underlying trends in the contour.

During synthesis, the model is "driven" by means of the input parameters u_j and b_j and also optionally by varying F_j and H_j. This can be done in a number of ways, but Ross and Osterndorf use simple equations that link the usual linguistic features (position within the phrase, stress value of the syllable etc.) to change the values of the parameters every syllable. During synthesis, the noise terms are ignored. Training the model is somewhat complicated, but in essence we attempt to generate F0 contours on a training corpus and adjust the parameters until the best match between the F0 contours produced by the model and the natural ones is found. This approach is very similar to the expectation-maximisation (EM) technique used to train hidden Markov models. This is explained in full in Section 15.1.8.

9.6.3 Hidden Markov models

The dynamic-system model is a natural choice for statistical generation of F0 contours since it is well suited to the job of generating continuous trajectories. If it has any weaknesses, we can point to the facts that the state trajectories are limited to being those of a first-order filter, the noise terms have to be Gaussian and the training process can be quite intricate. An alternative is to use hidden Markov models (HMMs) since these are in general easier to train and allow more complexity with regard to noise/covariance terms.

HMMs are, however, similar to dynamic-system models in certain ways, in that they have a hidden state space, within which the model dynamics operate, and a projection from this to an observation space. We have used HMMs before for various tasks including part-of-speech (POS) tagging and phrase-break prediction. In those cases, the output distributions were discrete and gave the probability of seeing (say) a word given a POS tag. It is also possible to have HMMs with continuous output distributions that give the probability of seeing a continuous value given a state. This sort of HMM can therefore be used for F0 generation. In the dynamic-system model, the hidden state space is a continuous trajectory, and this seems very natural for F0 contours. HMMs, by contrast, have a discrete state space, and, when we move from one state to another during generation, the observation probabilities suddenly change. In this mode an HMM will generate extremely unnatural contours in which we have a series of constant F0 values, a jump to a different series of constant F0 values and so on. No real F0 contours have behaviour anything like this. However, if we adopt the formulation of Tokuda *et al.* [452], we can generate from the HMM states while also taking into account the learned dynamics of F0 behaviour. Inanoglu [230] describes a system in which F0 for various emotions can be accurately generated by HMMs (diagrams of this are shown in Figures 15.13 and 15.14 later). This method of HMM synthesis is the subject of Chapter 15, so we leave further explanation of HMM F0 synthesis until then. We finish, however, with the observation that extremely natural F0 contours can successfully be generated with HMMs by using this formulation.

9.6.4 Functional models

Given the considerable difficulties in constructing a model of prosodic form, never mind any difficulties in using such a model, we can ask whether an explicit model of prosodic form is really necessary. While experience has shown that *some* notion of prosodic form is necessary (see Section 9.1.2), it need not in fact be necessary to model this explicitly; rather, we can learn a model of prosodic form and keep it as a hidden part of a more-general model [500], [501]. The **superposition of functional contours** (SFC) [31], [216], [217], [321] is a model in which the mapping from the meaning and linguistic structure to F0 is attempted in a unified fashion. The SFC model is based on the idea that there are identifiable prosodic **functions**[3] and that each has a characteristic prosodic pattern. Since the scope of each function is variable and we can have several independent functions operating on a particular part of the utterance (for instance syntax and emotion), we need to combine these effects to form a single prosodic pattern.

In the SFC model, the syllable is used as the unit of prosodic control and each syllable is described by three F0 values and a duration value representing the prosodic lengthening to be applied. Each prosodic function is assigned a generator, whose purpose is to generate the prosodic output for the particular sequence of syllables given when that prosodic function is active. A training database therefore comprises a set of utterances, each marked with the multiple prosodic functions and the four-parameter descriptions for each syllable. Each generator is then trained by an iterative procedure, which learns globally what prosodic patterns each generator should produce. In the original paper Bailly and Holm used neural networks as the generators, but, in principle, any trainable model can be used for this, including the dynamic-system models and HMMs described above. The advantage of this approach is that realistic and natural prosody can be generated without the need for determining which syllables receive pitch accents and what type they should be. All this is modelled and optimised implicitly in the model.

9.7 Timing

Many aspects of intonation, prominence and phrasing are manifested as differences in the **timing** of speech. It is therefore the task of the timing module in a TTS system to assign a temporal structure to the utterance. The first question then to ask is how exactly should we specify these timing values? Most commonly this is formulated in terms of unit **durations**, although other models assign timing values to significant points in the utterance [35]. The question then is what type of units should we assign these durations to? It is well known from experimental phonetics that the temporal properties of speech

---

[3] Here the word "function" means "purpose" and is a combination of affective and augmentative prosody.

are complex. If one talks more quickly, it is not the case that every part of the utterance contracts by a constant factor. Emphasis generally lengthens sections of speech, but in very particular ways. At a micro-level, we know that the articulators all have specific dynamic properties and, when a phone is spoken at a faster rate, some articulators may undershoot their target more than others.

A completely comprehensive approach to timing would therefore attempt to model all these factors, but in general it is very hard to determine any reliable anchor points at a level lower than the phone. In practical TTS, the complex sub-phone timing patterns are often ignored because firstly it is difficult to predict the interactions and secondly it is subsequently difficult to know what to do with the sub-phone information once obtained. In practice, then, it is most common to predict syllable durations or phone durations. Syllables are attractive since they seem to be the natural units of prosody; we have seen elsewhere that they are the units of prominence and therefore pitch-accent placement. In addition, we usually think of the syllable as being the natural unit of rhythm. Phones are attractive since they are the smallest units we can predict and therefore allow the most detail in prediction.

9.7.1 Formulation of the timing problem

Unlike intonation, timing synthesis can in fact be formulated as a regression algorithm directly, since our goal is to predict a single continuous value from a set of features. Regardless of whether we are predicting a phone duration or syllable duration, the problem can be formulated as a mapping from a set of features to a single value, expressed in seconds:

$$X(f_1, f_2, \ldots, f_N) \rightarrow d$$

Apart from the fact that we are predicting a continuous value instead of a class label, this formulation is the same as we have been using throughout the book for other classification problems, so it is natural to apply these approaches directly to the duration problem. In addition to these, certain specialised approaches have also been developed for use with this problem.

9.7.2 The nature of timing

The most common view of timing is that it is governed by both phonetics and prosody. Regardless of other factors, the fundamentals of articulatory speech production give a basic temporal pattern to speech. The articulation required to produce some vowels, such as /ii/, means that they are inherently longer that others such as /o/. In addition to these largely "automatic" aspects of timing, we also have phonetic timing factors that are part of the allophonic system of the language. The classic example in English is with words such as MELT and MELD where we see that the vowel in MELD is considerably longer than that in MELT. This is a consistent phenomenon (BENT versus BEND, TIGHT versus TIDE), but is *not* a physiological phenomenon alone; it is quite possible to speak these word pairs with equal durations, and some accents of English, most noticeably Scottish, do in fact do this. In general, however, it seems that the identity and interactions of phones have a strong bearing on the duration of each.

In addition to this we have prosodic factors. It is well observed that the durations at the ends of phrases are longer than elsewhere. The extent of this **phrase-final lengthening** is not completely understood, but roughy speaking we expect the last syllable in a phrase to be longer than otherwise and that the amount of lengthening in each phone increases towards the end of that syllable. It is important to realise that more than timing is affected by the presence of a phrase end; voice quality changes here too and in general the speech is spoken with considerably less vocal effort.

The other main prosodic factor affecting timing is prominence, with prominent syllables being considerably longer than non-prominent ones. Although it is not a proper prosodic factor, lexical stress behaves in the same way so that lexically stressed syllables are longer than unstressed syllables. It is hard to determine whether intonation has much of an effect on duration since all pitch accents lie on prominent syllables and so it is difficult to separate their effects. One observed phenomenon occurs when a complex intonation form is used with a very short, say single-syllable, utterance. If we say the name SUE? with a doubting intonation pattern, which has a low accent followed by a rise, we find that the length of the word seems to stretch to accommodate the complex pitch pattern.

9.7.3 Klatt rules

One of the most widely used sets of deterministic rules was developed by Dennis Klatt [10], [253]. Although designed for use with the MITalk formant synthesiser, these rules can be used with any synthesis technique.

In this, the basic properties of each type of phone are given by two values, a **minimum duration** M and an **inherent duration**. The inherent duration is a sort of average, but is best thought of as the duration of the phone spoken in a neutral canonical context. The final duration is then given by

$$\text{Duration} = [(\text{inherent duration} - \text{minimum duration}) \times A] + \text{minimum duration} \tag{9.8}$$

The term A is calculated by the successive application of a set of rules, whereby each firing rule multiples A by a factor. The factors are calculated as follows.

> **Clause-final lengthening** If a segment is the vowel or in the coda of a clause-final syllable, then $A = 1.4A$.
> **Non-phrase-final shortening** If a segment is not in a phrase-final syllable $A = 0.6A$. If the segment is a phrase-final post-vocalic liquid or nasal, $A = 1.4A$.
> **Non-word-final shortening** If a segment is not in a word-final syllable, $A = 0.85A$.
> **Polysyllabic shortening** If a vowel is in a polysyllabic word, $A = 0.80A$.
> **Non-initial-consonant shortening** If a consonant is in non-word-initial position, $A = 0.85A$.

Unstressed shortening Unstressed segments are half again more compressible, such that minimum duration = minimum duration/2. If the segment is unstressed or reduced, set A as follows:
- vowel in word-medial syllable: $A = 0.5A$
- other vowels: $A = 0.7A$
- pre-vocalic liquid or glide: $A = 0.1A$
- all others: $A = 0.7A$

Prominence If the segment is in a prominent syllable, $A = 1.4A$.

Post-vocalic context of vowels A vowel is modified by the consonant that follows it in the following ways:
- no following consonants, word-final position: $A = 1.2A$
- before a voiced fricative: $A = 1.6A$
- before a voiced plosive: $A = 1.2A$
- before a nasal: $A = 0.85A$
- before a voiceless plosive: $A = 0.7A$
- before all others: $A = A$

These effects are less in non-phrase-final positions, in which case set $A = 0.7 + 0.3A$.

Shortening in clusters
- vowel followed by a vowel: $A = 1.2A$
- vowel preceded by a vowel: $A = 0.7A$
- consonant surrounded by consonants: $A = 0.5A$
- consonant followed by a consonant: $A = 0.7A$
- consonant preceded by a consonant: $A = 0.7A$

In addition, Klatt also states that pauses should be 200 ms long and that non-reduced vowels preceded by voiceless plosives should be lengthened by 25 ms.

9.7.4 The sums-of-products model

Van Santen [411], [463] proposed the **sums-of-products model**, which can be seen as a semi-trainable generalisation of the Klatt model described above. Van Santen makes a few key observations regarding the nature of duration prediction. Firstly he notes that many possible feature combinations will never be observed in the training data (this of course is the sparse-data problem we have come across before). Crucially, however, he notes that duration modelling can be seen as a well-behaved regression modelling problem, in which certain properties of the problem allow us to make assumptions that can help ease the problem of data sparsity. Specifically, he states that to a large degree the factors influencing duration are **monotonic** (or "directionally invariant" in his terms). In practice this means that, while we know that emphasis and phrasing both affect the length of a phone, and that, strictly speaking, we cannot consider these features as being independent, their general influence on the duration is well behaved, such that emphasis always lengthens a phone, as does phrasing; it is not the case that the effects of these features ever reverse. This means that it is possible to build a model in which the interactions of these features do not have to be comprehensive.

In general, models of this type will not work because they don't model interactions between the features. In van Santen's approach, however, the process of building the model is to use knowledge to design the features. Hence the features actually used in the model will often be transformations of the ones generally available in the utterance structure. This process of hand design helps produce features that can operate more independently than the original features. The parameters can be set in a number of ways; how this is done is independent of the model itself. A sampling approach is one possibility.

9.7.5 The Campbell model

Campbell [83] proposed a model that uses the syllable as the fundamental unit of duration. In this a syllable duration is predicted from a set of linguistic features, after which the individual phone durations within the syllable are then calculated. This approach has the attraction that it is more modular, in that we have one component modelling the prosodic part of the duration and another modelling the phonetic part.

The prosodic part of the Campbell model uses a neural network to calculate a syllable duration. This is an attractive approach since the neural network, unlike the Klatt or sums-of-products model, can model the interactions between features. A somewhat awkward aspect of the model, however, is the fact that the syllable duration itself is of course heavily influenced by its phonetic content. If left as is, the phonetic variance in model duration may swamp the prosodic variance that the neural network is attempting to predict. To allow for this, Campbell also includes some phonetic features in the model. Campbell maps from syllable to phone durations using a model based on his **elasticity hypothesis**, which states that each phone in a syllable expands or contracts according to a constant factor, normalised by the variance of the phone class. This operates as follows. A "mean" syllable containing the correct phones with their mean durations is created. This duration is then compared with the predicted duration, and the phones are either expanded or contracted until the two durations match. This expansion/contraction is performed with a constant variance, meaning that, if the vowel is expanded by 1.5 standard deviations of its variance, the constant before it will be expanded by 1.5 standard deviations of *its* variance.

The problem of the absolute syllable duration can easily be solved by having the neural network predict this z-score instead of an absolute duration, an approach followed by [455]. This then frees the neural network from phonetic factors completely and allows it to use only prosodic features as input.

In general this model is quite effective since it solves the feature-explosion problem by positing a modular approach. One significant weakness, however, is that the elasticity hypothesis is demonstrably false. If it were true then we would expect the z-scores for all the phones in a syllable to be the same, but this is hardly ever the case; the z-scores for phones across a syllable vary widely, depending on context, position and other features. This is a problem with only the second component in the model and a more-sophisticated model of syllable/phone-duration interaction could solve this. In fact, there is no reason why a second neural network could not be used for this problem.

9.7.6 Other regression techniques

As we might expect by this stage in the book, most of the usual classification/regression algorithms have been applied to the duration-prediction problem. These include decision trees [372], neural networks for phone prediction [109], [157], genetic algorithms [319] and Bayesian belief networks [182]. Comparative studies of decision trees and neural networks found little difference in accuracy between these two approaches [72], [187], [472].

9.8 Discussion

In this chapter, we have considered techniques for synthesizing the acoustics of prosody. Any discussion on this topic will draw more or less the same conclusions as given in the discussion in Chapter 6, where we explained that there was an enormous range of models, theories and explanations for prosodic phenomena. The consequence of this is that developing engineering solutions to practical prosodic problems is considerably more difficult than with the verbal component of language. Hence the problem is complicated because there is no commonly agreed way of representing the linguistic form of the phenomena we are attempting to synthesize.

While many of the AM models and deterministic acoustic models provide useful and adequate representations for intonation, the trend is clearly towards the data-driven techniques described in Section 9.6. These have several advantages; besides bypassing the thorny theoretical issues regarding the "true nature" of intonation, they have the ability to analyse databases automatically, and in doing are also inherently robust against any noise that can occur in the data, whether it be from errors in finding F0 values or from other sources.

The purpose of all the prosody algorithms described in this chapter is to provide part of the specification which will act as input to the synthesizer proper. In the past, the provision of F0 and timing information was uncontested as a vital part of the synthesis specification and most of today's systems still use them. As we shall see in Chapters 15 and 16, however, some third-generation systems do not require any acoustic prosody specification at all, making used of higher-level prosodic representations instead. Rather than use F0 directly, stress, phrasing and discourse information are used. While such an approach completely bypasses all the problems described in this chapter, it does have the consequence of increasing the dimensionality of the feature space used in unit selection or HMM synthesis. It is therefore a practical question of tradeoff whether such an approach is better than the traditional approach.

Most work on the acoustics of prosody has focused on the modelling of F0 and timing. This has mainly been because these are the parts of prosody which are most readily identifiable and separable from the influence of verbal factors. In addition, these are the aspects of speech which signal-processing algorithms find easiest to modify. It is quite clear, however, that prosodic factors have a strong influence on a range of other acoustic aspects of the signal. For instance at the ends of phrases the vocal effort is considerably reduced and the overall voice quality is quite different from elsewhere. To date, only a few studies have been concerned with these effects [183], [417], [424]. If the prosodic influence on voice quality is not modelled within a TTS system the naturalness will definitely suffer, but it is unclear whether these affects should be modelled explicitly, or implicitly within a unit-selection or HMM synthesizer.

9.8.1 Further reading

Articles on all the models and techniques described here are readily available, but there are few that consider all models together and offer comparisons. The best account of the AM model is given by Ladd [269]. Ladd not only describes the model and its development, but also serves as a solid introduction to the whole field of intonational phonology and the practical phonetics associated with it. Accounts of the INTSINT model are given in Hirst [212], the Tilt model in Taylor [436], the Fujisaki model in Fujisaki [166], the SFC model in Bailly and Holm [31], the Dutch model in [446] and the British school in O'Connor and Arnold [333].

9.8.2 Summary

Prosodic form and acoustics
- The history of intonation research is full of attempts at finding a system of prosodic form that could act as the equivalent of phonemes in the verbal component of language.
- No consensus has been reached on this issue, which has led to the large number of quite different models that have been developed.
- F0/pitch and timing are the two main acoustic representations of prosody, but this is largely because these are quite easy to identify and measure; other aspects of the signal, most notably voice quality, are heavily influenced also.

Intonation models
- The British school describes intonation contours as comprising a pre-head, head, nucleus and tail, each of which has an inventory of shapes.
- The AM school (which includes ToBI) describes intonation in terms of abstract high and low tones. Diacritics (*, %, -) are used to specify which tones align with syllables and boundaries. The tones can be combined in various ways to form an inventory of pitch accents (e.g. H* + L).
- The INTSINT model provides a way of labelling intonation across theories and languages, and of labelling data for further analysis in much the same way as the IPA is used in phonetics.
- The Fujisaki model describes intonation as a sequence of commands with continuous parameters, which when fed into the appropriate filters generates an F0 contour. The model has an accent component and phrase component, which are superimposed to give the final contour

- The Tilt model describes intonation with a sequence of events, each of which is described by a set of continuous parameters.
- Many more intonation models exist, some of which are modifications of the above.

Intonation synthesis
- The acoustic models, such as the Tilt and Fujisaki models, have synthesis algorithms "built-in" and hence synthesis is fairly trivial with these.
- The phonological models usually have independently developed synthesis algorithms.
- In recent years, the notion of explicit models has been challenged by a number of data-driven techniques, which learn intonation effects from data.
- Some models, such as SFC, do away even with notions of explicit prosodic form and learn all the processes and representations of intonation from data.

Timing
- The purpose of a timing algorithm is to assign a time, usually in the form of a duration, to each linguistic unit.
- Nearly all algorithms take the form of a regression algorithm, wherein certain linguistic features are used as input to an algorithm that outputs a time.

10 Signals and filters

This chapter introduces the fundamentals of the field of **signal processing**, which studies how signals can be synthesised, analysed and modified. Here, and for the remainder of the book, we use the term signal in a more specific sense than before, in that we take it to mean a **waveform** that represents a pattern of variation against time. This material describes signals in general, but serves as a precursor to the following chapters which describe the nature of speech signals and how they can be generated, manipulated and modified. This chapter uses the framework of *digital signal processing*, a widely adopted set of techniques used by engineers to analyse many types of signals.

10.1 Analogue signals

A signal is a pattern of variation that encodes information. Signals that encode the variation of information over time can be represented by a **time waveform**, which is often just called a **waveform**. Figure 10.1 shows an example speech waveform. The horizontal axis represents time and the vertical axis represents amplitude, hence the figure shows how the amplitude of the signal varies with time. The amplitude in a speech signal can represent diverse physical quantities: for example, the variation in air pressure in front of the mouth, the displacement of the diaphragm of a microphone used to record the speech or the voltage in the wire used to transmit the speech. Because the signal is a continuous function of amplitude over time, it is called an **analogue signal**. By convention, analogue signals are represented as a continuous function x over time t. This means that, if we give a value for time, the function $x(t)$ will give us the amplitude at that point.

Signals can be classified as **periodic**, that is signals that repeat themselves over time, or **aperiodic**, that is signals that do not. Strictly speaking, no speech sound is ever exactly periodic because there is always some variation over time. Such signals are termed **quasi-periodic**. In Figure 10.1 it should be clear that the vowel sound is quasi-periodic, whereas the fricative is aperiodic. In the subsequent sections we will build a framework for describing signals, by first considering periodic signals and then moving on to considering any general signal.

10.1 Analogue signals

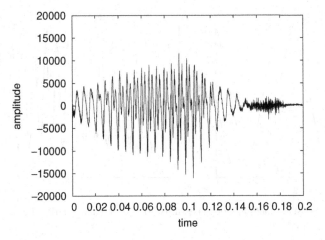

Figure 10.1 An example speech wave of /... ii th .../.

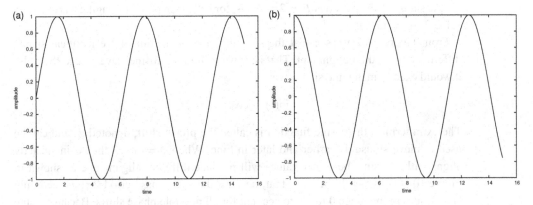

Figure 10.2 (a) Sine and (b) cosine waves.

10.1.1 Simple periodic signals: sinusoids

The **sinusoid** signal forms the basis of many aspects of signal processing. A sinusoid can be represented by either a **sine** function or a **cosine** function:

$$x(t) = \sin(t)$$
$$x'(t) = \cos(t)$$

It might not be immediately clear why the sine and cosine functions, which we probably first encountered in trigonometry, have anything to do with waveforms or speech. In fact it turns out that the sinusoid function has important interpretations beyond trigonometry and is found in many contexts in the physical world where oscillation and periodicity are involved. For example, both the movement of a simple pendulum and a bouncing spring are described by sinusoid functions.

Plots of these functions against time are shown in Figure 10.2, and from these it should be clear that both are periodic, that is that they exactly repeat over time. The **period** T

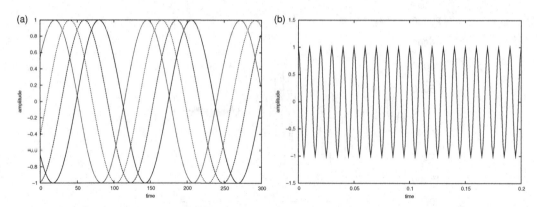

Figure 10.3 (a) The cosine function with phase shifts $-1, -2, -3$ and -4. (b) The cosine function with frequency 100 Hz.

is the length of time between any two equivalent points on successive repetitions. In the simple sinusoid, the period is $T = 2\pi$; that is, for any given point, the signal has the same value 2π seconds later.

From Figure 10.2 we see that the sine and cosine functions have identical shape, differing only in their alignment. That is, if we shifted the cosine wave $\pi/2$ to the right, it would exactly match the sine wave:

$$\sin(t) = \cos(t - \pi/2)$$

The extra term in the cosine function is called the **phase shift**, denoted ϕ, and can be used to move sinusoids earlier and later in time. When $\phi = -\pi/2$ the cosine exactly aligns with the sine, but other values will produce different alignments, as shown in Figure 10.3(a). It should be clear that, because the signal has period 2π, ϕ need only have values in the range 0 to 2π to account for all possible phase shifts. Because a sine can be represented by a cosine with an appropriate phase shift and vice versa, we can use either sines or cosines exclusively. By convention, cosines are generally used, giving the general formula

$$x(t) = \cos(t + \phi)$$

We define **frequency**, F, as the number of times the signal repeats in unit time, and this is clearly the reciprocal of the period:

$$F = \frac{1}{T} \tag{10.1}$$

The frequency is measured in cycles per second, **hertz**, so the frequency of our sinusoid is $1/(2\pi)$ Hz. A sinusoid of a different frequency can be generated by multiplying the variable t inside the sine function. For a frequency of 1 Hz, we multiply t by 2π, and so, for any frequency F expressed in Hz, we multiply t by $2\pi F$. Figure 10.3(b) shows a sinusoid with frequency 100 Hz,

$$x(t) = \cos(2\pi F t + \phi)$$

To save us writing 2π everywhere, a quantity called **angular frequency** is normally used, which is denoted by ω and has units **radians per second**:

$$\omega = 2\pi F = \frac{2\pi}{T} \tag{10.2}$$

The parameters ω and ϕ can thus be used to create a sinusoid with any frequency or phase. A final parameter, A is used to scale the sinusoid and is called the **amplitude**. This gives the general sinusoid function of

$$x(t) = A\cos(\omega t + \phi) \tag{10.3}$$

10.1.2 General periodic signals

Many non-sinusoidal waveforms are periodic: the test for a periodic waveform is that there exists some period T for which the following is true:

$$x(t) = x(t+T) = x(t+2T) = x(t+3T)\ldots \tag{10.4}$$

That is, at multiples of T the value is always the same. The lowest value for which this is true is called the **fundamental period**, denoted T_0. The frequency $F_0 = 1/T_0$ is called the **fundamental frequency**, and its angular equivalent is $\omega_0 = 1/(2\pi T_0)$. A **harmonic frequency** is any integer multiple of the fundamental frequency, $2F_0, 3F_0, \ldots$

It can be shown that *any* periodic signal can be created by summing sinusoids that have frequencies that are harmonics of the fundamental. If we have a series of harmonically related sinusoids numbered $0, 1, 2, \ldots, N$, then the complete signal is given by

$$x(t) = a_0\cos(0 \times \omega_0 t + \phi_0) + a_1\cos(1 \times \omega_0 t + \phi_1) + a_2\cos(2 \times \omega_0 t + \phi_2) + \cdots \tag{10.5}$$

where appropriate setting of the amplitudes a_k and phase shifts ϕ_k is used to generate the given waveform.

Since $0 \times \omega_0 t = 0$ and $a_1\cos(\phi_0)$ is a constant, we can combine $\cos(\phi_0)$ and a_0 into a new constant A_0, leaving the first term as simply this new A_0. This can be thought of as an "amplitude offset" (or "DC offset" in electrical terminology), so that the waveform does not have to be centred on the y-axis. In the most general case, the number of terms can be infinite, so the general form of Equation (10.5) is

$$x(t) = A_0 + \sum_{k=1}^{\infty} a_k\cos(k\omega_0 t + \phi_k) \tag{10.6}$$

Equation (10.6) is a specific form of the **Fourier series**, one of the most important concepts in signal processing. The process as just described is known as **Fourier synthesis**, that is the process of synthesising signals by using appropriate values and applying the Fourier series. It should be noted that, while the sinusoids must have frequencies that are harmonically related, the amplitudes and phases of each sinusoid can take on any value. In other words, the form of the generated waveform is determined solely by the amplitudes and phases.

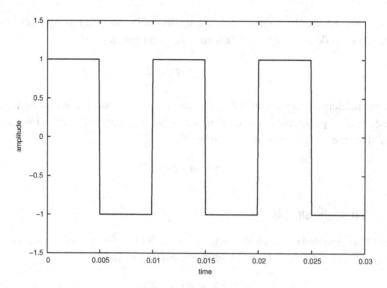

Figure 10.4 A square wave of fundamental frequency 100 Hz.

To demonstrate some of the properties of Fourier synthesis, consider the task of synthesising a square wave, as shown in Figure 10.4, and defined by

$$x(t) = \begin{cases} 1 & \text{for } 0 \le t \le \tfrac{1}{2}T_0 \\ -1 & \text{for } \tfrac{1}{2}T_0 \le t \le T_0 \end{cases} \qquad (10.7)$$

This signal can be generated by adding only the odd harmonics of the fundamental, using the same phase for every harmonic, and choosing appropriate amplitudes:

$$a_k = \begin{cases} 4/(k\pi) & k = 1, 3, 5, \ldots \\ 0 & k = 0, 2, 4, 6, \ldots \end{cases} \qquad (10.8)$$

Figure 10.5 shows the process of creating a square waveform by adding more and more harmonics to the fundamental sinusoid.

We have only been able to generate the square waveform because we knew the appropriate values for a_k and ϕ_k for every harmonic. The converse problem, that of finding unknown a_k and ϕ_k for a known waveform function $x(t)$, is termed **Fourier analysis** and will be described in Section 10.1.4.

10.1.3 Sinusoids as complex exponentials

While the general form of the sinusoid, $x(t) = A\cos(\omega_0 t + \phi)$, faithfully represents a sinusoid, it turns out that, for types of operations we consider below, use of a different form of the sinusoid, known as the **complex exponential**, greatly simplifies the mathematics and calculations. The basis of this representation is **Euler's formula**, which states

$$\boxed{e^{j\theta} = \cos\theta + j\sin\theta} \qquad (10.9)$$

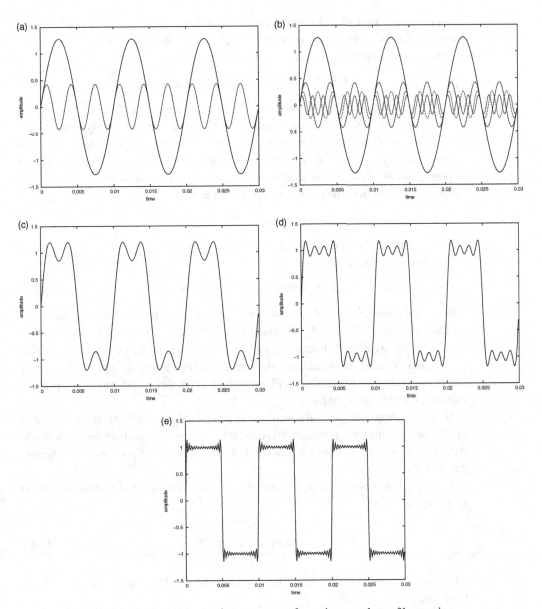

Figure 10.5 Fourier synthesis of a square wave for various numbers of harmonics: (a) superimposed sine waves of the first and third harmonics, (b) superimposed sine waves of the first to seventh harmonics, (c) synthesised wave of the first and third harmonics, (d) synthesised waves of the harmonics 1 to 7 and (e) synthesised waveform of harmonics 1 to 29.

where θ is any number and $j = \sqrt{-1}$. The **inverse Euler formulas** are

$$\cos\theta = \frac{e^{j\theta} + e^{-j\theta}}{2} \qquad (10.10)$$

$$\sin\theta = \frac{e^{j\theta} - e^{-j\theta}}{2j} \qquad (10.11)$$

From here on, we will make frequent use of **complex numbers**, which comprise a real part, x, and an imaginary part, jy, where $j = \sqrt{-1}$, such that $z = x + jy$.

If we add an amplitude and set $\theta = \omega t + \phi$, we get

$$A e^{j\omega t + \phi} = A\cos(\omega t + \phi) + jA\sin(\omega t + \phi) \qquad (10.12)$$

where A, ω and ϕ are all real and have the same meanings as before. At first glance, this seems insane – why on Earth should we replace our previous, easily interpretable representation of a sinusoid, $A\cos(\omega t + \phi)$, with a different equation, that includes an extra, imaginary term, $jA\sin(\omega t + \phi)$, and then rewrite it as an exponential? The reason is purely that it makes manipulation much easier. As a demonstration, consider the following, which allows us to represent phase differently:

$$x(t) = A e^{j\omega t + \phi} = A e^{j\phi} e^{j\omega t} \qquad (10.13)$$

$e^{j\phi}$ is a constant, not dependent on t, and we can combine this with the amplitude A to form a new constant, X, giving

$$x(t) = X e^{j\omega t} = A e^{j\omega t + \phi} = A e^{j\omega t} e^{j\phi} \qquad (10.14)$$

Here, the new quantity, X, is a *complex* amplitude and describes both the original amplitude and the phase. The purely sinusoidal part, $e^{j\omega t}$, is now free of phase information. This is highly significant, because it turns out that, in general, we want to describe, modify and calculate amplitude and phase together. (From here on, amplitude terms in complex exponential expressions will generally be complex – if the pure amplitude is required, it will be denoted $|A|$.)

While the complex exponential form has an imaginary part, $j\sin(\omega t)$, to the waveform, in physical systems such as speech, this does not exist and the signal is fully described by the real part. When requiring a physical interpretation, we simply ignore the imaginary part of $e^{j\omega t}$.

Using the complex exponential form, we can write a more general form of the Fourier synthesis equation:

$$\boxed{x(t) = \sum_{-\infty}^{\infty} a_k e^{jk\omega_0 t}} \qquad (10.15)$$

where a_k is a complex amplitude representing the amplitude and phase of the kth harmonic:

$$a_k = A_k e^{j\phi_k} \qquad (10.16)$$

Equation (10.15) is the general form of Fourier synthesis which sums harmonics from $-\infty$ to $+\infty$. It can be shown that, for all real-valued signals (that is, ones like speech with no imaginary part), the complex amplitudes are *conjugate symmetric* such that $a_{-k} = a_k$. In this case, the negative harmonics don't add any new information and the signal can be faithfully constructed by summing from 0 to ∞:

$$x(t) = \sum_{0}^{\infty} a_k e^{k\omega_0 t} \qquad (10.17)$$

In fact, the summation from $-\infty$ to $+\infty$ keeps the mathematics easier, so we often use the complete form.

10.1.4 Fourier analysis

What if we know the equation for the periodic signal $x(t)$ and wish to find the complex amplitudes a_k for each harmonic? The **Fourier-analysis** equation gives this:

$$a_k = \frac{1}{T_0} \int_0^{T_0} x(t) e^{-jk\omega_0 t} \, dt \quad (10.18)$$

It should be clear that integrating a sinusoid over a single period will give 0 for all sinusoids[1]

$$\int_0^{T_0} \sin(\omega_0 t) \, dt = 0$$

This holds for any harmonic also – while we have more positive and negative areas, we have equal numbers of them, so the sum will always be zero. In exponential form, we can therefore state

$$\int_0^{T_0} e^{k\omega_0 t} \, dt = 0$$

Now consider the calculation of the **inner product** of two harmonically related sinusoids, such that one is the complex conjugate of the other

$$\int_0^{T_0} e^{jk\omega_0 t} e^{-jl\omega_0 t} \, dt$$

This can be written

$$\int_0^{T_0} e^{\omega_0(k-l)t} \, dt$$

When $k \neq l$, this quantity $k - l$ is another integer, which will represent one of the harmonics. But we know from Equation (10.1.4) that this will always evaluate to 0. However, when $k = l$, then $k - l = 0$ so the integral is

$$\int_0^{T_0} e^{j\omega_0 0 t} \, dt = \int_0^{T_0} 1 \, dt = T_0$$

Using these results, we can state the **orthogonality property of sinusoids**:

$$\int_0^{T_0} e^{jk\omega_0 t} e^{-jl\omega_0 t} \, dt = \begin{cases} 0 & \text{if } k \neq l \\ T_0 & \text{if } k = l \end{cases} \quad (10.19)$$

Imagine that we have a signal $x(t) = e^{jl\omega_0 t}$ of known fundamental period but unknown harmonic (that is, we know ω_0 but not l). We can calculate the inner product of this

---

[1] This can be seen by using the interpretation that integration over an interval is finding the area between the curve and the x-axis. From the sine wave in Figure 10.2 we see that the area between 0 and π will cancel out the area between π and 2π.

signal with each harmonic $k = 1, 2, 3 \ldots$ of the fundamental. All the answers will be zero except the one for when the harmonics are the same ($k = l$). Since a general periodic signal is just the sum of a set of harmonics, we can use this method to find the value of every harmonic in a signal. Mathematically, we show this by starting with the Fourier series:

$$x(t) = \sum_{-\infty}^{\infty} a_k e^{jk\omega_0 t}$$

If we multiply both sides by the complex-conjugate exponential and integrate over period T_0 we get

$$\int_0^{T_0} x(t) e^{-jl\omega_0 t} dt = \int_0^{T_0} \left(\sum_{-\infty}^{\infty} a_k e^{jk\omega_0 t} \right) e^{-jl\omega_0 t} dt$$

$$= \sum_{-\infty}^{\infty} a_k \left(\int_0^{T_0} e^{jk\omega_0 t} e^{-jl\omega_0 t} dt \right)$$

$$= \sum_{-\infty}^{\infty} a_k \left(\int_0^{T_0} e^{-j(k-l)\omega_0 t} dt \right)$$

$$= a_l T_0 \qquad (10.20)$$

and if we replace l with k we get the Fourier-analysis definition.

Fourier analysis can easily be demonstrated graphically. Figure 10.6 shows the multiplication of a square wave by sinusoids of various harmonics. It can be seen from the graphs that the first and third harmonics have a non-zero area, so their Fourier coefficients will be non-zero. The multiplication of the square wave and second harmonic produces a curve that is symmetrical about the x-axis, so the area beneath the x-axis will exactly match the area above, resulting in a total area of 0.

One interesting point about the Fourier synthesis and analysis equations is that, while they are of similar form, one is a sum whereas the other is an integral. This is solely because periodic signals are composed of an integer number (possibly infinite) of discrete harmonics, necessarily enforcing a discrete frequency description. The waveform, however, is a continuous function. As we shall see, there are several variations on these equations in which frequency can be continuous or time discrete.

10.1.5 The frequency domain

The previous sections have shown how arbitrary periodic signals can be composed of harmonics with appropriate amplitudes and phases. It is often very useful to study the amplitude and phases directly as a function of frequency. Such representations are referred to as **spectra** and are described as being in the **frequency domain**. By contrast, waveforms as a function of time are said to be in the **time domain**.

Because the harmonic coefficients are complex, we can plot the spectrum in either **Cartesian** or **polar** form. The Cartesian form has a **real spectrum** and **imaginary spectrum** as shown in Figures 10.7(a) and (b). Alternatively, we can plot the

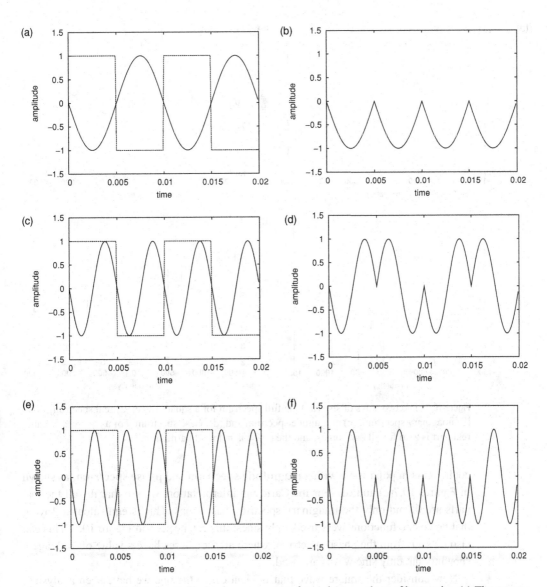

Figure 10.6 Fourier synthesis of a square wave for various numbers of harmonics. (a) The square wave and sine waves of the first harmonic. (b) Multiplication of square wave and sine wave of the first harmonic. (c) The square wave and sine wave of the second harmonic.
(d) Multiplication of square wave and sine wave of the second harmonic. (e) The square wave and sine wave of the third harmonic. (f) Multiplication of square wave and sine wave of the third harmonic. We can see that the curve in part (d) has symmetry about the x-axis and so there are equal amounts of area above and below the x-axis. The total area is therefore 0; hence the contribution of the second harmonic to a square wave is 0. Parts (b) and (f) have more area beneath the x-axis than above; hence the total area will sum to a non-zero value.

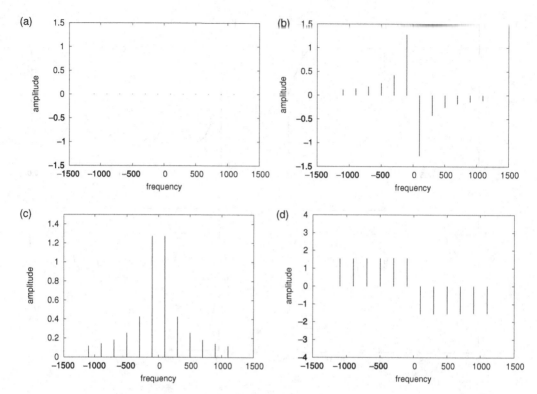

Figure 10.7 The two ways of showing the full spectrum for a square wave: (a) real spectrum, (b) imaginary spectrum, (c) magnitude spectrum and (d) phase spectrum. For a square wave, the real part is zero for all harmonics, and the phase is π for all harmonics.

coefficients in polar form with a **magnitude spectrum** and **phase spectrum**, as shown in Figures 10.7(c) and (d). For the Cartesian interpretation, we see that the all the real parts are 0, and only the imaginary spectrum has values. This is equivalent to saying that no cosine functions were used, only sines: this can be seen in Figure 10.5. This can also be seen from the phase spectrum, where all the phases have a value of $\pi/2$, again showing that only sine waves are used.

Now consider the square wave that is identical to the one we have been analysing, except that it is shifted a quarter of a period to the left. If we subject this to Fourier analysis, we get the spectra shown in Figure 10.8. This shows that the imaginary part is now zero, and the real-part spectrum now contains the non-zero coefficients. The phase spectrum is also different, while the magnitude spectrum is the same as for the original square wave. This result is highly significant since it shows that only the magnitude spectrum is unaffected by a delay in the signal, so we can say that the magnitude spectrum best captures the essential characteristics of the shape of the square wave.

The difference between the two square waves is in their relative position in time – one is slightly later than the other. This agrees with what we expect from human hearing: if we play a note on a piano and then play the same note at some other time, the notes sound exactly the same, even though their position with respect to some reference time $t = 0$

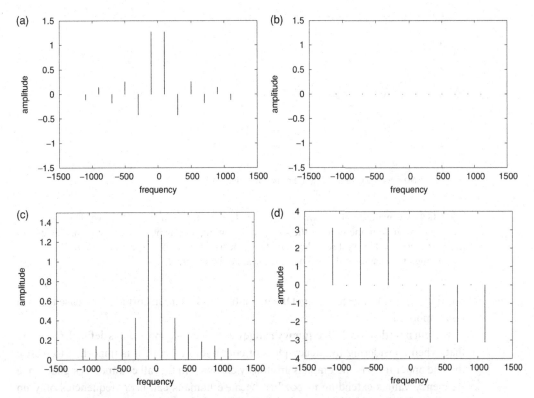

Figure 10.8 Spectra for a square wave with delay: (a) real spectrum, (b) imaginary spectrum, (c) magnitude spectrum and (d) phase spectrum. The real and imaginary spectra are clearly different from Figure 10.7. The magnitude spectrum is the same, and the phase spectrum is the same apart from the additional shift.

will generally be different. In fact, from extensive studies, we know that the human ear is largely insensitive to phase information in discerning speech and other sounds, such that, if two signals have the same magnitude spectrum but different phase spectra, in general they will be perceived as sounding the same.[2] This fact applies not only to the square wave as a whole, but also to the individual harmonics – if we synthesise a waveform with the magnitude spectrum of a square wave and *any* phase spectrum, it will sound exactly the same as the normal square wave. Since the square wave shape is created only by careful alignment of the sinusoids, it will generally be the case that a wave created with an arbitrary phase spectrum will not form a square waveform at all; nevertheless, **any two waveforms with the same magnitude spectra will sound the same regardless of phase or the shape of the waveform.** Figure 10.9 shows a waveform created with the magnitude values of a square wave.

This is an important fact for speech analysis and synthesis. For analysis, we generally need to study only the magnitude spectra and can ignore phase information. For synthesis,

---

[2] Phase information is, however, used to **localise** sounds, that is, we use the phase difference between the signals arriving at our left and right ears to estimate where the source of the signal is.

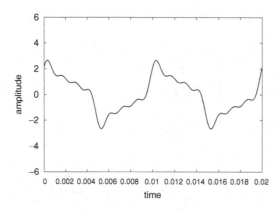

Figure 10.9 A waveform created from harmonics 1, 3, 5 and 7 of a "square" wave. The magnitude values are the same as the standard definition, but the phase values for harmonics 3 and 9 are set to 0. This will sound exactly the same as the waveform shown in Figure 10.5(d), even though the shapes of the two waveforms are quite different.

our signal generator need only get the magnitude information correct; the phase is much less important.

We will now discuss the **frequency range** over which the spectrum is defined. First note that, when considering frequency plots of continuous signals, it is important to realise that the upper limit in the graph is arbitrarily chosen – in fact all continuous signals have frequency values extending to ∞. Just because humans can hear frequencies only up to say 20 000 Hz doesn't mean that they stop being physically present above this level. Secondly, note that the spectra for the square wave in Figures 10.7 and 10.8 are plotted with negative as well as positive values for frequency. Our initial definition of frequency stated that it was the inverse of the period, and, since we can't have a negative period, why negative frequency? The use of negative frequency comes from the fact that a true continuous periodic signal is defined in time from −∞ to ∞. As we shall see later, time and frequency share a duality, with neither being more "real" than the other. Since time extends from −∞ to ∞, so does frequency.

If we take the cosine sinusoid definition and write it in terms of complex exponentials using the inverse Euler Equation (10.11),

$$x(t) = \cos(\omega t) = \frac{e^{j\omega t} + e^{-j\omega t}}{2}$$

$$x'(t) = \sin(\omega t) = \frac{e^{j\omega t} + e^{-j\omega t}}{2j}$$

we see that this signal actually has *two* frequencies, a positive ω and a negative ω. This is a general property of all *real* signals: they will have conjugate symmetry around $\omega = 0$, i.e. the vertical axis in the spectrum. Conjugate symmetry is a complex property, in which the real part of the spectrum is reflected in the vertical axis and the imaginary part is reflected in the vertical axis and again in the horizontal axis. This is clearly seen in Figures 10.7 and 10.8.

Signals that are not purely real, such as $e^{j\omega t}$ itself, have only one frequency value. Because of this, it can be argued that complex exponentials really are a simpler form than the sine or cosine form.

10.1.6 The Fourier transform

In general, we need to analyse non-periodic signals as well as periodic ones. Fourier synthesis and analysis cannot deal with non-periodic signals, but a closely related operation, the **Fourier transform**, is designed to do just this. By definition, the Fourier transform is

$$X(j\omega) = \int_{-\infty}^{\infty} x(t)e^{-j\omega} \, dt \qquad (10.21)$$

Its similarity to Fourier analysis (Equation (10.18)) should be clear:

$$a_k = \frac{1}{T_0} \int_0^{T_0} x(t)e^{-jk\omega_0 t} \, dt$$

The first difference is that we are now computing an integral from $-\infty$ to ∞ rather than over a single period: as our signal now may not have a period we can not of course integrate over it. Secondly, the result is a *continuous* function over frequency, not simply the complex amplitudes for a set of discrete harmonics. For non-periodic signals we have to abandon the idea of discrete harmonics: the magnitude and phase are instead expressed as continuous functions of frequency.

The Fourier transform can be explained as follows. Imagine that we have a **rectangular pulse** waveform $x(t)$, which has values for a finite duration T, defined as follows:

$$x(t) = \begin{cases} 1 & \text{for } -\frac{1}{2}T \leq t \leq \frac{1}{2}T \\ -1 & \text{otherwise} \end{cases}$$

This is equivalent to just a single period of a square wave, centred at $t = 0$, with values -1 everywhere else. This is clearly not periodic. But imagine that we artificially make a periodic version by replicating its shape as shown in Figure 10.10(a). This gives us our familiar square wave and by Fourier analysis we get the spectrum, the real part of which is shown in Figure 10.10(b). Now imagine that we perform the same trick, but move the repetitions further apart – in effect we are increasing the period of our waveform (Figure 10.10(c)). After performing Fourier analysis, we get the real spectrum shown in Figure 10.10(d). It should be immediately apparent that the overall shape (the **envelope**) is the same, but the harmonics are now much closer together, as we would expect of a waveform of longer period. If we extend the period further still (Figure 10.10(e)), we get the real spectrum shown in Figure 10.10(f) and here the harmonics are closer still. As $T_0 \to \infty$, the "harmonics" in the spectrum become infinitely dense, resulting in a continuous function.

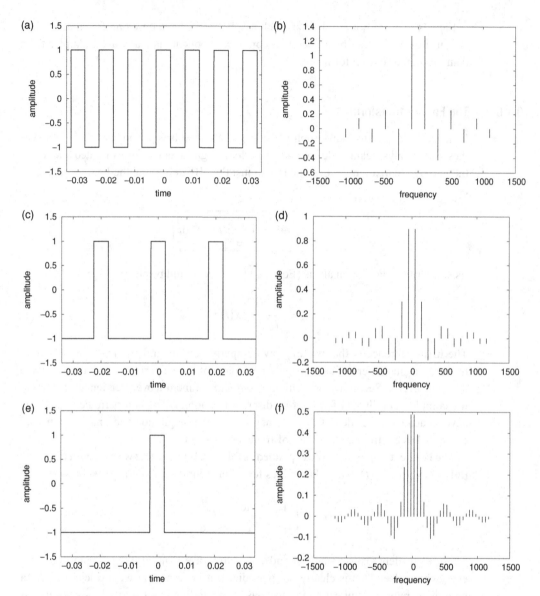

Figure 10.10 Real spectra for a signal of constant shape but decreasing period. As the period decreases, the harmonics get closer, but the envelope of the spectrum stays the same. In the limit, period → ∞, the harmonics will be infinitely close and the spectrum will be be a continuous function of frequency.

Let us now calculate the continuous spectrum of the rectangular pulse directly. From its definition, the Fourier transform of the waveform is

$$X(j\omega) = \int_{-T/2}^{T/2} e^{-j\omega t} \, dt$$

$$= \frac{e^{-j\omega T/2}}{-j\omega} - \frac{-e^{j\omega T/2}}{-j\omega}$$

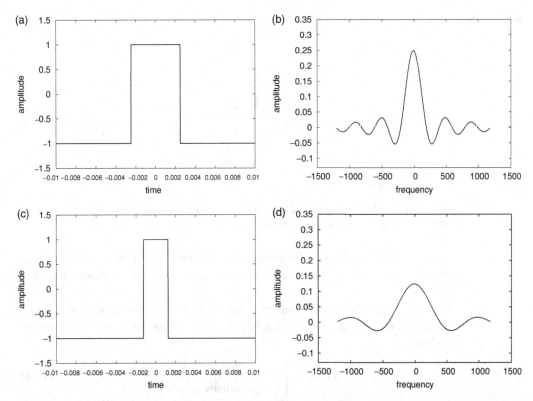

Figure 10.11 The waveform and real part of the spectrum for two rectangular pulses of different durations. These figures demonstrate the scaling principle whereby compression in one domain gives rise to spreading in the other.

We can use the inverse Euler formula to simplify this to

$$X(j\omega) = \frac{\sin(\omega T/2)}{\omega/2} \qquad (10.22)$$

Any function of the form $\sin(x)/x$ is known as a **sinc function**. The real spectrum of this is shown in Figure 10.11(b), and we can see that this has the exact shape suggested by the envelope of the harmonics in Figure 10.10. Figures 10.11(c) and (d) show the waveform and Fourier transform for a pulse of shorter duration. The Fourier transform is now more "spread-out", and this demonstrates an important aspect of the Fourier transform known as the **scaling property**.

The **inverse Fourier transform** is

$$\boxed{x(t) = \frac{1}{2\pi} \int_{-\infty}^{\infty} X(j\omega) e^{j\omega t} \, dt} \qquad (10.23)$$

and when we compare this with the Fourier-synthesis equation (Equation (10.18))

$$x(t) = \sum_{-\infty}^{\infty} a_k e^{jk\omega_0 t}$$

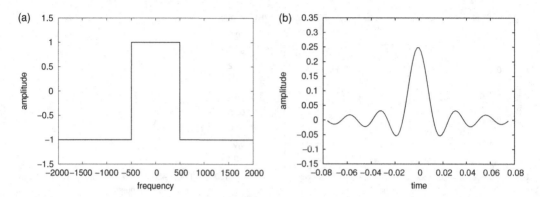

Figure 10.12 Part (a) is the real part of a spectrum; part (b) is its inverse Fourier transform. Together with Figure 10.11, these figures demonstrate the duality principle.

we can see that the discrete frequency representation of a fundamental frequency multiplied by a harmonic index has now been replaced with a continuous frequency variable ω.

Let us now use the inverse Fourier transform to find the waveform for a defined spectrum. Consider a spectrum which is 1 within a certain frequency region and -1 elsewhere:

$$X(j\omega) = \begin{cases} 1 & \text{for } -\omega_b \leq t \leq \omega_b \\ -1 & \text{otherwise} \end{cases} \quad (10.24)$$

The inverse Fourier transform of this is

$$x(t) = \frac{1}{2\pi} \int_{-\omega_b}^{\omega_b} e^{j\omega t} dt$$

$$= \frac{1}{2\pi} \frac{e^{-j\omega_b t}}{-j2\pi t} - \frac{-e^{-j\omega_b t}}{-j2\pi t}$$

$$= \frac{\sin(\omega_b t)}{\pi t} \quad (10.25)$$

This is shown in Figure 10.12. When we compare Equation (10.25) with Equation (10.22) and Figure 10.11 with Figure 10.12 we see that the Fourier transform of a rectangular pulse is a sinc spectrum and the inverse Fourier transform of a rectangular spectrum is a sinc waveform. This demonstrates another special property of the Fourier transform known as the **duality principle**. These and other general properties of the Fourier transform are discussed further in Section 10.3.

10.2 Digital signals

We now turn our attention to **digital signals**. All the signals we have discussed upto now are *analogue signals* in which the amplitude and time dimensions are represented by real numbers. By contrast, digital signals are represented by a sequence of integers.

In the "real world", most signals are analogue, including speech signals, for which the amplitude is a measure of sound pressure level or some other physical quantity. Digital signals are mostly found in computers.

10.2.1 Digital waveforms

A digital signal is represented by a sequence of evenly spaced points, the value at each point being the amplitude, represented by an integer. This integer amplitude value can be interpreted as the digital manifestation of sound pressure level, voltage or some other such analogue quantity. Formally, a digital sequence is represented as follows:

$$x[n] = \ldots, x_{-2}, x_{-1}, x_0, x_1, x_2, \ldots \qquad (10.26)$$

For digital sequences, it is usual to use n and write $x[n]$, to distinguish from continuous time t in analogue signals. Each point in the sequence is called a *sample*, and the distance between one sample and the next is called the *sample period*, T_s. The **sample rate** or **sample frequency** is given by $F_s = 1/T_s$. The sample rate limits the frequencies which the digital signal can describe. The highest frequency that a digital signal can contain is called the **Nyquist** frequency and is exactly half the sample rate.

The choice of sample rate is a compromise between wanting to have the highest-possible quality (high sample rate) and the lowest storage or transmission cost (low sample rate). Practical data from human perception studies show that the limit of human hearing is about 20 000 Hz and this figure would lead us to use sampling rates of >40 000 Hz. Indeed, compact discs and digital audio tapes have sampling rates of 44 100 Hz and 48 000 Hz, respectively. However, it can be shown that, for speech signals, very little useful speech energy is found in the higher frequencies, so sample rates of 16 000 Hz are often used and found to be perfectly adequate for accurate representation of speech signals. Where space is tight, lower sampling rates are used, most commonly in the telephone, which has a standardised sample rate of 8000 Hz. While speech at this sample rate is normally quite intelligible, it is clear that it sounds much poorer than when higher sample rates are used.

The **bit range** describes the **dynamic range** of a digital signal, and is given by the range of the integer used to store a sample. Widely used values are 16 bits, which gives an absolute dynamic range of 2^{16}, meaning that the amplitude of the sample can range from $-32\,768$ to $32\,767$. Dynamic range is normally expressed in the logarithmic decibel scale. In digital signals, we take the reference as 1 and therefore the dynamic range of a 16-bit signal is

$$10 \log(2^{16}) \approx 96 \text{ dB} \qquad (10.27)$$

To save on space and transmission bandwidth, telephones and other systems frequently use 8 bits (48 dB) and 12 bits (72 dB). At 8 bits, speech quality is noticeably worse than at 16 bits; 12 bits is a reasonable compromise, but, if space is not an issue (and with most types of speech synthesis it is not), 16 bits is normally used.

10.2.2 Digital representations

We use somewhat different notation when dealing with digital signals. First, an integer n is used instead of t for the time axis, where n is related to t by the sample period:

$$n = tT_s \tag{10.28}$$

From this, we can relate the continuous signal $x(t)$ and digital signal $x[n]$ by $x[n] = x(n/T_s)$. We use square brackets "[ ]" rather than parentheses (round brackets) "( )" when describing digital functions. Strictly speaking, a signal $x[n]$ is in the **n-domain** but we shall informally refer to it as being in the time domain.

Secondly, it is common to use *normalised frequency*, rather than natural frequency. From (10.28) we see that frequency, angular frequency and normalised angular frequency are related by

$$\hat{\omega} = \frac{\omega}{T_s} = \frac{2\pi F}{T_s} \tag{10.29}$$

Using n and normalised angular frequency $\hat{\omega}$ makes calculations easy since the equations are the same regardless of the sample rate. To convert to real frequency, the normalised frequency is simply multiplied by the sample rate. In general, we use the same terms for amplitude and phase for both analogue and digital signals.

10.2.3 The discrete-time Fourier transform

For a continuous signal, we can move from the time domain to the frequency domain via the Fourier transform, but, since we are now dealing with digital signals, this formula cannot be used directly. We will now derive an equivalent transform for digital signals. Starting with the previously defined Fourier transform (Equation (10.21))

$$X(j\omega) = \int_{-\infty}^{\infty} x(t)e^{-j\omega t}\, dt$$

we can use this on a digital sequence with the relation $x[n] = x(n/T_s)$. The Fourier transform is calculated as an integral over a range, which can be interpreted as finding the area under the curve over this range. The numerical method of the Riemann sum says that we can approximate a continuous area function by adding up the areas of separate rectangles. The area of each rectangle is the value of the function at that point multiplied by the width of the rectangle. In our case, the width of the rectangle is the sample period T_s. This gives a numerical version of the Fourier transform:

$$X(j\omega) = \sum_{-\infty}^{\infty} (x(nT_s)e^{-j\omega t})T_s$$

T_s is independent of the integral, so we move it outside, and use $x[n]$ in place of $x(nT_s)$, giving

$$X(j\omega) = T_s \sum_{-\infty}^{\infty} x[n]e^{-j\omega n T_s}$$

We can get rid of the T_s by making use of normalised frequency, $\hat{\omega} = \omega/T_s$. The result is the **discrete-time Fourier transform** or **DTFT**:

$$X(e^{j\hat{\omega}}) = \sum_{-\infty}^{\infty} x[n] e^{-j\hat{\omega}n} \qquad (10.30)$$

It should be clear that the DTFT is a transform in which time is discrete but frequency is continuous. As we shall see in later sections, the DTFT is a very important transform in its own right but, for computational purposes, it has two drawbacks. Firstly, the infinite sum is impractical, and secondly we wish to obtain a representation for frequency that is useful computationally, and such a representation must be finite.

10.2.4 The discrete Fourier transform

For the first problem, we choose a sufficiently large interval L that the samples outside $0 < n < L$ are zero. For the second problem, we will use a similar operation to time-domain sampling and "sample" the spectrum at discrete intervals. Just as $n = tT_s$, we define the spacing between each discrete frequency and the next as $F_s/N = 1/(NT_s)$, where N is the total number of frequency values we want. Each angular frequency therefore lies at

$$\omega_k = \frac{2\pi}{NT_s}$$

giving the transform

$$X(e^{j\hat{\omega}_k}) = \sum_{n=0}^{L} x[n] e^{-j\hat{\omega}_k n}$$

If we set $L = N$ (discussed below), and use the above definition of ω_k, we can write this in a more compact usable form:

$$X[k] = \sum_{n=0}^{N-1} x[n] e^{-j\frac{2\pi}{N}kn} \qquad (10.31)$$

which gives us the standard definition of the **discrete Fourier transform** or **DFT**.

The interpretation is the same as for the Fourier transform defined in Section 10.1.6; for each required frequency value, we multiply the signal by a complex exponential waveform of that frequency, and sum the result over the time period. The result is a complex number that describes the magnitude and phase at that frequency. The process is repeated until the magnitude and phase have been found for every frequency.

The DFT states that the length of the analysed waveform and the number of resultant frequency values are given by N. This makes intuitive sense: if we analyse just a few samples, we can expect to obtain only a somewhat crude version of the spectrum; by analysing more samples, we acquire more information "to go on" and can fill in more detail in the spectrum: that is, the points in the spectrum (ω_k) get closer and closer together. As $N \to \infty$, we analyse all the waveform values, and the distance between the frequency points in the spectrum $\to 0$. This is exactly the case that the DTFT describes. Both the DTFT and the DFT have inverse transforms:

$$x[n] = \frac{1}{2\pi} \int_{-\pi}^{\pi} X(e^{j\hat{\omega}}) e^{j\hat{\omega}n} d\hat{\omega} \qquad (10.32)$$

$$x[n] = \frac{1}{N} \sum_{n=0}^{N-1} x[n] e^{j\frac{2\pi}{N}kn} \qquad (10.33)$$

In practice an algorithm called the **fast Fourier transform** or **FFT** is commonly used to calculate the DFT. This is not another *type* of transform, merely a fast way of calculating a DFT: the result for the FFT is exactly the same as for the DFT. The FFT works only for values of N that are powers of 2; hence it is common to find N set to 256 or 512. If a DFT of a different N is required, a time signal of the next highest power of 2 is chosen, and the extra values are set to 0. So, for example, if we have a signal of length 300, we set $N = 512$, "pad" the extra 212 values with 0, and then calculate a 512-point FFT. The use of the FFT is pervasive in practice, and one often hears references to "the FFT of the signal" etc. We repeat that, in terms of the outcome, the FFT and DFT are exactly the same and hence taking "the FFT of the signal" is no different from "taking the DFT". An overview of the FFT can be found in Oppenheim and Schafer [342].

10.2.5 The z-transform

We now introduce another important transform, the **z-transform**.[3] The z-transform is perhaps the most useful of all transforms in practical digital signal processing. It is defined as

$$\boxed{X(z) = \sum_{n=-\infty}^{\infty} x[n] z^{-n}} \qquad (10.34)$$

This should look familiar – if we use $z = e^{j\hat{\omega}}$, we get

$$X(e^{j\hat{\omega}}) = \sum_{-\infty}^{\infty} x[n] e^{-j\hat{\omega}n}$$

which is the discrete-time Fourier transform, Equation (10.30). Just like the DTFT, the z-transform is a transform that operates on a *digital* signal but produces a *continuous* function.

An important property of the z-transform is its effect on a delay. Define a signal

$$y[n] = x[n - n_d]$$

---
[3] What? You can't be serious – *another* transform!

that is, a delayed version of $x[n]$. By use of a dummy variable $m = n - n_d$, we can find its z-transform as follows:

$$Y(z) = \sum_{n=-\infty}^{\infty} y[n]z^{-n}$$

$$= \sum_{n=-\infty}^{\infty} x[n-n_d]z^{-n}$$

$$= \sum_{m=-\infty}^{\infty} x[m]z^{n_d-m}$$

$$= \sum_{m=-\infty}^{\infty} x[m]z^{-m}z^{n_d}$$

The summation in m is just the z-transform of x defined in m rather than n, so we can write

$$Y(z) = X(z)z^{n_d} \tag{10.35}$$

Hence delaying a signal by n_d multiplies its z-transform by z^{n_d}. This important property of the z-transform is discussed further in Section 10.3.5.

Why use yet another transform? The reason is that by setting z to $e^{j\hat{\omega}}$ we greatly simplify the DTFT, and this simplification allows us to perform sophisticated operations in analysis and synthesis with relatively simple mathematical operations. The key to the ease of use of the z-transform is that Equation (10.34) can be rewritten as

$$X(z) = \sum_{n=-\infty}^{\infty} x[n](z^{-1})^n \tag{10.36}$$

That is, the transform can be written as a *polynomial* in z^{-1}, so that the techniques of polynomial solving can be brought to bear.

A final transform, the **Laplace transform**, is frequently used in continuous-time signal processing, and is the continuous-time version of the z-transform. The z-transform and the Laplace transform can be thought of as generalisations of the DTFT and Fourier transform, respectively. The Laplace transform is not required in this book, since all our signal processing is performed on digital signals. We mention it here for completeness only.

10.2.6 The frequency domain for digital signals

For continuous signals, we showed that the frequency domain extends from $-\infty$ to ∞. However, for digital signals we stated that a frequency of half the sampling rate, the Nyquist frequency, was the highest possible frequency that could be represented by a digital signal. What then is the range of the frequency domain? As for continuous signals the range extends from $-\infty$ to ∞, but no *new* information is found outside the Nyquist range. Rather, the spectrum repeats itself at multiples of the Nyquist frequency. In other words, the spectrum of a digital signal is **periodic**. It is for this reason that the spectrum of digital signals calculated from the DTFT is described as $X(e^{j\omega t})$, rather that $X(j\omega)$ as is normal for the Fourier transform. The "period" of the spectrum is twice the Nyquist frequency, since the spectrum is defined from $-F_N$ to $+F_N$. In terms of normalised frequency, the frequency range is just $-\pi$ to $+\pi$ and so the "period" is 2π.

Just as with real continuous signals, real digital signals exhibit conjugate symmetry in the frequency domain. Hence in a 512-point DFT the first 256 values will be conjugate symmetric with the second 256. For this reason only half a digital spectrum is normally plotted, and this will have half the number of values used to calculated the DFT.

10.3 Properties of transforms

At first sight the number of transforms can seem unnecessarily large. (Recall that each has an inverse transform also.) Why do we need so many? The differences are simply attributable to us wanting to use both continuous and discrete representations for time and frequency, and to make analysis easier (in the case of the z-transform and Laplace transform). The differences between them can be clearly shown in the following table:

	Discrete time	Continuous time
Discrete frequency	Discrete Fourier transform (DFT)	Fourier analysis, periodic waveform
Continuous frequency	Discrete Fourier transform (DTFT)	Fourier transform
Continuous variable	z-transform	Laplace transform

We now describe various general properties of the transforms we have introduced. For demonstration purposes, we will describe these using the Fourier transform, but in general the following hold for all our transforms.

10.3.1 Linearity

All the transforms are **linear**. If we have an input $x(t) = \alpha s_1(t) + \beta s_2(t)$, then $X(j\omega) = \alpha X_1(j\omega) + \beta X_2(j\omega)$. The proof of this for the Fourier transform is

$$X(j\omega) = \int_{-\infty}^{\infty} (\alpha x_1(t) + \beta x_2(t)) e^{-j\omega t}$$
$$= \alpha \int_{-\infty}^{\infty} x_1(t) e^{-j\omega t} + \beta \int_{-\infty}^{\infty} x_2(t) e^{-j\omega t}$$
$$= \alpha X_1(j\omega) + \beta X_2(j\omega) \qquad (10.37)$$

10.3.2 Time and frequency duality

The Fourier transform and inverse Fourier transform have very similar forms. If we have a function $g(.)$ and define a signal as $g(t)$ then its Fourier transform will be

$G(j\omega) = \mathcal{F}\{g(t)\}$ (here we use the notation $\mathcal{F}$ to denote the operation of the Fourier transform). If we define the spectrum in terms of our function as $g(\omega)$ and take the inverse Fourier transform, we get $G(t) = 2\pi \mathcal{F}\{g(\omega)\}$. It should be clear that the form is the same apart from the 2π term, which is required because of our use of angular frequency:

$$g(t) \xrightarrow{\mathcal{F}} G(w) \qquad (10.38)$$

$$g(\omega) \xleftarrow{\mathcal{IF}} G(t) \qquad (10.39)$$

Figure 10.12 demonstrates this property.

10.3.3 Scaling

We have already described the Fourier transform of a pulse of various durations in Section 10.1.6. The shorter-duration pulse has a spectrum that is more spread out, and the longer-duration pulse has a more-contracted spectrum. This is known as the **scaling property**, which states that **compressing a signal will stretch its Fourier transform and vice versa**. This can be formally stated as

$$x(at) \xrightarrow{\mathcal{F}} \frac{1}{|a|} X(j(w/a)) \qquad (10.40)$$

10.3.4 Impulse properties

From the scaling property, we know that, as the duration of the pulse approaches zero, the spectrum spreads out more and more. In the limit, we reach the **unit impulse**, which has zero width. This is a particularly important signal. For digital signals, this is simply defined as a signal, $\delta[n]$, which has a value 1 at time 0 and a value 0 elsewhere. For continuous-time signals, the definition is a little trickier, but in this case the unit impulse is described as a signal that has values of zero everywhere apart from at time 0 (i.e. zero width), at which time it is described by an integral that defines its *area* (not height) as 1:

$$\delta(t) = 0 \qquad \text{for } t \neq 0 \qquad (10.41)$$

$$\int_{-\infty}^{\infty} \delta(t)dt = 1 \qquad (10.42)$$

As expected from the scaling property, the Fourier transform of an impulse is a function that is "infinitely stretched", that is, the Fourier transform is 1 at all frequencies. Using the duality principle, a signal $x(t) = 1$ for all t will have a Fourier transform of $\delta(\omega)$, that is, an impulse at time $\omega = 0$. This is to be expected – a constant signal (a DC signal in electrical terms) has no variation and hence no information at frequencies other than 0:

$$\delta(t) \xrightarrow{\mathcal{F}} 1 \qquad (10.43)$$

$$1 \xrightarrow{\mathcal{F}} \delta(\omega) \qquad (10.44)$$

10.3.5 Time delay

A signal $x(t)$ that is delayed by time t_d is expressed as $x(t - t_d)$. If we denote the Fourier transform of $x(t)$ as $X(j\omega)$ and the Fourier transform of $x(t - t_d)$ as $X_d(j\omega)$, we see from the Fourier-transform definition that $X_d(j\omega)$ is

$$X_d(j\omega) = \int_{-\infty}^{\infty} x(t - t_d) e^{-j\omega t}\, dt \tag{10.45}$$

To simplify the integration, we use a dummy variable v and make the substitution $v = t - t_d$:

$$X_d(j\omega) = \int_{-\infty}^{\infty} x(v) e^{-j\omega(v + t_d)}\, dv$$

$$= e^{-j\omega t_d} \int_{-\infty}^{\infty} x(v) e^{-j\omega v}\, dv$$

$$X_d(j\omega) = e^{-j\omega t_d} X(j\omega)$$

So delaying a signal by t_d seconds is equivalent to multiplying its Fourier transform by $e^{-j\omega t_d}$. If we apply this to a unit impulse, $x(t - t_d) = \delta(t - t_d)$, we get

$$X_d(j\omega) = \int_{-\infty}^{\infty} \delta(t - t_d) e^{-j\omega t}\, dt$$

$$= e^{-j\omega t_d} \int_{-\infty}^{\infty} \delta(v) e^{-j\omega v}\, dv$$

and, from the impulse property, we know that the Fourier transform of $\delta(t)$ is 1, leaving

$$X_d(j\omega) = e^{-j\omega t_d} \tag{10.46}$$

At first sight, this may seem strange since the Fourier transform for a shifted impulse seems very different from that for a normal impulse, which simply had a Fourier transform of 1. Recall, however, that the magnitude of $e^{-j\omega t_d}$ will be 1, so the magnitude spectrum will be the same as the delta function. The phase of $e^{-j\omega t_d}$ is simply a linear function of t_d – this is as we should expect; the longer the delay t_d the more the phase spectrum will be shifted. It should be noted that the above result is the Fourier-transform result of the z-transform delay derived in Equation (10.35).

10.3.6 Frequency shift

From the duality principle, it then follows that, if we have a signal $x(t) = e^{j\omega_0 t}$ (i.e. a sinusoid), the Fourier transform will be $2\pi\delta(\omega - \omega_0)$, that is a single impulse at the frequency ω_0. This is entirely to be expected – the frequency-domain impulse can be interpreted as a harmonic and is consistent with the Fourier-analysis result.

More generally, if we multiply a signal by $e^{j\omega_0 t}$ we cause a **frequency shift** of ω_0 in the Fourier transform:

$$x(t) e^{j\omega_0 t} \leftarrow X(j(\omega - \omega_0)) \tag{10.47}$$

Shift in the frequency domain corresponds to **modulation** in the time domain.

10.3.7 Convolution

Convolution is defined as the overlap between two functions when one is passed over the other, and is given by

$$f(t) = g(t) \otimes h(t) = \int h(\tau)g(t-\tau)d\tau \qquad (10.48)$$

If we take the Fourier transform of this, we get

$$F(j\omega) = \mathcal{F}\left\{\int_{\tau=-\infty}^{\infty} h(\tau)g(t-\tau)d\tau\right\}$$

$$= \int_{t=-\infty}^{\infty} \left(\int_{\tau=-\infty}^{\infty} h(\tau)g(t-\tau)\right) e^{-j\omega t} dt$$

$$= \int_{\tau=-\infty}^{\infty} h(\tau)d\tau \int_{t=-\infty}^{\infty} g(t-\tau)e^{-j\omega t} dt \qquad (10.49)$$

From the time-delay property (Equation (10.46)) we know that

$$\int_{\tau=-\infty}^{\infty} g(t-\tau)e^{-j\omega t} = e^{-j\omega\tau}G(j\omega)$$

and substituting this into (10.49) gives

$$F(j\omega) = \int_{\tau=-\infty}^{\infty} h(\tau)e^{-j\omega\tau}G(j\omega)$$

The $\int h(\tau)e^{-j\omega\tau}$ term is simply the Fourier transform of $h(t)$ defined in terms of τ rather than t, so we can state

$$\boxed{F(\omega) = G(\omega)H(\omega)} \qquad (10.50)$$

Hence, **convolution in the time domain corresponds to multiplication in the frequency domain**. It can also be proved that multiplication in the time domain corresponds to convolution in the frequency domain.

10.3.8 Analytical and numerical analysis

The main practical difference between the transforms is whether an analytical or numerical method is used in the calculation. For transforms and inverse transforms that contain integrals, it should be clear that we must be able to find the analytical form of the integral. This can be a blessing in that we can use analytical techniques to find and succinctly describe the transforms of some signals, but it can also be a difficulty, because first we have to be able to represent the signal as an analytical function, and second we have to be able to find its integral. When dealing with arbitrary signals, say from a recording, it is nearly certain that we will not know the form of the signal, so a numerical method must be used.

10.3.9 Stochastic signals

All the transforms we have examined either integrate or sum over an infinite sequence. This will give a meaningful result only if the sum of the signal over that range is finite; and it can easily be shown that this is the case for all periodic and many non-periodic signals. There are, however, signals for which this is not so, and in such cases we say that the Fourier (or other transform) does not exist for these signals. One important class of such signals is constituted by **stochastic signals**, which are signals generated by some random process. The "noisy" signals (for example those found in fricatives) that we have been mentioning up until now are such signals.

In trying to characterise these signals we face two problems. Firstly, it is difficult to describe them in the time domain due to the very nature of the random process. Secondly, since the transforms we have introduced cannot be used as defined, we cannot express these signals directly in the frequency domain either. We get round these issues by focusing on *averages* within the signal; specifically, we use a measure of the **self-similarity** of a signal called its **autocorrelation function**. For a signal $x[n]$, this is defined as

$$R(j) = \sum_{n=-\infty}^{\infty} x[n]x[n-j] \qquad (10.51)$$

which is the expected value of the product of a stochastic signal with a time-shifted version of itself. This function *does* have a Fourier transform, and this gives the **power spectral density** of the signal.

10.4 Digital filters

A **filter** is a mathematically defined **system** for modifying signals. In signal processing filters are put to many uses; as their name might suggest, they can "filter out" some unwanted portion of a signal. In speech synthesis, we are interested in them mainly because of the acoustic theory of speech production (Section 11.1), which states that speech production is a process by which a glottal *source* is modified by a vocal tract *filter*. This section explains the fundamentals of filters and gives insight into how they can be used as a basis for modelling the vocal tract. **Digital filters** are of course filters that operate on digital signals. Here, we describe a specific but powerful kind of filter, called the **linear time-invariant (LTI)** filter.

Using the notation $x[n] \mapsto y[n]$ to describe the operation of the filter ("$\mapsto$" is read "maps to"), we state the time-invariance property as

$$\boxed{x[n-n_0] \mapsto y[n-n_0] \quad \text{for any } n_0} \qquad (10.52)$$

which states that the system behaviour is not dependent on time. A **linear system** has the properties of **scaling**, such that if $x[n] \mapsto y[n]$ then

$$\alpha x[n] \mapsto \alpha y[n]$$

and **superposition**, such that if $x_1[n] \mapsto y_1[n]$ and $x_2[n] \mapsto y_2[n]$ then

$$x[n] = x_1[n] + x_2[n] \mapsto y[n] = y_1[n] + y_2[n]$$

These combine to form the **linearity** principle:

$$\boxed{x[n] = \alpha x_1[n] + \beta x_2[n] \mapsto y[n] = \alpha y_1[n] + \beta y_2[n]} \quad (10.53)$$

The linearity property combines with the principle of Fourier synthesis – since the filter linearly combines the components in a waveform, we can construct and analyse a filter for components separately, without having to worry about their interactions.

10.4.1 Difference equations

Digital LTI filters are often described in terms of their time-domain **difference equation**, which relates input samples to output samples. The **finite-impulse-response** or **FIR** filter has the difference equation

$$\boxed{y[n] = \sum_{k=0}^{M} b_k x[n-k]} \quad (10.54)$$

The **infinite-impulse-response** or **IIR** filter has the difference equation

$$\boxed{y[n] = \sum_{k=0}^{M} b_k x[n-k] + \sum_{l=1}^{N} a_l y[n-l]} \quad (10.55)$$

As well as operating on the input, the IIR filter operates on previous values of the output and for this reason these filters are sometimes referred to as **feedback** systems or **recursive** filters. The FIR is a special case of an IIR filter, formed when the coefficients $\{a_l\}$ are set to zero. (Note that the $\{a_l\}$ coefficients do not have a value for 0 since the output for $y[n]$ must be calculated before it can be used on the right-hand side of the equation.)

Consider a simple FIR filter with $M = 3$ and $b_k = \{3, -1, 2, 1\}$. Since $M = 3$, this is known as a **third-order** filter. From the FIR definition equation, we can see that this will modify any input sequence x as follows:

$$y[n] = 3x[n] - x[n-1] + 2x[n-2] + x[n-3] \quad (10.56)$$

Figure 10.13 shows an example signal $x = \{0, 1, 2, 1, 0, -1, -2, -1, 0, 0\}$ and the output of the filter for this example signal.

10.4.2 The impulse response

While the above clearly showed how our filter created an output signal $y[n]$ from our given input signal $x[n]$, it isn't immediately clear how the filter will modify other, different input signals. Rather than having to examine the output with respect to a wide range of different input signals, it would be useful to have a characterisation of the effect of

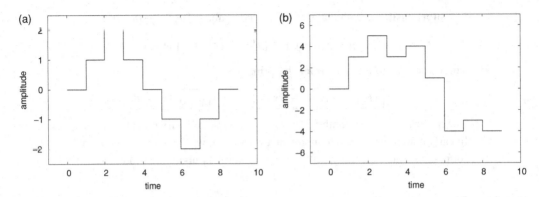

Figure 10.13 Operation of an FIR filter on an example finite waveform: (a) input signal and (b) output signal.

the filter that is independent of the specific form of the input. Such a characterisation is called the **response** of the filter, and a particularly useful response is provided by using a **unit impulse** as input. The output is then the **impulse response**, denoted $h[n]$.

A unit impulse, δ, is a special function, which has value 1 at $n = 0$ and 0 elsewhere:

$$\delta[n] = \begin{cases} 1 & n = 0 \\ 0 & n \neq 0 \end{cases} \tag{10.57}$$

The output of the example FIR filter for an impulse is given in Figure 10.14(a), which shows that the impulse response is simply the coefficients in the filter. Because of this, the duration of the impulse will always be the order of the filter. Since the order is finite, the impulse response is finite, which gives rise to the name of the filter.

Consider now an example first-order IIR filter, with $a_1 = -0.8$ and $b[0] = 1$. Its difference equation is therefore

$$y[n] = x[n] - 0.8y[n-1] \tag{10.58}$$

If we give this filter the unit impulse as an input, then we can calculate the output for any n after 0. The output is shown in Figure 10.14(b). First of all we observe that the output is not of a fixed duration: in fact it is infinite as the name of the filter implies. Secondly we observe that the response takes the form of a decaying exponential. The rate of decay is governed by the filter coefficient, a_1. (Note that if $a_1 > 1$ the output will grow exponentially: this is generally undesirable and such filters are called **unstable**.)

Consider the impulse response of the second-order IIR filter $a_1 = -1.8$, $a_2 = 1$, shown in Figure 10.15(a). Just as in the first-order case, this filter has an infinite response, but this time takes the form of a sinusoid. This shows the power of the IIR filter – with only a few terms it can produce a quite complicated response. The filter $a_1 = -1.78$, $a_2 = 0.9$ has its impulse response shown in Figure 10.15(b). We can see that it is a decaying sinusoid; we have in effect combined the characteristics of the examples in Figures 10.14(b) and 10.15(a). The key point to note is that in all cases the input is the same, i.e. a unit impulse. The characteristics of the output are governed solely by the

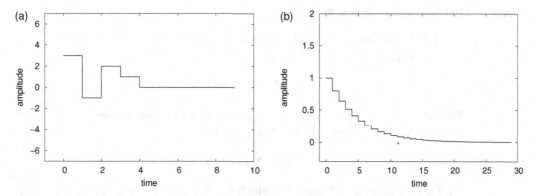

Figure 10.14 Example impulse responses for FIR and IIR filters: (a) FIR filter and (b) IIR filter. Note that, while the number of coefficients in the IIR filter is actually less than for the FIR filter, the response will continue indefinitely (only values up to 30 samples are shown).

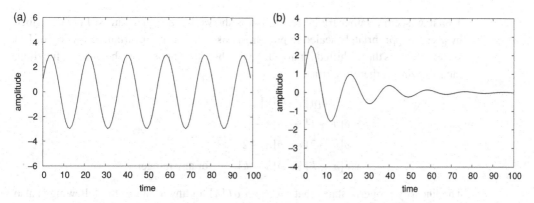

Figure 10.15 Impulse responses for two IIR filters, showing a sinusoid (a) and a damped-sinusoid (b) response.

filter – after all, the input has no periodicity or decay factor built in. If we choose a slightly different set of coefficients, we can create an output with a different period and decay factor.

Different filters can produce a wide range of impulse responses; the decaying sinusoid patterns here come only from a specific type of second-order IIR filter. In general though, the more terms in the filter, the more complicated the impulse response will be. The impulse response is important for a number of reasons. Firstly, it provides us with a simple way of examining the characteristics of the filter. Because the input lasts for only a single sample, the output is not "cluttered" with further interference from subsequent input samples. Impressionistically, exciting the filter with an impulse response can be thought of as "giving the filter a nudge and seeing what happens". Secondly, and more importantly, we can use the principle of *linearity* and *time invariance* to describe the responses of more-general signals by using the impulse response, which is explained next.

10.4.3 The filter convolution sum

The response for a non-unit impulse is simply the unit impulse response scaled. So, using the linearity principle, we can state that

$$3\delta[n] \mapsto 3h[n]$$

Since the unit impulse is simply a single sample, a full input sequence $x[n]$ can be characterised by a sequence of scaled unit impulses:

$$x[n] = \sum w[k]\delta[n-k]$$

What are the values of w? Since the delta function will set everything outside $n-k$ to 0, it should be clear that $w[n] = x[n]$ for a given n. Therefore, we can state that

$$x[n] = \sum_{k=0}^{k} x[k]\delta[n-k] \tag{10.59}$$

This may seem a rather trivial result, but it shows that a signal can be fully described by a set of appropriately scaled impulse responses. The time-invariance property of LTI systems tells us that a shifted input will give a shifted response. For the impulse function, the following is therefore true:

$$\delta[n] \mapsto h[n]$$
$$\delta[n-1] \mapsto h[n-1]$$
$$\delta[n-2] \mapsto h[n-2]$$
$$\delta[n-k] \mapsto h[n-k] \quad \text{for any integer } k$$

The linearity property states that $\alpha\delta[n] \mapsto \alpha h[n]$ for any α and so the following is also true:

$$x[0]\delta[n] \mapsto x[0]h[n]$$
$$x[1]\delta[n-1] \mapsto x[1]h[n-1]$$
$$x[2]\delta[n-2] \mapsto x[2]h[n-2]$$
$$x[k]\delta[n-k] \mapsto x[k]h[n-k] \quad \text{for any integer } k$$

We can use these two properties on Equation (10.59) to find the output of the filter

$$\sum_k x[k]\delta[n-k] \mapsto \sum_k x[k]h[n-k]$$

which gives us

$$\boxed{y[n] = \sum_k x[k]h[n-k]} \tag{10.60}$$

which is known as the **convolution sum**. It shows that we can calculate the output for a filter for **any** input, provided that we know the impulse response $h[n]$.

10.4.4 The filter transfer function

We will now examine the frequency-domain description or **frequency response** of the LTI filter by means of the z-transform. The IIR difference Equation (10.55), expanded to a few terms, is

$$y[n] = b_0x[n] + b_1x[n-1] + \cdots + b_Mx[n-M] + a_1y[n-1] + \cdots + a_Ny[n-N]$$

By using the notation $\mathcal{Z}\{.\}$ to denote the z-transform, we can take the z-transform of both sides as follows:

$$Y(z) = \mathcal{Z}\{b_0x[n] + b_1x[n-1] + \cdots + b_Mx[n-M] + a_1y[n-1] + \cdots + a_Ny[n-N]\}$$

The coefficients are not affected by the z-transform, so, using the linearity of the z-transform, this simplifies to

$$Y(z) = b_0\mathcal{Z}\{x[n]\} + b_1\mathcal{Z}\{x[n-1]\} + \cdots + b_{n-M}\mathcal{Z}\{x[n-M]\} + a_1\mathcal{Z}\{y[n-1]\} + \cdots + a_N\mathcal{Z}\{y[n-N]\}$$

The z-transform time-delay property (Equations (10.35) and (10.46)) states that $\mathcal{Z}\{x[n-k]\} = X(z)z^{-k}$ and so our equation becomes

$$Y(z) = b_0X(z) + b_1X(z)z^{-1} + \cdots + b_{n-M}X(z)z^M + a_1Y(z)z^{-1} + \cdots + a_NY(z)z^N$$

If we group the $Y(z)$ terms on one side, and factor the common $Y[z]$ and $X[z]$, we get

$$Y(z) = \frac{b_0 + b_1z^{-1} + \cdots + b_Mz^{M-1}}{1 - a_1z^{-1} - \cdots - a_Nz^{N-1}} X(z)$$

The term is known as the **transfer function** of the filter,

$$H(z) = \frac{Y(z)}{X(z)} = \frac{b_0 + b_1z^{-1} + \cdots + b_Mz^{M-1}}{1 - a_1z^{-1} - \cdots - a_Nz^{N-1}} \tag{10.61}$$

$$= \frac{\sum_{k=0}^{M} b_k z^{-k}}{\sum_{l=0}^{N} a_l z^{-l}} \tag{10.62}$$

and fully defines the characteristics of the filter in the z-domain. From this, the frequency-domain characteristics of the filter can easily be found (as explained below).

10.4.5 The transfer function and the impulse response

We have previously shown that the impulse response, which completely characterises the filter, allows us to calculate the time-domain output for any input by means of the convolution sum:

$$y[n] = \sum_{k=-\infty}^{\infty} h[k]x[n-k] \tag{10.63}$$

Section 10.3 showed how to derive the Fourier transform for a convolution and we can apply this to a digital signal by using the z-transform in the same way:

$$Y(z) = \mathcal{Z}\left\{\sum_{k=-\infty}^{\infty} h[k]x[n-k]\right\}$$

$$= \sum_{k=-\infty}^{\infty} h[k] \sum_{n=-\infty}^{\infty} x[n-k]z^{-n}$$

$$= \sum_{k=-\infty}^{\infty} h[k]z^{-k}X(z) \qquad (10.64)$$

The $\sum h[k]z^{-k}$ term is simply the z-transform of $h[n]$ defined in terms of k rather than n, so (10.64) simplifies to

$$\boxed{Y(z) = H(z)X(z)} \qquad (10.65)$$

which is the same result as Equation (10.62). This shows that **the transfer function $H(z)$ is the z-transform of the impulse response** $h[n]$. This gives us the following two important properties:

1. An LTI filter is completely characterised by its impulse response, $h[n]$, in the time domain. The z-transform of the impulse response is the transfer function, $H(z)$, which completely characterises the filter in the z-domain.
2. The transfer function is a ratio of two polynomials, with the same coefficients as used in the difference equation.

10.5 Digital filter analysis and design

Section 10.4 fully described the relationship between the time-domain difference equations and the z-domain transfer function of an LTI filter. For first-order filters, the coefficients are directly interpretable, for example as the rate of decay in an exponential. For higher-order filters this becomes more difficult, and, while the coefficients $\{a_l\}$ and $\{b_k\}$ fully describe the filter, they are somewhat hard to interpret (for example, it was not obvious how the coefficients produced the waveforms in Figure 10.15). We can, however, use **polynomial analysis** to produce a more easily interpretable form of the transfer function.

10.5.1 Polynomial analysis: poles and zeros

Consider a simple quadratic equation defined in terms of a real variable x:

$$f(x) = 2x^2 - 6x + 1$$

This equation can be **factorised** into

$$f(x) = G(x - q_1)(x - q_2)$$

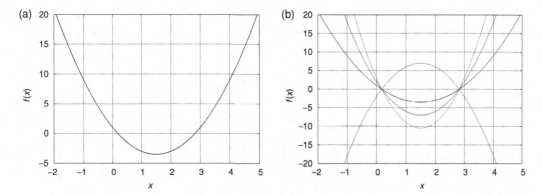

Figure 10.16 Plots of polynomial: (a) $2x^2 - 6x + 1$ and (b) $g \times (2x^2 - 6x + 1)$ for various values of g. The points where the curve intersects the x-axis are called "zeros".

and then the **roots** q_1 and q_2 of the quadratic can be found by setting $f(x)$ to 0 and solving for x (in this case the roots are $x = 0.38$ and $x = 2.62$). G is the **gain**, which simply scales the function.

We can plot the polynomial as shown in Figure 10.16(a). The roots of this expression (the points at which $f(x) = 0$) are marked on the graph. Although the roots mark only two points, they do in fact describe the general curve. Figure 10.16(b) shows the effect of varying the gain – the curve can become steeper or shallower, or even inverted if the gain is set to a negative number, but in all cases the general shape of the curve is the same. Because setting x to one of the roots results in $f(x) = 0$, we call these roots **zeros**. Now consider the same polynomial used as a denominator expression, where by convention we keep the gain as a numerator term:

$$f(x) = \frac{1}{2x^2 - 6x + 1} = \frac{G'}{(x - p_1)(x - p_2)}$$

The plot of this is shown in Figure 10.17(a). The shape of this curve is very different, but is again completely characterised by the roots plus a gain factor. This time values at the roots cause the function to "blow up", $x \to p_1$, $f(x) \to \pm\infty$. Such roots are called **poles** because they create a "pole-like" effect in the absolute (magnitude) version of the function, shown in Figure 10.17(b).

We will now show how polynomial analysis can be applied to the transfer function. A polynomial defined in terms of the complex variable z takes on just the same form as when defined in terms of x. The z form is actually less misleading, because in general the roots will be complex (e.g. $f(z) = z^2 + z - 0.5$ has roots $0.5 + 0.5j$ and $0.5 - 0.5j$.). The transfer function is defined in terms of negative powers of z – we can convert a normal polynomial into one in negative powers by multiplying by z^{-N}. So a second-order polynomial is

$$H(z) = \frac{1}{z^2 - a_1 z - a_2} = \frac{z^{-2}}{1 - a_1 z^{-1} - a_2 z^{-2}} = G \frac{z^{-2}}{(1 - p_1 z^{-1})(1 - p_2 z^{-1})}$$

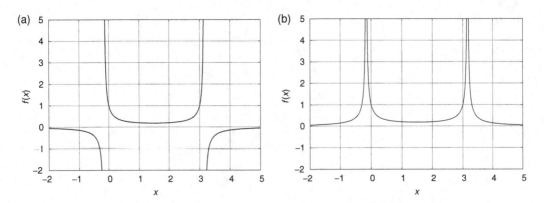

Figure 10.17 Plots of functions with a polynomial as the denominator: (a) $1/(2x^2 - 6x + 1)$ and (b) $|1/(2x^2 - 6x + 1)|$. The points where the function $\to \infty$ are called "poles".

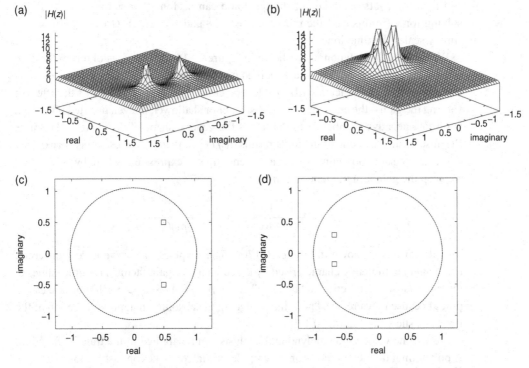

Figure 10.18 z-Domain and pole–zero plots for two filters, A and B: (a) z-domain, filter A; (b) z-domain, filter B; (c) pole–zero plot, filter A; and (d) pole–zero plot, filter B.

Figure 10.18(a) shows a three-dimensional plot of the above transfer function with $a_1 = 1.0$ and $a_2 = -0.5$. The numerator has two roots at $z = 0$, which form zeros. The roots of the denominator are complex ($0.5 + 0.5j$ and $0.5 - 0.5j$) and it can be seen that the poles do not now lie on the x-axis (which would happen only if they were completely real).

Since the poles and zeros completely characterise the filter (apart from the gain), we can dispense with the need to draw three-dimensional graphs and simply plot where the poles and zeros lie on the z-plane. This is shown in Figure 10.18(c). Figures 10.18(b) and 10.18(d) show the three-dimensional plot and pole–zero plot for a different second-order filter.

This analysis can be extended to any digital LTI filter. Since the general-form transfer function, $H(z)$ is defined as the ratio of two polynomials

$$H(z) = \frac{b_0 + b_1 z^{-1} + \cdots + b_M z^{-M}}{1 - a_1 z^{-1} - \cdots - a_N z^{-N}}$$

this can always be expressed in terms of its factors:

$$H(z) = g \frac{(1 - q_1 z^{-1})(1 - q_2 z^{-1}) \cdots (1 - q_M z^{-1})}{(1 - p_1 z^{-1})(1 - p_2 z^{-1}) \cdots (1 - p_N z^{-1})}$$

from which pole–zero analysis can be performed. Furthermore, a factored filter can be split into smaller filters:

$$H(z) = g \underbrace{\frac{1 - q_1 z^{-1}}{1 - p_1 z^{-1}}}_{} \cdot \underbrace{\frac{1 - q_2 z^{-1}}{1 - p_2 z^{-1}}}_{} \cdots \underbrace{\frac{1 - q_M z^{-1}}{1 - p_M z^{-1}}}_{}$$

$$= H_1(z) H_2(z) \ldots H_N(z) \tag{10.66}$$

which is often a useful step when constructing or analysing a complicated filter.

10.5.2 Frequency interpretation of the z-domain transfer function

Recall that we showed the relationship between the z-transform

$$X(z) = \sum_{n=-\infty}^{\infty} x[n] z^{-n}$$

and the discrete-time Fourier transform by setting $z = e^{j\hat{\omega}}$:

$$X(e^{j\hat{\omega}}) = \sum_{-\infty}^{\infty} x[n] e^{-j\hat{\omega}n}$$

Given any z-domain expression, we can generate the equivalent frequency-domain expression by simply replacing z with $e^{j\omega}$. In Section 10.2.6, we explained that the spectrum $X(e^{j\hat{\omega}})$ of a digital signal is periodic with period 2π for normalised frequency. This can be thought of as a function that sweeps out a circle of radius 1 in the z-domain. The frequency domain is in effect the particular "slice" on the z-domain that this circle describes.

Let us return to the three-dimensional representation of the magnitude of $H(z)$, for an example second-order filter, as shown in Figure 10.19(a). This is as before, but the unit circle is now superimposed. The value at the unit circle rises and falls with the shape of the plot, and, when we "unfold" this, we get a two-dimensional graph of magnitude against frequency as shown in Figure 10.19(b). The phase spectrum can be generated in the same

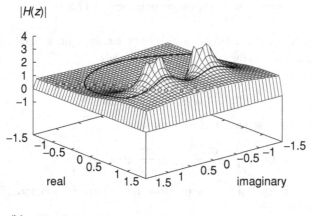

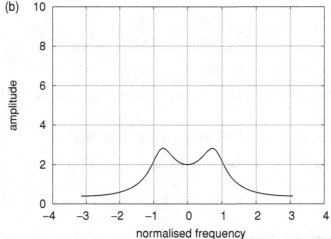

Figure 10.19 The relationship between the z-domain (a) and the frequency domain (b). The unit circle drawn on the z-domain diagram can be unfolded and straightened to produce the frequency response shown in part (b).

way, that is, by taking the values from the unit circle from the z-domain phase function. As previously mentioned, we can dispense with three-dimensional graphs because the pole–zero diagram fully characterises the transfer function. The frequency domain can easily be obtained from the z-domain by use of the poles and zeros.

10.5.3 Filter characteristics

We will now use the above results on pole–zero analysis and frequency-domain interpretation to examine the properties of first-order and second-order filters, and show how these can be extended to the general case of all digital LTI filters.

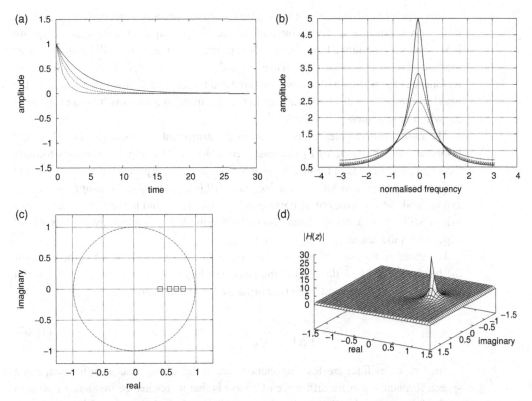

Figure 10.20 Plots of a first-order IIR filter, with $a_1 = 0.8, 0.7, 0.6$ and 0.4: (a) time-domain impulse response, (b) magnitude spectrum of frequency response, (c) pole–zero plot and (d) magnitude z-domain transfer function. As the length of decay increases, the frequency response becomes sharper. Because only a single coefficient is used, there will be one pole, which will always lie on the real axis. As $a_1 \to 1$, the impulse response will have no decay and the pole will lie at 1.0. Because the pole lies on the unit circle, it will lie in the frequency response, and hence there will be an infinite value for the frequency at this point in the spectrum.

Consider again a first-order IIR filter:

$$h[n] = b_0 x[n] - a_1 y[n-1]$$

$$H(z) = \frac{b_1}{1 - a_1 z^{-1}}$$

$$H(e^{j\omega}) = \frac{b_1}{1 - a_1 e^{-j\omega}}$$

Figure 10.20 shows the time-domain impulse response, z-domain pole–zero plot and frequency-domain magnitude spectrum for this filter with $b_0 = 1$ and a_1 taking the values $0.8, 0.7, 0.6$ and 0.4. Figure 10.20(d) shows a three-dimensional z-domain magnitude plot for a single value, $a_1 = 0.7$.

It is clear that the value of the coefficient a_1 completely describes the shape of the three functions. For a single-term polynomial, the root of $1 - a_1 z^{-1}$ is simply a_1. Since the filter coefficient is real, so is the root, and hence the pole will always lie on the real axis in the pole–zero plot.

This type of filter is commonly called a **resonator** because it accurately models the types of resonances commonly found in nature. The peak in the the frequency response is known as a **resonance** because signals at or near that frequency will be amplified by the filter. For our purposes, it provides a good model of a single vocal-tract **formant**. Resonances are normally described by three properties: **amplitude** – the height of the peak, **frequency** – where the peak lies in the spectrum, and **bandwidth** – a measure of the width or sharpness of the peak.

The radius of the pole, r, controls both the **amplitude** and the **bandwidth** of the resonance. As the value of a_1 decreases, its pole moves along the real axis towards the origin, and hence the frequency response has a lower amplitude at $\hat{\omega} = 0$. From the difference equations, we know that a first-order IIR filter with $a_1 < 1$ is simply a decaying exponential. Small values of a_1 correspond to slow decay and narrow bandwidth, large values (still <1) to a steeper decay and wider bandwidth. This is exactly what we would expect from the scaling property of the Fourier transform.

The standard measure of bandwidth is the **3-dB down point**, which is the width of the resonance at $1/\sqrt{2}$ down from the peak. For larger values of $r = |a_1|$, two useful approximations of the bandwidth (in normalised frequency) are

$$\hat{B} \approx \begin{cases} -2\ln(r) \\ 2(1-r)/\sqrt{r} \end{cases} \quad 0.5 < r < 1.0 \tag{10.67}$$

The first-order filter creates a resonance pattern that already has a similar shape to a speech formant; the main difference of course is that its resonance frequency is at zero on the frequency axis. The resonance can easily be shifted from $\hat{\omega} = 0$ by moving its pole off the real axis, and this is shown in Figure 10.21. The effect of the pole position on the frequency response can be seen more clearly by expressing the pole in polar form, $p_1 = re^{j\theta}$. The resonance frequency is given by θ, the angle of the pole, and the bandwidth is controlled as before by r.

The first-order resonator as just described has one serious drawback: because its pole is not on the real axis, its coefficient a_1 will be complex also, which is a violation of the condition that all the coefficients of the polynomial should be real. This problem can be countered by the use of **complex-conjugate pairs** of poles. If we have a pole with the desired values $re^{j\theta}$, we simply create a second pole with value $re^{-j\theta}$. This will ensure that the coefficients are always real, and, because the pole has a negative value for θ, its resonance will occur in the negative frequency range, leaving a single pole in the positive range. If we have two poles, we naturally now have a second-order filter:

$$h[n] = b_0 x[n] - a_1 y[n-1] - a_2 y[n-2]$$

$$H(z) = \frac{b_1}{1 - a_1 z^{-1} - a_2 z^{-2}}$$

$$H(z) = \frac{1}{(1 - p_1 z^1)(1 - p_2 z^1)}$$

$$H(e^{j\omega}) = \frac{b_1}{1 - a_1 e^{-j\omega} - a_2 e^{-2j\omega}}$$

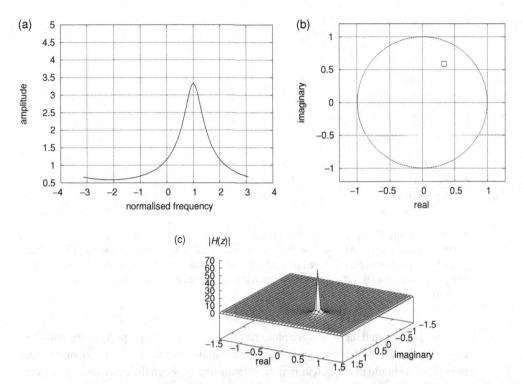

Figure 10.21 Plots of a first-order IIR filter, with the single pole at $0.37 + 0.58j$: (a) the magnitude spectrum of the transfer function, (b) pole–zero plot and (c) the magnitude z-domain transfer function. Since the pole is not on the real axis, the resonance occurs at a non-zero frequency.

The full frequency response can be seen more clearly by expressing the poles in polar form:

$$p_1 = re^{j\theta}$$
$$p_2 = p_1^* = re^{-j\theta}$$

Using this we can write the transfer function and frequency response as

$$H(z) = \frac{1}{(1 - re^{j\theta}z^{-1})(1 - re^{-j\theta}z^{-1})}$$

$$= \frac{1}{1 - 2r\cos(\theta)z^{-1} + r^2 z^{-2}}$$

$$H(e^{j\omega}) = \frac{1}{1 - 2r\cos(\theta)e^{-j\omega} + r^2 e^{-j2\omega}}$$

The following table gives four sets of example poles and coefficients for this filter, and the pole–zero plot and frequency responses for these are shown in Figure 10.22:

r	θ	p_1	p_2	a_1	a_2
0.9	1.0	$0.48 + 0.75j$	$0.48 - 0.75j$	0.97	−0.81
0.8	1.0	$0.43 + 0.67j$	$0.43 - 0.67j$	0.86	−0.64
0.7	1.0	$0.38 + 0.59j$	$0.38 - 0.59j$	0.75	−0.48
0.6	1.0	$0.32 + 0.51j$	$0.32 - 0.51j$	0.65	−0.36

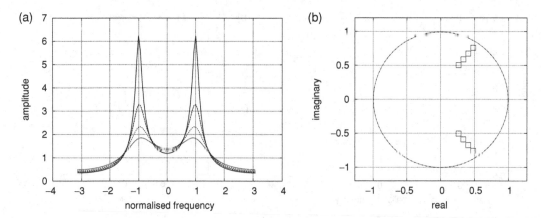

Figure 10.22 Plots of a second-order IIR filter for a series of poles: (a) pole–zero plot and (b) frequency response. All have the same angle, but differ in distance from the unit circle. The poles in the pole–zero diagram are conjugate symmetric, meaning that they are symmetric in the real axis. Because of this the frequency response will have peaks at the same positive and negative values.

From the table and the pole–zero plot, the complex-conjugate property of the poles can clearly be seen. This property gives rise to the symmetry in the frequency domain about the y-axis. It should also be clear that the relationship between the poles and coefficients is less obvious. These example responses were defined by keeping the angle θ constant, while varying the distance of the pole from the origin, r. It can be seen that as r approaches 1, and the position of the pole nears the unit circle, the frequency response becomes sharper. Exactly the same effect was observed with the first-order filter. Moving the poles off the real axis has the effect of moving the peaks away from the origin in the frequency response. For larger values of r (>0.5) the resonance frequency is approximated by θ, as for the first-order complex pole. For smaller values, the approximation breaks down because the **skirts** of each pole (that is, the roll-off areas on either side of the actual resonance) interact and cause a shifting of frequency.

Figure 10.23 and the table below show the effect of keeping r constant while varying the angle θ:

r	θ	p_1	p_2	a_1	a_2
0.8	0.75	0.58 + 0.54j	0.58−0.54j	1.17	−0.64
0.8	1.0	0.43 + 0.67j	0.43−0.67j	0.86	−0.64
0.8	1.25	0.25 + 0.76j	0.25−0.76j	0.50	−0.64
0.8	1.5	0.056 + 0.78j	0.05−0.78j	0.11	−0.64

As we might expect, the peaks now all have similar shapes, but their locations in the frequency response are different.

In general it is the poles which define the most important characteristics of a resonator, namely the resonance amplitude, frequency and bandwidth. Zeros do have a role to play and can be used to achieve other effects in the spectrum. Figure 10.24 shows the effect

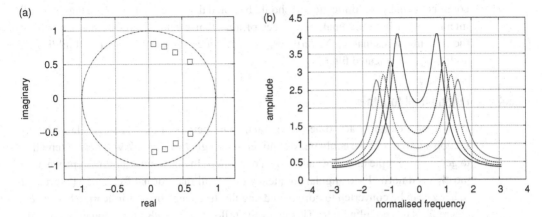

Figure 10.23 Plots of a second-order IIR filter for a series of poles with the same radius but different angles: (a) pole–zero plot and (b) frequency response. The positions of the peaks vary with the angle of the poles. The bandwidth of the peaks is constant, but the amplitudes differ because as the poles near the frequency axis their skirts affect each other, giving each peak a lift.

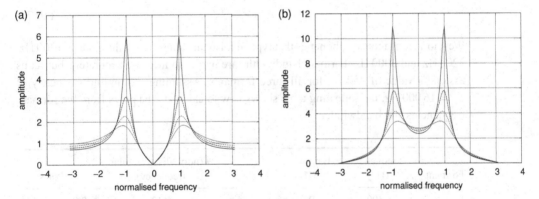

Figure 10.24 Plots of a second-order IIR filter with poles and zeros for various pole positions: (a) pole–zero plot and (b) frequency response. Part (a) has a zero on the unit circle on the real axis. Part (b) has a zero on the unit circle on the imaginary axis.

of using a term $b_1 = 1$, which will place a zero at the origin, and $b_1 = -1$, which will place a zero at the ends of the spectrum. Extra zeros can be used to generate notches or **anti-resonances** in the spectrum.

We know that the vocal tract has multiple formants. Rather than developing more and more complicated models to relate formant parameters to transfer functions directly, we can instead make use of the factorisation of the polynomial to simplify the problem. Recall from Equation (10.66) that any transfer-function polynomial can be broken down into its factors. We can therefore build a transfer function of any order by combining simple first- and second-order filters:

$$H(z) = H_1 H_2 \ldots H_N \qquad (10.68)$$

This greatly simplifies the construction of a complicated filter since we can make simple resonators in isolation using pole–zero analysis to set their parameters, and afterwards combine them by multiplication. It should be noted that poles that are close together will interact, and hence the final resonances of a system cannot always be predicted from the constituent resonators. Nevertheless, this provides a powerful model for building an arbitrarily complicated filter.

10.5.4 Putting it all together

We will now demonstrate some of the practical properties of the LTI filter. The previous section showed how we could build a model of a single formant. We will now extend this to generate a simple three-formant model of a vowel. In discussing digital filters, we have mostly used normalised angular frequency $\hat{\omega}$. Recall that normal frequency (expressed in hertz) can be converted to normalised angular frequency by multiplication by $2\pi/F_s$, where F_s is the sampling rate. This gives the following formulas for estimating the pole radius and angle for a single formant of peak frequency f and bandwidth B:

$$\theta = \frac{2\pi F}{F_s} \qquad (10.69)$$

$$r = e^{-2\pi B/F_s} \qquad (10.70)$$

We know from acoustic phonetics that typical formant values for an /ih/ vowel are 300 Hz, 2200 Hz and 3000 Hz. Formant bandwidths are harder to measure accurately, but let us assume a value of 250 Hz for all three formants. Assuming a sampling frequency of $F_s = 16\,000$ Hz, the following table shows how to calculate the poles from the formant frequencies and bandwidths:

Formant	Frequency (Hz)	Bandwidth (Hz)	r	θ (normalised angular frequency)	Poles
F1	300	250	0.95	0.12	$0.963 \pm 0.116j$
F2	2200	250	0.95	0.86	$0.619 \pm 0.719j$
F3	3000	250	0.95	1.17	$0.370 \pm 0.874j$

We construct a second-order IIR filter for each formant, F_n:

$$H_n(z) = \frac{1}{(1 - p_n z^{-1})(1 - p_n^* z^{-1})}$$

where p_n is the pole and p_n^* is complex conjugate of the pole for the formant n. We create the complete transfer function $H_{/ih/}$ for the /ih/ vowel by simply multiplying each formant's transfer function:

$$H_{/ih/}(z) = H_1(z)H_3(z)H_3(z)$$

$$= \frac{1}{(1 - p_1 z^{-1})(1 - p_1^* z^{-1})(1 - p_2 z^{-1})(1 - p_2^* z^{-1})(1 - p_3 z^{-1})(1 - p_3^* z^{-1})}$$

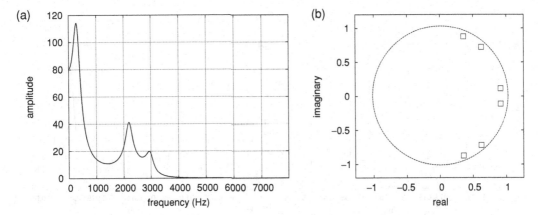

Figure 10.25 A three-formant, six-pole model of the /ih/ vowel: (a) frequency response and (b) pole–zero plot.

Figure 10.25 shows the pole–zero plot and frequency response of the transfer function. From the plot, we can see that the formant frequencies do indeed appear at approximately the desired locations.

By definition, H gives the transfer function and frequency response for a unit impulse. In reality, of course, the vocal-tract input for vowels is the quasi-periodic glottal waveform. For demonstration purposes, we will examine the effect of the /ih/ filter on a square wave, which we will use as a (very) approximate glottal source. We can generate the output waveforms $y[n]$ by using the difference equation, and find the frequency response of this vowel from $H(e^{j\omega})$. The input and output in the time domain and frequency domain are shown in Figure 10.26. If the transfer function does indeed accurately describe the frequency behaviour of the filter, we should expect the spectrum of $y[n]$ calculated by use of the DFT to match $H(e^{j\omega})X(e^{j\omega})$. We can see from Figure 10.26 that indeed it does.

Filtering a periodic signal demonstrates an important point of all LTI filters. While they amplify and change the phase of components of an input signal, they do not change the fundamental period of the signal itself. In Figure 10.26, it is clear that the output waveform has the same period as the input, and the harmonics have the same spacing in the input and output. Again, this clearly models speech behaviour – if we generate a voiced sound with our glottis, we can move our mouth into a number of configurations, each of which will produce a different sound, but in all cases the fundamental frequency of the sound is unchanged.

10.6 Summary

Periodic signals

- A complex exponential, $e^{j\omega t}$, describes a sinusoidal waveform.
- A harmonic is any integer multiple of the fundamental frequency of a periodic waveform.

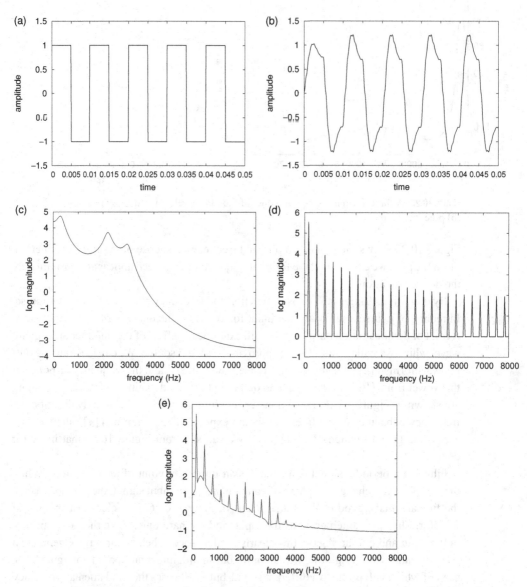

Figure 10.26 Operation of a simulated /ih/ vowel filter on a square wave. Part (b) shows the time-domain output for the square-wave input (a). Although the shape of the output is different, it has exactly the same period as the input. Part (d) is the magnitude spectrum of the square wave. Part (e) is the magnitude spectrum of the output, calculated by use of the DFT. The harmonics are at the same intervals as in the input spectrum, but each has had its amplitude changed. Part (e) demonstrates that the effect of the filter is to multiply the input spectrum by the frequency response, $H(e^{j\omega})$, part (c).

- Any periodic waveform can be made from a sum of suitably scaled and phase-shifted sinusoids. This is the Fourier series.
- Fourier analysis can be used to find the amplitudes and phases of the component harmonics in a periodic waveform.

The frequency domain

- A representation in the frequency domain is called a spectrum.
- Spectra are complex, and are usually represented by the magnitude spectrum and phase spectrum.
- The ear is relatively insensitive to phase, so the magnitude spectrum is normally sufficient to characterise the frequency-domain representation of a signal.
- For all real signals, the spectrum is conjugate symmetric.

Digital signals

- A digital signal can be obtained from a continuous signal by sampling at even intervals, given by the sampling period.
- The sampling rate is the reciprocal of the sampling period.
- A digital signal has an upper bound on the frequencies it can represent. This is called the Nyquist frequency and is exactly half the sample rate.
- Use of normalised frequency, $\hat{\omega} = \omega/F_s$, avoids the need to keep track of the sampling rate in digital-signal-processing expressions.
- Spectra obtained from digital signals are periodic, with normalised period 2π.

Transforms

- The Fourier transform converts a continuous waveform into a continuous spectrum:

$$X(j\omega) = \int_{-\infty}^{\infty} x(t) e^{-j\omega t} \, dt$$

- The discrete-time Fourier transform converts a digital waveform into a continuous spectrum:

$$X(e^{j\hat{\omega}}) = \sum_{-\infty}^{\infty} x[n] e^{-j\hat{\omega} n}$$

- The discrete Fourier transform converts a digital waveform into a discrete spectrum:

$$X[k] = \sum_{n=0}^{N-1} x[n] e^{-j\frac{2\pi}{N} kn}$$

- The z-transform converts a digital waveform into a continuous z-domain representation:

$$X(z) = \sum_{n=-\infty}^{\infty} x[n] z^{-n}$$

- All transforms have the properties of linearity, scaling, time delay, frequency shift and convolution.

Digital filters

- Digital, time-invariant linear filters operate on an input signal $x[n]$ and produce an output signal $y[n]$.
- A filter is fully defined by two sets of real coefficients $\{a_l\}$ and $\{b_k\}$.
- The operation of the filter in the time domain is given by the difference equation

$$y[n] = \sum_{k=0}^{M} b_k x[n-k] + \sum_{l=1}^{N} a_l y[n-l] \qquad (10.71)$$

- The time-domain characteristics of a filter are fully captured by the response $h[n]$ to a unit impulse $\delta[n]$.
- The operation of the filter in the z-domain is given by the transfer function

$$H(z) = \frac{b_0 + b_1 z^{-1} + \cdots + b_M z^{-M}}{1 - a_1 z^{-1} - \cdots - a_N z^{-N}}$$

- The transfer function $H(z)$ is the z-transform of the impulse response $h[n]$.
- The frequency response of the filter is given by setting $z = e^{j\hat{\omega}}$, which gives

$$H(e^{-j\hat{\omega}}) = \frac{b_0 + b_1 e^{-j\hat{\omega}} + \cdots + b_M e^{-jM\hat{\omega}}}{1 - a_1 e^{-j\hat{\omega}} - \cdots - a_N e^{-jN\hat{\omega}}}$$

- The z-domain transfer function can be factored into two polynomials, the roots of which are called poles and zeros:

$$H(z) = g \frac{(1 - q_1 z^{-1})(1 - q_2 z^{-1}) \ldots (1 - q_M z^{-1})}{(1 - p_1 z^{-1})(1 - p_2 z^{-1}) \ldots (1 - p_N z^{-1})}$$

- Poles have an easily understood interpretation, and, in many cases, formant frequency and bandwidth can be directly related to pole values. By suitable choice of pole, a frequency response with appropriate resonances can usually be constructed.
- Complicated filters can be created by combining simpler filters.
- Filters change the amplitude and phase of the components in the input. They do not change the fundamental period or fundamental frequency.

11 Acoustic models of speech production

The speech-production process was qualitatively described in Chapter 7. There we showed that speech is produced by a source, such as the glottis, which is subsequently modified by the vocal tract acting as a filter. In this chapter, we turn our attention to developing a more-formal quantitative model of speech production, using the techniques of signals and filters described in Chapter 10.

11.1 The acoustic theory of speech production

Such models often come under the heading of the **acoustic theory of speech production**, which refers both to the general field of research in mathematical speech-production models and to the book of that title by Fant [158]. Although considerable previous work in this field had been done prior to its publication, this book was the first to bring together various strands of work and describe the whole process in a unified manner. Furthermore, Fant backed his study up with extensive empirical studies with X-rays and mechanical models to test and verify the speech-production models being proposed. Since then, many refinements to the model have been made, as researchers have investigated trying to improve the accuracy and practicalities of these models. Here we focus on the single most widely accepted model, but conclude the chapter with a discussion on variations on this.

As with any modelling process, we have to reach a compromise between a model that accurately describes the phenomena in question and one that is simple, effective and suited to practical needs. If we tried to capture every aspect of the vocal organs directly, we would have to account for every muscle movement, articulator shape and tissue absorption characteristic directly, and then determine how each of these affected speech production. This would be a huge undertaking, and, even if an accurate description of all these factors could be determined, the result could be too complicated for useful analysis. We therefore have to make quite a number of simplifying assumptions in order to obtain a usable model.

11.1.1 Components in the model

Our first task is to build a model in which the complex vocal apparatus is broken down into a small number of independent components. One way of doing this is shown in

Acoustic models of speech production

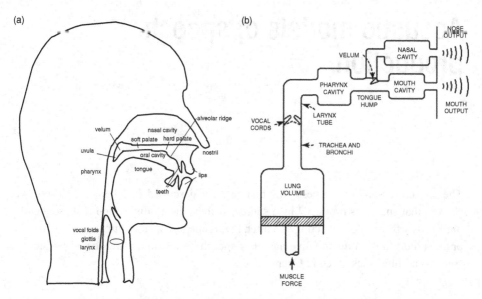

Figure 11.1 A diagram and a model of the vocal organs: (a) a mid-sagittal drawing of vocal organs and (b) a model of vocal organs with discrete components identified.

Figure 11.1(b), where we have modelled the lungs, glottis, pharynx cavity, mouth cavity, nasal cavity, nostrils and lips as a set of discrete, connected systems. If we make the assumption that the entire system is linear (in the sense described in Section 10.4), we can then produce a model for each component separately and determine the behaviour of the overall system from the appropriate combination of the components. While of course the shape of the vocal tract will be continuously varying in time when speaking, if we choose a sufficiently short time frame, we can consider the operation of the components to be constant over that short period of time. This, coupled with the assumption of linearity, then allows us to use the theory of linear time-invariant (LTI) filters (Section 10.4) throughout. Hence we describe the pharynx cavity, mouth cavity and lip radiation as LTI filters, and so the speech-production process can be considered as the operation of a series of z-domain transfer functions on the input.

For example, in the case of vowels, speech is produced by the glottal source waveform travelling through the pharynx, and, as the nasal cavity is shut off, the waveform progresses through the oral cavity and is radiated into the open air via the lips. Hence, since filters connected in series are simply multiplied in the z-domain, we can write the system equation for vowels as

$$Y(z) = U(z)P(z)O(z)R(z) \tag{11.1}$$

where $U(z)$ is the glottal source, with $P(z)$, $O(z)$ and $R(z)$ representing the transfer functions of the pharynx, the oral cavity and the lips, respectively. Since $P(z)$ and $O(z)$ linearly combine, it is normal to define a single **vocal-tract transfer function** $V(z) = P(z)O(z)$, such that Equation (11.1) is written

$$Y(z) = U(z)V(z)R(z) \tag{11.2}$$

which can be represented in the (discrete) time domain as

$$y[n] = u[n] \otimes v[n] \otimes r[n] \qquad (11.3)$$

The configuration of the vocal tract governs which vowel sound is produced, and by studying this we can gain an understanding of how the physical properties of air movement relate to the transfer function and of the sounds produced. In a similar fashion, transfer functions can be determined for the various types of consonants. To determine the form these transfer functions take, we must investigate the physics of sound, so this is dealt with next.

11.2 The physics of sound

11.2.1 Resonant systems

We have informally observed that the vocal-tract filter acts as a resonator; that is, it amplifies certain frequencies and attenuates others. How does this behaviour arise?

The resonant nature of systems is often demonstrated by considering the motion of a mass and spring as shown in Figure 11.2. Imagine that this system is set in motion by displacing the mass to the right and then letting go. The extended spring will try to return to its original position and in doing so will set the mass in motion. Initially, the speed of movement will be slow because the inertia of the mass must be overcome. As the spring reaches its original position, it will no longer exert any force, but, since the mass is now moving, it will continue to move and pass the original position, moving to the left. As it does so, the spring will again be displaced from its original position. The more this is done, the stronger the force will become, until eventually the mass stops and the spring pulls the mass towards the at-rest position again. From Newton's laws we also know that the system will continue moving until acted upon by an external force.

The frequency at which the system oscillates is determined by the **inertia** of the mass and the **compliance** of the spring,

$$\text{resonance frequency} = \frac{1}{2\pi}\sqrt{\frac{k}{m}}$$

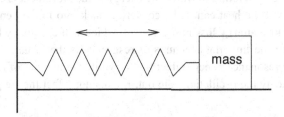

Figure 11.2 A simple mass/spring oscillator. If the mass is displaced from its original, at-rest position, it will indefinitely oscillate horizontally.

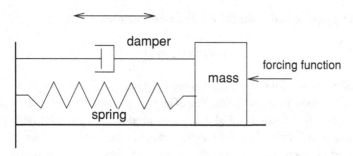

Figure 11.3 A mass/spring/damper system driven by an external force. Here a periodic force is applied, and this will determine the frequency at which the system oscillates. As the driving frequency nears the natural resonance frequency (determined by the mass and spring) the size of the oscillations will increase.

Hence a heavier mass will mean a slower oscillation and a stiffer spring will mean a quicker oscillation. The important point is that the system has a single resonance frequency determined by the physical properties of the mass and spring.

The above equates to a source–filter system wherein a single impulse has been given to the system: we saw in Chapter 10 that the impulse response from a system is a good way to characterise the system itself independently of the input. Now let us consider the case when a periodic **forcing function** is exerted on the mass, as shown in Figure 11.3. Assuming that the force is of sufficient strength, the mass will move with the periodicity of the driving (source) force. Again, this is what we would expect from Chapter 10, where we showed that the source (not the filter) largely determines the output frequency of a system. Imagine now that we send sources of different frequencies into the system. We will find that at frequencies near to the resonance frequency the size of the oscillations will increase. In fact, at the resonance frequency itself, the oscillations will get larger and larger, and, if no other factors are involved, will eventually become so large as to break the system.[1] Hence a mass–spring system with a driving force acts as a linear filter fed with a source.

The above is of course somewhat idealised because we know that in general systems don't shake themselves apart, and, in the unforced case, when a mass and spring are set in motion, eventually the motion will stop. This is due to the third property, known as **resistance**. In the mechanical world, resistance is opposition to velocity, and is often described as a **damping** effect on the system shown by the damper in Figure 11.3. Normally resistance is caused by **friction** in which the energy in the movement is converted into heat. Since in general the heat cannot be converted back into motion energy, the effect of friction is to cause **energy loss** in the system, which will eventually bring the system to a halt. One of the important features of resistance is that usually its effect on the frequency of the resonance is negligible; that is, while the amount of resistance will determine how quickly the oscillations die out, it does not affect the frequency at

---

[1] This is what happens when things "shake themselves apart", most famously the Tacoma Narrows bridge.

which those oscillations occur. This is an important point for our model, because it often allows us to study the frequencies of resonances without needing to consider the damping effects.

All these effects were shown in our examination of linear filters in Chapter 10. For example, Figure 10.15(a) shows the displacement of the mass against time for the undamped case, whereas Figure 10.15(b) shows the same system with damping. Figures 10.22 and 10.23 show the separation of frequency and damping effects on a pole–zero plot. There, the angle of the pole or zero determined the frequency, whereas the radius of the pole determined the damping, with poles on the unit circle meaning that the damping was zero. In terms of a frequency response, the damping manifests itself as amplitude and bandwidth of the resonance peak, with high amplitude and narrow bandwidth arising from cases of low damping.

The three factors of inertia (mass), resistance and capacitance (the reciprocal of spring compliance) are collectively known as **impedance** because, intuitively, the system "impedes" the motion in the system.

11.2.2 Travelling waves

The behaviour of the vocal tract is in many ways similar to the behaviour of the mass/spring/damper system just described. The input "forcing function" is the glottal source, which is used to drive the system (the vocal tract) whose physical properties determine the resonance characteristics, which then modify the input signal so as to amplify some frequencies and attenuate others. One crucial difference however, is that the mass/spring/damper system is what is known as a **lumped-parameter** system in that it comprises easily identifiable, discrete components such as the mass and spring. By contrast, the vocal tract is a **distributed system**, in which the properties of inertia, capacitance and resistance are evenly distributed throughout the system. We can formulate the properties of a distributed system by modelling it as a large number of small connected discrete systems. Using calculus, we can determine the behaviour of a distributed system by considering the behaviour of the system as the size of the elements tends to zero. This steps involved in this are fully described in Rabiner and Schafer [368]; for our purposes though, we will simply take the results of this analysis and apply it to the problem of the propagation of sound in the vocal tract.

Before considering this directly, we will appeal again to an equivalent physical system, since this will let us examine an aspect of resonance behaviour that is central to our study, namely that of **travelling waves**. Consider first the easy-to-visualise example of waves travelling in a rope, one end of which is in a person's hand and the other is tied to a wall. This is shown in Figures 11.4 and 11.5. If the person introduces some energy into the rope by giving the rope a jerk (i.e. an impulse), a wave will form at that end and travel down the rope. When the wave reaches the tied end at the wall, it will be reflected, and will travel backwards towards the person. When it reaches the held end, the wave will be reflected back again towards the wall and so on, until all the energy has eventually been lost from the system.

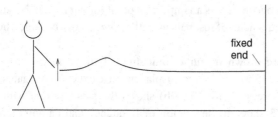

Figure 11.4 If the held end of the rope is given a jerk, this creates an impulse, which will travel along the rope towards the wall.

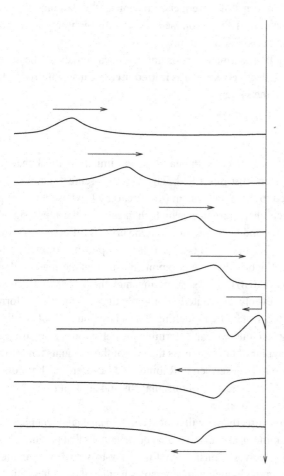

Figure 11.5 The impulse in the rope travels towards the wall, where it is reflected out of phase with the original pulse. The reflected pulse travels backwards along the rope.

The person can create a forcing function (i.e. source) input to the system by moving their hand in a repetitive fashion. In doing so, a series of pulses will form and travel forwards down the rope towards the wall, and these pulses will create a **travelling wave**, the frequency of which is dictated by the hand movement. When this forward travelling wave is reflected, a backward travelling wave will be created. The forward wave and backward wave will **interfere** – if the two have a peak at a given point, a large peak will occur, whereas if one has a negative peak while the other has a positive peak then these will cancel out, leaving the rope at that point in its neutral position. In general, the observed shape of the rope will not form a simple pattern. However, at some frequencies, the reflected wave will exactly reinforce the forward wave, and the rope will settle down into a fixed pattern. In such cases, the wave does not look like it is travelling at all, so this is termed a **standing wave**. The term is a little misleading because it might be taken to mean that there is no motion – this is not the case. It merely appears that the waves are not moving. Standing waves are created only at certain frequencies. At these frequencies, the wave is amplified, that is the overall displacement in the rope at a peak is greater than the input. Thus the rope acts as a resonator, amplifying certain frequencies and attenuating others. The frequencies at which the resonances occur are governed by the length of the rope and the speed of wave travel.

The above demonstrates some of the fundamental properties of travelling waves. First, note that some of the behaviour is determined by the motion of the hand, while some is determined by the properties of the rope itself. For example, the frequency of the waves is determined purely by the rate of movement of the hand; the rope itself has no influence on this at all. In contrast, the rate at which the wave travels along the rope is not influenced by the hand at all; no matter how quickly or slowly the hand moves, the wave always travels at the same speed, and this speed is determined by the properties of the rope. This shows why it is appropriate to use the theory of signals and filters where an initial source (the wave created by the moving hand) has its behaviour modified by a filter (the rope). Furthermore, we should also clearly see the parallels between this and the behaviour of the glottis source and vocal-tract filter in speech production.

What then determines the behaviour of the rope? Firstly, we have the physical properties of the material of the rope; its inherent stiffness, its length, its mass and so on. Secondly, we have what are known as the **boundary conditions**, that is, the specifics of how the rope is terminated at either end. In the above example, one end of the rope was attached to a wall; there are other possibilities, such as the end of the rope being free – in such a case reflection also occurs, but in a different way from that in the fixed case.

11.2.3 Acoustic waves

The properties of sound waves travelling in a **tube** are very similar to those of waves travelling in a rope. Instead of the wave travelling by means of moving the rope, a sound wave travels in air by a process of displacement of the particles of air. In the absence of any particular sound, the air has certain ambient properties, such as pressure and temperature. The effect of a sound source in the air causes the particles to move backwards and forwards, and the wave thereby spreads from the source. This is shown in Figure 11.6, where a forcing function at one end of the tube causes a disturbance, which travels along the tube, causing **compressions**, whereby the particles come closer together,

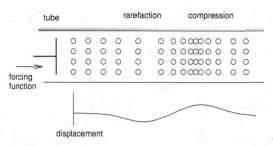

Figure 11.6 A longitudinal sound wave travelling along a tube. The wave travels by means of compression and rarefaction of the air particles. The displacement of a slice of air particles against position in the tube is shown below.

and **rarefaction**, whereby they move further apart. This is similar to the situation with the rope, in which the wave travels along the rope, but the rope itself is only temporarily displaced; the rope doesn't end up travelling with the wave.

One difference is that sound waves are **longitudinal** whereas waves in a rope are **transverse**. This means that, rather than the particles of sound moving up and down relative to the direction of wave motion, they move backwards and forwards along the axis of wave motion. Despite this significant physical difference, the behaviour can be modelled in exactly the same way. As with the rope, the specifics of how the wave travels in the air are determined by the physical properties of the air and, in the case of standing waves, by the boundary conditions. The physics of general sound-wave propagation is complicated and forms an extensive area of study. For our purposes, however, we need consider only one specific type of sound propagation, that of sound propagation in a tube. Furthermore, we can assume that the tube has constant cross-sectional area along its length, since this can be generalised to cases in which the tube varies in area.

An important property of the air is its **speed of sound**, denoted by c. For a given pressure and temperature, this is constant, and, although it can be calculated from more fundamental properties, it is easily measured empirically. A typical value is that a sound wave travels 340 metres in one second at room temperature and pressure. By speed of sound we mean the distance travelled by one part of the wave in unit time. Note the similarity to the rope; since the speed of sound propagation is constant, it doesn't matter what the source does, all waves travel at exactly the same speed.

In sound propagation in general, the signal or wave is manifested as **particle velocity**. This is the pattern of movement of air particles which makes up the sound wave. In tubes, however, the mathematics is considerably simplified if we use a related quantity, **volume velocity**, which is just the particle velocity multiplied by the area. Since the area of the tube is constant, this doesn't complicate our analysis. The air in the tube has an impedance Z, and the effect of this is to "impede" the volume velocity. This brings us to **pressure**, which can be thought of as the work required to move the air particles through a particular impedance. The fundamental equation linking the three is

$$P = ZU \qquad (11.4)$$

Hence for a constant pressure a high impedance will mean a reduced particle velocity, or, conversely, to maintain a constant particle velocity, more pressure will be needed if the impedance increases. This can easily be visualised: imagine pushing something in a pipe. If the pipe is unconstricted (nothing "impeding", low impedance) a small amount of pressure (effort) will move something down the tube. If the tube is blocked, considerably more effort will be required to move the object.

Acoustic impedance is formulated in a similar fashion to the mechanical impedance described in Section 11.2.1. First we have **acoustic inductance**, meaning that, when a sound wave passes through the air, the inertia in the air impedes any acceleration of the air particles. Next we have a property called **acoustic capacitance**, which is equivalent to the effect of the spring in the mechanical system. This arises because the air is compressible, and the capacitance is a measure of how strongly the air resists this compression. Unlike inertia, this is a property that reacts to displacement, that is, distance moved, rather than acceleration. Finally we have **acoustic resistance**, which dampens the system and is responsible for energy loss.

In our uniform tube, the impedance is distributed evenly along the entire length of the tube rather than existing at a single point, as in a single mass or spring. Hence, the length of the tube is an important factor and the distributed nature means that determining the resonance characteristics is somewhat more complicated. Since length is a factor, a standard measure of impedance in such systems is called the **characteristic impedance**, which is the impedance that would occur if the tube were infinitely long. This is termed Z_0 and is given by

$$Z_0 = \rho c \qquad (11.5)$$

where ρ is the density of the air. In air, the inductance is given by density divided by area, ρ/A, and capacitance is given by $A/(\rho c^2)$. For now, we shall assume that the resistance in air is in fact zero. We do this firstly because it helps simplify the derivation of the full model, but in fact it is further justified because the friction due to the air itself is relatively minor, and in reality other resistive factors (such as the contact with the tube walls) provide a greater (and more-complicated) contribution to the overall energy loss. Tubes that have zero resistance component are termed **lossless tubes** for obvious reasons.

11.2.4 Acoustic reflection

Consider again the properties of waves travelling in a rope. There we saw that, at the termination of the rope, the travelling wave was reflected, and the reflected backward travelling wave interfered with the forward wave to determine the behaviour of the complete system. Travelling sound waves move along the tube in a similar way. Depending on the boundary conditions at the end of the tube, some portion of the acoustic wave is reflected back, and this interferes with the wave travelling forwards. At specific frequencies, governed by the length of the tube and the speed of sound, the backward and forward waves will reinforce each other and cause resonances.

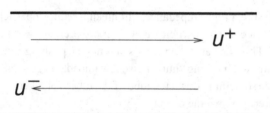

Figure 11.7 A uniform tube with forward travelling wave, u^+, and backward travelling wave u^-.

In such a system, the volume velocity at a given point and time can be expressed as

$$u(x,t) = u^+(t-x) - u^-(t+x) \qquad (11.6)$$

where $u^+(t)$ is the forward travelling wave and $u^-(t)$ is the backward travelling wave. This is shown in Figure 11.7. The minus sign between the terms indicates that the two waves can cancel each other out; when the forward wave meets an equally sized volume velocity of the backward wave, the overall effect is to have no motion.

Equation (11.5) allows us to state an expression for pressure in terms of the forward and backward volume velocities at a point in space and time in a similar way:

$$p(x,t) = \frac{\rho c}{A}\left(u^+(t-x) + u^-(t+x)\right) \qquad (11.7)$$

Here $\rho c / A$ is the characteristic impedance Z_0 of the tube. Note that in this case the terms are additive; an increase in pressure from the forward wave meeting an increase in pressure from the backward wave results in an even higher combined total pressure. If the area of the tube remains constant, the wave simply propagates through the tube. If, however, the area changes, then the impedance changes, which causes reflection. The reflections set up standing waves and these cause resonances. In this way, the impedance patterns of the tube govern the resonance properties of the model.[2]

11.3 The vowel-tube model

In vowel production, the glottal source creates a wave, which passes through the pharyngeal cavity and then the oral cavity, and finally radiates through the lips. The nasal cavity is blocked off, so the pharyngeal cavity and mouth can be combined and modelled as a single tube. This tube varies in cross-sectional area along its length, and it is the ability of a speaker to vary the configuration of this tube that gives rise to the different vowel sounds. Modelling a continuously varying tube is complicated, but we can approximate this by considering a series of short uniform tubes connected in series, as shown in Figure 11.8. If the number of tubes is sufficiently large, a continuous tube can be modelled

---

[2] Note that the above expressions can be derived from a first-principles solution of the wave equations and the properties of air. This is beyond the scope of our study, but several good accounts are available [364], [368].

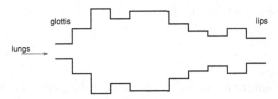

Figure 11.8 A vocal-tract model composed of a sequence of joined uniform tubes.

with arbitrary accuracy using this method (we return to the issue of just how many tubes are needed in Section 11.3.6).

Recall that we described the overall process of vowel production as one whereby a signal from the glottis $u[n]$ travels through and is modified by the vocal tract and lips to produce the speech waveform $y[n]$. In the z-domain, this is represented by Equation (11.2):

$$Y(z) = U(z)V(z)R(z)$$

where $U(z)$ is the z-transform of the glottal source, $V(z)$ is the vocal-tract filter and $R(z)$ is the radiation effect from the lips. Initially, we wish to find $V(z)$ since this is the part of the expression that characterises the phonetic nature of the sound. From $V(z)$ we can find the spectral envelope, formants and other linguistically useful information. To find $V(z)$, we study the evolution of the volume-velocity signal from the glottis to the lips. We will denote the volume velocity at the lips $U_L(z)$ and, to avoid confusion, denote the volume-velocity glottal source $U_G(z)$ rather than simply $U(z)$. Hence our expression is

$$U_L(z) = V(z)U_G(z) \tag{11.8}$$

After this, we can determine the final speech signal $Y(z)$ by applying the radiation characteristic $R(z)$ to $U_L(z)$. Rather than attempt to find $V(z)$ from the tube properties directly, we will use the tube model to find $U_L(z)$ in terms of $U_G(z)$, and divide the former by the latter to find $V(z)$.

11.3.1 Discrete time and distance

In practice we wish to use our analysis on digital signals, so our first task is to convert the expressions for volume velocity and pressure in a tube into a digital form, so that we can perform the analysis independently of the sampling rate.

Recall from Section 10.2.1 that we can convert time t to a discrete-representation time variable n by

$$n = tF_s$$

where F_s is the sampling rate. As is standard in discrete signal processing, we will use normalised frequency ($\hat{F}$) and normalised angular frequency ($\hat{\omega}$), which are calculated from frequency (F) and angular frequency (ω) by

$$\hat{F} = \frac{\hat{\omega}}{2\pi} = \frac{F}{F_s}$$

In addition, it is convenient to normalise our measure of distance from the start of the tube, x, with respect to the sampling frequency. It also helps simplify the analysis if we normalise this with respect to the speed of sound. As with time and frequency normalisation, this allows us to derive general expressions for the system independently of the sampling frequency used and the speed of sound in the medium. Hence normalised distance d is given by

$$d = \frac{xF_s}{c} \tag{11.9}$$

Using this result we can state the volume-velocity and pressure functions, given in Equations (11.6), (11.7) and (11.9), in discrete normalised time and distance as

$$u[d, n] = u^+[n - d] - u^-[n + d] \tag{11.10a}$$

$$p[d, n] = \frac{\rho c}{A}(u^+[n - d] - u^-[n + d]) \tag{11.10b}$$

In the next section we will develop a general expression for the reflection and transmission of a wave at the junction of two tubes and, in the subsequent sections, show how this result can be used to determine the behaviour of the whole system.

11.3.2 Junction of two tubes

If we consider a single tube k in our model, then from Equations (11.6), (11.7) and (11.9) we can state the volume-velocity and pressure functions in discrete normalised time and distance as

$$u_k[d, n] = u_k^+[n - d] - u_k^-[n + d] \tag{11.11a}$$

$$p_k[d, n] = \frac{\rho c}{A_k}(u_k^+[n - d] - u_k^-[n + d]) \tag{11.11b}$$

Now we consider the junction between a tube k with normalised length D_k and its neighbouring tube $k + 1$, as shown in Figure 11.9. Since this tube is also uniform, it can also be described by Equations (11.11a) and (11.11b).

Joining two tubes creates a boundary condition at the junction, and it is this junction which determines how the sound propagates as a whole. In considering behaviour at a junction, we can use an important principle of physics, namely that pressure and volume velocity cannot change instantaneously anywhere. So, despite the sudden change in cross-sectional area, the volume velocity and pressure cannot abruptly change, and it follows from this that at the point of the junction the pressure and volume velocity must

11.3 The vowel-tube model

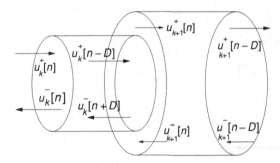

Figure 11.9 Behaviour of forward and backward travelling waves at a tube junction.

be equal. Hence

$$u_k[D_k, n] = u_{k+1}[0, n]$$
$$p_k[D_k, n] = p_{k+1}[0, n]$$

Using these boundary conditions we can relate the forward and backward volume-velocity equations of tube k and tube $k + 1$ at the junction. That is, at the end ($d = D_k$) of tube k and the start of tube $k + 1$ ($d = 0$)

$$u_k^+[n - D_k] - u_k^-[n + D_k] = u_{k+1}^+[n] - u_{k+1}^-[n] \quad (11.12)$$

and likewise with the pressure:

$$\frac{\rho c}{A_k}(u_k^+[n - D_k] + u_k^-[n + D_k]) = \frac{\rho c}{A_{k+1}}(u_{k+1}^+[n] + u_{k+1}^-[n]) \quad (11.13a)$$

$$\frac{1}{A_k}(u_k^+[n - D_k] + u_k^-[n + D_k]) = \frac{1}{A_{k+1}}(u_{k+1}^+[n] + u_{k+1}^-[n]) \quad (11.13b)$$

We can substitute these equations into one another and eliminate one of the volume-velocity terms. For example, from Equation (11.12) we know that $u_k^-[n + D_k] = u_{k+1}^+[n] + u_{k+1}^-[n] + u_k^+[n - D_k]$, which we can substitute into Equation (11.13b) to give an expression for $u_{k+1}^+[n]$:

$$u_{k+1}^+[n] = \left(\frac{2A_{k+1}}{A_k + A_{k+1}}\right) u_k^+[n - D_k] + \left(\frac{A_{k+1} - A_k}{A_k + A_{k+1}}\right) u_{k+1}^-[n] \quad (11.14)$$

This equation is interesting in that it shows the make up of the forward travelling wave in tube $k + 1$. As we would expect, this is partly made up of the forward travelling wave from tube k and partly from some of the backward travelling wave being reflected at the junction. The coefficient of u_k^+ in Equation (11.14) is the amount of u_k^+ that is **transmitted** into the next tube, tube $k + 1$, and is termed the **transmission coefficient**. The coefficient of u_{k+1}^- is the amount of u_{k+1}^- that is **reflected** back into tube $k + 1$ at the junction and is termed the **reflection coefficient** (Figure 11.10). Since they are simply related, only one is needed in order to describe the junction and by convention reflection coefficients are used. The reflection coefficient is often denoted r_k (meaning

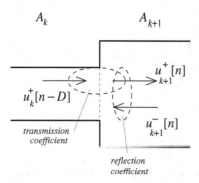

Figure 11.10 A tube junction showing how wave u^+ is made from reflection of u^- in tube $k+1$, and transmission of u^+ from tube k.

the reflection between tubes k and $k+1$) and from (11.14) is given by

$$r_k = \frac{A_{k+1} - A_k}{A_k + A_{k+1}} \qquad (11.15)$$

Of particular importance is the fact that the boundary condition is completely described by the reflection coefficient, which is governed entirely by the areas of the tubes; the speed-of-sound and density terms have been eliminated. We shall see that reflection coefficients are a simple and useful way to define the characteristics between two tubes and we will henceforth use them instead of tube areas. Using reflection coefficients, we can find expressions that relate the forward wave or the backward wave in one tube to the forward and negative waves in the other tube:

$$u_k^+[n - D_k] = \frac{1}{1 + r_k} u_{k+1}^+[n] + \frac{r_k}{1 + r_k} u_{k+1}^-[n] \qquad (11.16a)$$

$$u_k^-[n + D_k] = \frac{-r_k}{1 + r_k} u_{k+1}^+[n] + \frac{1}{1 + r_k} u_{k+1}^-[n] \qquad (11.16b)$$

11.3.3 Special cases of junctions

The above result can be used to help set boundary conditions for the tube at the lips and the glottis. To see this, let us consider three special cases resulting from changing areas in the tubes.

Firstly, if two connected tubes have the same cross-sectional area we see from (11.15) that the reflection coefficient is 0 and the transmission coefficient is 1. That is, there is no reflection and all of the wave is transmitted past the boundary. This is entirely what we would expect from our uniform-tube model. Next, let us consider the situation when the tube is completely closed at one end. A closed termination can be modelled by letting the second tube have an infinitely small area (that is, the size of the junction between the tubes disappears to nothing and hence creates a solid wall). The effect of this can be found from Equation (11.15). As the area of the second tube $A_{k+1} \to 0$, the reflection coefficient $r_k \to -1$. Now consider a completely open tube. This can again be modelled with the tube-junction model, but this time the area of the second tube is infinitely big.

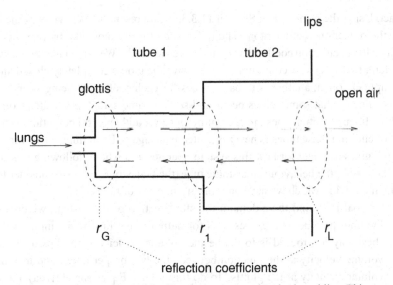

Figure 11.11 A two-tube model with terminations at the glottis and lips. This system has three reflection coefficients. In the middle we have r_1, the only "real" reflection coefficient, whose value is given by the area equation (11.15). The lips and glottis reflection coefficients are artificial values designed to ensure that there are some losses from the system. Note that there is no backward wave entering at the lips and no forward wave (save the source) entering at the glottis.

Hence, from Equation (11.15), as $A_{k+1} \to \infty$ the reflection coefficient $r_k \to 1$. Hence both a closed termination and an open termination completely reflect the wave.

There is a crucial difference between the two situations, however, and this can be seen from the sign of the reflection coefficient. In the closed termination case, the tube ends in a solid wall and the reflection coefficient is -1. From Equation (11.16b) we can see that in such a case the volume velocity will be 0. If we find the equivalent pressure expressions for Equation (11.16b), we see that the pressure terms add, and hence the pressure at this point is $\rho c / A_1$. Intuitively, we can explain this as follows. The solid wall prevents any particle movement and so the volume velocity at this point must be 0. The pressure, however, is at its maximum, because the wall stopping the motion is in effect an infinite impedance, meaning that, no matter how much pressure is applied, no movement will occur. In the open-end situation, when the reflection coefficient is 1, we see from Equation (11.16b) that the volume velocity is at a maximum and now the pressure is 0. Intuitively, this can be explained by the fact that now the impedance is 0, and hence no pressure is needed to move the air particles. Hence the pressure $p(L, t)$ is 0, and the volume velocity is at its maximum.

11.3.4 The two-tube vocal-tract model

Here we will consider a model in which we have two connected tubes of different cross-sectional areas. In this formulation, the tubes have areas A_1 and A_2, and normalised lengths D_1 and D_2, such that $D_1 + D_2 = D$, the total normalised length of the vocal tract. Using the results from Section 11.3.3, we represent the behaviour at the boundary by the reflection coefficient r_1. The lips and glottis are modelled by two special tubes, which have reflection coefficients r_L and r_G respectively. We can set these so as to produce different effects – one configuration is to have the glottis completely closed and the lips completely open; another is to have values for each coefficient being slightly less than this value so that some losses occur and so that some sound propagates from the lips. These lip and glottis tubes are special because we add an extra condition such that any wave entering these tubes is never reflected back again. Examining a two-tube model in depth gives us a feel for how this system operates as a whole; following this section we will consider the behaviour of a single tube, the behaviour when we consider losses and a final model in which we have an arbitrary number of tubes.

Our goal is to find the z-domain transfer function of the system, which we will do by dividing the z-domain expressions for output at the lips by the input at the glottis. The best way to proceed is to define the volume velocity at the lips, use this to find the volume velocity at the junction between the two proper tubes, and use this to find the volume velocity at the glottis. To do this we use Equations (11.16a) and (11.16b), which relate the volume velocity in one tube to the volume velocity in the next. Since our end point is a z-domain expression, we will make repeated use of the z-transforms of Equations (11.16a) and (11.16b), which are

$$U_k^+(z) = \frac{z^{D_k}}{1+r_k} U_{k+1}^+(z) - \frac{r_k z^{D_k}}{1+r_k} U_{k+1}^-(z) \qquad (11.17a)$$

$$U_k^-(z) = \frac{-r_k z^{-D_k}}{1+r_k} U_{k+1}^+(z) + \frac{z^{-D_k}}{1+r_k} U_{k+1}^-(z) \qquad (11.17b)$$

We start by defining $U_L(z)$ as the output at the lips. We now feed this into Equations (11.17a) and (11.17b) and use the fact that there is no backward wave ($U_L^- = 0$) to set the volume velocities at junction $k = 2$ as

$$U_2^+(z) = \frac{z^{D_2}}{1+r_L} U_L(z) \qquad (11.18a)$$

$$U_2^-(z) = \frac{-r_L z^{-D_2}}{1+r_L} U_L(z) \qquad (11.18b)$$

By feeding these values into Equations (11.17a) and (11.17b) again, we can calculate the volume velocities U_1^+ and U_1^- at the junction $k = 1$:

$$U_1^+(z) = \frac{z^{D_1}}{1+r_1} U_2(z) - \frac{r_1 z^{D_1}}{1+r_1} U_2(z)$$

$$= \frac{z^{D_1} z^{D_2}}{(1+r_1)(1+r_L)} U_L(z) + \frac{r_1 r_L z^{D_1} z^{-D_2}}{(1+r_1)(1+r_L)} U_L(z)$$

$$U_1^-(z) = \frac{-r_1 z^{-D_1}}{1+r_1} U_2(z) + \frac{z^{-D_1}}{1+r_1} U_2(z)$$

$$= \frac{r_1 z^{-D_1} z^{D_2}}{(1+r_1)(1+r_L)} U_L(z) - \frac{z^{-D_1} z^{-D_2}}{(1+r_1)(1+r_L)} U_L(z)$$

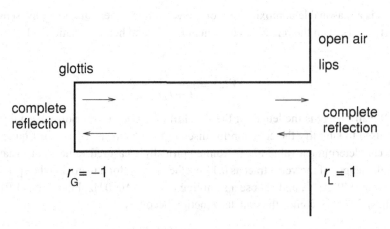

Figure 11.12 A uniform-tube model of the vocal tract when the glottis end is closed and the lips are open.

We have now reached the glottis, which is modelled by a special tube of length 0 and reflection coefficient r_G. We take the length of this tube to be 0, thus, since $z^0 = 1$,

$$U_G(z) = \frac{1}{1+r_G} U_1^+(z) - \frac{r_G}{1+r_G} U_1^-(z)$$

To find U_G^+ we simply repeat the above procedure. Note that we don't need to know U_G^- since we are assuming that it gets fully absorbed into the lungs. Thus

$$U_G^+(z) = \frac{z^{D_1}z^{D_2} + r_1 r_L z^{D_1} z^{-D_2} + r_1 r_G z^{-D_1} z^{D_2} + r_L r_G z^{-D_1} z^{-D_2}}{(1+r_G)(1+r_1)(1+r_L)} U_L(z)$$

By rearranging to get U_L/U_G and by dividing top and bottom by $z^{D_1}z^{D_2}$ we get the standard form of the transfer function as a rational function in z^{-1}:

$$V(z) = \frac{U_L}{U_G} = \frac{(1+r_G)(1+r_1)(1+r_L)z^{-(D_1+D_2)}}{1 + r_1 r_L z^{-2D_2} + r_1 r_G z^{-2D_1} + r_L r_G z^{-2(D_1+D_2)}}$$

The terms containing r in the numerator sum to form the gain, G. The normalised tube lengths D_1 and D_2 must sum to the overall length $D_1 + D_2 = D$, which gives

$$V(z) = \frac{U_L}{U_G} = \frac{Gz^{-D}}{1 + r_1 r_L z^{-2(D-D_1)} + r_1 r_G z^{-2D_1} + r_L r_G z^{-2D}} \tag{11.19}$$

Note that the lengths of the tubes D_1 and D_2 must be integers; this is a constraint imposed by the discrete modelling of the vocal tract. The discrete modelling also dictates the number of poles: a higher sampling rate requires more poles and hence the granularity of the tube lengths increases.

11.3.5 The single-tube model

A special case of the two-tube model is the single-tube or uniform-tube model in which the cross-sectional area is constant along the entire length of the tube (Figure 11.12).

This is a reasonable approximation of the vocal tract when producing the schwa vowel. If $A_1 = A_2$, then the reflection coefficient $r_1 = 0$ and hence Equation (11.19) simplifies to

$$V(z) = \frac{Gz^{-D}}{1 + r_L r_G z^{-2D}} \quad (11.20)$$

Recall that D is the length of the vocal tract in units of the normalised distance that we used to simplify the equation for discrete time analysis. By use of Equation (11.9), we can determine a value for D from empirically measured values. For instance, if we set the length of the vocal tract as 0.17 m (the average for a man) and the speed of sound in air as 340 m s^{-1}, and choose a sampling rate of 10 000 Hz, Equation (11.9) gives the value of D as 5. Hence the transfer function becomes

$$V(z) = \frac{Gz^{-5}}{1 + r_L r_G z^{-10}} \quad (11.21)$$

that is, an IIR filter with five zeros and ten poles. It can be shown that a polynomial $1 + r_L r_G z^{-10}$ can be factorised to a set of ten complex roots, such that in polar terms the radius of each root is $r_L r_G$ and the angle is given by

$$\theta_n = \frac{\pi(2n-1)}{10} \quad (11.22)$$

The numerator of the transfer function tells us that the output is delayed by a factor that is a function of the length of the tube. All these zeros lie at 0, and can be ignored for purposes of determining the frequency response of the transfer function. The poles are evenly spaced on the unit circle at intervals of $\pi/5$. The first is at $\pi/10 = 0.314$, which, when converted to a real frequency value with Equation (10.29), gives 500 Hz. Subsequent resonances occur every 1000 Hz after that, i.e. at 1500 Hz, 2500 Hz and so on.

Recall that the losses in the system come from the wave exiting the vocal tract at the glottis and lips, and the degree to which this occurs is controlled by the special reflection coefficients r_L and r_G. Let us first note that, since these are multiplied, the model is not affected by where the losses actually occur, and in fact in some formulations the losses are taken to occur only at the glottis or only at the lips. Secondly, recall from Section 11.3.3 that the special case of complete closure at the glottis would produce a reflection coefficient of 1, whereas the special condition of complete openness at the lips would produce a reflection coefficient of -1. This is a situation in which there are no losses at all in the system (which is unrealistic of course, for the simple reason that no sound would escape and hence a listener would never hear the speech). For these special cases the product $r_L r_G$ is 1, which means that the denominator term is

$$V(z) = \frac{Gz^{-5}}{1 - z^{-10}}$$

It can be shown that, when this is factorised, all the poles are spaced as before, but lie on the unit circle. This is of course exactly what we would expect from a lossless case.

11.3 The vowel-tube model

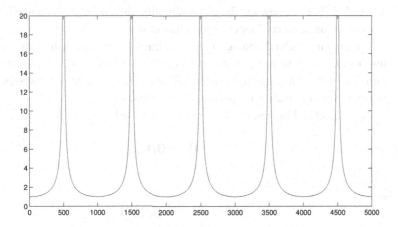

Figure 11.13 A magnitude spectrum for a uniform tube, showing that the resonances are sharp spikes (because there is no damping) that lie at even intervals along the spectrum.

We can find the frequency response of the vocal tract by setting $z = e^{j\hat{\omega}}$. For the special lossless case, then,

$$V(e^{j\hat{\omega}}) = \frac{2e^{j5\hat{\omega}}}{1 + e^{j10\hat{\omega}}}$$

Making use of the inverse Euler expression, Equation (10.11) can also be written as

$$V(e^{j\hat{\omega}}) = \frac{1}{\cos(5\hat{\omega})} \qquad (11.23)$$

This frequency response is shown in Figure 11.13, and it can be seen that the formants are very sharp (there is no damping) and lie at evenly spaced intervals, as predicted by the transfer function.

11.3.6 The multi-tube vocal-tract model

We will now develop a general expression for the transfer function of a tube model for vowels, in which we have an arbitrary number of tubes connected together.

We saw in Section 11.3.5 that the order of the final transfer function is determined by the normalised length of the tube D. This is turn is a function of the sampling rate and vocal-tract length, so longer vocal tracts or higher sampling rates will require higher-order filters. Staying with the case of a 0.17-m vocal tract with speed of sound 340 m s^{-1} and sampling rate 10 000 Hz, we find that we need a tenth-order transfer function as shown in the cases of the uniform and two-tube models. The difference between the transfer functions for these two models is simply the inclusion of the extra non-zero coefficients in the denominator. On extending this principle, we find that the most powerful transfer function is one with a non-zero coefficient for every term in z^{-1}. Such a transfer function can be created by a model with exactly ten equal-length tubes, each of which has length

$D_N = 1/2$. Importantly, for the given sampling rate, no extra accuracy is achieved by increasing the number of tubes beyond this value.

The principle behind finding the N-tube transfer function is the same as for the two-tube model – start at the lips and apply Equations (11.18a) and (11.18b) for each tube in turn. This is tedious to perform in the manner used for the two-tube system, but we can use matrix multiplication to speed up the process.

Equations (11.17a) and (11.17b) can be expressed in matrix form as

$$\mathbf{U}_k = \mathbf{Q}_k \mathbf{U}_{k+1}$$

where

$$\mathbf{U}_k = \begin{bmatrix} U_k^+(z) \\ U_k^-(z) \end{bmatrix}$$

and, because $D = 1/2$, then

$$\mathbf{Q}_k = \begin{bmatrix} \dfrac{z^{1/2}}{1+r_k} & \dfrac{-r_k z^{1/2}}{1+r_k} \\ \dfrac{-r_k z^{-1/2}}{1+r_k} & \dfrac{z^{-1/2}}{1+r_k} \end{bmatrix}$$

We can use the special reflection coefficients at the lips and glottis to model the terminating boundary conditions as before. In the matrix formulation, this gives

$$\mathbf{U}_{N+1} = \begin{bmatrix} U_L(z) \\ 0 \end{bmatrix} = \begin{bmatrix} 1 \\ 0 \end{bmatrix} U_L(z) \tag{11.24}$$

for the lips and

$$\mathbf{U}_G = \begin{bmatrix} \dfrac{1}{1+r_G} & \dfrac{r_G}{1+r_G} \end{bmatrix} U_1(z) \tag{11.25}$$

for the glottis. We can use these boundary conditions to formulate an expression for the transfer function for an arbitrary number of tubes

$$\frac{1}{V(z)} = \frac{U_G}{U_L} = \begin{bmatrix} \dfrac{1}{1+r_G} & \dfrac{r_G}{1+r_G} \end{bmatrix} \prod_{k=1}^{N} \mathbf{Q}_k \begin{bmatrix} 1 \\ 0 \end{bmatrix}$$

which can be solved by iteratively calculating the terms. The result of this will always take the form

$$V(z) = \frac{1+r_0}{2} \frac{z^{-N/2} \prod_{k=1}^{N}(1+r_k)}{1 - a_1 z^{-1} - a_2 z^{-2} - \cdots - a_N z^{-N}} \tag{11.26}$$

which, as we saw before, is simply a delay in the numerator, followed by an all-pole expression in the denominator. This is an important result, because it shows that a vocal tract without internal losses can be accurately modelled by an all-pole filter.[3]

It is possible to simplify the calculation by making the further assumption that the glottis is a completely closed end, with infinite impedance and reflection coefficient 1. If we take

$$A_N(z) = \sum_{k=1}^{N} a_k z^{-k}$$

it is possible to show that in the case $r_G = r_0 = 1$ the following equation holds

$$A_N(z) = A_{N-1}(z) + r_k z^{-k} D_{k-1}(z^{-1}) \quad (11.27)$$

That is, the Nth-order polynomial in z^{-1} can be determined from the $(N-1)$ order polynomial in z^{-1}, z and the reflection coefficient r_k. If we therefore start at $N = 0$ and set $A_0(z) = 1$ we can calculate the value for A_1, and then iteratively for all $A(z)$ up to N. This is considerably quicker than carrying out the matrix multiplications explicitly. The real value in this, however, will be shown in Chapter 12, when we consider the relationship between the all-pole tube model and the technique of linear prediction.

11.3.7 The all-pole resonator model

This general result is important because it shows that, for any number of tubes, the result is always an all-pole filter. From Section 10.5.1, we know that this can be factorised into an expression containing individual poles, which can often be interpreted directly as formant resonances. Here we see that we have come full circle: in Section 10.5.3 we showed that a second-order IIR filter could faithfully model the shape of a single formant; in Section 10.5.4 we showed that several of these could be cascaded to model a number of formants. In this chapter, we saw in Section 11.2.1 that a physical mass/spring/damper system can be modelled by the same equations, and that a single mass, spring and damper create a resonance pattern of exactly the same type as a second-order IIR formant model. In the preceding section, we have just shown that a connected tube model has an all-pole transfer function, with the number of poles equalling the number of tubes. From this result, it is clear that an all-pole tube model can faithfully create any formant pattern if the parameters attributed to the tubes (i.e. the cross-sectional areas) are set correctly. Furthermore, the transfer function for the tube model can be factorised so as to find the poles and possibly identify individual formants.

An important point to realise is that the formant patterns that arise from this all-pole model are a function of the whole model; while the transfer function can be factorised, it is *not* appropriate to ascribe individual poles or formants to individual tubes in the model. The situation is more complex than this, and the formant patterns are created from the properties of all the tubes operating together.

---

[3] Rabiner and Schafer [368] show that the above can be calculated more easily, and without direct matrix multiplication, if we assume that the glottal impedance is infinite.

11.4 Source and radiation models

11.4.1 Radiation

So far, we have developed a transfer function $V(z)$, which is defined as the ratio of the volume velocity of the lips over the volume velocity at the glottis. In practice, however, when we measure sound, we normally in fact measure pressure signals, since this is what most microphones are designed to respond to. Most microphones operate in the **far field**, that is at some distance from the lips, and hence the signal is influenced by a **radiation** impedance from the lips. This can be modelled by another transfer function, $R(z)$, which, from Equation (11.4), is defined as

$$R(z) = \frac{P(z)}{U_L(z)} \qquad (11.28)$$

Modelling radiation accurately soon becomes very complicated, so, for our purposes, we use a very simple model that has been shown to do a reasonable job at approximating the radiation. This takes the form of a function that differentiates the signal, and this can be modelled by a FIR filter with a single zero:

$$R(z) = 1 - \alpha z^{-1} \qquad (11.29)$$

where α is a value less than but quite near to 1 (e.g. 0.97). This creates a high-frequency boost of about 6 dB per octave.

11.4.2 The glottal source

We qualitatively described in Section 7.1.2 the process of how the glottis produces sound. In voiced speech, the vocal folds undergo a cycle of movement that gives rise to a periodic source sound. At the start of this cycle the vocal folds are closed. Pressure is exerted from the lungs, which causes a build up in pressure beneath the folds until eventually the pressure forces open the folds. When the folds open, air moves through them and, as it does so, the pressure beneath the folds reduces. Eventually the pressure is low enough that the tension closes the vocal folds, after which the cycle repeats. The rate at which this cycle occurs determines the fundamental frequency of the source, and ranges from about 80 Hz to 250 Hz for a typical man, and from about 120 Hz to 400 Hz for a typical woman or child. A plot of volume velocity against time for the glottal source is shown in Figure 11.14. Here we see the three main parts of the cycle: the **open phase**, during which the glottis is opening due to the pressure from the lungs, the **return phase**, during which the glottis is closing due to the vocal-tract tension, and the **closed phase**.

Modelling the glottis is a tricky problem in that building a simple model seems relatively easy, but developing a highly accurate model that mimics glottal behaviour in all circumstances is extremely difficult and has not in any way been achieved. We will return to the difficulties in accurate glottal modelling in Section 13.3.5, but for now we will demonstrate a simple model that at least gives us a flavour of how this works.

11.4 Source and radiation models

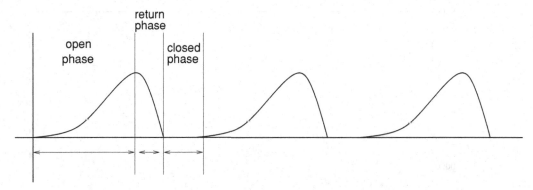

Figure 11.14 A plot of an idealised glottal source, as volume velocity over time. The graph shows the open phase, when air is flowing through the glottis, the short return phase, when the vocal folds are snapping shut, and the closed phase, when the glottis is shut and the volume velocity is zero. The start of the closed phase is the instant of glottal closure.

An extensive study into glottal modelling is given in Flanagan [164], which describes various mass/spring/damper systems. These models can be somewhat difficult to model in discrete-time systems, so instead we adopt models that simply generate a time-domain function that has the properties described above. One such model [368], [376] is given by

$$u[n] = \begin{cases} \frac{1}{2}(1 - \cos(\pi n/N_1)) & 0 \leq n \leq N_1 \\ \cos(\pi(n - N_1)/(2N_2)) & N_1 \leq n \leq N_2 \\ 0 & \text{otherwise} \end{cases} \quad (11.30)$$

where N_1 marks the end of the open phase and N_2 marks the end of the return phase/start of the closed phase. Another quite widely used model is the Lijencrants–Fant model [159] (sometimes known as the LF model). This describes the derivative of the glottal waveform and is given by

$$u[n] = \begin{cases} 0 & 0 \leq n < T_0 \\ E_0 e^{\alpha(n-T_0)} \sin[\Omega_0(n - T_0)] & T_0 \leq n < T_e \\ -E_1[e^{\beta(n-T_e)} - e^{\beta(T_c-T_e)}] & T_e \leq n < T_c \end{cases} \quad (11.31)$$

where T_0 is the instant of glottal opening, T_e is the position of the negative minimum, T_c is the instant of glottal closure, α, β and Ω control the shape of the function, and E_0 and E_1 determine the relative heights of the positive and negative parts of the curve. The LF function and its derivative, which is the normal glottal flow function, are shown in Figure 11.15. Quatieri [364] gives a third model, in which $g(n)$ is created by two time-reversed exponentially decaying sequences:

$$u[n] = (\beta^{-n}u[-n])(\beta^{-n}u[-n]) \quad (11.32)$$

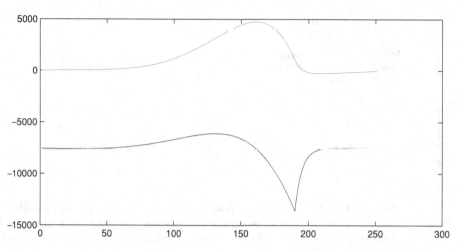

Figure 11.15 A plot of the Lijencrants–Fant model for glottal flow and the derivative of glottal flow for a single period.

which have the nice property of having a simple z-transform

$$U(z) = \frac{1}{(1-\beta z)^2} \qquad (11.33)$$

The most important control element in the glottal source is of course the rate at which the cycles occur. In Equation (11.30), this is determined by the positions of N_1 and N_2; in Equation (11.32) by T_0, T_e and T_c; and in Equation (11.31) by the impulse function $u[n]$. In the last case, the glottal volume-velocity function can be thought of as a low-pass filtering of an impulse stream. From these expressions and empirical measurements, it is known that this low-pass filter creates a roll off of about -12 dB per octave. It is this, combined with the radiation effect, that gives all speech spectra their characteristic spectral slope.

While Equation (11.32) gives a reasonable approximation of the glottal volume-velocity signal, it does not model the secondary effects of **jitter, shimmer** and **ripple**. More realism can be added to the glottal signal by the addition of zeros into the transfer function, giving the expression

$$U(z) = \frac{\prod_{k=1}^{M}(1-u_k z^{-1})}{(1-\beta z)^2} \qquad (11.34)$$

We have defined $U(z)$ as a volume-velocity signal, mainly for purposes of developing the vocal-tract transfer function. While, in reality, the radiation $R(z)$ occurs after the operation of $V(z)$, we aren't restricted to this interpretation mathematically. As we shall see, it is often useful to combine $U(z)$ and $R(z)$ into a single expression. The effect of this is to have a system wherein the radiation characteristic is applied to the glottal-flow waveform before it enters the vocal tract. This is equivalent to measuring the pressure waveform at the glottis, and, if we adopt the radiation characteristic of Equation (11.29), the effect of this is to differentiate the glottal-flow waveform. We call this signal the **glottal-flow derivative**, denoted $\dot{s}[n]$ and $\dot{S}(z)$, and this is shown together with the normal glottal flow in Figure 11.15.

The interesting feature of the glottal-flow derivative is that it shows that the primary form of excitation into the vocal tract is a large negative impulse. This is useful in that the output of the vocal tract immediately after this should approximate well to the true impulse response of the system.

11.5 Model refinements

11.5.1 Modelling the nasal cavity

When the velum is lowered during the production of nasals and nasalised vowels, sound enters via the velar gap, propagates through the nasal cavity and radiates through the nose. Hence, for a more complete model, we have to add a component for the nasal cavity. This in itself is relatively straightforward to model; for a start, it is a static articulator, so it doesn't have anywhere near the complexity of shapes that occur in the oral cavity. By much the same techniques as we employed above, we can construct an all-pole transfer function for the nasal cavity.

The complication arises because of the behaviour of the oral cavity. For the case of vowels, the sound wave moves from the glottis through the oral cavity and radiates from the lips. In the case of nasalised vowels, the sound propagates from the glottis and through the pharynx as before, but at the velum some sound enters the nasal cavity while the remainder propagates through the mouth. The sound entering through the velum is filtered by the nasal cavity and radiates through the nose. The sound continuing through the oral cavity radiates through the lips in the manner described above. One complication to this is that some sound in the oral cavity is reflected backwards and will enter the velar gap, and likewise some sound in the nasal cavity will be reflected backwards and will enter the oral cavity. If we ignore this interaction for now, the system for nasalised vowels can be modelled as shown in Figure 11.16 with a simple splitting operation at the velum. Because we now have two filters operating in parallel, it is not possible to derive a single transfer function for the whole system. Rather we construct a transfer function for the oral cavity and a different one for the nasal cavity and then simply add their outputs to find the final signal.

The case of nasal consonants is more complicated. As with nasalised vowels, the sound wave starts at the glottis and then moves through the pharynx and partly through the mouth. Some sound now travels through the velum into the nasal cavity and radiates through the nose. The remainder continues travelling through the oral cavity, but, instead of being radiated at the lips, now meets a closure at the lips. Hence a full reflection occurs, and so the sound travels backwards through the mouth, and some continues down towards the glottis and some enters the nasal cavity and radiates from the nose. In addition to this, there are of course the forward and backward reflections being caused all the time by the changing cross-sectional areas of the nose and mouth. An equivalent way of looking

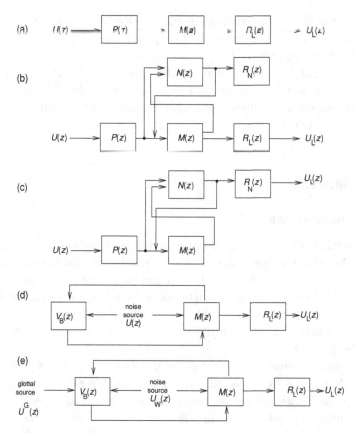

Figure 11.16 LTI filter models for various types of sound production: (a) vowel and approximant, (b) nasalised vowel, (c) nasal, (d) unvoiced obstruent and (e) voiced obstruent.

at this is that the pharynx, nasal cavity and nose operate for nasals in the same way as the pharynx, oral cavity and lips operate for vowels. In addition, there is now a **side resonator** of the front of the mouth. In phonetic terms the effect of this oral side-cavity is very important; since the nasal cavity is static, it is in fact the ability of the tongue to create different oral side-cavity configurations which is responsible for the nasal sounds /m/, /n/ and /ng/ all sounding different.

Acoustically, the effect of the side branch is to "trap" some of the sound, which creates **anti-resonances**. These can be modelled by the inclusion of zeros in the transfer function. As with the case of nasalised vowels, the parallel nature of the system means that we can't use a single transfer function; rather we have a system with an all-pole transfer function for the pharynx and back of the mouth, a splitting operation, an all-pole function for the nose and a pole-and-zero function for the oral cavity.

11.5.2 Source positions in the oral cavity

Many consonant sounds, such as fricatives and stops, have a sound source located in the oral cavity. This is created by the tongue nearing another surface (the roof of the mouth, the teeth etc.) so as to cause a tight constriction. In the case of unvoiced obstruents, the glottis is open, and air flows through it until it reaches the constriction, at which point turbulence occurs. For voiced obstruents, the glottis behaves as with vowels, but again this air flow is made turbulent by the constriction. The difference in placement of the constriction is what gives rise to the difference between the various fricative and stop sounds of the same voiced sound. Hence /s/ has an alveolar constriction whereas /sh/ has a palatal constriction. It is important to realise, however, that the constrictions themselves do not generate different-sounding sources; it is the fact that the constrictions configure the vocal tract into differently shaped cavities through which the source resonates that is responsible for the eventual difference in the sounds.

The effect of a sound source in the middle of the vocal tract is to split the source such that some sound travels backwards towards the glottis while the remainder travels forwards towards the lips. The vocal tract is thus effectively split into a backward and a forward cavity. The forward cavity acts a tube resonator, similarly to the case of vowels but with fewer poles because the cavity is considerably shorter. The backward cavity also acts as a further resonator. The backward-travelling source will be reflected by the changes in cross-sectional area in the back cavity and at the glottis, creating a forward-travelling wave that will pass through the constriction. Hence the back cavity has an important role in the determination of the eventual sound. This back cavity acts as a side resonator, just as with the oral cavity in the case of nasals. The effect is to trap sound and create anti-resonances. Hence the back cavity should be modelled with zeros as well as poles in its transfer function.

11.5.3 Models with vocal-tract losses

The tube model described in Section 11.3 is termed "lossless" because no energy losses occur during the propagation of sound through the tube itself. Any energy losses that do occur happen through the lips and glottis, where the signal travels outwards but is not reflected back. There have been several studies on modelling the losses which of course do occur in the tubes themselves. In terms of the discussion given in Section 11.2.3, this means that the impedance now is not just composed of an inductive and a capacitance component, but also has a resistive component.

Rabiner and Schafer [368] give an extensive review of this issue and summarise that there are three main ways in which losses can occur. First, the passage of the wave will cause the walls of the vocal tract (e.g. the cheeks) to vibrate. This can be modelled by refinements to the transfer function. The effect is to raise the formant centre frequencies slightly, dampening the formants and especially the lower formants; we should expect this since the mass of the vocal-tract walls will prevent motion at the higher frequencies. The effects of friction and thermal conduction can also be modelled by adding resistive terms to the transfer function. The overall effect of these is the opposite of the above; the formant centre frequencies are somewhat lowered, and the effects of this are more pronounced at higher frequencies.

Apart from the overall contribution to damping, these effects tend to cancel each other out, so the overall contribution of these resistive terms is quite small. Since the degree of loss from the lips and glottis is considerably higher, these loss terms are often ignored altogether because they add considerable complication to the model with little gain in modelling accuracy. It should be noted that it is the particular form of the IIR expression that we derived that gives rise to a lossless tube; other IIR filter expressions can easily model losses.

11.5.4 Source and radiation effects

Our model of the glottis and lips was extremely simple in that we used special reflection coefficients r_G and r_L to generate reflections and losses (for values other than $|1|$). As we might expect, in reality the situation is much more complicated. Firstly, there is no such thing as an "ideal source" where we can simply add volume velocity into a system without any further effect. The source in fact is always to a greater or lesser degree coupled to the rest of the system, and has to be modelled if more accuracy is desired. In general this is modelled by having a real impedance connecting the source to the rest of the system. Again, assuming that sound can just propagate from the lips and "disappear" with no further effect is quite simplistic. A more-accurate model can be created by modelling the radiation as a complex impedance connected as part of the system.

In this chapter we have concentrated on sound propagation through a tube, and saw that even this, the simplest of all acoustic systems, can be quite complicated. Another entire branch of the field of acoustics is devoted to sound propagation through three-dimensional spaces, and this can be drawn on to produce a better model of radiation through the lips. As with the other refinements mentioned in this section, modelling both the source and lip losses with more realistic expressions does increase the accuracy of the overall model, but at considerable modelling expense.

11.6 Discussion

In developing our model, we have attempted to balance the needs of realism with tractability. The all-pole vocal-tract model that described in Section 11.3 will now be adopted for the remainder of the book as the model best suited to our purposes. In subsequent chapters, we shall in fact see that this model has some important properties that make its use particularly attractive.

One of the key jobs in making assumptions in a model is to recognise that such assumptions have indeed been made. So long as this is done, and these assumptions are borne in mind, we should not run into too many problems when we rely on the model to deliver faithful analyses. With this in mind, we will now note the main assumptions we have made and discuss any potential problems that might arise because of them. The following list is in approximate increasing order of severity.

> **Linear filter** Throughout we have assumed that the system operates as a time-invariant linear (LTI) filter of the type described in Chapter 10. While it is well known that many non-linear processes are involved in vocal-tract sound propagation, in general the linear model provides a very good approximation.

Time invariance During normal speech, the vocal tract is constantly moving, but, so long as the rate of change is very slow compared with the sampling rate, it is quite adequate to model the system as a sequence of piecewise-stationary configurations. In Chapter 12, we will discuss further the issue of quantising the time dimension so as to produce stationary "snapshots" of speech, but for now let us say that a good approximation to a continuously varying tract is to generate a new stationary vocal-tract configuration about every 10 ms.

Straight tube One of the first assumptions we made was that we could model the vocal tract by a straight tube or sequence of straight tubes. The vocal tract is of course curved; very roughly it starts as a vertical tube at the glottis, takes a 90-degree turn at the top of the pharynx and then continues horizontally through the mouth. In fact it is one of the main principles of acoustic modelling that, so long as any bend is not too severe, there is little difference between the propagation in straight and curved tubes. This can be seen clearly in music, where a modern trumpet is actually a quite long (10 m) tube that has been wrapped many times to make its playing and carrying more convenient. The bends in the trumpet have little effect on the overall sound.

Discrete-tube model The number of tubes determines the granularity at which we model the real, continuously varying tube. As we saw in Section 11.3.6, the number of tubes is determined by the sampling rate, and, so long as we have the correct number of tubes for the sampling rate being used, there will be no loss of accuracy. A rule of thumb is that we use one tube for every 1 kHz of sampling rate, so that 10 kHz requires ten tubes and so on. For a true continuous signal (equivalent to an infinite sampling rate) we see that we require an infinite number of zero-length tubes, i.e. a continuous tube. For all practical purposes, the discrete nature of the tube has no effect on accuracy.

All-pole modelling Only vowel and approximant sounds can be modelled with complete accuracy by all-pole transfer functions. We will see in Chapter 12 that the decision on whether to include zeros in the model really depends on the application to which the model is put, and mainly concerns tradeoffs between accuracy and computational tractability. Zeros in transfer functions can in many cases be modelled by the addition of extra poles. The poles can provide a basic model of anti-resonances, but cannot model zero effects exactly. The use of poles to model zeros is often justified because the ear is most sensitive to the peak regions in the spectrum (which are naturally modelled by poles) and less sensitive to the anti-resonance regions. Hence using just poles can often generate the required spectral envelope. One problem, however, is that as poles are used for purposes other than their natural one (to model resonances) they become harder to interpret physically, and this will have knock-on effects on, say, determining the number of tubes required, as explained above.

Lossless tube The real losses in the vocal tract are slight, but, since the difference between a pole being on the unit circle (no loss whatsoever) and slightly within the circle (say at a radius of 0.95) is quite noticeable, any loss factors will affect the overall system. That said, so long as any ignored losses in the vocal tract are compensated for by increased loss at the lips and glottis, the effect should not be too great. If only the lips and glottis are used for controlling losses, the effect is that a single damping factor is applied to all the formants, and hence it is not possible to generate formants with independent bandwidths.

Radiation characteristics The use of a reflection coefficient to model the radiation from the lips is really quite crude. In reality, radiation is very complicated and affected by the exact shape of the lips and head. In our simple example we effectively modelled the region outside the lips as an additional tube – for the case of complete reflection this is an infinitely wide tube, that is, the vocal tract opens into a wall that extends infinitely far in all directions. For the cases of less than total reflection, an additional tube of finite area was used. Since a real head is actually concave with respect to the direction of radiation, rather than convex in the additional-tube case, we see that this in fact is a very crude model. Rabiner and Schafer [368] and Flanagan [164] have developed more sophisticated expressions for lip radiation. An important radiation effect that we have not touched upon is that the radiation determines the directional characteristics of the sound propagation. One commonly observed effect is that high frequencies are more narrowly directed, so that, if a speaker is facing away from the listener, the high frequencies are not heard – the speech sounds "muffled".

Source–filter separation This is one area where our model starts to break down significantly. Firstly we saw that modelling the glottis as a completely isolated source is quite unrealistic. In addition, however, there are many physiological effects that make a complete source–filter separation unrealistic. For example, it is clear that, when speaking with a high fundamental frequency, the entire set of muscles in the throat and mouth is much tenser than in low-pitched speech.

Glottal source Our model of the source itself is perhaps the crudest and least-accurate part of our model. We know for example that the shape of the volume-velocity waveform changes with frequency – it is not the case that a fixed low-pass filter will do. The model described above also says nothing about **vocal effort**. It is clear that shouting is not just the same as speaking with a louder source input – the entire characteristic of the glottal source has changed. Furthermore, the dynamics of glottal signals are very complicated. In particular, when real glottal signals are studied in the regions where voicing starts and stops, we see many irregular pitch periods, isolated periods of very short duration and glottal signals that seem very noisy. The glottal model described in Section 11.4 models none of these effects. A significant amount of research has been conducted into building more sophisticated models, and progress has been made. However, many of these more-sophisticated models are difficult to implement and control and hence the simple model explained above is often used. We shall return to this issue in Section 13.3.5, where we shall see that difficulties in modelling source characteristics well are in fact the most difficult issue in speech synthesis, to the extent that they have been one of the main driving factors in determining the path of speech synthesis in recent years.

11.6.1 Further reading

Many books deal extensively with modelling the vocal tract acoustically, including Flanagan [164], Rabiner and Schafer [368] and of course Frant [158].

11.6.2 Summary

Model of the vocal organs
- The vocal apparatus can be modelled as a set of discrete, interconnected components.
- Each component functions as either a source component, which generates a basic sound, or a filter component, which modifies this sound.
- The filter components are joined in various configurations (Figure 11.16) to form tubes.
- For vowels, nasals and approximants, the source component is the glottis. The glottis is located at the end of the vocal-tract tube.
- Radiation occurs at the lips and nose, and this can also be modelled by a filter.
- The filter components are the pharynx, oral cavity and nasal cavity.
- All of the above are modelled as lossless tubes.
- All of the above act as linear time-invariant filters.
- Each component tube can be modelled, without loss of accuracy, as a series of connected shorter uniform tubes. The number of uniform tubes required is a function of the required length of the main tube and the sampling rate.

Lossless-tube models
- The sound wave generated by the source travels along the tube and in so doing is modified.
- The tubes are assumed lossless; hence all the impedance terms are reactive, meaning that they alter the frequency characteristics of the travelling wave, but do not cause energy losses.
- The main source of this is reflection, which occurs when the tube changes in cross-sectional area.
- A tube that has a closed termination and a tube that has a completely open termination cause complete reflection. At a closed termination, the pressure is at a maximum and the volume velocity is zero; at an open termination the pressure is zero and the volume velocity is at a maximum.
- At other changes in area, some part of the sound is transmitted and some is reflected backwards.
- The backward waves interfere with the forward waves, such that they reinforce at certain frequencies. This causes resonances and hence creates formants.
- While each tube in itself is lossless, losses occur in the system because some sound escapes from the lips and glottis.

The transfer function
- By iterative application of the equations relating volume velocity in one tube to volume velocity in the next, we can determine a transfer function for the whole tube.

- It is shown (Equation (11.26)) that all such models produce a transfer function containing only poles (i.e. a polynomial expression in z^{-1} in the denominator).
- We know, from Section 10.5.4, that such transfer functions naturally produce resonance patterns and hence formants.
- The polynomial expression can be factorised to find the individual poles.
- The values of the coefficients in the denominator polynomial are determined by the set of reflection coefficients in the tube. There is no simple relationship between the reflection coefficients and the poles, but they can be calculated if need be.
- The bandwidths of the resonances are determined by the losses in the system, and, since these occur only at the lips and glottis (or even just the lips or just the glottis), there are only one or two parameters in the system controlling all of these.
- Modelling nasals and obstruents is more complicated because of the inclusion of side branches in the model. These "trap" the sound, and are best modelled with filters that also contain zeros.
- Since nasals and obstruents involve parallel expressions, they cannot be expressed as a single transfer function.

12 Analysis of speech signals

In this chapter we turn to the topic of **speech analysis**, which tackles the problem of deriving representations from recordings of real speech signals. This book is of course concerned with speech *synthesis* – and at first sight it may seem that the techniques for generating speech "bottom-up" as described in Chapters 10 and 11 may be sufficient for our purpose. As we shall see, however, many techniques in speech synthesis actually rely on an analysis phase, which captures key properties of real speech and then uses these to generate new speech signals. In addition, the various techniques here enable useful characterisation of real speech phenomena for purposes of visualisation or statistical analysis. Speech analysis then is the process of converting a speech signal into an alternative representation that in some way better represents the information which we are interested in. We need to perform analysis because waveforms do not usually directly give us the type of information we are interested in.

Nearly all speech analysis is concerned with three key problems. First, we wish to remove the influence of phase; second, we wish to perform source/filter separation, so that we can study the spectral envelope of sounds independently of the source that they are spoken with. Finally, we often wish to transform these spectral envelopes and source signals into other representations that are coded more efficiently, have certain robustness properties, or more clearly show the linguistic information we require.

All speech signals in the real world are continuous signals, which describe the pattern of air pressure variation over time. These signals can be recorded with a variety of analogue means, but, for computer analysis, we require our signals to be **digitised** such that the continuous signal is converted into a discrete signal. For our purposes, we need not go into the details of analogue sound capture and recording since it is most likely that any reader will either come across signals that are already digitised, or be able to "record" any acoustic signal directly into a digitised form by using a standard computer sound card, without the need for any interim analogue recording step.[1]

12.1 Short-term speech analysis

A common starting point in speech analysis is to find the magnitude spectrum from a speech signal. We want this to be in a discrete form that is easy to calculate and store in

---

[1] Many standard books explain the process of analogue-to-digital conversion, for instance see Rabiner and Schafer [368].

a computer, so we use the discrete Fourier transform (DFT) as our principal algorithm. In practice the fast-Fourier-transform (FFT) implementation is nearly always used due to its speed (see Section 10.2.4).

12.1.1 Windowing

When we speak, the glottis and vocal tract are constantly changing. This is problematic for most of the techniques we introduced in Chapter 10 because these were designed to work on stationary signals. We can get around this problem by assuming that the speech signal is in fact stationary if considered over a sufficiently short period of time. Therefore we model a complete speech waveform as a series of short-term **frames** of speech, each of which we consider as a stationary time-invariant system.

A frame of speech $x[n]$ is obtained from the full waveform $s[n]$ by multiplication by a **window** $w[n]$ in the time domain:

$$x[n] = w[n]s[n] \qquad (12.1)$$

Three common windows are the **rectangular window**, **hanning window** and **hamming window**:

$$\text{rectangular } w[n] = \begin{cases} 1 & 0 \le n \le L-1 \\ 0 & \text{otherwise} \end{cases} \qquad (12.2)$$

$$\text{hanning } w[n] = \begin{cases} 0.5 - 0.5\cos(2\pi n/L) & 0 \le n \le L-1 \\ 0 & \text{otherwise} \end{cases} \qquad (12.3)$$

$$\text{hamming } w[n] = \begin{cases} 0.54 - 0.46\cos(2\pi n/L) & 0 \le n \le L-1 \\ 0 & \text{otherwise} \end{cases} \qquad (12.4)$$

Figure 12.1 shows a sinusoid after windowing by the three windows just described. The rectangular window abruptly cuts off the signal at the window edges, whereas the hanning tapers the signal so that the discontinuities are absent. The hamming window is just a raised version of the hanning, and hence has a similar shape except for a slight discontinuity at the window edge.

While it is an essential part of speech analysis, the windowing process unfortunately introduces some unwanted effects in the frequency domain. These are easily demonstrated by calculating the DFT of the sinusoid. The "true" frequency-domain representation for this, calculated in continuous time with a Fourier transform, is a single delta peak at the frequency of the sinusoid. Figure 12.2 shows the DFT and log DFT for the three window types. The DFT of the rectangular window exhibits a spike at the sinusoid frequency, but also has prominent energy on either side. This effect is much reduced in the hanning window. In the log version of this, it is clear that there is a wider **main lobe** than in the rectangular case – this is the key feature since it means that more energy is allowed to pass through at this point in comparison with at the neighbouring frequencies. The hamming window is very similar to the hanning, but has the additional

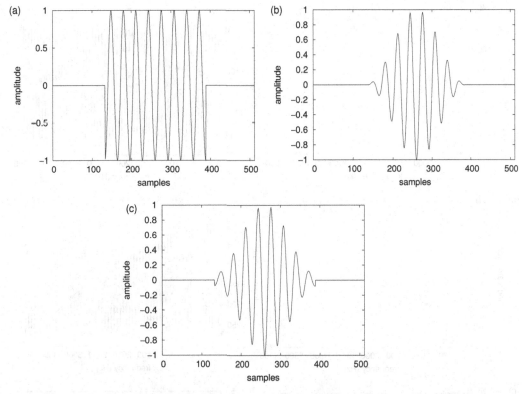

Figure 12.1 Effects of windowing in the time domain: (a) rectangular window, (b) hanning window and (c) hamming window.

advantage that the side lobes immediately neighbouring the main lobe are more suppressed. Figure 12.3 shows the effect of the windowing operation on a square-wave signal and shows that the harmonics are preserved far more clearly when the hamming window is used.

12.1.2 Short-term spectral representations

Using windowing followed by a DFT, we can generate **short-term spectra** from speech waveforms. The DFT spectrum is complex and and can be represented by its real and imaginary parts or its magnitude and phase parts. As explained in Section 10.1.5, the ear is not sensitive to phase information in speech, so the magnitude spectrum is the most suitable frequency-domain representation. The ear interprets sound amplitude in an approximately logarithmic fashion – so a doubling in amplitude produces only an additive increase in perceived loudness. Because of this, it is usual to represent amplitude logarithmically, most commonly on the decibel scale. By convention, we normally look at the **log power spectrum**, that is the log of the square of the magnitude spectrum. These operations produce a representation of the spectrum with which one attempts to

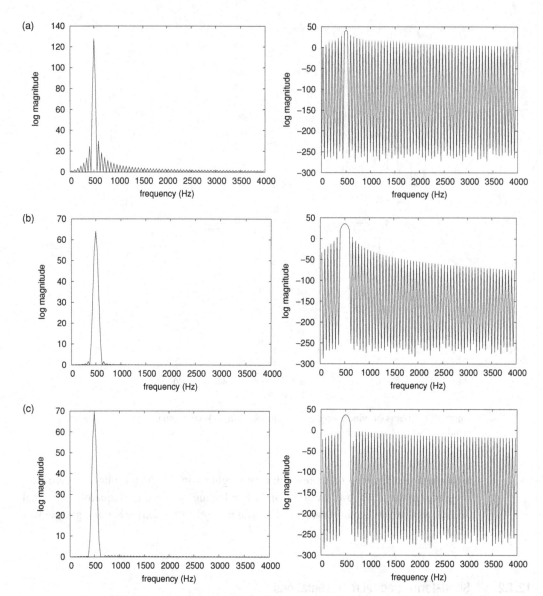

Figure 12.2 Effects of windowing in the frequency domain, with magnitude spectra on the left-hand side and log magnitude spectra on the right-hand side: (a) rectangular window, (b) hanning window and (c) hamming window.

match human perception: because phase is not perceived, we use the power spectrum; and because our response to signal level is logarithmic, we use the log power spectrum.

Figures 12.4 and 12.5 show some typical power spectra for voiced and unvoiced speech. For the voiced sounds, we can clearly see the harmonics as the series of spikes. They are evenly spaced, and the fundamental frequency of the source, the glottis, can be estimated by taking the reciprocal of the distance between the harmonics. In addition,

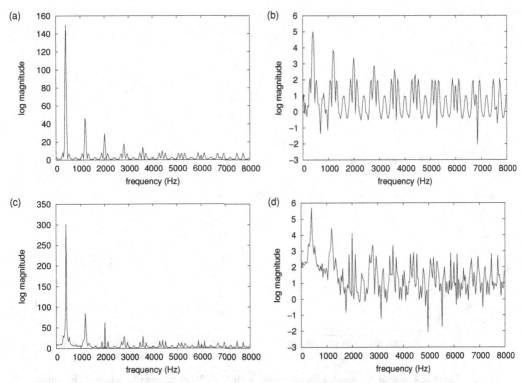

Figure 12.3 A comparison of effects of rectangular and hamming windows on a square wave of fundamental frequency 400 Hz and DFT length 512. Both spectra show significant deviation from the ideal of a single spike for each harmonic, but it is clear that the general pattern is that of a square wave, i.e. there are only odd harmonics, which are exponentially decaying. More distortion occurs for the weaker harmonics because their own energy is swamped by the side lobes of the other harmonics. This effect is much worse for the rectangular-window case, for which the higher harmonics are barely discernible. (a) Magnitude spectrum, hamming window. (b) Log magnitude spectrum, hamming window. (c) Magnitude spectrum, rectangular window. (d) Log magnitude spectrum, rectangular window.

the formants can clearly be seen as the more general peaks in the spectrum. The general shape of the spectrum ignoring the harmonics is often referred to as the **envelope**.

12.1.3 Frame lengths and shifts

While the spectra of the hamming and hanning windowed sinusoids give a truer frequency representation than the rectangular window, the spike has a noticeable width, and is not the ideal delta function that the continuous Fourier transform predicts. The width of the spike is an inverse function of window length – as the window lengthens the spike becomes narrower, as demonstrated in Figure 12.6. This of course is predicted by the scaling property (see Section 10.3.3). While longer windows produce a more accurate spectrum, the amount of speech required to generate (say) a window of length 2096 is

Analysis of speech signals

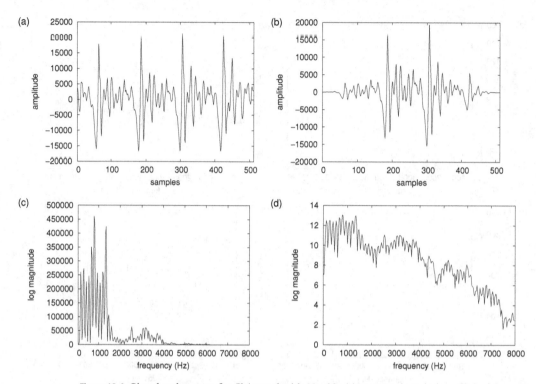

Figure 12.4 Signal and spectra for /ih/ vowel with $N = 32$: (a) rectangular window, (b) hanning window, (c) magnitude spectrum and (d) log magnitude spectrum.

such that the frame-based assumption of stationarity will have been violated, because the vocal tract shape and glottal period may have changed substantially within this time.

Hence we encounter the problem referred to as the **time–frequency tradeoff** in speech. The speech spectrum is constantly varying as we move from saying one sound to another. In fact it is a rare case when two consecutive frames of speech share exactly the same frequency characteristics. From the definition of the DFT we know that to achieve a high frequency resolution we need a large number of waveform samples. However, as the frame length increases, we include speech from different vocal-tract or glottis configurations. For example, if our window is very long we might have several different phonemes in the window, all with completely different characteristics. If, on the other hand, we use a very short window, in which we are sure to be capturing only a single speech sound, we will have only a few terms from the DFT, resulting in too coarse a representation of the spectrum. Hence there is a tradeoff. If we require information about the fundamental frequency of the signal, it is important that the frame length be long enough to include one or more periods of speech. Figure 12.7 shows waveforms and spectra for a signal for a variety of window lengths.

In addition to the length of window, we also have to consider how often we should calculate the spectrum. The **frame shift** is defined as the distance between two consecutive frames. Again, there is no single correct value for this. In some applications (e.g.

12.1 Short-term speech analysis

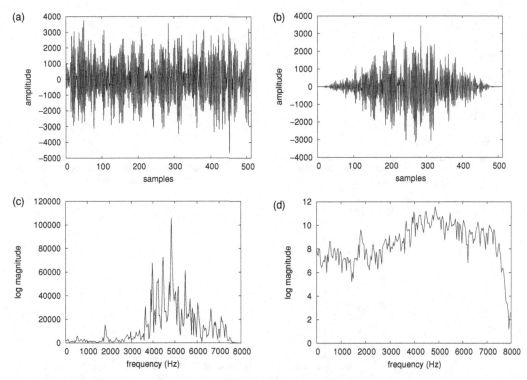

Figure 12.5 Signal and spectra for /s/ vowel with $N = 32$: (a) rectangular window, (b) hanning window, (c) magnitude spectrum and (d) log magnitude spectrum.

when calculating a spectrogram, as described below) a very short frame shift (e.g. 1 ms) is desirable because we wish to represent every possible transient effect in the evolution of the spectrum. For other uses, this is considered overkill, and, since spectra take time to calculate and space to store, a longer frame shift of about 10 ms is often used. In general we decide frame shifts on the basis of the rate of evolution of the vocal tract. A good rule of thumb is that the vocal tract can be considered stationary for the duration of a single period, so values in this range are often chosen. Typical values of frame length and frame shift are about 25 ms and 10 ms, respectively. These can be expressed in terms of numbers of samples by multiplication by the sample rate being used, and frame lengths are often chosen to be powers of two (128, 256 etc.) because of the use of the FFT.

In speech synthesis, it is common to perform **pitch-synchronous** analysis, in which the frame is centred around the pitch period. Since pitch is generally changing, this makes the frame shift variable. Pitch-synchronous analysis has the advantage that each frame represents the output of the same process, that is, the excitation of the vocal tract with a glottal pulse. In unvoiced sections, the frame rate is calculated at even intervals. Of course, for **pitch-synchronous** analysis to work, we must know where the pitch periods actually are: this is not a trivial problem and will be addressed further in Section 12.7.2. Note that fixed-frame-shift analysis is sometimes referred to as **pitch-asynchronous** analysis.

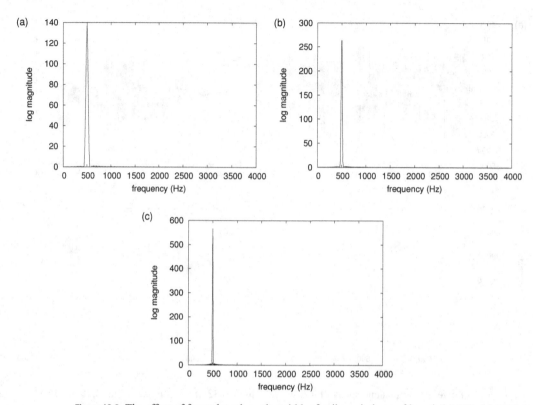

Figure 12.6 The effect of frame length on the width of spike: windows of length (a) 512, (b) 1024 and (c) 2048.

Figure 12.7 shows the waveform and power spectrum for five different window lengths. To some extent, all capture the envelope of the spectrum. For window lengths of less than one period, it is impossible to resolve the fundamental frequency and so no harmonics are present. As the window length increases, the harmonics can clearly be seen. At the longest window length, we have a very good frequency resolution, but, because so much time-domain waveform is analysed, the position of the vocal tract and pitch have changed over the analysis window, leaving the harmonics and envelope to represent an average over this time rather than a single snapshot.

12.1.4 The spectrogram

A spectrum shows only the frequency characteristics for a single frame at a single point in time – often this is referred to as a spectral **slice**. We often wish to observe the evolution of the spectrum over time – this can be displayed as a **spectrogram**. A spectrogram is a three-dimensional plot as shown in Figure 12.8. The horizontal axis denotes time, the vertical axis frequency. The level of darkness represents amplitude – the darker the plot the higher the amplitude at this point. The spectrogram is formed by calculating the DFT

12.1 Short-term speech analysis

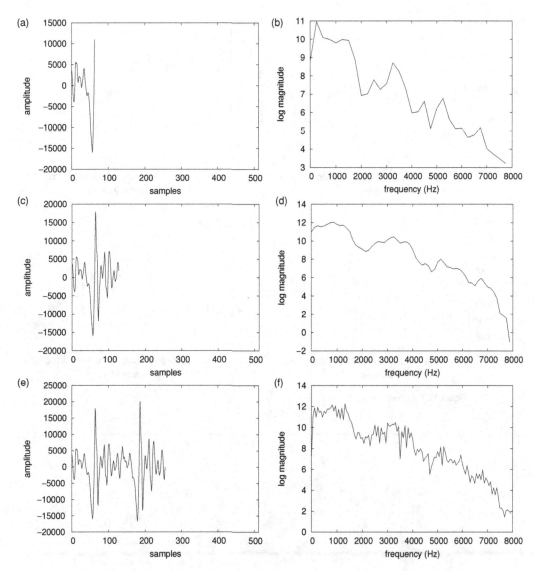

Figure 12.7 Signals and spectra for a vowel with various window lengths: (a) 64 sample frame, (b) log magnitude spectrum of 32 sample frame, (c) 128 sample frame, (d) log magnitude spectrum of (c), (e) 256 sample frame, (f) log magnitude spectrum of (e), (g) 512 sample frame, (h) log magnitude spectrum of (g), (i) 1024 sample frame and (j) log magnitude spectrum of (i). For the shorter frames, only a coarse spectrum can be discerned. The harmonics are clearest for length 512; above this length the changes in fundamental frequency across the window mean that there is no single harmonic spacing.

at successive points along the waveform. For a high-definition spectrogram, the frame shift is normally small, say 1 ms.

The choice of window length is important to the appearance of the spectrogram – Figure 12.9 shows a **narrow-band** spectrogram, which was calculated with a DFT length

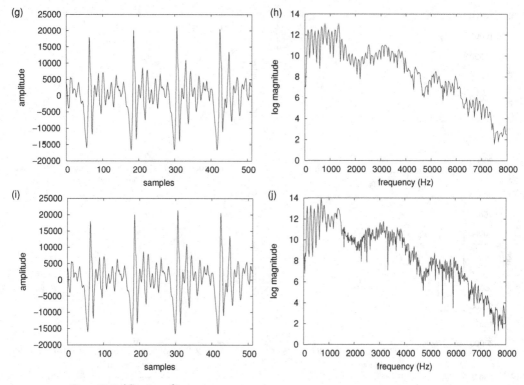

Figure 12.7 (*Continued*)

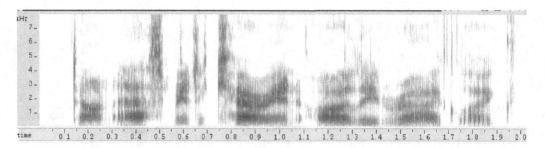

Figure 12.8 A wide-band spectrogram.

of 400; Figure 12.8 shows a **wide-band** spectrogram calculated with a DFT length of 40. The narrow-band spectrogram has a detailed spectrum, so the harmonics can clearly be seen as horizontal stripes in the voiced sounds. It should be clear that the distance between adjacent stripes for a particular time value is constant, as we would expect of harmonics. In fact an estimate of the fundamental frequency can be taken by measuring the inter-harmonic distance. The wide-band spectrogram has a very short DFT and so the frequency information is course and the harmonics are not discernible. The formants, however, clearly are – these are seen as the wide horizontal stripes. Since the window length is short, the time resolution is high, and hence several frames are needed to cover a

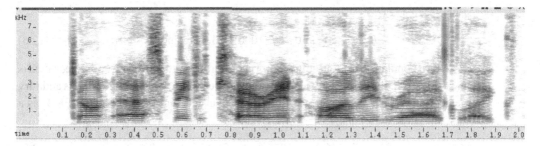

Figure 12.9 A narrow-band spectrogram.

single period. The effect of this can be seen in the fine vertical streaks: frames that contain a glottal impulse will be dark; those which do not will be lighter. The distance between successive dark streaks can be used to gain an estimate of the pitch. Because the frames are much longer in the narrow-band case, it is not possible to discern pitch in this way. The narrow-band and wide-band spectrograms clearly demonstrate the time–frequency tradeoff. In general, the wide-band spectrogram is used for finding information about phones, since it shows the formant structure more clearly.

Most speech-analysis computer packages have spectrogram-display software, so it is easy to visualise and examine speech. The reader should be made aware, however, that many of these packages perform a degree of graphical post-processing, which brings out the salient features in the spectrogram. This is often achieved by setting saturation levels that specify cut-offs beyond which everything is either white or black. It is important to be aware of these post-processing steps because they can give rise to different-looking spectrograms, and, if caught unaware, one might mistakenly conclude that two speech signals are more dissimilar than they really are.

12.1.5 Auditory scales

It is well known that human sensitivity to the frequency scale is not linear; for instance we know that musical relationships are clearly logarithmic. Studies into the low-level perception of sounds have resulted in a number of **auditory scales**, which define a new frequency range that is more in line with the human sensitivity to sounds at different frequencies.

The **mel-scale** was the product of experiments with sinusoids in which subjects were required to divide frequency ranges into sections. From this, a new scale was defined, in which one **mel** equalled one thousandth of the pitch of a 1-kHz tone [415]. The mapping from linear frequency to this scale is given by

$$m = 2595 \log_{10}(1 + f/700) \qquad (12.5)$$

Further studies based on loudness (i.e. the perception of amplitude) found that a more accurate representation was one that had a linear response for lower frequencies, becoming logarithmic for higher frequencies. One popular scale based on this is the

Bark-scale [521]:

$$B = 13\arctan(0.76f/1000) + 3.5\arctan((f/7500)^2) \tag{12.6}$$

A third popular scale is the **equivalent rectangular bandwidth** or **ERB** scale, which is measured from the ability to detect sinusoids in the presence of noise. From this auditory perception is described in terms of an "equivalent rectangular bandwidth" (ERB) as a function of centre frequency [316]:

$$B = 6.23 \times 10^{-6} f^2 + 9.339\,10^{-2} f + 28.52 \tag{12.7}$$

12.2 Filter-bank analysis

We will now turn to the important problem of **source–filter separation**. In general, we wish to do this because the two components of the speech signal have quite different and independent linguistic functions. The source controls the pitch, which is the acoustic correlate of intonation, while the filter controls the spectral envelope and formant positions, which determine which phones are being produced. There are three popular techniques for performing source–filter separation. First we will examine **filter-bank analysis** in this section, before turning to **cepstral analysis** and **linear prediction** in the next sections.

If we visually examine a short-length DFT such as that in Figure 12.4, or in fact any spectrogram, we can see the general shape of the spectral envelope. That is, if we "squint" and ignore the harmonics, in many cases the shape of the spectral envelope is discernible independently of the particular positions of the harmonics. When doing this we are effectively averaging out the influence of individual harmonics and creating a blurring effect. This is the principle behind filter-bank analysis.

Filter-bank analysis is performed by first creating a series of **bins**, each centred on a particular frequency. Time-domain filter-bank analysis can be performed by creating a **band-pass** filter (either FIR or IIR), which allows frequencies at or near the bin frequency to pass, but attenuates all other frequencies to zero. After the filtering, the amount of energy is calculated and assigned to that bin. Once this has been done for every bin, we have a representation of how energy varies according to frequency. The effect of the filtering operation is to blur the effect of individual harmonics, such that the final representation is largely that of the vocal tract irrespective of the source. Alternatively, we can perform filter-bank analysis on spectrum. Here, we take a magnitude spectrum and multiply it by a windowing function centred on a bin frequency. This has the effect of setting all energy outside the window to zero, while allowing us to measure the energy inside the window. This is show in Figure 12.10(a). Often the bins are not spaced at equal frequencies, but defined logarithmically or according to one of the frequency scales of Section 12.1.5. This is shown in Figure 12.10(b).

Filter-bank analysis is a simple and robust technique for finding the spectral envelope, but it is only a partial solution to source–filter separation. First, the amount of "blurring" required to eliminate the harmonics may be too severe and may eliminate some of the

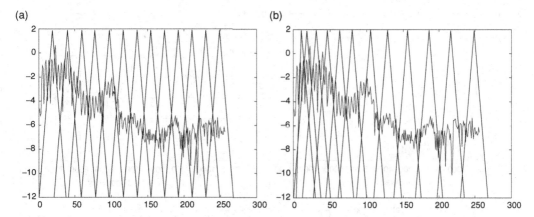

Figure 12.10 Filter-bank analysis on magnitude spectra: (a) with evenly spaced bins and (b) with bins spaced according to the mel-scale.

detail we require; alternatively, especially with high-pitched speech, we can find that peaks in the envelope fall between two harmonics and therefore are ignored. Secondly, filter-bank analysis gives us only the spectral envelope derived from the vocal-tract transfer function; it does not give us the source. For many applications (e.g. ASR) this is not a problem, but in TTS we generally require both components of the signal, so other techniques are normally used.

12.3 The cepstrum

12.3.1 Cepstrum definition

The cepstrum is (most commonly) defined as the **inverse DFT of the log magnitude of the DFT of a signal**:

$$c[n] = \mathcal{F}^{-1}\{\log|\mathcal{F}\{x[n]\}|\} \tag{12.8}$$

where $\mathcal{F}$ is the DFT and $\mathcal{F}^{-1}$ is the inverse DFT. Figure 12.11 shows the process of calculating the DFT, log DFT and inverse DFT on a single frame of speech. We will now look at how this operation performs source–filter separation.

12.3.2 Treating the magnitude spectrum as a signal

Consider the magnitude spectrum of a periodic signal, such as that shown in Figure 12.11. As we have proved and seen in practice, this spectrum will contain harmonics at evenly spaced intervals. Because of the windowing effects, these harmonics will not be delta-function spikes, but will be somewhat more "rounded". For most signals, the amplitude of the harmonics tails off quite quickly as frequency increases. We can reduce this effect by compressing the spectrum with respect to amplitude; this can easily be achieved by calculating the log spectrum. It should be clear in the many examples of log spectra

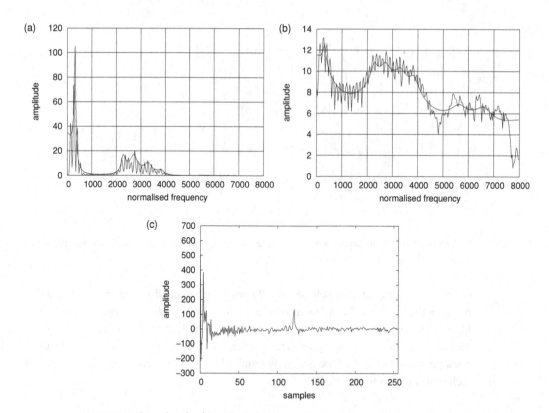

Figure 12.11 Steps involved in calculating the cepstrum: (a) magnitude spectrum and envelope, and (b) log magnitude spectrum and envelope and (c) cepstrum. For demonstration purposes, an estimate of the spectral envelope has been overlaid on the two DFT spectra. The "periodicity" of the harmonics in the log spectrum can clearly be seen. This gives rise to the spike at point 120 in the cepstrum. The low values (<30) in the cepstrum describe the envelope. It should be clear that these low values are separated from the spike – because of this it is a simple task to separate the source and filter.

in this chapter that the relative amplitudes of the harmonics are much closer in the log spectra than they are in the absolute spectra. Figure 12.5(b) shows this effect.

Now, forget for a moment what the log spectrum actually represents, and consider it a *waveform* instead. If we were told that Figure 12.11(b) was a waveform, we would informally describe this as some sort of quasi-periodic signal (after it, it has a repeating pattern of harmonics), with a time-evolving "amplitude effect" that makes some "periods" higher than others. It is clear that the rate at which the "periods" change is much greater than the rate of evolution of the "amplitude effect". If we wished to separate these, we would naturally turn to our standard tool for analysing the frequency content of waveforms, namely the Fourier transform. If we calculate the DFT of our "waveform" we get a new representation, which we will designate the **cepstrum**. This is shown in Figure 12.11(c). As we would expect, the periodicity of the "waveform" is shown by a large spike at about position 120. Because we haven't performed any windowing operations that cause blurring, we would expect this to be close to a delta function.

Again, as we would expect, because the "periods" were not sinusoidal, this has harmonics at multiples of the main spike. The "amplitude effect" will be present in the cepstrum also, but, because this is varying much more slowly than the "periods", it will be in the lower range of the cepstrum. Since the "amplitude effect" is the spectral envelope and the spikes represent the harmonics (i.e. the pitch), we see that these operations have produced a representation of the original signal in which the two components lie at different positions. A simple filter can now separate them.

12.3.3 Cepstral analysis as deconvolution

Recall from Section 11.1 that Equation (11.2) described the speech-production process as

$$y[n] = u[n] \otimes v[n] \otimes r[n]$$

where $u[n]$ is the glottal or other source, $v[n]$ is the vocal-tract filter and $r[n]$ is the radiation. In the time domain, these expressions are convoluted and so are not easily separable. If we don't know the form of the vocal-tract filter, we can't perform any explicit inverse-filtering operation to separate these components.

We can, however, perform separation as follows. Let us subsume the radiation filter $r[n]$ into $v[n]$ to give a single filter $v'[n]$. We can then write the time-domain expression of Equation (11.2) as

$$y[n] = u[n] \otimes v'[n] \qquad (12.9)$$

Taking the Fourier transform of Equation (12.9) gives

$$Y(e^{j\omega}) = U(e^{j\omega})V'(e^{j\omega})$$

If we now take the log of the magnitude of this we get

$$\log(|Y(e^{j\omega})|) = \log(|U(e^{j\omega})||V'(e^{j\omega})|)$$
$$= \log(|U(e^{j\omega})|) + \log(|V'(e^{j\omega})|)$$

in which the source spectrum and filter spectrum are now just added together. We can now return to the time domain by taking an inverse DFT, giving us

$$c[n] = c_u[n] + c_v[n] \qquad (12.10)$$

that is, a time domain representation created by addition of source and filter. Unlike in Equation (12.9), the source and filter are now added, rather than convoluted, in the time domain.

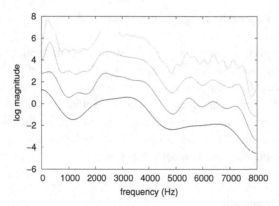

Figure 12.12 Series of spectra calculated from cepstra. The 512-point cepstrum has all its values after a cut-off point K set to 0 and then a 512-point DFT is applied to generate a spectrum. The figure shows this spectrum for four values of K, namely 10, 20, 30 and 50. (In the figure each has been graphically separated for clarity, with $K = 10$ as the bottom spectrum and $K = 50$ as the top. In reality all lie with in the same amplitude range.) As the number of coefficients increases, so does the level of spectral detail, until effects of harmonics start appearing. The choice of optimal cut-off point depends on whether spectral precision or elimination of harmonics is the top priority.

12.3.4 Cepstral analysis discussion

If we take the DFT of a signal and then take the inverse DFT of that, we of course get back to where we started. The cepstrum calculation differs in two important respects. First, we are only using the magnitude of the spectrum – in effect we are throwing away the phase information. The inverse DFT of a magnitude spectrum is already very different from the inverse DFT of a normal (complex) spectrum. The log operation scales the harmonics, which emphasises the "periodicity" of the harmonics. It also ensures that the cepstrum is the sum of the source and filter components rather than their convolution.

The cepstrum is useful because it splits the signal into an envelope, given by the first few coefficients, and a source, given by the spike. Subsequent analysis usually throws away one of these parts: if we are interested in the vocal tract we use only the low coefficients; if we are interested in the pitch and behaviour of the glottis, we study the peak. We can demonstrate this by calculating a spectrum in which the spike is eliminated – this is done by simply setting a cut-off point K above which all the cepstral coefficients are set to zero. From this, we can create a log magnitude spectrum by applying a DFT on the modified cepstrum. Figure 12.12 shows this for four different values of K. The figure shows that a reasonably accurate spectral envelope can be generated from about 30 coefficients, but the number chosen really depends on the degree of precision required in the spectrum.

For many cases in which the spectral envelope is required, it is advantageous to keep the low part of the cepstrum as it is, instead of converting it back to a frequency-domain spectrum. The low cepstral coefficients form a very compact representation of the envelope, and have the highly desirable statistical modelling property of being independent so that only their means and variances, not their covariances, need be stored to provide an accurate statistical distribution. It is for this reason that cepstra are the feature representation of choice in most speech-recognition systems. Despite this, cepstra are quite difficult to interpret visually, so we often find that, while systems use cepstra internally, we still use spectra when we wish to inspect a signal visually. The higher part of the cepstrum contains the pitch information, and is often used as the basis for pitch-detection algorithms, which will be explained in Section 12.7.1

12.4 Linear-prediction analysis

Linear prediction (LP) is another technique for source–filter separation, in which we use the techniques of LTI filters to perform an explicit separation of source and filter. In LP we adopt a simple system for speech production, where we have an input source $x[n]$, which passes through a linear time-invariant filter $h[n]$, to give the output speech signal $y[n]$. In the time domain this is

$$y[n] = x[n] \otimes h[n] \qquad (12.11)$$

The convolution makes it difficult to separate the source and filter in the time domain, but this is easily achieved if we transform our system to the z-domain:

$$H(z) = \frac{X(z)}{Y(z)} \qquad (12.12)$$

In Section 12.12, we showed that the lossless-tube model was a reasonable approximation for the vocal tract during the production of a vowel. If we assume for now that $H(z)$ can therefore be represented by an all-pole filter, we can write

$$Y(z) = X(z)H(z) \qquad (12.13)$$

$$= \frac{1}{1 - \sum_{k=1}^{P} a_k z^{-k}} X(z) \qquad (12.14)$$

where $\{a_k\}$ are the filter coefficients. By taking the inverse z-transform[2] we see that in the time domain this becomes

$$y[n] = \sum_{k=1}^{P} a_k y[n-k] + x[n] \qquad (12.15)$$

---

[2] We can most easily prove this in reverse; if

$$y[n] = \sum_{k=1}^{P} a_k y[n-k] + x[n]$$

then taking the z-transform of both sides results in

$$Y(z) = \sum_{n=-\infty}^{n=\infty} \left(\sum_{k=1}^{P} a_k y[n-k] \right) z^{-n} + X(z)$$

$$= \sum_{k=1}^{P} a_k \left(\sum_{n=-\infty}^{n=\infty} y[n-k] z^{-n} \right) + X(z)$$

The importance of this equation is that it states that the value of the signal y at time n can be determined by a linear function of the previous values of y, apart from the addition of the current value of the input $x[n]$. Because of this, the technique is known as **linear prediction**. In essence, this states that an output signal is closely approximated by a linear sum of the previous samples:

$$\tilde{y}[n] = \sum_{k=1}^{P} a_k y[n-k] \qquad (12.16)$$

and this is identical to the difference equation (12.15) except for the input term. In linear predication, the input $x[n]$ is usually referred to as the **error**, and is often written $e[n]$ to show this. This may seem a little harsh, but the concept comes from the fact that $y[n]$ is the actual signal and $\tilde{y}[n]$ is its approximation by the model, the only difference being the contribution from $x[n]$. Since this accounts for the difference between the predicted and actual signals it is called the error. We now turn to the issue of how to find automatically the predictor coefficients of $H(z)$ for a frame of speech.

12.4.1 Finding the coefficients: the covariance method

The goal of linear prediction is to find the set of coefficients $\{a_k\}$ which generates the smallest error $e[n]$, over the course of the frame of speech. We will now examine two closely related techniques for finding the a_k coefficients, both of which are based on error minimisation. We can state Equation (12.15) in terms of the error for a single sample:

$$e[n] = y[n] - \sum_{k=1}^{P} a_k y[n-k] \qquad (12.17)$$

We wish to find the values $\{a_k\}$ which minimise the squared error, summed over the frame of speech. Using E to denote the sum squared error we can state

$$E = \sum_{n=0}^{N-1} e^2[n]$$

$$= \sum_{n=0}^{N-1} \left(y[n] - \sum_{k=1}^{P} a_k y[n-k] \right)^2$$

Using the dummy substitution $m = n - k$ gives

$$Y(z) = \sum_{k=1}^{P} a_k \left(\sum_{m=-\infty}^{m=\infty} y[m] z^{-m} z^{-k} \right) + X(z)$$

$$= \sum_{k=1}^{P} a_k Y(z) z^{-k} + X(z)$$

Rearranging the $Y(z)$ terms gives the all-pole transfer function of Equation (12.14).

We can find the minimum of E by differentiating with respect to *each* coefficient a_k, and setting to zero:

$$\frac{\delta E}{\delta a_j} = 0 = \sum_{n=0}^{N-1}\left(2\left(y[n] - \sum_{k=1}^{P} a_k y[n-k]\right) y[n-j]\right) \quad \text{for } j = 1, 2, 3, \ldots, P,$$

$$= -2 \sum_{n=0}^{N-1} y[n]y[n-j] + 2 \sum_{n=0}^{N-1} \sum_{k=1}^{P} a_k y[n-k]y[n-j]$$

which gives

$$\sum_{n=0}^{N-1} y[n-j]y[n] = \sum_{k=1}^{P} a_k \sum_{n=0}^{N-1} y[n-j]y[n-k] \quad \text{for } j = 1, 2, 3, \ldots, P, \quad (12.18)$$

If we define $\phi(k, j)$ as

$$\phi(j, k) = \sum_{n=0}^{N-1} y[n-j]y[n-k] \quad (12.19)$$

Equation (12.18) can be written more succinctly as

$$\phi(j, 0) = \sum_{k=1}^{P} \phi(j, k) a_k \quad (12.20)$$

the full form of which is

$$\phi(1, 0) = a_1 \phi(1, 1) + a_2 \phi(1, 2) + \cdots + a_P \phi(1, P)$$
$$\phi(2, 0) = a_1 \phi(2, 1) + a_2 \phi(2, 2) + \cdots + a_P \phi(2, P)$$
$$\phi(3, 0) = a_1 \phi(3, 1) + a_2 \phi(3, 2) + \cdots + a_P \phi(3, P)$$
$$\vdots$$
$$\phi(P, 0) = a_1 \phi(P, 1) + a_2 \phi(P, 2) + \cdots + a_P \phi(P, P)$$

We can easily calculate all the quantities $\phi(1, 0)$, $\phi(2, 0)$, so it should be clear that this is a set of P equations with P unknown terms, namely the a_k coefficients. This can be written in matrix form as

$$\begin{bmatrix} \phi(1,0) \\ \phi(2,0) \\ \phi(3,0) \\ \vdots \\ \phi(P,0) \end{bmatrix} = \begin{bmatrix} \phi(1,1) & \phi(1,2) & \cdots & \phi(1,P) \\ \phi(2,1) & \phi(2,2) & \cdots & \phi(2,P) \\ \phi(3,1) & \phi(3,2) & \cdots & \phi(3,P) \\ \vdots & & & \vdots \\ \phi(P,1) & \phi(P,2) & \cdots & \phi(P,P) \end{bmatrix} \begin{bmatrix} a_1 \\ a_2 \\ a_3 \\ \vdots \\ a_P \end{bmatrix} \quad (12.21)$$

and expressed as

$$\phi = \psi \mathbf{a}$$

We can find the value of $\mathbf{a}$ (that is, the coefficients a_k) by solving this equation. This can be done simply, by solving for a_1 in terms of the other coefficients and substituting again, but this is extremely inefficient. Because the square matrix ϕ is symmetric a much faster technique known as **Cholskey decomposition** can be used [295], [302].

This method is called the **covariance method** (for somewhat obscure reasons). It works by calculating the *error* in the region of samples from $n = 0$ to $n = N - 1$, but in fact uses *samples* in the region $n = -P$ to $n = N - 1$ to do so. This is because the method requires a full calculation of the error at time $n = 0$, and hence must use some previous samples to do this calculation. The calculation over many frames can be found either by simply using the main signal itself, and moving the start position n to the designated frame start each time, or by creating a separate windowed frame of speech, of length $N + P$ samples, and performing the calculation starting P samples into the frame. The choice is mainly determined by implementation issues (e.g. a system may have already performed the windowing operation before we start). No special windowing functions are required with this method: either we can perform the calculations directly on the main waveform, or a rectangular window can be used if windowing must be performed.

12.4.2 The autocorrelation method

A slightly different method known as the **autocorrelation method** can also be used. Since this has many practical advantages over the covariance method, it is the method most often used in practice.

Recall from Section 10.3.9 that the autocorrelation function of a signal is defined as

$$R[n] = \sum_{m=-\infty}^{\infty} y[m]y[n-m] \qquad (12.22)$$

This is similar in form to Equation (12.19), and, if we can calculate the autocorrelation, we can make use of important properties of this signal. To do so, in effect we have to calculate the squared error from $-\infty$ to ∞. This differs from the covariance method, in which we just consider the speech samples from a specific range. To perform the calculation from $-\infty$ to ∞, we window the waveform using a hanning, hamming or other window, which has the effect of setting all values outside the range $0 \leq n < N$ to 0.

While we are calculating the error from $-\infty$ to ∞, this will always be 0 before the windowed area, because all the samples are 0 and hence multiplying them by the filter coefficients will just produce 0. This will also be the case significantly after the window. Importantly, however, for P samples after the window, the error will not be 0 since it will still be influenced by the last few samples at the end of the region. Hence the calculation of the error from $-\infty$ to ∞ can be rewritten as

$$\phi(j,k) = \sum_{-\infty}^{\infty} y[n-j]y[n-k]$$
$$= \sum_{n=0}^{N-1+P} y[n-j]y[n-k]$$

which can be rewritten as

$$\phi(j,k) = \sum_{n=0}^{N-1-(j-k)} y[n]y[n+j-k] \qquad (12.23)$$

Because this is a function of one independent variable $j - k$, rather than the two of (12.23), we can rewrite it as

$$\phi(j,k) = \sum_{n=0}^{N-1-k} y[n]y[n+k] \tag{12.24}$$

Since the limits of this expression equate to $-\infty$ to ∞, this is the same as the autocorrelation function (12.22), and we can write

$$R(j-k) = \phi(j,k)$$

Now we can rewrite the matrix form of (12.20) in terms of the autocorrelation function:

$$\phi(j,0) = \sum_{k=1}^{P} \phi(j,k)a_k \tag{12.25}$$

$$R(j) = \sum_{k=1}^{P} R(j-k)a_k \tag{12.26}$$

This significance of all of this is that the autocorrelation function (of any signal) is even, that is $R(j-k) = R(j+k) = R(|j-k|)$, which means that, when we now write the matrix in Equation (12.21) in terms of R, we obtain

$$\begin{bmatrix} R(1) \\ R(2) \\ R(3) \\ \vdots \\ R(P) \end{bmatrix} = \begin{bmatrix} R(0) & R(1) & R(2) & \cdots & R(P-1) \\ R(1) & R(0) & R(1) & \cdots & R(P-2) \\ R(2) & R(1) & R(0) & \cdots & R(P-3) \\ \vdots & & & & \\ R(P-1) & R(P-2) & R(P-3) & \cdots & R(0) \end{bmatrix} \begin{bmatrix} a_1 \\ a_2 \\ a_3 \\ \vdots \\ a_P \end{bmatrix}$$
(12.27)

Unlike the square matrix in Equation (12.21), this matrix is symmetric and all the elements in its diagonals are symmetric. This is known as a **Toeplitz** matrix, and, because of the properties just mentioned, it is significantly easier to invert.

12.4.3 Levinson–Durbin recursion

The matrix given in Equation (12.23) can of course be solved by any matrix-inversion technique. Such techniques can be slow, however (usually of the order of p^3, where p is the dimensionality of the matrix), and hence faster techniques have been developed to find the values of $\{a_k\}$ from the autocorrelation functions $R(k)$. In particular, it can be shown that the **Levinson–Durbin** recursion technique can solve Equation (12.27) in time p^2. For our purposes, analysis speed is really not an issue, so we will forgo a detailed discussion of this and other, related techniques. However, a brief overview of the technique is interesting in that it sheds light on the relationship between linear prediction and the all-pole tube model discussed in Chapter 11.

The Levinson–Durbin recursion operates by considering a set of initialisation conditions, and using these to find the coefficients of a first-order polynomial (that is, just a single term in z^{-1}) which minimise the mean squared error. From this we find the coefficients of a second-order polynomial, third-order polynomial and so on, using the previous polynomial and minimisation of the error, until we have reached a polynomial of the required filter order P.

The algorithm initialises by setting the error and predictor coefficients of a (contrived) zeroth-order polynomial as

$$a_0^0 = 0$$
$$E^0 = R(0)$$

Next we define a weighting factor for the ith model as

$$k_i = \frac{-R(i) - \sum_{j=1}^{i-1} a_j^{i-j} R(i-j)}{E^{i-1}}$$

Using this, we find the predictor coefficients for the ith model:

$$a_i^i = k_i \qquad (12.28)$$
$$a_j^i = a_j^{i-1} - k_i a_{i-j}^{i-1} \qquad 1 \leq j \leq i-1 \qquad (12.29)$$

and from this, we can update the minimum mean-squared prediction error for this step as

$$E^i = \left(1 - k_i^2\right) E^{i-1}$$

These steps are repeated until $i = p$, whereupon we have a polynomial and hence set of predictor coefficients of the required order. We have just seen how the minimisation of error over a window can be used to estimate the linear-prediction coefficients. Since these are in fact the filter coefficients that define the transfer function of the vocal tract, we can use these in a number of ways to generate other useful representations.

12.5 Spectral-envelope and vocal-tract representations

The preceding sections showed the basic techniques of source–filter separation using first cepstral and then linear-prediction analysis. We now turn to the issue of using these techniques to generate a variety of representations, each of which by some means describes the spectral envelope of the speech.

12.5.1 Linear-prediction spectra

The main motivation in LP analysis is that it provides an automatic and explicit means for separating the source and filter components in speech. Recall the basic equation (11.2) of the vocal tract,

$$Y(z) = U(z)V(z)R(z)$$

Here $Y(z)$ is the speech, $U(z)$ is the source, $V(z)$ is the vocal tract and $R(z)$ is the radiation. Ideally, the transfer function $H(z)$ found by linear-prediction analysis would be $V(z)$, the vocal-tract transfer function. In the course of doing this, we could then find $U(z)$ and $R(z)$. In general $H(z)$ is a close approximation to $V(z)$ but is not exactly the same. The main reason for this is that the LP minimisation criterion means that the algorithm attempts to find the lowest error for the *whole* system, not just the vocal-tract component. In fact, $H(z)$ is properly expressed as

$$H(z) = G(z)V(z)R(z)$$

for voiced speech, where $G(z)$ represents the glottal transfer function, and

$$H(z) = V(z)R(z)$$

for unvoiced speech. The distinction that the LP coefficient represents $H(z)$ instead of $V(z)$ has most significance for source modelling, so we will delay a full discussion of this until Section 12.6. For most purposes, the fact that $H(z)$ models more than just the vocal tract isn't a big problem; the upshot is that some additional poles are included in the transfer function, but the effect of these can often be ignored, and a good estimate of the vocal-tract envelope can be found by converting $H(z)$ directly to the frequency domain. If a purer estimate of $V(z)$ is required, one of the techniques described in Section 12.6 can be used instead.

The transfer function $H(z)$ that we have been estimating can be converted to the frequency domain by simply setting $z = e^{j\omega t}$ (see Section 10.5.2) to give $H(e^{j\omega t})$. Doing this gives us the spectral envelope of the LP filter, which, apart from the extra poles, we can interpret as the vocal-tract spectral envelope. A plot of this LP envelope overlayed on the DFT is shown in Figure 12.13(a). It is quite clear that the LP spectrum is indeed a good estimate of the spectral envelope.

Figure 12.14 shows spectral envelopes for a range of sounds. Here we can see a much-noted "weakness" of LP envelope estimation, which is that the spectrum can appear very "peaky". This arises because, in many speech sounds, particularly vowels, the poles lie close to, but not on, the unit circle. At such locations, even a very small difference in their distance from the unit circle can result in much-sharper formant peaks. For example, for a sampling rate of 10 000 Hz, a pole with radius 0.96 will produce a formant bandwidth of 130 Hz, whereas a pole of 0.98 produces a formant of bandwidth 65 Hz, so any slight error in determining the radius of the pole can produce a spectrum with exaggerated formants. This problem does not occur with pole frequencies, so the LP envelope can normally be relied upon to find the positions of resonances quite accurately.

Recall that in Chapter 11 we showed that the number of tubes required is directly related to the sampling rate, length of tube and speed of sound (see Section 11.3.5). For a speed of sound of 340 m s^{-1} and vocal-tract length of 17 cm, we showed that ten tube sections were required for a sampling rate of 10 000 Hz. Assuming these values for the speed of sound and vocal-tract length, we see that we require one tube section per 1000 Hz sampling rate. Since each tube section gives rise to one pole, we conclude that $V(z)$ should be a tenth-order all-pole filter for a sampling rate of 10 000 Hz a 16th-order filter for 16 000 Hz and so on. The source ($G(z)$) and radiation ($R(z)$) filters are considerably simpler, and can be adequately modelled by four to six additional poles for both. It is highly advantageous to keep the number of poles even, so an overall value of

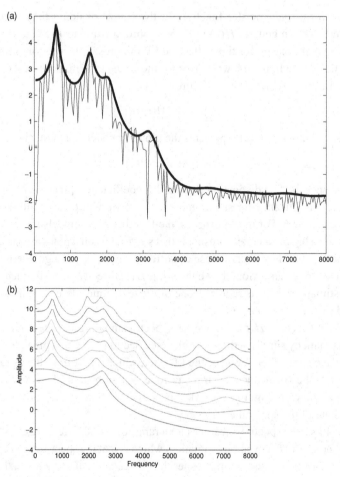

Figure 12.13 Examples of linear-prediction spectra. (a) LP spectrum superimposed on the DFT. (b) LP spectra for analysis of order 4, 6, 8, 10, 12, 14, 16 and 18. In reality the plots lie on top of one another, but each has been separated vertically for purposes of clarity. The bottom plot is from LP analysis of order 4; the top plot is for order 18.

14 poles or 16 poles is normally adopted for a sampling rate of 10 000 Hz, and 20 or 22 poles for a sampling rate of 16 000 Hz.

Figure 12.13(b) shows the spectra predicted from LP filters for a range of orders. It is quite clear that, as the order increases, so does the detail in the spectrum. It is interesting to compare this with Figure 12.12, which shows the same investigation but with cepstra. It is clear that the cepstral spectra are much more heavily influenced by the harmonics as the order of analysis increases.

12.5.2 Transfer-function poles

As with any filter, we can find the poles by finding the roots of the transfer-function polynomial.

12.5 Spectral-envelope and vocal-tract representations

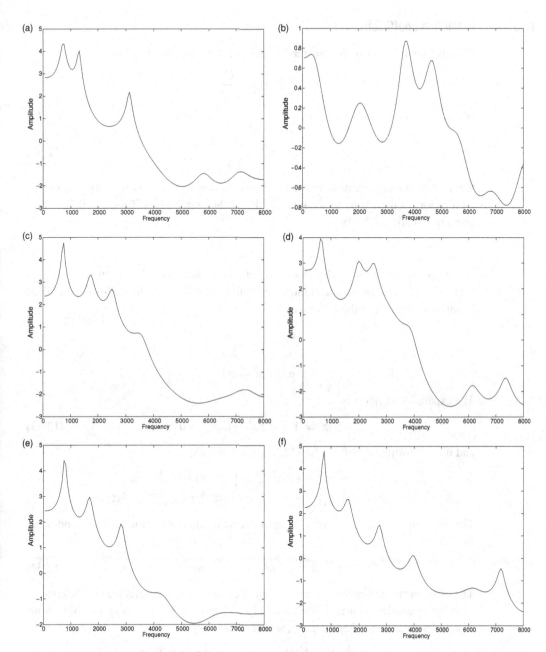

Figure 12.14 Examples of spectra calculated from LP coefficients for various sounds: (a) /ow/, (b) /s/, (c) /eh/, (d) /iy/, (e) /ae/ and (f) /ae/. Note how all the spectra are devoid of any harmonic influence, and that all seem to describe the spectral envelope of the speech. Note the characteristic formant patterns and roll-off for all vowels, and compare this with the spectra for /s/. The same vowel is present in examples (e) and (f). Note that, while the formant positions are the same in both examples, the amplitudes and bandwidths are quite different, especially in the fourth formant, which is barely noticeable in example (e).

12.5.3 Reflection coefficients

Consider again the all-pole transfer function we have been using all along,

$$V(z) = \frac{G}{A(z)}$$

where $A(z)$ is

$$A(z) = 1 - \sum_{k=1}^{P} a_k z^{-k} \qquad (12.30)$$

Recall that the key equation in the Levinson–Durbin recursion is the step that relates a predictor coefficient a_j to the quantity k and the previous predictor coefficient a_{j-1} given in Equation (12.29).

$$a_j^i = a_j^{i-1} - k_i a_{i-j}^{i-1} \qquad (12.31)$$

Let us now examine what happens when we substitute Equation (12.30) into Equation (12.31). For demonstration we shall choose the case for a third-order filter, in which case we have three expressions:

$$\begin{aligned} a_3^3 &= k_3 \\ a_2^3 &= a_2^2 - k_3 a_1^2 \\ a_1^3 &= a_1^2 - k_3 a_2^2 \end{aligned} \qquad (12.32)$$

The expansion of $A^3(z)$ is

$$A^3(z) = 1 - a_1^3 z^{-1} - a_2^3 z^{-2} - a_3^3 z^{-3} \qquad (12.33)$$

and if we substitute equations for (12.32) into this we get

$$\begin{aligned} A^3(z) &= 1 - [a_1^2 - k_3 a_2^2]z^{-1} - [a_2^2 - k_3 a_1^2]z^{-2} - k_3 z^{-3} \\ &= 1 - a_1^2 z^{-1} - a_2^2 z^{-2} + k_3 a_2^2 z^{-1} + k_3 a_1^2 z^{-2} - k_3 z^{-3} \end{aligned}$$

The first terms, $1 - a_1^2 z^{-1} - a_2^2 z^{-2}$, are of course just the expansion of the second-order filter $A^2(z)$,

$$A^3(z) = A^2(z) - k_3\left[-a_2^2 z^{-1} - a_1^2 z^{-2} + z^{-3}\right] \qquad (12.34)$$

The other terms are similar to the polynomial expansion, except that the coefficients are in the reverse order to normal. If we multiply each term by z^3 we can express these terms as a polynomial in z rather than z^{-1}, which gives

$$\begin{aligned} A^3(z) &= A^2(z) - k_3 z^{-3}\left[-a_2^2 z^2 - a_1^2 z + 1\right] \\ A^3(z) &= A^2(z) - k_3 z^{-3} A^2(z^{-1}) \end{aligned} \qquad (12.35)$$

In a similar fashion, the above relation can shown for any order of filter.

The significance of all this can be seen if we compare Equation (12.35) with Equation (11.27). Recall that Equation (11.27) was used to find the transfer function for the tube model from the reflection coefficients, in the special case in which losses occurred only at the lips. The equations are in fact identical if we set $r_k = -k_i$. This means that, as a by-product of Levinson–Durbin recursion, we can in fact easily determine the reflection coefficients which would give rise to the tube model having the same transfer function as that of the LP model. This is a particularly nice result; we have shown not only that the tube model is all-pole and that the LP model is all-pole, but also that the two models are in fact equivalent and hence we can find the various parameters of the tube directly from LP analysis.

Reflection coefficients are a popular choice of LP representation because they are easily found, are robust with respect to quantisation error and can in fact be used directly without having to be converted into predictor coefficients (see Rabiner and Schafer [368] for how this is done). Furthermore, and perhaps most significantly for our purposes, they have a physical interpretation, which means that they are amenable to interpolation. For example, if we had two different vocal-tract configurations and wished to generate a set of vocal-tract configurations in between these, we could do it by interpolating the reflection coefficients, in the knowledge that by doing so we were creating a set of reasonable vocal-tract shapes. This process could not be done with the predictor coefficients themselves, since this would very probably lead to unnatural or unstable filters.

12.5.4 Log area ratios

Recall from our tube model that the reflection coefficients were originally defined in terms of the ratio of the areas between two adjacent tubes. Equation (11.15) stated that the amount of the forward travelling wave that was reflected back into the tube was given as

$$r_k = \frac{A_{k+1} - A_k}{A_{k+1} + A_k}$$

Rearranging this gives

$$\frac{A_k}{A_{k+1}} = \frac{1 - r_k}{1 + r_k} \quad (12.36)$$

Hence it is possible to find the ratio of two adjoining tubes from the reflection coefficients. It is not, however, possible to find the absolute values of the areas of the tubes (this is independent, just as the gain is independent of the filter coefficients). A common way of representing (12.36) is in terms of **log area ratios**, which are defined as

$$g_k = \ln\left(\frac{A_k}{A_{k+1}}\right) = \ln\left(\frac{1 - r_k}{1 + r_k}\right) \quad (12.37)$$

Such log area ratios come in useful in a variety of ways, for example as a means of smoothing across joins in a manner that reflects physical reality.

12.5.5 Line-spectrum frequencies

Recall from the tube model of Section 11.3 that losses in the model are caused exclusively by wave propagation from the ends of the tube; no losses occur within the tube itself. Recall also from Section 11.5.3 that the resistance factors which cause the losses have virtually no effect on the location of the resonance peaks. In using the Levinson–Durbin solution to the LP equations, we have simplified the loss situation further by assuming that losses occur only from one end of the tube. **Line-spectrum frequencies (LSFs)** are found from an adaptation of the tube model whereby we eliminate the losses from the glottis. This can be thought of as a process of creating a perfectly lossless tube and calculating its transfer function. Because the resistive factors have a minor impact on the locations of the formants in the frequency domain, this modelling assumption helps us find good estimates for the formant positions.

If the tube is completely lossless, all the poles will lie on the unit circle; the formants will not be damped and hence have infinite amplitude. The spectrum of this will therefore be extremely "spiky" and it is this property that gives rise to the name, in that the spectrum appears as a pattern of lines rather than a smooth function.

We start the LSF analysis by constructing two new transfer functions for the lossless tube. We do this by adding a new tube at the glottis, which reflects the backward-travelling wave (whose escape would otherwise cause the loss) into the tube again. As explained in Section 11.3.3, this can be achieved by having either a completely closed or a completely open termination. In the closed case, the impedance is infinite, which can be modelled by a reflection coefficient value of 1; by contrast the open case can be modelled by a coefficient of value -1.

Recall that we showed in Section 12.5.3 that the operation of the Levinson–Durbin calculation was in fact equivalent to the determination of the transfer function for the tube model from its reflection coefficients. In both cases, we made use of an expression whereby we can formulate a new transfer function of order $P+1$, from an existing transfer function of order P and a new reflection coefficient r_{p+1}:

$$A^{p+1}(z) = A_p(z) + r_{p+1} z^{p+1} A_p(z^{-1})$$

We can therefore use this to construct the transfer function for the LSF problem, where we wish to simulate a completely lossless system by addition of one extra tube and a reflection coefficient. We do this for both the open and the closed case, which gives us two new transfer functions, by convention called $P(z)$ and $Q(z)$:

$$P(z) = A_p(z) + z^{-(p+1)} A_p(z^{-1}) \qquad (12.38)$$

$$Q(z) = A_p(z) - z^{-(p+1)} A_p(z^{-1}) \qquad (12.39)$$

As with any polynomial, these can be factorised to find the roots, but, when we do so in this case, we find that because the system is lossless all the roots lie on the unit circle. If we solve for $P(z)$ and $Q(z)$ we find that the roots are interleaved on the unit circle. In essence, the LSF representation converts the LP parameters from a set of $P/2$ pairs of complex-conjugate poles, which contain frequency and bandwidth information, into a set of $P/2$ pairs of frequencies, one of which represents the closed termination and the other the open termination. Note that, while the LSF tube is lossless, the loss information in the original LP model is still present. In the process of generating $A(z)$ back from $P(z)$ and $Q(z)$, the extra ± 1 reflection coefficient is removed and hence the loss and the bandwidths can be fully recovered.

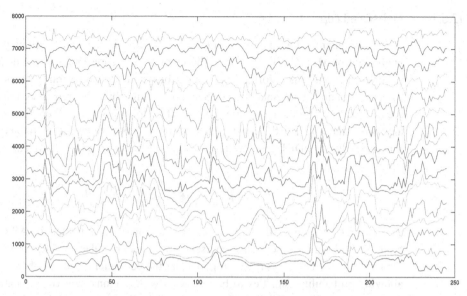

Figure 12.15 An example speech of KSF line-spectrum frequencies for an entire utterance. When the lines bunch together we take this as evidence of a formant being present.

The LSF representation is attractive for our purposes because it is in the frequency domain and hence amenable to many types of analysis. This is true of the original LP poles as well of course; we can convert these into a set of frequencies by taking their angle from the pole–zero plot. The LSFs have been shown to be superior, however, in that they can be robustly estimated and interpolated, whereas the original pole frequencies can be quite sensitive to noise and quantisation. A final property of the LSFs is that, when two occur close together, the overall value of the transfer function $H(z) = 1/a(z)$ is high. This implies that the original poles responsible for these frequencies form a formant, rather than being simply poles "making up the shape" of the spectral envelope. See Figure 12.5 for an example of line-spectral frequencies.

12.5.6 Linear-prediction cepstra

The cepstral representation has benefits beside its implicit separation of the source and filter. One often-cited property is that the components of a cepstral vector are to a large extent uncorrelated, which means that accurate statistical models of such vectors can be built using only means and variances and so covariance terms are not needed (see Section 15.1.3). Because of this and other properties, it is often desirable to transform the LP coefficients into cepstral coefficients. Assuming that we wish to generate a cepstrum of the same order as the LP filter, this can be performed by the following recursion:

$$c_k = \begin{cases} \ln(G) & k = 0 \\ a_k + \dfrac{1}{k} \sum_{i=1}^{k-1} i c_i a_{k-1} & 1 < k \leq P \end{cases} \quad (12.40)$$

12.5.7 Mel-scaled cepstra

A very popular representation in speech recognition is the **mel-frequency cepstral coefficient** or **MFCC**. This is one of the few popular represenations that does not use linear prediction. This is formed by first performing a DFT on a frame of speech, then performing a filter-bank analysis (see Section 12.2) in which the frequency-bin locations are defined to lie on the mel-scale. This is set up to give say 20–30 coefficients. These are then transformed to the cepstral domain by the **discrete cosine transform** (we use this rather than the DFT since we require only the real part to be calculated):

$$c_i = \sqrt{\frac{2}{N}} \sum_{j=1}^{N} m_j \cos\left(\frac{\pi i}{N}(j - 0.5)\right) \qquad (12.41)$$

It is common to ignore the higher cepstral coefficients, and often in ASR only the bottom 12 MFCCs are used. This representation is very popular in ASR, for two reasons. Firstly it has the basic desirable property that the coefficients are largely independent, allowing probability densities to be modelled with diagonal covariance matrices (see Section 15.1.3). Secondly, the mel-scaling has been shown to offer better discrimination between phones, which is an obvious help in recognition.

12.5.8 Perceptual linear prediction

It is possible to perform a similar operation with LP coefficients. In the normal calculation of these, spectral representations aren't used, so scaling the frequency domain (as in the case of the mel-scaled cepstrum) isn't possible. Recall, however, that in the autocorrelation technique of LP we used the set of autocorrelation functions to find the predictor coefficients. In Section 10.3.9 we showed that the power spectrum is in fact the Fourier transform of the autocorrelation function, and hence the autocorrelation function can be found from the inverse transform of the power spectrum.

So, instead of calculating the autocorrelation function in the time domain, we can calculate it by first finding the DFT of a frame, squaring it, and then performing an inverse DFT. In this operation it is possible to scale the spectrum as described above. The final LP coefficients will therefore be scaled with respect to the mel or Bark scale, such that more poles are used for lower frequencies than for higher ones.

12.5.9 Formant tracking

The above representations are all what can be termed "bottom-up" representations in that they are all found from the original speech frame by means of a mathematical function or process. We can also consider other representations in which we move away from these techniques and move towards representations that we believe more accurately represent the properties of the signal that are most perceptually important. In Chapter 7, we showed that formants are considered one of the most important representations in traditional phonetics, but in addition they are used in formant synthesis, which will be

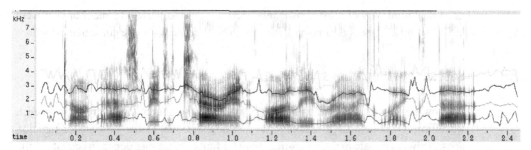

Figure 12.16 An example spectrogram with formants superimposed, as calculated by a formant tracking algorithm.

explained later in Section 13.2, and they have also been used as cost functions in unit selection, as explained in Chapter 16.

In traditional algorithms, which operated by a mixture of signal processing and rules, a three-stage approach is generally followed. First the speech is pre-processed, for example to downsample the speech to the frequency range where the formants occur; second candidate formants are found, and finally a heuristic step that determines the formants from the most likely candidates is applied. As we have just seen, in many cases the poles which result from LP analysis can be considered to give rise to formants, so it is natural that many formant-tracking algorithms use LP poles as their initial candidates. An alternative is to find peaks in either the DFT or the LP spectrum and assign these as candidates. Yet another technique is that of **multiple-centroid analysis**, which is based on the properties which define the overall shape of the spectrum [116]. Often the algorithms fix the number of formants to be found (usually three), and the post-processing stage uses dynamic time warping to find the best formant paths (formant tracks) across the course of the utterance. See Figure 12.16.

More recently, statistical processing has been brought to bear on this problem. In essence, formant tracking is no different from any statistical problem in which we have evidence extracted from features and constraints determined from a knowledge of the problem. Given such an approach, it is possible to put this in statistical terms in which the candidates form a likelihood, and the known dynamics of the formants (e.g. their number and how much they can move from frame to frame) can form a prior. With this, a Viterbi search can be performed, which finds the most likely formants for a sentence in a single step. Algorithms of this kind make use of hidden Markov models [4], non-linear prediction algorithms [129] and graphical models [297]. These statistical algorithms are significantly more accurate and robust than their heuristic counterparts.

In the early days of speech research, formants held the centre stage as *the* representation for speech. Today, however, it is extremely rare to find formants used in speech recognition, synthesis or other systems. This is partly because robust formant trackers have always proved hard to develop, so researchers moved towards cepstra and LP representations that could be robustly found from speech, even if their form didn't explicitly show the type of information that was thought most important. Furthermore, formants are really only a powerful representation for vowel sounds; they are not particularly good discriminators of most consonants. Hence, even if a perfect formant tracker were available, other types of representations would be needed for some consonant sounds.

12.6 Source representations

We now turn to the topic of finding and analysing the source signal from the speech waveform. Of course LP attempts to do just this, but, just as the LP transfer function doesn't always equate to the vocal-tract transfer function, neither does $x[n]$ equate to the source signal. This is basically because in LP analysis $H(z)$ over-extends itself due to the minimisation criteria, and so subsumes some of the characteristics of the true glottal source. With a little effort, however, we can in fact extract a good representation of the source by using a number of techniques, which we will now examine. As with all our other modelling problems, we have a range of tradeoffs between modelling complexity and ease of use, and we will examine a few of these now.

12.6.1 Residual signals

In the LP formulation of Equation (12.15)

$$y[n] = \sum_{k=1}^{P} a_k y[n-k] + x[n]$$

we termed $x[n]$ the **error signal**, because it is the difference between the predicted and actual signals. Hence, for a single frame of speech, if we perform LP analysis and note the values of $x[n]$ we will have a complete error signal for that frame. Alternatively, we can find this by a process of **inverse filtering**. From simple rearrangement of Equation (12.15), we can express $x[n]$ as follows:

$$x[n] = y[n] - \sum_{k=1}^{P} a_k y[n-k]$$

$$= \sum_{k=0}^{P} a_k y[n-k] \quad \text{where } a_0 = 1$$

Here we have expressed $x[n]$ purely in terms of the coefficients $\{a_k\}$ and the current and previous values of $y[n]$: if we compare this with Equation (10.54), we see that this is just a standard FIR filter. Hence, if we have the speech signal and the coefficients for a single frame, we can find $x[n]$ by a standard FIR filter operation. Apart from a gain normalisation, this will be identical to the error signal $x[n]$. The inverse-filtering technique is useful because often we are required to find the error signal after the LP analysis has been performed, or for cases (explained below) where we wish to find the residual for an interval different from that used for analysis. If the residual signal is then fed through the normal LP (IIR) filter, the original speech is reconstructed. This process is shown in Figure 12.17.

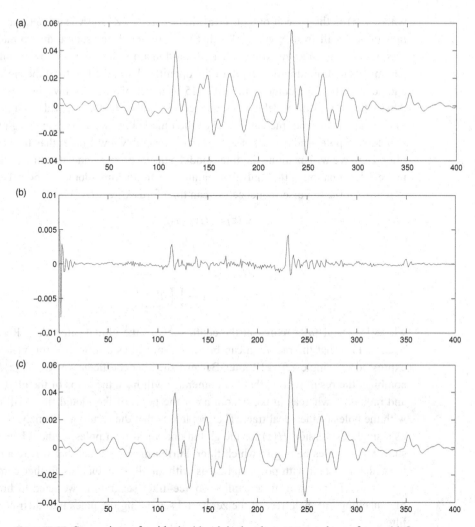

Figure 12.17 Comparison of residual with original and reconstructed waveforms. (a) One windowed frame of voiced speech. (b) The residual for the frame of speech, calculated with an inverse filter of the predictor coefficients. Note that the error is highest at the start of the frame, where there are no proper previous values on which to base the predication, and at the instant of glottal closure. The prediction error is highest at the latter point because it is here that the glottal derivative flow is at its highest. (c) Reconstructed waveform, created by passing the residual through a filter. This is virtually identical to the original speech.

If we apply the inverse-filtering technique to each frame in the entire speech waveform, we create a sequence of residual signals, one for each frame. A single residual waveform for the whole utterance can be made from these residual frames; but, since the details of this are somewhat subtle, a description of this is left until Section 12.6.5.

How do we now find the glottal-source signal from the residual $x[n]$? Recall that in the z-domain the main filter expression is

$$Y(z) = U(z)V(z)R(z)$$

where $Y(z)$ is the speech output, $U(z)$ the source, $V(Z)$ the vocal tract and $R(z)$ the radiation. Recall from Section 11.4 that we can model this source in two main ways. Firstly, we can model it explicitly as a time-domain function, which has a number of parameters determining the length of the opening phase, the length of the closed phase and so on. The Lijencrants–Fant (LF) [159] model of this is shown in Figure 11.15. Alternatively, we can model it as a sequence of impulses that are passed through a glottal filter $G(z)$. From experimentation and explicit modelling, we know that this glottal filter will contain poles and zeros, but will be of considerably lower order than the vocal-tract filter. Here we will examine the filter model, since a proper understanding of this helps in the determination of the equivalent parameters for the time-domain model. Using $I(z)$ to represent the impulse sequence, we can therefore express $U(z)$ as

$$U(z) = G(Z)I(z) \tag{12.42}$$

$$U(z) = \frac{\prod_{k=0}^{M} b_k^G z^{-k}}{1 - \prod_{l=1}^{N} a_l^G z^{-l}} I(z) \tag{12.43}$$

where $\{b_k^G\}$ and $\{a_l^G\}$ represent the coefficients of the glottal filter $G(z)$. Recall from Section 11.4 that the radiation can be approximated as a differentiator, which can be expressed as a single-zero FIR filter. Because the filter coefficients of $G(z)V(z)R(z)$ will combine, the result will be that the numerator will have the zeros of the glottal source and radiation, whereas in the denominator the poles of the glottal filter will combine with the poles of the vocal tract. The problem is that the result of LP analysis will give us a single expression $H(z)$ for the glottal and vocal-tract poles, without being able to show which poles belong to which filter. Furthermore, the fact that we are attempting to model a system with poles and zeros with an all-pole analysis further complicates the matter. To perform more explicit source–filter separation, we need to find a way of cancelling out the effect of the zeros and separating the poles into their component filters.

12.6.2 Closed-phase analysis

Figure 12.18 shows a diagram of a glottal-flow input, the derivative of glottal flow and the output speech. From this, we see that during each pitch period there is an interval when the glottal flow is zero because the glottis is shut. Since there is no contribution from the glottis here, if we estimate $H(z)$ over this interval, we will be able to equate this to a combined $V(z)$ and $R(z)$ expression, free of influence from $G(z)$. Calculation of the LP coefficients over this interval is known as **closed-phase analysis**.

To perform closed-phase analysis properly, we need to determine when the glottis shuts; these points are called the **instants of glottal closure** and can be found automatically by the epoch-detection algorithms described below in Section 12.7.1. A further benefit to performing closed-phase analysis is that during these intervals the closed glottis obeys the model assumptions of being completely reflective and having an infinite

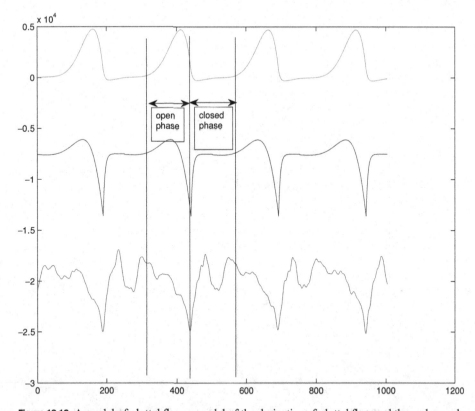

Figure 12.18 A model of glottal flow, a model of the derivative of glottal flow and the real speech waveform over time.

impedance; by contrast, during the open phase there are losses at the glottis, but, worse, there will definitely be interaction with the sub-glottal system (that is, the assumption made in Section 11.3.3 where we assume that no signal is reflected back from the lungs will not hold).

So, given that $G(z)$ has negligible effect during the closed phase, all we need now do is remove the effect of $R(z)$ to find $V(z)$. From experimentation and knowledge of radiation properties, we can approximate $R(z)$ as a differentiator of the signal. This can be modelled by a first-order FIR filter of the form

$$y[n] = x[n] - \alpha x[n-1]$$

Hence the effect of this can be removed by multiplying the signal by the inverse equivalent of this, which is an IIR filter, acting as an integrator:

$$x[n] = y[n] + \alpha x[n-1]$$

For reasons of stability, it is best to have α slightly less than the "true" integrating/differentiating value of 1; values in the range 0.95 to 0.99 are often used. This process is known as **pre-emphasis**.

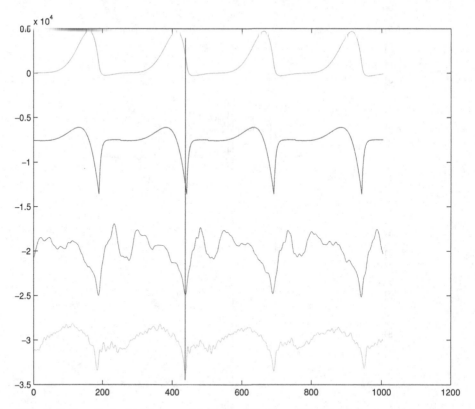

Figure 12.19 From top to bottom, a model of glottal flow, a model of the derivative of glottal flow, the real speech waveform and the real residual.

Now we have a reasonably accurate expression for the vocal-tract transfer function, we can inverse filter the speech to find the glottal-flow signal. The trick here is to implement inverse filtering of the signal over a *whole* pitch period (or longer) rather than just over the interval within which the LP coefficients were calculated. By doing this, we inverse filter over the region where the glottis is open, and from this we can find the form of the input, glottal signal. This process is shown in Figure 12.19, together with the shapes of the glottal flow and its derivative from the Lijencrants–Fant model. The figure clearly shows that the derivative of glottal flow is highly similar to the residual signal, especially with regard to the large negative spike at the instant of glottal closure. Figure 12.19 also shows that the assumptions made about the impulse excitation in the basic LP model are also quite justified; the excitation in the closed phase coming from the large negative-going spike in the glottal-derivative signal can be approximated by an impulse quite easily.

In many cases, especially with female or other high-pitched speech, the length of the closed phase can be very small, perhaps only 20 or 30 samples. In the autocorrelation method, the initial samples of the residual are dominated by the errors caused by calculating the residual from the zero signal before the window. The high (and erratic) error in the residual can be seen in the first few samples of the residual in Figure 12.17(b). For short analysis windows this can lead to a residual dominated by these terms, and for this reason covariance analysis is most commonly adopted for closed-phase analysis.

It should be noted that in many real-life cases the pre-emphasis term is not applied, and hence the calculated residual is in fact a representation of the derivative of glottal flow. This is the case for most speech-synthesis purposes, and henceforth we will assume that all residuals are calculated in this way, unless stated otherwise.

12.6.3 Open-phase analysis

It is not always possible or desirable to used closed-phase analysis. Sometimes other constraints in an application mean that a fixed frame spacing is required, and sometimes it is simply not possible to find the instant of glottal closure with any degree of accuracy. What then happens if the LP analysis is performed over an arbitrary section of speech? The main upshot of this is that the LP analysis will attempt to model the effect of the zeros in the glottal transfer function, and in doing so include the effect of the zeros in its poles.

From the basic Maclaurin expansion of

$$\sum_{n=0}^{\infty} x^n = 1 + x + x^2 + x^3 + \cdots = \frac{1}{1-x}$$

we can show that

$$\sum_{n=0}^{\infty} b^n z^{-1} = \frac{z}{z-b} = \frac{1}{1-bz^{-1}}$$

Upon inverting this, we get

$$\frac{1}{\sum_{n=0}^{\infty} b^n z^{-1}} = 1 - bz^{-1}$$

which shows that a single zero can be exactly modelled by an infinite number of poles. It can be shown that, so long as the number of poles is much larger than the number of zeros, a reasonable approximation of the effect of zeros can in fact be modelled by the poles. In doing so, however, we are in effect modelling these zeros with *all* the poles in the transfer function, so the poles will all be at locations slightly different from those in the case of closed-phase analysis. Luckily, this doesn't seem to have too great an influence on the frequency response of the LP transfer function, but it does have a knock-on effect when the LP filter is used to find the residual. Since we have modelled the glottal zeros with poles, the residual now does not closely resemble the glottal derivative, rather it is significantly "noisier" and "spikier". This means first that the spectrum of the residual will be flat and secondly that the main excitation will be in the form of isolated impulses rather than the shape of the glottal-derivative function of closed-phase analysis. This can be seen by comparing the open-phase residual of Figure 12.17 with the closed-phase residual of Figure 12.19. In general, fixed-frame LP analysis is the norm, so the open-phase residual signals such as that shown in Figure 12.17 are most commonly observed. Note that this ability to model zeros with poles also allows us to provide a means to represent vocal-tract transfer functions for nasals and obstruents with all-pole functions. While the accuracy of these all-pole expressions is not as high as for vowels, the convenience in being able to use a single method and expression for all speech sounds usually outweighs any issues arising from the inability to model vocal-tract zeros properly.

It is worth making a final point about residuals when they are used as sources. Recall from the original LP equation that, as well as the residual being the outcome when the speech is inverse filtered, it is also simply the error in the LP analysis. Hence, if a residual is used to excite the filter from which it came, the result should be a perfect reconstruction of the original signal. As we shall see in Chapter 14, there are many implementation issues to do with frame edges and so on, but it is worth noting that, no matter how crude any of the assumptions we have made with regard to LP analysis in general, a powerful feature of the model is that the speech can be exactly reconstructed if excited with the correct input.

12.6.4 Impulse/noise models

Linear prediction was in fact primarily developed for use in speech-coding applications. As we have just seen, performing LP analysis allows us to deconvolute the signal into a source and filter, which can then be used to reconstruct the original signal. Simply separating the source and filter doesn't actually save any space; the trick comes in replacing the residual by a much simpler (and more economical) source signal.

In what we shall henceforth call **classical linear prediction** the source is modelled by two systems; one for voiced sounds and one for obstruent sounds. In Figure 12.19, we saw that the integrated residual signal was a reasonably close approximation to the glottal-flow-derivative signal. The primary feature in such a signal is the negative spike caused by the glottal closure. Since this is reasonably close to an impulse, the classical LP model simply uses a single impulse as the excitation for voiced sounds. To do this properly, one impulse is generated for every pitch period. Thus, in analysis, an estimation of the pitch period has to be found and stored/transmitted. All obstruent sounds have a frication source, which is generated by white noise in our model from Section 11.5.2. In the classical LP system, all we need store/transmit is the fact that we have an obstruent sound, upon which a noise source is used during reconstruction. For voiced obstruents, both the impulse and the noise source are used.

The impulse/noise model provides a very efficient means of encoding the residual, but does so at some cost, the effects of which can clearly be heard in the reconstructed signal as a buzz or metallic sound. Apart from the space saving, having an impulse-excited system has the additional benefit that we are free to generate the impulses when we like; in other words, we are not limited to generating the impulses at the same spacing as in the analysis. This provides a means of changing the pitch of a signal and in fact forms the basis of the synthesis technique described in Section 13.3.

12.6.5 Parameterisation of glottal-flow signals

Performing the above analysis allows us to find a reasonable estimate of the glottal flow or its derivative. In fact, we can go one stage further and parameterise this by further analysis. One technique described by Quatieri [364] involves fitting the Lijencrants–Fant model to the residual by means of least-squares minimisation. In philosophy, this is of course similar to LP analysis itself, whereby we attempt to find the set of predictor coefficients which will give the lowest error. In practice, the LF model is non-linear and so the LP technique is not available to us. Instead, numerical techniques such as the Gauss–Newton method are used to iterate to a set of optimum model parameters. Dutoit [148] describes a different technique whereby the glottal-source parameters are actually found at the same time as the predictor coefficients.

Although we can find the optimal parameters for any given frame for a model such as LF, we never actually achieve a perfect fit. In fact, we can determine a sample-by-sample difference between the model and the residual. This, of course, is another "residual" or "error" signal, which now represents the modelling error after the LP model and the LF model have been taken into account. An interesting study by Plumpe *et al.* [356] describes how this signal is in fact highly individual to speakers and can be used as the basis of speaker recognition or verification.

12.7 Pitch and epoch detection

12.7.1 Pitch detection

The process of finding the fundamental frequency (F0) or pitch from a signal is known as **pitch detection** or **pitch tracking**. Recall from Chapter 7 that the fundamental frequency is defined as the frequency at which the glottis vibrates during voiced speech, whereas pitch is the perceived rate of periodicity in the speech signal heard by the listener. In general, we ignore this distinction, mainly because in most cases any difference between these two values is less than the known accuracy of any detection algorithm. One point that is important, though, is that F0 tracks are in fact used for many quite different purposes. In some of the signal-processing operations described below, an accurate frame-by-frame value is required; however, in intonational analysis, we are usually interested only in the general shape of the track, so small perturbations are often smoothed for such needs (Figure 12.20).

As with formant tracking, traditional **pitch-detection algorithms (PDAs)** often have three stages: a pre-processing stage, where for instance the speech is downsampled or low-pass filtered; a bottom-up candidate-finding step; and a decision-making step whereby the best candidates are picked.

A wide variety of techniques has been proposed for pitch tracking, including the following.

- Using the cepstrum. Since cepstral analysis generates a representation in which the pitch information appears as a spike, it is possible to find this spike and determine the pitch from that.

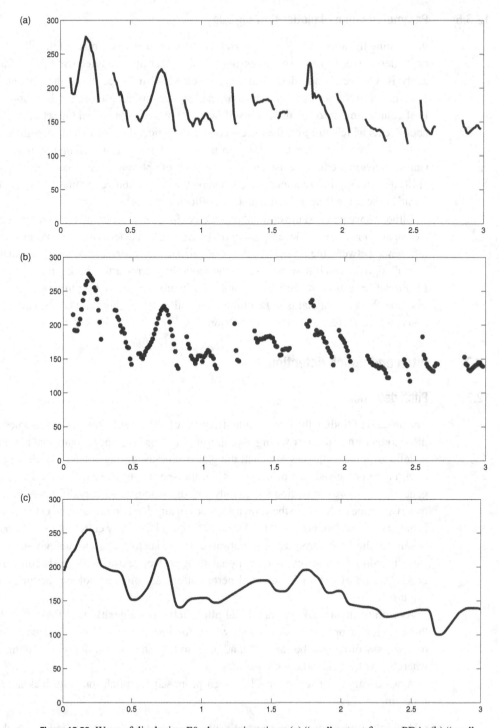

Figure 12.20 Ways of displaying F0 plots against time: (a) "raw" output from a PDA; (b) "raw" output, but plotted with symbols rather than lines; and (c) smoothed and interpolated output from a PDA.

- Using the autocorrelation function. In the time domain, successive pitch periods are usually similar to one another. Since the autocorrelation function measures how similar a signal is to itself when shifted, it will have a high value when a pitch period is most closely overlapped with its neighbour. We can hence find the period by finding these points where the autocorrelation function is at a maximum. In a similar fashion, the normalised cross-correlation function can be used.
- Working in the frequency domain. Since the harmonics in the DFT are spaced at multiples of the period, they can be used to find the period. One strength in this technique is that there are multiple harmonics and hence more than one measurement can be made. A drawback is that they can often be hard to measure accurately, especially at higher frequencies.
- Residual analysis. Since the residual is mostly devoid of vocal-tract influence, it represents the source, and from this the period can be seen. In practice this technique is rarely used because the artefacts in LP analysis often make the determination of the period quite difficult.

The classic work on pitch detection is that of Hess [202], who gives a thorough overview of all the PDA techniques just mentioned. With PDAs, the basic techniques go only so far and an important component of the algorithm is the use of many heuristic or empirically determined factors. Bagshaw [28] conducted an objective test on six popular techniques and claims that the time-domain super-resolution pitch-determination algorithm [309] is the best. Other well-known pitch-trackers include Talkin's get_f0 and RAPT algorithms [431] (these were not included in Bagshaw's test, but are generally acknowledged to be very accurate). As with formant tracking, traditional PDAs suffer somewhat from their heuristic rather than probabilistic formulation. Recently, more interest in probabilistic techniques has arisen [141], and for instance Li *et al.* [283] cite accuracy results for a probabilistic technique that is significantly more accurate than get_f0.

12.7.2 Epoch detection: finding the instant of glottal closure

We have just seen that closed-phase LP requires that we analyse each pitch period separately. This type of speech analysis is called **pitch-synchronous analysis** and can be performed only if we are in fact able to find and isolate individual periods of speech. We do this by means of a **pitch-marking** or **epoch-detection** algorithm (EDA).

The idea behind an EDA is to locate a single instant in each pitch period that serves as an "anchor" for further analysis. These positions are often known as **pitch marks**, **pitch epochs** or simply **epochs**. In general, they can refer to any reference point, but are often described in terms of salient positions on the glottal-flow signal, such as the peak of the flow for each pitch period. For many types of analysis (such as TD-PSOLA, which will be described in Section 14.2) it doesn't really matter where the anchor point is chosen, so long as it is consistent from period to period. Often a time-domain reference point such as the peak of the highest excursion in each period is used, or the trough of the lowest excursion. That said, many analysis techniques focus on one particular point known as

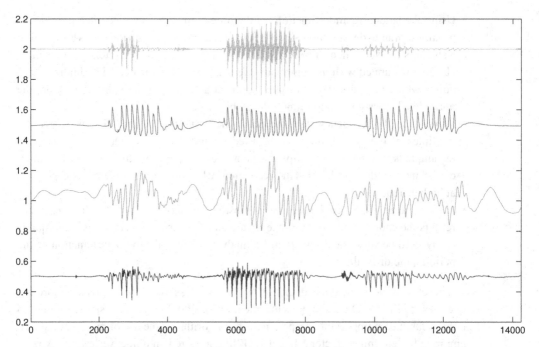

Figure 12.21 A comparison of (from top to bottom) waveform, laryngograph signal, integrated residual and residual.

the **instant of glottal closure** (IGC). This is manifested as a large negative-going spike in the derivative of glottal flow, as shown in Figure 12.21. If we view this as the input to the vocal tract (which we do on including the radiation term), we see that it closely resembles a large negative impulse. Such an impulse excites the filter in such a way that we can examine the filter properties without interference from the ongoing effects of the glottis. As we saw above, this is the basis on which closed-phase LP analysis is performed.

Locating the instant of glottal closure (and epoch detection in general) is widely acknowledged as a thorny problem. In the past, most pitch-synchronous applications used a rough estimate of the ICG, because the technique was not particularly sensitive for determining these instants exactly. Indeed, the closed-phase LP technique described above does not need a particularly accurate estimate – the window used by the technique does not have to start exactly at the ICG. With the increasingly widespread adoption of pitch-synchronous speech-synthesis techniques (described in Chapter 14) it becomes more important to locate the ICG, or some other anchor point, more exactly. Dutoit [148] gives a review of some of the difficulties involved, and describes a process common in the early 1990s whereby researchers would actually mark epochs by hand. This was seen as being an acceptable burden when speech databases were small and when the only alternative was to use other much-poorer-sounding synthesis techniques. Dutoit describes some of the difficulties involved in this, most specifically the fact that, because of various phase effects, it is not always the largest negative peak in the waveform

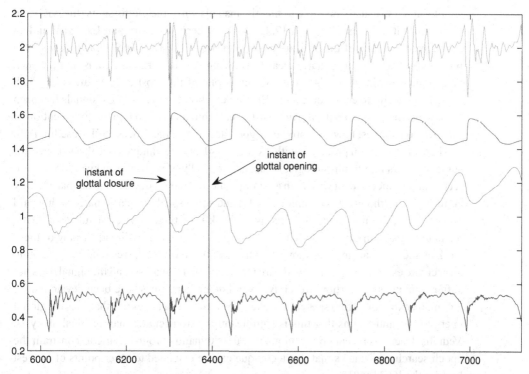

Figure 12.22 Fine-grained comparison of (from top to bottom) waveform, laryngograph signal, integrated residual and residual. The vertical lines show the instants of glottal closure and glottal opening. Here we can see that, while the general periodicity of the form of the Lx and residual glottal flow are similar, the functions are not simply related in shape. Note that, when the glottis is open, the Lx signal is at a minimum because the current flow is at a minimum. The integrated residual signal approximates the glottal volume velocity. Ideally the closed phase in this signal should be flatter, but the many modelling errors combine to give this part of the waveform a gradual upward slope. A small sharpening in the gradient of this can be seen at the instant of glottal opening.

which corresponds to the IGC. Furthermore, it is known that phase distortion can make time-domain marking of waveforms virtually impossible for either machine or human.

These difficulties led many to a radical alternative, which was to measure the glottal behaviour by means other than from the speech waveform. A common technique was to use a device known as an **electroglottograph** or a **laryngograph**. This consists of two metal discs, which are placed on each side of the protruding part of the larynx (the "Adam's apple"). A very-high-frequency alternating current is passed through the system, such that the current passes from one disc, through the larynx, to the other disc, and from this the electrical impedance and flow of current across the larynx can be measured. If the glottis is open, the impedance is high and the current is low; if the glottis is shut the impedance is low and the current is high. Laryngograph signals (known as Lx signals) do not, however, generate the glottal-flow signal itself because the electrical impedance and flow are not simply related. The Lx signal is, however, quite simple, and it is usually possible to find the instants when the glottis shuts (the IGC) and opens from it. Figures 12.21 and 12.22 show a speech waveform, an Lx signal and an LP-derived estimate of the derivative of glottal flow. The Lx signal clearly shows each pitch period, and is particularly clear at showing exactly when the transition between voiced and unvoiced sounds occurs, which is one of the most difficult areas in which to perform accurate epoch detection. Epoch detection by means of Lx signals has been quite widely used in TTS during the last 15 years, but it is rare to find explicit mention of this. Acero [3] does, however, provide a good account of the process of Lx use for TTS.

There are two main problems with Lx use. Firstly, the technique doesn't work well for some speakers due to physiological reasons. Secondly, it is simply awkward to have to record all the speech data with both a microphone and the Lx device. It is not particularly comfortable (although it is completely safe) and some speakers, particularly highly paid voice talents, are not happy with its use. Thankfully, there now exist fully automatic techniques that can extract the IGCs from the speech waveform directly. Many of these work on the residual, and it is known that the negative pulse of the residual can give quite an accurate estimation of the IGC if care is taken with the analysis and the signal does not suffer from phase distortion. Recently several other techniques have been developed for automatic extraction from speech. Brookes [68] describes a technique that uses the glottal energy flow, and claims that this is significantly more accurate than residual analysis. Vedhius describes a general technique that uses dynamic programming to constrain the epoch search, and shows that this technique can be extended to other points of interest besides the IGC [468].

12.8 Discussion

For those approaching these topics for the first time, some of the formulations, the reasoning and the sheer number of representations can make the subject of speech analysis seem quite daunting. It is in fact virtually impossible to learn this subject purely from the mathematics: speech analysis is a practical subject and requires practical study, and by far the best means to study it is by interactively playing with real speech signals. Luckily there exist packages that allow intimate examination of speech signals. Using these, it is possible to examine magnitude spectra, LP envelopes, spectrograms of various types and so on. In fact, a good understanding of the speech representations described here can be gained from interactive study alone without any understanding of the equations involved (although this of course is not recommended).

Speech analysis is a broad topic and is used in many applications. Historically, most of the techniques described here were first developed for use in speech coding, where the aim is to transmit a signal down a wire as efficiently as possible. With any general type of data, there is often considerable redundancy, and by studying the patterns in which the data naturally occurs we can study this redundancy and eliminate it. These general techniques go only so far, and it soon became apparent that special-purpose coders designed specifically for speech signals could be much more efficient, delivering more than a factor of ten compression rate. In fact, LP was developed for just this purpose.

Linear predictive coding (LPC), as it is known, separates the source and filter for each frame, but sends only the filter coefficients, normally encoded as reflection coefficients or line-spectral frequencies. The residual is not transmitted; rather an estimate of the pitch is sent for voiced signals, together with a flag stating whether the source is noisy (i.e. is an obstruent sound). When the speech is to be reconstructed, the pitch information is used to generate an impulse sequence for voiced signals, and white noise is added if the noise flag is set. The quality of LPC speech is considerably worse than the original, but it has proved a popular choice, especially for situations with user acceptance and requirements for very low transmission rates (usually military). More recently, LPC coders with more sophisticated residual-reconstruction techniques have been used in normal commercial applications.

Automatic speech-recognition (ASR) systems use speech analysis as their first stage. All ASR systems are essentially pattern recognisers; so they work well when presented with input features with certain properties. These include that the representation should be compact, so that the number of parameters to be learned is low; and this shares obvious common ground with the speech-coding applications just mentioned. A good example is that many ASR systems use multi-variate Gaussians to model phone distributions (see Chapter 15). For a vector with 10 components, doing so fully requires a set of 10 means and a set of 10 by 10 (100) covariances. If, however, the components are statistically independent of each other, we can ignore most of the covariance terms and model the distribution with ten means and just ten variances. In addition, it is desirable that the representations should be invariant for a single class; for this reason, recognition is never performed on the waveform itself, since the phase aspect makes signals with the same envelope look quite different.

Currently, mel-scale cepstral coefficients, and perceptual LP coefficients transformed into cepstral coefficients, are popular choices for the above reasons. Specifically, they are chosen because they are robust with respect to noise, can be modelled with diagonal covariance, and, with the aid of the perceptual scaling, are more discriminative than would otherwise be the case. From a speech-synthesis point of view, these points are worth making, not because the same requirements exist for synthesis, but rather to make the reader aware that the reason why MFCCs and PLPs are so often used in ASR systems is to do with the above factors, not because they are intrinsically better in any general-purpose sort of way. This also helps explain why there are so many speech representations in the first place; each has strengths in certain areas, and will be used as the application demands. In fact, as we shall see in Chapter 16, the application requirements which make, say, MFCCs so suitable for speech recognition are almost entirely absent for our purposes. We shall leave a discussion of what representations really are suited for speech-synthesis purposes until Chapter 16.

12.8.1 Further reading

There are several good books on speech analysis, most of which adopt an application-neutral attitude. The classic work is Rabiner and Schafer [368]. Even though it was written several decades ago, this book remains an excellent reference for all the fundamentals of speech analysis. It assumes that the reader is familiar with basic signal processing, and so can be a little difficult for those with no prior experience. Given an understanding of the topics covered in Chapter 10 of this book, the reader should have little difficulty. An excellent more-recent book is Quatieri [364]. The subject matter doesn't differ much from that of Rabiner and Schafer, but the book covers many of the techniques that have been developed since. Other good references include Furui [168], Huang et al. [224], Kleijn and Paliwal [256], Deller and Proakis [128] and Gold and Morgan [176].

12.8.2 Summary

- Speech analysis is in general concerned with three main problems.
  1. Eliminate phase; the magnitude DFT does this.
  2. Separate source and filter; this can be done with cepstral analysis or linear prediction.
  3. Transform the representation into a space that has more desirable properties: log magnitude spectra follow the ear's dynamic range; mel-scaled cepstra scale according to the frequency sensitivity to the ear; log area ratios are amenable to simple interpolation; and line-spectral frequencies show the formant patterns robustly.
- The continuously varying signal is divided by a process called windowing into a sequence of stationary smaller signals called frames.
- Windowing affects the signal, so a perfect representation of the original waveform is not possible.
- The commonest spectral representation is achieved by applying a discrete Fourier transform (DFT) to the frame of speech.
- Successive application of DFTs can be used to generate a spectrogram.
- The cepstrum is the inverse Fourier transform of the log magnitude spectrum. It separates variation in frequency across the range and so implicitly separates source and filter components in the spectrum.
- Linear prediction (LP) performs source–filter separation by assuming that an IIR system represents the filter. This allows the filter coefficients to be found by a process of minimising the error predicted from the IIR filter.
- The LP equations can be solved by the autocorrelation or covariance methods. Each involves solving the equations by means of efficient matrix inversion.
- The direct LP, IIR, coefficients are inherently unrobust with regard to quantisation and interpolation. Various derived representations, including reflection coefficients, log area ratios and line-spectral frequencies, avoid these problems.
- The source signal, called the residual, can be found by inverse filtering the speech signal with the LP coefficients.
- With appropriate care during analysis and filtering, the commonalities between this residual and the glottal-flow-derivative signal can be shown.

13 Synthesis techniques based on vocal-tract models

In this chapter we introduce the three main synthesis techniques which dominated the field up until the late 1980s, collectively known as **first-generation** techniques. Even though these techniques are used less today, it is still useful to discuss them because, apart from simple historical interest, they give us an understanding of why today's systems are configured the way they are. As an example, we need to know why today's dominant technique of unit selection is used rather than the more-basic approach which would be to generate speech waveforms "from scratch". Furthermore, modern techniques have been made possible only by vast increases in processing power and memory, so in fact, for applications that require small footprints and low processing cost, the techniques explained here remain quite competitive.

13.1 Synthesis specification: the input to the synthesiser

First-generation techniques usually require a quite-detailed, low-level description of what is to be spoken. For purposes of explanation, we will take this to be a phonetic representation for the verbal component, together with a time for each phone and an F0 contour for the whole sentence. The phones will have been generated by a combination of lexical lookup, G2T rules and post-lexical processing (see Chapter 8), while the timing and F0 contour will have been generated by a classical prosody algorithm of the type described in Chapter 9. It is often convenient to place this information in a new structure called a **synthesis specification**. Hence the specification is the input to the synthesiser and the waveform is the output.

Initially, we will assume a simple model where the specification is a list $S = \langle s_1, s_2, \ldots, s_N \rangle$, in which each item s_j contains the phoneme identity, a duration and one or more F0 values. Recall from Chapter 9 that the F0 contour is necessarily dependent on the timing information, since the F0 generator must produce an F0 contour that matches the length of the utterance. One option is to specify the F0 contour on a frame-by-frame basis, so that if we had, say, a phoneme of duration 50 ms and a frame shift of 10 ms, we would have five F0 values in the specification. A phoneme of duration 70 ms would have seven frames and so on:

$$[F0] = \langle 121, 123, 127, 128, 126 \rangle \quad \text{(five-frame example)}$$
$$[F0] = \langle 121, 123, 127, 128, 126, 121, 118 \rangle \quad \text{(seven-frame example)}$$

So an example of a complete specification item would look like

$$s_j = \begin{bmatrix} \text{PHONEME} & /ah/ \\ \text{DURATION} & 70 \\ \text{F0} & \langle 125, 123, 119, 117, 115, 115, 114 \rangle \end{bmatrix}$$

Having different numbers of F0 values per item can be unwieldy, so sometimes the F0 contour is sampled at the mid point of the unit or other key positions to give a simpler specification:

$$s_j = \begin{bmatrix} \text{PHONEME} & /ah/ \\ \text{DURATION} & 50 \\ \text{F0 START} & 121 \\ \text{F0 MID} & 123 \\ \text{F0 END} & 127 \end{bmatrix}$$

Once the specification has been created, the synthesis process itself can begin.

13.2 Formant synthesis

Formant synthesis was the first genuine synthesis technique to be developed and was the dominant technique until the early 1980s. Formant synthesis is often called **synthesis by rule**; a term invented to make clear at the time that this was synthesis "from scratch" (at the time the term "synthesis" was more commonly used for the process of reconstructing a waveform that had been parameterised for speech-coding purposes). As we shall see, most formant-synthesis techniques do in fact use rules of the traditional form, but data-driven techniques have also been used.

Formant synthesis adopts a **modular, model-based, acoustic-phonetic** approach to the synthesis problem. The formant synthesiser makes use of the acoustic-tube model, but does so in a particular way so that the control elements of the tube are easily related to acoustic-phonetic properties that can easily be observed. A typical basic layout of a formant synthesiser is shown in Figure 13.1. Here we see that the sound is generated from a source, which is periodic for voiced sounds and white noise for obstruent sounds. This basic source signal is then fed into the vocal-tract model. In virtually all formant synthesisers, the oral and nasal cavities are modelled separately as parallel systems. Hence the signal passes into the component which models the oral cavity, but can also pass into the component for modelling the nasal cavity if required for a nasal or nasalised sound. Finally, the outputs from these components are combined and passed through a radiation component, which simulates the load and propagation characteristics of the lips and nose.

The first point of note regarding the formant synthesiser is that it is *not* an accurate model of the vocal tract. Even taking into account the crude assumptions regarding

13.2 Formant synthesis

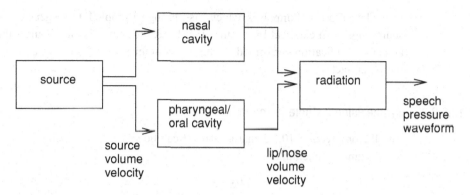

Figure 13.1 A block diagram of a basic formant synthesiser.

all-pole modelling, losses and so on, the formant-synthesiser structure differs significantly in that it allows separate and independent control of each formant. In the vocal-tract model developed in Chapter 11, the formants emerged as a result of the whole, concatenated-tube model; it was not possible to point to just one section of tube and say that that section was responsible for the behaviour of a particular formant. Hence the formant synthesiser is based on a **lumped-parameter** model of speech generation, rather than the **distributed model** developed in Chapter 11. The reason for modelling the formants as individual components is that, this allows the system designer to gain direct control of the formants and the reason for wishing to do this is that, at the time when formant synthesisers were developed, it was much easier to read real formant values from spectrograms than to determine real vocal-tract configurations. Even today, with MRI and EMA, the logistics of acquiring formant values is considerably simpler than that of determining vocal-tract shapes. Hence constructing a synthesis paradigm based on formant control is one of pragmatics (being able to control what we can observe) and modularity (whereby each component can be separately controlled).

That said, formant synthesis does share much in common with the all-pole vocal-tract model. As with the tube model, the formant synthesiser is modular with respect to the source and vocal-tract filter. The oral-cavity component is formed from the connection of between three and six individual formant resonators in series, as predicted by the vocal-tract model, and each formant resonator is a second-order filter of the type discussed in Section 10.5.3.

13.2.1 Sound sources

The source for vowels and voiced consonants can be generated either as an explicit time-domain periodic function, of the type described in Section 11.4.2, or by an impulse sequence that is fed into a glottal LTI filter. For obstruent sounds a random-noise generator is used. Sounds such as voiced fricatives use both sources. If we adopt the impulse/filter approach, we see that in fact these sources are equivalent to those of the classical impulse/noise LP model (Section 12.6.4). One difference between the impulse/glottal-filter source and LP models is that in the impulse/glottal-filter model the voiced source is formed by a series of impulses, which pass through a simple filter to generate a periodic source signal. In classical LP, on the other hand, the source is the impulses themselves and any modification performed on these is combined with the use of a filter to model the vocal tract.

13.2.2 Synthesising a single formant

Recall from Section 10.5.3 that a formant can be created by a second-order IIR filter whose transfer function is

$$H(z) = \frac{1}{1 - a_1 z^{-1} - a_2 z^{-2}}$$

This can be factorised as

$$H(z) = \frac{1}{(1 - p_1 z^{-1})(1 - p_2 z^{-1})}$$

To generate filter coefficients a_1 and a_2 with real values, the poles have to form a complex-conjugate pair. Recall that a second-order expression like this generates a spectrum with two formants, but, because only one of these lies in the positive-frequency range, the above expression does in fact generate a single usable formant with the full range of amplitude, bandwidth and frequency position.

The poles can be represented in polar form. Since they form a complex-conjugate pair, they have the same magnitude r and angles θ and $-\theta$:

$$H(z) = \frac{1}{(1 - re^{j\theta} z^{-1})(1 - re^{-j\theta} z^{-1})} \qquad (13.1)$$

As explained previously, poles can be visualised on a pole–zero plot, such as that shown in Figure 10.22. The magnitude of the pole determines the bandwidth and amplitude of the resonance; for low values of r, the formant will be very flat; as r increases the formant will become sharp, until a pole with a value of 1 will have an infinite amplitude (which should not occur in practice). The angle θ determines the normalised angular-frequency position of the pole and is expressed in radians; it can be converted into a frequency in hertz by multiplication by $2\pi F_s$, where F_s is the sampling frequency. If we multiply Equation (13.1) out again, we can write the transfer function in terms of the radius and angle of the poles:

$$H(z) = \frac{1}{1 - r(e^{j\theta} + e^{-j\theta}) z^{-1} + r^2 z^{-2}}$$

and from Euler's formula (Equation (10.9)) we can see that the complex terms cancel out to give a value for the first coefficient expressed in cosine terms:

$$H(z) = \frac{1}{1 - 2r \cos(\theta) z^{-1} + r^2 z^{-2}}$$

While we can generate a formant of any desired frequency directly by appropriate use of θ, controlling the bandwidth directly is a little more difficult. The position of the formant will change the shape of the spectrum somewhat, so a precise relationship for all cases is not possible. Note also that, when two poles (from two separate second-order filters) are close together, they have the effect of combining into a single resonance peak and this again makes bandwidth calculation somewhat difficult. However, it can be shown that a reasonable approximation linking the normalised bandwidth of a formant to the radius of the pole is

$$\hat{B} = -2\ln(r)$$

Using this, we can write the transfer function in terms of the normalised frequency $\hat{F}$ and normalised bandwidth $\hat{B}$ of the formant:

$$H(z) = \frac{1}{1 - 2e^{-2\hat{B}}\cos(2\pi\hat{F})z^{-1} + e^{-2\hat{B}}z^{-2}}$$

As before, values for the normalised frequency $\hat{F}$ and bandwidth $\hat{B}$ can be determined from the frequency F and bandwidth B by dividing by F_s, the sampling rate. Figures 10.23 and 10.24 show the spectrum for a single formant with different values for frequency and bandwidth.

13.2.3 Resonators in series and parallel

A transfer function that creates multiple formants can be formed by simply multiplying several second-order filters together. Hence the transfer function for the vocal tract is given as

$$H(z) = H_1(z)H_2(z)H_3(z)H_4(z) \tag{13.2}$$

$$= \underbrace{\frac{1}{1 - 2e^{-\hat{B}_1}\cos(2\pi\hat{F}_1)z^{-1} + e^{-2\hat{B}_1}z^{-2}}}_{H_1} \cdots \underbrace{\frac{1}{1 - 2e^{-\hat{B}_4}\cos(2\pi\hat{F}_4)z^{-1} + e^{-2\hat{B}_4}z^{-2}}}_{H_4}$$

$$\tag{13.3}$$

where F_1 is the frequency of the first formant, B_1 is the bandwidth of the first formant and so on. The terms in the denominator are multiplied out to give the standard all-pole filter transfer function

$$H(z) = \frac{1}{1 - a_1 z^{-1} - a_2 z^{-2} - \cdots - a_8 z^{-8}} \tag{13.4}$$

and time-domain difference equation

$$y[n] = x[n] + a_1 y[n-1] + a_2 y[n-2] + \cdots + a_8 y[n-8] \tag{13.5}$$

These equations show the main advantage of the formant technique, namely that, for a given set of formant values, we can easily create a single transfer function and difference equation for the whole oral tract. In a similar fashion, a nasal system can be created, which likewise links the values of the nasal formant to a transfer function and a difference equation.

This formulation is known as a **serial** or **cascade** formant synthesiser and was the first type to be developed. Note that the relative amplitudes of the formants are determined as part of this calculation rather than being set explicitly. This works because each formant sits on the skirt of the others, and this naturally generates formants of the correct amplitude. Hence the spectrum generated from such a formulation will always seem "natural" in terms of its shape. Counter to these advantages is the problem that the interactions of the formants when cascaded can sometimes have some unwanted effects. In particular, when two formants are close together, the resultant values as measured from the spectrum can often differ from the input formant parameters and hence a lack of control appears.

An alternative formulation is the **parallel** formant synthesiser, in which each formant is modelled in isolation, and, during speech generation, the source signal is fed through each separately. After this, the signals from all the formants are combined. The advantage is that the frequency of each formant can be more accurately specified; the disadvantage is that now the amplitude of each formant must be carefully controlled; they do not appear on their own, unlike in the cascading method. In addition, there is no neat single transfer function or difference equation for the parallel case; rather each formant has its own transfer function

$$H_1(z) = \frac{1}{1 - a_{11}z^{-1} - a_{12}z^{-2}} \qquad (13.6)$$

$$H_2(z) = \frac{1}{1 - a_{21}z^{-1} - a_{22}z^{-2}} \qquad (13.7)$$

$$\vdots \qquad (13.8)$$

and difference equation

$$y_1[n] = x[n] + a_{11}y_1[n-1] + a_{12}y_1[n-2] \qquad (13.9)$$

$$y_2[n] = x[n] + a_{21}y_2[n-1] + a_{22}y_2[n-2] \qquad (13.10)$$

$$\vdots \qquad (13.11)$$

which are added in the time domain to give the final signal

$$y[n] = y_1[n] + y_2[n] + \cdots \qquad (13.12)$$

The two configurations are shown in Figure 13.2. Note that the parallel configuration has an amplitude (gain) control for each formant.

13.2.4 Synthesising consonants

The source characteristics of consonants differ depending on the class of sound being produced. All unvoiced sounds use only the noise source; nasals and approximants use only the periodic source whereas voiced obstruents (i.e. voice fricatives, affricates and stops) use both sources. Approximants are generated in much the same way as vowels. Some consonants (such as [h]) can be synthesised in the same way as vowels; that is, by sending a sound source through the oral cavity resonators. Most other consonants are,

13.2 Formant synthesis

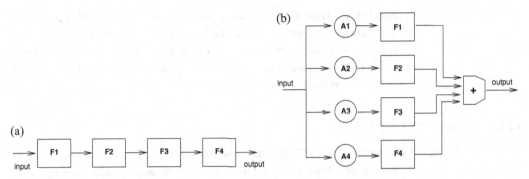

Figure 13.2 Formant configurations: (a) cascade or serial configuration and (b) parallel configuration.

however, more complicated than /h/ because their source is not at one end of the vocal tract.

Recall from Section 11.5 and Figure 11.16 that we used a number of different tube configurations for each consonant type. The same approach is used in formant synthesis. In the case of nasals, the nasal cavity acts as a resonator and the nose is the only place where sound escapes. The oral cavity does, however, still have an effect – even with the velum lowered sound from the glottis still passes into the oral cavity, but now is reflected at the lips rather than escaping. The oral cavity still operates as a resonator and some of the glottal source wave that it modifies enters the nasal cavity through the lowered velum. The oral cavity therefore has a significant effect on the sound of the nasal and this explains why we can have different-sounding nasals even though the nasal cavity itself has only a single, unmoving shape.

In real speech obstruents have their frication source generated at a place of constriction, not by the glottis. Normally this place of constriction is caused by the tongue nearing the roof of the mouth. In these cases, some of the sound created at the place of constriction travels back towards the glottis, is reflected there and travels forwards, through the constriction and towards the lips. Hence the vocal tract is effectively split into separate front and rear cavities, which have separate resonance properties.

While details differ from system to system, in formant synthesis these more-complex vocal organ configurations can be modelled by the use of parallel resonators. In the case of nasals, one set of resonators can be used to model the nasal cavity while another models the oral cavity. The outputs of these are then simply added to generate the output. While this obviously differs from the real vocal tract (where the output of the oral cavity is further modified by the nasal cavity), with appropriate choices of parameters a convincing effect can be achieved. Fricative sounds are generated in a similar fashion, with one combination of resonators being used for the forward cavity and another for the back cavity. Again this is simpler that the real vocal tract, but this limitation can be overcome with appropriate choice of parameters.

Further refinement can be achieved by the use of zeros. These can be used to create anti-resonances, corresponding to a notch in the frequency response. Here the format-synthesis model again deviates from the all-pole tube model, but recall that we adopted the all-pole model only to make the derivation of the tube model easier. While the all-pole model has been shown to be perfectly adequate for vowel sounds, the quality of nasal and fricative sounds can be improved by the use of some additional zeros. In particular, it has be shown [254] that the use of a single-zero anti-resonator in series with the normal resonators can produce realistic nasal sounds.

13.2.5 A complete synthesiser

The **Klatt synthesiser** [10], [254] is one of the most sophisticated formant synthesisers developed. It combines all the components just described, and includes both a parallel and a cascade resonator system. A diagram of this system is shown in Figure 13.3. Recall that, from the tube model, we determined that we require about one formant per 1000 Hz. The Klatt synthesiser was set up to work at 10 kHz and six main formants are used. It is worth noting that most literature on formant synthesis uses a sampling rate of 8 kHz or 10 kHz; not because of any principled quality reason but mostly because space, speed and output requirements at the time prevented higher sampling rates being used. If a higher sampling rate is required, this number can easily be extended. That said, it has been shown that only the first three formants are used by listeners to discriminate sounds, so the higher formants are there simply to add naturalness to the speech.

When generating speech, in principle all the parameters (that is, the formant values, bandwidth, voicing, F0 etc.) could be changed on a sample-by-sample basis. In principle, however, a slower rate of parameter update, such as every 5 ms, is used.

13.2.6 The phonetic input to the synthesiser

The formant-synthesis technique just described of course deals with only half the problem; in addition to generating waveforms from formant parameters, we have to be able to generate formant parameters from the discrete pronunciation representations of the type represented by the synthesis specification. It is useful therefore to split the overall process into separate **parameter-to-speech** (i.e. the formant synthesiser just described) and **specification-to-parameter** components.

Before going into this, we should ask the following question: how good does the speech sound if we give the formant synthesiser "perfect" input? The specification-to-parameter component may produce errors and, if we are interested in assessing the quality of the formant synthesis itself, it may be difficult to do this directly from the specification. Instead we can use the technique of **copy synthesis**, whereby we forget about automatic text-to-speech conversion, and instead artificially generate the best possible parameters for the synthesiser. This test is in fact one of the corner stones of speech-synthesis research; it allows us to work on one part of the system in a modular fashion, but more importantly it acts as a proof of concept as to the synthesiser's eventual suitability for inclusion in the full TTS system. The key point is that, if the synthesis sounds bad with the best possible input, then it will only sound worse when potentially error-full input is given instead. In effect copy synthesis sets the upper limit on expected quality from any system.

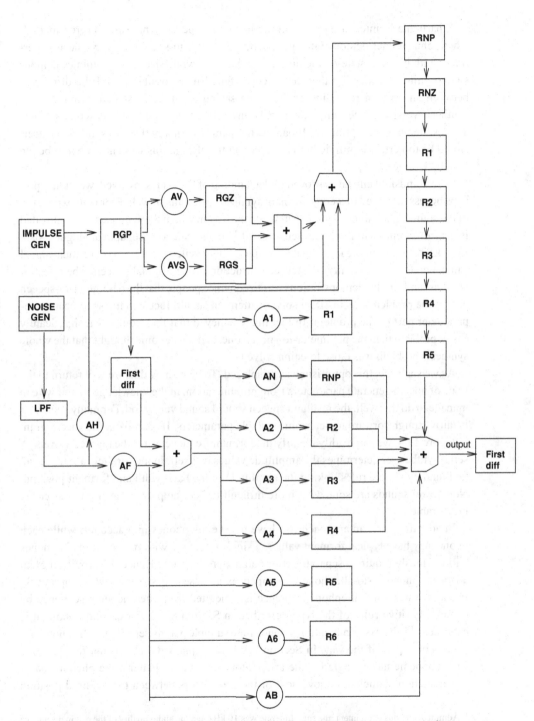

Figure 13.3 The complete Klatt synthesiser. The boxes R1, R2 etc. correspond to resonators generating formants. The A boxes control amplitudes at those points. The diagram shows both the serial and the parallel configurations of the formants.

A particularly interesting event in the history of speech synthesis occurred in 1972 when John Holmes demonstrated just such a parameter-to-speech copy-synthesis experiment with his then-state-of-the-art parallel formant synthesiser [218]. Holmes demonstrated quite convincingly that most people found it impossible to tell the difference between an original recording of his voice saying a particular sentence and a copy-synthesis version of the same sentence. Holmes' demonstration was met with considerable excitement and optimism. It had been assumed by many that the synthesiser itself would be too crude to mimic human speech faithfully, but this was shown not to be the case by Holmes.[1]

Given hindsight and an improved understanding of the issues involved, we should perhaps be less surprised that copy-formant synthesis performs so well. From our exposition of the source/filter model we see that, while the way in which the synthesiser is controlled is very different from human articulation, it is in essence a reasonable analogue. Also, from LP resynthesis, we know that an all-pole filter excited by the correct residual signal can generate a perceptually perfect reconstruction of the original speech. The effect of Holmes' and similar demonstrations was to convince people that the parameter-to-speech part of the problem was in effect solved; attention should focus on the specification-to-parameter part of the problem. In fact many believed that this would be a significantly easier problem than the parameter-to-speech one, and hence some thought that the whole synthesis problem was close to being solved.

As we shall see, this optimism was misplaced. To help show this, we will return to the issue of how to generate parameters from pronunciation, in the knowledge that, if we can manage to do this well, the resultant speech should sound very good. For many sounds, it is fairly straightforward to determine suitable parameters. If we look at the spectrogram in Figure 13.4, we can readily identify the formant frequencies for the marked vowels. A serial synthesiser determines the amplitude values by itself; in the parallel case these can be found by simple rules linking the amplitude to the bandwidth and formant position. Non-vowel sounds are somewhat more difficult in that the parameters are not as easily measurable.

In adopting such an approach, we have to take allophones into account; while each phone may have typical formant values, a single phoneme with two or more allophones will obviously require a separate set of parameters for each. Hence the provision of an adequate phoneme-to-allophone component is an essential step of such an approach. In most systems, an allophone sequence is generated from the phoneme sequence by context-sensitive rules of the type described in Section 8.1.2. For each allophone it is often possible by experimentation to determine suitable parameter values. This, however, gets us only part of the way. In Section 7.3.4, we explained that it is not the case that each phoneme has a single unique articulatory or acoustic pattern; the phenomena of coarticulation, namely the movement of the articulators between the required position

---

[1] Admittedly, in this experiment, the sampling rate was 10 kHz and the audio quality of the original sentence wasn't perfect (which would help cover up some slight differences). Furthermore, it appears that Holmes spent an enormous amount of time carefully adjusting the parameters; it wasn't just the case that he measured a few values from a spectrogram and the results sounded fine. These issues aside, the demonstration was still impressive.

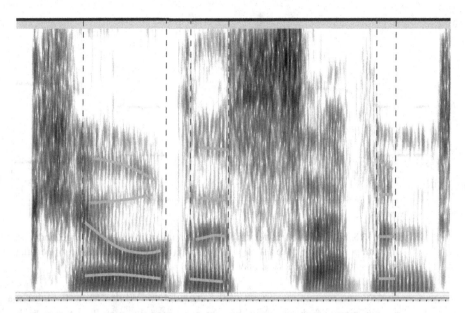

Figure 13.4 An example spectrogram with formant values marked.

for one phone and that for the next, ensures that a huge variety of acoustic patterns is observed for each phoneme.

One technique, originally developed by Holmes, Mattingly and Shearme [219], is to define a **target** that represents the steady-state portion of each phone. Depending on the duration of the phone, this will be longer or shorter. Between the steady-state portion of one phone and that of the next, a **transition** region occurs, where the parameter values are interpolated between the steady states of the first phone and the second. The rate of change of the transition is constant; hence for different durations of the same phone sequence the transition portion will look the same whereas the steady-state portion will change. An enhancement to the model allows for the possibility that the speech is spoken so quickly that the steady-state portion is never reached; in such cases the parameters **undershoot** the target and continue on, trying to reach the target values for the next phoneme. By this means, an attempt to mimic the more casual speaking style of fast speech is made. This process is shown in Figure 13.5.

13.2.7 Formant-synthesis quality

The general assessment of formant synthesis is that it is intelligible, often "clean" sounding, but far from natural. In looking for reasons for this, we can say firstly that the source model is just too simplistic; we shall, however, leave discussion of this until Section 13.3.5 since problems with this have much in common with the LP technique explained next. Secondly, the target–transition model is also too simplistic, and misses many of the subtleties really involved in the dynamics of speech. While the shapes of the formant trajectories are measured from a spectrogram, the underlying process is one of motor

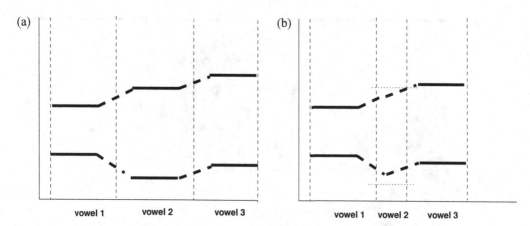

Figure 13.5 Formant patterns for different speech rates: (a) schematic target and transition formant positions for three vowels; and (b) as in (a), but here vowel 2 is so short that the steady-state target positions are not reached.

control and muscle movement of the articulators. While each articulator may move in a fairly simple fashion, when they are combined the overall system is highly complex. Furthermore, this complexity is simply that of the vocal tract; since this acts as a recursive filter, the effect on the passage of the source waveform is even more complicated. Finally, the assumptions made about the nature of the vocal-tract model do have some effect; even though each of these assumptions is valid on its own, the lack of precision adds up and affects the overall model. Significantly, although it can be shown that most of the bad effects caused by these assumptions can be bypassed, this can be done only by manipulating the parameter values away from their "natural" interpretation. In other words, while by appropriate manipulation the formant synthesiser can produce very natural speech, this is at a cost of having to use the parameters in strange ways that don't relate to their observable and easy-to-understand interpretations. Since this is the main motivation for using formant synthesis, we see that there is a real conflict in having a model that has easy control and one that produces high-quality speech.

That said, the main criticism of formant synthesis is that it doesn't sound *natural*. It can in fact produce quite *intelligible* speech, and is still competitive in that regard with some of the best of today's systems. The reason for this goes back to our initial discussion on the nature of communication in Chapter 2. There we showed that a primary factor in successful communication was that of contrast between symbols (phonemes in our case). Hence the success of the message-communicating aspect of speech (i.e. the intelligibility) can be achieved so long as the signal identifies the units in a contrasting fashion such that they can successfully be decoded. In the target–transition model, the targets are of course based on typical or canonical values of the formants for each phoneme; and as such should be clear and distinct from one another. The fact that the transitions between them don't accurately follow natural patterns only slightly detracts from the ability to decode the signal. In fact, it could be argued that good formant synthesis could produce speech *more* intelligible than slurred natural speech insofar the targets are harder to identify in the latter.

Given that one of the main problems lies with generating natural dynamic trajectories of the formants, we now turn to a series of techniques designed to do just that by means of determining these dynamics from natural speech.

13.3 Classical linear-prediction synthesis

One of the main difficulties in building a more-sophisticated target–transition model is that it is not always straightforward to find the formant values directly from speech. Although formant values can often be determined by visually scanning a spectrogram, this can be time-consuming and prone to human error. Use of an automatic formant tracker can bypass these practical difficulties, but in the years when formant synthesis was at its height accurate formant trackers had yet to be developed. In addition, both visual and automatic techniques suffer from the problem that in some cases the formants simply aren't easily discernible. The vocal-tract transfer function at that point is not in a position where it generates clear resonance peaks; rather the poles are well within the unit circle and hence the spectral envelope is not particularly "peaky". Hence, while formants may be the primary means of distinguishing between different certain phonemes, their values are not always clear for *every* frame of speech, so having formants as the primary means of control can lead to difficulties. Finally, even if we could find the formant values, it is certainly not the case that their dynamics would conform to simple target–transition models; in reality the movements are complex and hard to predict by a simple formula.

An alternative to using formants as the primary means of control is to use the parameters of the vocal-tract transfer function directly. The key here is that, if we assume the all-pole tube model, we *can* in fact determine these parameters automatically by means of LP, performed by the covariance or autocorrelation technique described in Chapter 12. In the following section we will explain in detail the commonality between LP and formant synthesis, where the two techniques diverge and how LP can be used to generate speech.

13.3.1 Comparison with formant synthesis

With the series formant synthesiser, we saw that the transfer function for a single formant with specified frequency and bandwidth could be created by a second-order filter,

$$H_n(z) = \frac{1}{1 - 2e^{-2\hat{B}} \cos(2\pi \hat{F})z^{-1} + e^{-2\hat{B}}z^{-2}}$$

and that a single transfer function for the whole oral cavity could then be made by cascading several of these second-order filters together,

$$H(z) = H_1(z)H_2(z)H_3(z)H_4(z) \tag{13.13}$$

$$= \underbrace{\frac{1}{1 - 2e^{-\hat{B}_1} \cos(2\pi \hat{F}_1)z^{-1} + e^{-2\hat{B}_1}z^{-2}}}_{H_1} \cdots \underbrace{\frac{1}{1 - 2e^{-\hat{B}_4} \cos(2\pi \hat{F}_4)z^{-1} + e^{-2\hat{B}_4}z^{-2}}}_{H_4}$$

$$\tag{13.14}$$

which when multiplied out can be expressed in terms of a polynomial of filter coefficients $A = \{a_1, a_2, \ldots, a_P\}$:

$$H(z) = \frac{1}{1 - a_1 z^{-1} - a_2 z^{-2} - \cdots - a_p z^{-p}} \qquad (13.15)$$

For an input $X(z)$ and output $X(Z)$, we write

$$Y(z) = \frac{X(z)}{1 - a_1 z^{-1} - a_2 z^{-2} - \cdots - a_p z^{-p}} \qquad (13.16)$$

and the inverse z-transform of this gives the time-domain difference equation;

$$y[n] = x[n] + a_1 y[n-1] + a_2 y[n-2] + \cdots + a_p y[n-p] \qquad (13.17)$$

$$= \sum_{k=1}^{P} a_k y[n-k] + x[n] \qquad (13.18)$$

Recall that Equation (13.18) is exactly the same as the LP equation (12.15), where $A = \{a_1, a_2, \ldots, a_P\}$ are the predictor coefficients and $x[n]$ is the "error" signal $e[n]$. This shows that the result of LP gives us the same type of transfer function as the serial formant synthesiser, and hence LP can produce exactly the same range of frequency responses as the serial formant synthesiser. The significance is of course that we can derive the LP coefficients automatically from speech and don't have to perform manual or potentially errorful automatic formant analysis. This is not, however, a solution to the formant-estimation problem itself; reversing the set of Equations (13.14)–(13.18) is not trivial, meaning that, while we can accurately estimate the all-pole transfer function for arbitrary speech, we can't necessarily decompose this into individual formants.

Beyond this the formant synthesiser and LP model start to diverge. Firstly, with the LP model, we use a single all-pole transfer function for all sounds. In the formant model, there are separate transfer functions in the formant synthesiser for the oral and nasal cavities. In addition, a further separate resonator is used in formant synthesis to create a voiced source signal from the impulse train; in the LP model the filter that does this is included in the all-pole filter. Hence the formant synthesiser is fundamentally more modular in that it separates these components. This lack of modularity in the LP model adds to the difficulty in providing physical interpretations for the coefficients.

13.3.2 The impulse/noise source model

One of the commonalities with the formant model is that LP synthesis maintains a source–filter separation. This means that, for a sequence of frames, we can resynthesise this with a different fundamental frequency from that of the original. The benefit is that, for a given transition effect that we wish to synthesise, we need analyse only one example of it; we can create the full range of fundamental-frequency effects by virtue of the separate control of the source.

In performing LP analysis, we are trying to find the set of predictor coefficients which minimises $e[n]$ over the analysis period. It can be shown that in doing so we are in fact attempting to find the set of predictor coefficients which gives rise to the error signal with the flattest spectrum. This is why the glottal filter is subsumed within the LP all-pole filter; otherwise the spectrum of the error signal wouldn't be flat. From this we conclude that the proper form of the source for the LP filter then is also a signal with a flat spectrum. For periodic sounds this can be provided by a sequence of impulses; for obstruent sounds it is provided by a noise source. When creating the source, then, we first pick the type of source that is appropriate to the phoneme class of the coefficients: an impulse train for voiced speech, a noise source for obstruents and both for voiced obstruents. Then, for voiced sounds, we generate an impulse train with period defined by the F0 values given in the specification.

It is worth noting that this is the same as the source-generation method used in formant synthesis. The main difference is that there the impulse sequence was fed into a special glottal filter in an attempt to produce a realistic glottal volume-velocity signal, whereas in the LP synthesis the impulse train is passed directly into the filter. The reason for this difference has been explained before, namely that LP analysis generates the coefficients for a filter that will model both the glottal and the vocal-tract characteristics. The fact that we use an impulse/noise model is why we call this **classical linear-prediction** synthesis; this distinguishes it from the other source techniques that use linear prediction, which we will described in Chapter 14.

13.3.3   Linear-prediction diphone-concatenative synthesis

By using LP analysis over a number of frames, we can exactly capture the vocal-tract dynamics of those frames. We therefore solve the transition-modelling problem implicitly: since we can measure the dynamics directly, we simply do so for all the types of transitions we are interested in and store the results. By then synthesising from these patterns, we can exactly recreate the original speech dynamics that we analysed. Independently, we generate a source signal using the noise source for obstruent sounds and the impulse source for voiced sounds. We control the fundamental frequency of the source by controlling the rate of impulses, which allows us to generate speech with a range of fundamental frequencies for a single set of stored frames.

This use of the LP model leads us to the technique called **concatenative synthesis**. Here, no explicit vocal-tract modelling is performed; rather, an exhaustive set of **units** is acquired and synthesis is performed by **concatenating** these units in appropriate order. The targets and transitions are stored within the units, and, so long as suitable units and good original speech are used, it should be possible to generate realistic parameter patterns via the concatenation technique.

One of the most-popular choices of unit is that of the **diphone**. A diphone is defined as a unit that starts in the middle of one phone and extends to the middle of the next phone. The justification for using diphones follows directly from the target–transition model, in which we have a stable "target" region (the phone middle), which then has a transition period to the middle of the next phone. Because the diphones start and stop in these middle, stable, target regions, they should have similar vocal-tract shapes at these points and therefore join together well. By comparison, concatenating units at phone

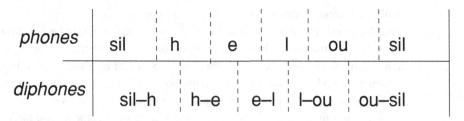

Figure 13.6 Phones and diphones.

boundaries is much less likely to be successful because this is where the variance in vocal-tract shapes is greatest.

A diphone synthesiser requires a different type of synthesis specification from the phone-centred one used previously. For a phoneme string of length N, a diphone specification of phone pairs of length $N - 1$ is required. Each item in the specification is comprised of two **states** known here as **half-phones**,[2] named STATE 1 and STATE 2. Each of these half-phones has its own duration. We can define the start and end points of the diphones in a number of ways, but the simplest is to place them half way between the start and end of each phone. Several options are available for the F0 component as before, but one neat way is to have a single F0 value for each half-phone, measured from the middle of that half-phone. This is shown graphically in Figure 13.6. A single item in a diphone specification would then for example look like

$$s_t = \begin{bmatrix} \text{STATE 1} \begin{bmatrix} \text{PHONE} & n \\ \text{F0} & 121 \\ \text{DURATION} & 50 \end{bmatrix} \\ \text{STATE 2} \begin{bmatrix} \text{PHONE} & t \\ \text{F0} & 123 \\ \text{DURATION} & 70 \end{bmatrix} \end{bmatrix}$$

For convenience, we often denote a single diphone by both its half-phones joined by a hyphen. Hence the phone string /h eh l ow/ gives the diphone sequence /h-eh eh-l l-ow/ or /h-eh/, /eh-l/, /l-ow/.

A key task is careful determination of the full set of possible diphones. This is dealt with in full in Section 14.1, but for now it suffices to say that we should find a database in which there is a high degree of consistency, to ensure good continuity and joins, and a broad range of phonetic contexts, to ensure that all diphone combinations can be found. While it is possible to use existing databases, it is more common to design and record specifically a speech database that meets these particular requirements.

---

[2] The terminology regarding unit names is unfortunately complicated by historical naming quirks. We'll go into this more thoroughly in Chapter 16.

Once the diphone tokens have been identified in the speech database, we can extract the LP parameters. This is normally done with fixed-frame analysis. First we window the section of speech into a number of frames, with, say, a fixed frame shift of 10 ms and a window length of 25 ms. The coefficients can be stored as is, or converted to one of the alternative LP representations. When storing the units in the database, it is important that the durations and F0 values of the two half-phones are recorded, since these are needed in the process of prosodic modification.

13.3.4 A complete synthesiser

From the diphone specification, we find the matching units in our database, and concatenate them into a **unit sequence**. From this we extract the sequence of LP parameters for each unit and concatenate them into a single sequence of parameters. In doing this we should have created a sequence of frames with appropriate targets and transitions for the phone sequence we are attempting to synthesise. Next we need to ensure that the speech we generate has the correct duration and F0 as given in the specification. From now on, it is easiest to think of the task in terms of half-phones; so, for each half-phone in the specification, we have a specified duration and specified F0. These can be compared with the database durations and F0 values of each half-phone in our unit sequence, and from this we can determine the amount of prosodic modification to be performed for each half-phone.

The unit durations can be made to match the specification durations in a number of ways. Firstly, we can simply stretch and compress the frame rate. Imagine we have a unit that is originally 120 ms long. This will comprise 12 frames at intervals of 10 ms. Imagine now that the target duration tells us to create a unit 90 ms long. This can be created by changing the synthesis frame rate to 7.5 ms, in which case the 12 frames will have a duration of 12×7.5 ms $= 90$ ms. If a longer duration is required, the synthesis frame rate can be lengthened. An alternative is to shorten the unit by keeping the frame rate the same but leaving out every fourth frame – this will result in nine frames of 10 ms $= 90$ ms. To lengthen the unit, appropriate frames can be duplicated. The leaving-out/duplication strategy is cruder than the compression/stretching strategy, but so long as the change is not too large (say within a factor of two) it works surprisingly well. By using one of these duration techniques we can generate a single sequence of LP frames for the entire utterance.

Given the durations, we can calculate the source signal for the utterance. We use an impulse source for sonorant phones, a noise source for unvoiced consonants, and a combined impulse and noise source for voiced obstruents. The source characteristics are switched at phone boundaries. For voiced sounds, the impulse sequence is created by placing impulses at a separation distance determined by 1/F0 at that point. Finally we feed the source signal into the filter coefficients to generate the final speech waveform for the sentence.

There are some minor modifications that we can make to this basic model. Firstly, it is possible to lessen the switching character of the sound source at phone boundaries by appropriate use of amplitude control on the noise and impulse source. This is motivated from measurements of glottal activity, which show that it may take a frame or two for the glottis to change modes. At boundaries between frames, we will encounter the situation that at one point we are having our synthetic speech signal $s[n]$ being generated by the coefficients of frame j, but as we reach the end of the frame we will have to start using the coefficients of the next frame, $j + 1$. During this cross-over, at a point K less than the filter order P, we will be using the low-valued ($<K$) coefficients of frame $j + 1$ and the high-valued ($>K$) coefficients of the previous frame, j. This can occasionally lead to unwanted artefacts, so it is often sensible to perform some sort of filter interpolation at this point. The simplest way is simply to smooth and interpolate the filter coefficients at the P samples around the frame boundary. Alternatively, one can adopt a technique of constant interpolation whereby in effect *every* sample has its own unique set of coefficients interpolated from the two nearest frames. This equates to a constantly moving vocal tract rather than the piecewise-stationary model assumed in normal frame-based analysis. While interpolation can be performed on the other representations as before, experimentation has shown that often the LP coefficients are sufficiently similar from one frame to the next to allow direct interpolation on the coefficients themselves.

13.3.5 Problems with the source

Despite its ability to mimic the target and transition patterns of natural speech faithfully, standard LP synthesis has a significant unnatural quality to it, often being impressionistically described as "buzzy" or "metallic" sounding. Recall that, while we measured the vocal-tract model parameters directly from real speech, we still used an explicit impulse/noise model for the source. As we will now see, it is this, and specifically the interaction of this with the filter, which creates the unnaturalness.

We can demonstrate that this is so with objective and subjective evidence. Recall from Section 12.6.1 that we can synthesise a signal $y'[n]$ that is exactly the same as the original speech signal $y[n]$ that we analysed provided that we excite the LP filter with the residual. If we look at a real residual signal such as that shown in Figure 12.17, we can clearly see a major spike for each pitch period. This is what we are modelling with our single impulse. We can, however, see that there are many other smaller spikes and patterns in the residual, and, since these are being completely ignored in the impulse/noise model, we can see that we are considerably over-simplifying the source.

Subjectively, we can analyse the source of the unnatural "buzzy" sound by again performing a copy-synthesis experiment. This time we take a complete, natural sentence (no diphones or other units), and synthesise it under various conditions. The first is to use the standard LP resynthesis model whereby we analyse the sentence into frames and resynthesise with a noise source and sequence of impulses that match the original pitch. The resultant speech sounds reasonably good, but clearly different (degraded) from the original, especially if we do this with good audio conditions and a high sampling rate (the good audio conditions allow us to hear more detail in the signal, and from this we can therefore detect errors more easily). If we now change the pitch but keep everything else the same (by changing the impulse sequence), we find that the speech becomes noticeably more buzzy. Thirdly, we can keep the original frames and pitch patterns, but change the durations by one of the two techniques described above. This sounds reasonably good, and for small-duration modifications the speech does not sound noticeably worse than it did under the first condition. Finally, we resynthesise, not using the impulse/noise source, but by taking the actual residual error signal (details are described below). That is, during analysis we record the error at every sample, and make a complete residual waveform. When the speech is resynthesised using this as a source, the speech sounds perfect and is in fact completely indistinguishable from the original. From this, we can safely conclude that the buzzy nature of the speech is indeed caused by use of an overly simplistic sound source.

13.4 Articulatory synthesis

Perhaps the most obvious way to synthesise speech is to try a direct simulation of human speech production. This approach is called **articulatory synthesis** and is actually the oldest in the sense that the famous talking machine of von Kempelen can be seen as an articulatory synthesizer [477]. This machine was a mechanical device with tubes, bellows and pipes, which, with a little training, could be used to produce recognisable speech. The machine was "played" in much the same way as one plays a musical instrument. The device is of course mimicking the vocal tract using sources and filters, it is just that this time the physical nature of these is acoustic. While it may seem amazing that a device developed as long ago as this can produce half-reasonable speech, we should now realise that a major reason for this is that the "specification-to-parameter" part of this device is controlled by a human in real time, which adds many advantages to the system in that the human can use mechanisms such as feedback to control the playback and hence mimic the natural speech-production process. In modern times, articulatory synthesis is tackled from a different perspective since it is obviously incomplete and impractical to have someone "play" the synthesiser in this way.

Many modern articulatory synthesisers are extensions of the acoustic-tube models described in Chapter 11 [11], [164], [288], [382]. As we saw there, a complex tube shape can be approximated by a number of smaller uniform tubes, and, since we know the sound-propagation properties of such systems, we can use these to build a more-complex general model. Hence the difference between articulatory synthesis and our other techniques is that with formant synthesis we are implicitly using the tube model, but in fact using the resultant behaviour of the tube model as our parameter space rather than the tubes themselves. With LP synthesis, we are using one form of the tube model (i.e. the all-pole model), but acquiring the tube parameters from data rather than attempting to build a model of how the tubes themselves behave. The attractive part of articulatory synthesis is that, since the tubes themselves are the controls, this is a much easier and more-natural way to generate speech; small, "natural" movements in tubes can give rise to the complex patterns of speech, thus bypassing the problems of modelling complex formant trajectories explicitly. Often articulatory-synthesis models have an interim stage, in which the motion of the tubes is controlled by some simple process (such as mechanical damping or filtering) intended to model the fact that the articulators move with a certain inherent speed. This **motor-control** space is then used to provide the parameters for the specification-to-parameter component.

The two difficulties in articulatory synthesis are, firstly, deciding how to generate the control parameters from the specification, and secondly finding the right balance between a highly accurate model that closely follows human physiology and a more-pragmatic model that is easy to design and control. The first problem is similar to that of formant synthesis. There, however, in many cases (but not all) it is straightforward to find the formant values from real speech; we simply record the speech, calculate the spectrogram and determine the formant values. The problem is considerably more difficult with articulatory synthesis in that the correct articulatory parameters cannot be found from recordings; rather we must resort to more-intrusive measures such as X-ray photography, MRI or EMA imaging (explained in Section 7.1.7). Not only is this sort of data inherently more difficult to collect, but also many of these techniques are relatively recent developments, meaning that during the early days of articulatory synthesis acquiring such data was particularly difficult.

The second problem concerns just how accurate our model of articulation should be. As we saw in our discussion on tube models, there is always a balance between the desire to mimic the phenomenon accurately and the need to do so with a simple and tractable model. The earliest models were more or less those described in Chapter 11, but since then numerous improvements have been made, many along the lines described in Section 11.5. These have included modelling vocal-tract losses, source–filter interaction, radiation from the lips, and of course improvements to glottal-source characteristics. In addition many authors have attempted to develop models of both the vocal tract itself and the controls within it, such that many approaches have models for muscle movement and motor control.

Both these problems present considerable difficulty, such that the best articulatory synthesis is quite poor compared with the best synthesis from other techniques (however, some good copy synthesis has been demonstrated). Because of this, articulatory synthesis has largely been abandoned as a technique for generating high-quality speech for engineering purposes. However, while this approach might not provide a good engineering solution for text-to-speech, it still arouses interest in a number of related disciplines. Firstly, there is considerable interest in the scientific field of speech production. Many have argued that, because the articulatory domain is the natural and true domain for speech production, it does in fact help explain the systematic organisation of higher-level aspects of speech. For instance the **articulatory phonology** of Browman and Goldstein [69] is based on the idea of **articulatory gestures** as the phonological primitives rather than the segment-based features described in Section 7.4.3. Boersma [60] is also notable for developing a theory of phonology coupled with an articulatory synthesiser. A second, related, field of interest is what we can term **articulatory physiology**, where the goal is to create fully accurate models of articulator movement. The emphasis here is somewhat different in that we are attempting to model specific articulators or effects precisely, rather than build a complete (but necessarily approximate) model, or to link this with a linguistic/phonetic model [467], [489]. Finally, articulatory synthesis is implicitly connected to the field of **audio-visual synthesis** or **talking-head synthesis**, where the idea is to build a complete *visual* animated model of the head while talking. Talking heads can be built in a number of ways, ranging from systems that attempt a bottom-up reconstruction of the head by modelling the articulators directly to more data-driven approaches that take photos or videos of real heads and then use morphing techniques to create animation [32], [188], [304].

13.5 Discussion

In order to gain a better understanding of contemporary techniques, it is important to understand the strengths and weakness, and ultimate shortcomings, of three techniques described here. We have seen a number of factors that enable us to draw up a "wish list" for a perfect synthesiser.

Modularity It is advantageous to have a modular system, so that we can control components separately. All three techniques provide a source/filter modularity, with the formant and articulatory techniques scoring better in this regard in that the glottal waveform itself is fully separable. The formant and articulatory synthesisers have further modularity, in that they allow separate control of the oral and nasal cavities. Beyond this, the formant synthesiser allows individual control of each formant, giving a final level of modularity that greatly outweighs those of the other techniques.

Ease of data acquisition Irrespective of whether the system is "rule-driven" or "data-driven", *some* data have to be acquired, even if this is just to help the rule-writer determine appropriate values for the rules. Here LP clearly wins, because its parameters can easily be determined from any real speech waveform. When formant synthesisers were mainly being developed, no fully reliable formant trackers existed, so the formant values had to be determined either manually or semi-manually. While better formant trackers now exist, many other parameters required in formant synthesis (e.g. zero locations or bandwidth values) are still somewhat difficult to determine. Articulatory synthesis is particularly interesting in that in the past it was next to impossible to acquire data. Now, various techniques such as EMA and MRI have made this much easier, so it should be possible to collect much bigger databases for this purpose. The inability to collect accurate articulatory data is certainly one of the main reasons why articulatory synthesis never really took off.

Effectiveness of model All three techniques are based on related but different models of speech, but some are more inherently "natural", meaning that the way the parameters change over time can be specified by simpler means.

It should be clear from our exposition that each technique has inherent tradeoffs with respect to the above wish list. For example, we make many assumptions in order to use the lossless all-pole LP model for all speech sounds. In doing so, we achieve a model whose parameters we can measure easily and automatically, but find that these are difficult to interpret in a useful sense. While the general nature of the model is justified, the assumptions we make to achieve automatic analysis mean that we can't modify, manipulate and control the parameters in as direct a way as we can with formant synthesis. Following on from this, it is difficult to produce a simple and elegant phonetics-to-parameter model, since it is difficult to interpret these parameters in higher-level phonetic terms.

On the other hand, with formant synthesis, we can in some sense relate the parameters to the phonetics in that we know for instance that the typical formant values for an /iy/ vowel are 300 Hz, 2200 Hz and 3000 Hz. However, because we can't find the required parameters easily for some arbitrary section of speech, we have to resort to a top-down fully specified model for transitions, which is often found severely lacking when compared with real transitions. Because parameter collection is so hard, it is difficult to come up with more-accurate phonetics-to-parameter models since the development-and-testing cycle is slow and inaccurate.

Moving on from these issues, we see problems that are common to all techniques. Firstly, there is the issue of source quality. In effect, all use the same or similar sources. While these perform their *basic* job of generating a glottal-waveform signal of specified F0 quite well, they fall far short of producing a genuinely natural-sounding one. In fact, our understanding of source characteristics now is such that we know that this part of the speech apparatus is highly individual, being something that gives individual speakers their identity. Having a single, "mean" model for all speakers is very crude. Furthermore, the shape of the signal changes with frequency, with vocal effort and with other speaking factors. We know that the source and filter aren't in fact completely decoupled, and this has an impact. Finally, the source has its own version of coarticulation, such that the source characteristics can become very complex over a phone boundary, especially when the voicing is switched on and off. More-sophisticated source models of the type described in Section 11.4 improve matters somewhat, but even these are not close to generating accurately the complex source patterns observed in real speech.

A second issue is that, as we demand ever-greater fidelity in the output speech, the various assumptions we have made in our models start having a more-noticeable negative effect. We have already shown this for the source modelling, but these effects are present too in the various filter model variants. In Section 11.6 we listed several assumptions in the source/tube model, namely assumptions made with losses, simple radiation characteristics, linearity of the filters and so on. As we wish to create speech of ever-higher quality, each of these starts to have an effect.

The final further issue concerns the specification-to-parameters problem. In going from formant synthesis to LP synthesis the key move was to abandon an explicit specification-to-parameters model of vocal-tract configurations and instead measure the required parameters from data. From this we can use at synthesis time a lookup table that simply lists the parameters for each unit (typically a diphone). The cost in doing so is firstly that we lose explicit control of the phenomenon and secondly of course that we incur significant extra storage costs. If we now, however, look at how closely the LP parameters for one diphone follow parameters for lots of examples of that same diphone in many different real situations, we see that in fact even this purely data-driven approach is severely lacking, and in many cases the set of parameters for our diphone will match only one specific linguistic context and can be quite different from that diphone in other linguistic contexts.

We have presented LP synthesis from one particular viewpoint; specifically one in which we show the similarity between this and formant and articulatory synthesis. It is quite common to see another type of explanation of the same system. This alternative explanation is based on the principle that in general we wish to record speech and play it back untouched; in doing so we have of course exactly recreated the original signal and hence the quality is perfect. The problem is of course that we can't collect an example of everything we wish to say; in fact we can collect only a tiny fraction. This leads us to the idea that we should perform synthesis from natural data, which has had the minimum, "non-intrusive" modification required to make it obey the synthesis specification.

These two final points are in fact the essence of the modern synthesis problem as currently defined. Firstly, we wish to create speech with a set of parameters that can faithfully generate the patterns of as wide a range of naturally occurring speech sounds as possible. Secondly, and because we might not be able to collect everything we wish, we will then modify those generate parameters to deliver speech that matches the specification. These two problems, natural-parameter generation and speech modification, are the subjects of the next chapters.

13.5.1 Further reading

Klatt's 1987 article "Review of text-to-speech conversion for English" [255] is an excellent source for further reading on first-generation systems. Klatt documents the history of the entire TTS field, and then explains the (then) state-of-the-art systems in detail. While his account is heavily biased towards formant synthesis, rather than LP or articulatory synthesis, it nonetheless remains a very solid account of technology before and at the time.

In the article Klatt gives a very thorough and fascinating account of text-to-speech from the early days until the mid 1980s, documenting the development of the first-generation techniques explained in this chapter. Klatt starts his discussion with an examination of what we might now term "pre-first-generation" systems, which were not intended to be used in text-to-speech, but rather as basic demonstrations that machines could generate speech. The first widely recognised modern speech synthesiser is the **vodor** developed at the AT&T telephone company by Dudley [143]. This was in many ways an electronic counterpart to von Kempelen's talking machine in that the vodor was "played" by human operators, who with a little practice could generate somewhat intelligible speech. The next notable system was the "pattern playback" developed at Haskins labs in the USA [106]. This device worked by reading parameters from a roll of paper that was fed into the machine. The marks on the roll could be measured from spectrograms or painted on by hand.

The development of the source–filter model and the general acoustic theory of speech production by Fant and others (see Chapter 11) was a key breakthrough step and allowed the development of the type of formant synthesisers described here. The Parametric

Artificial Talker (PAT) developed by Walter Lawrence and colleagues (including Peter Ladefoged) at Edinburgh University [278] worked in a similar fashion to the Haskins system except that the roll of paper now specified formant values. Fant developed the OVE I and OVE II synthesis systems, which were fully electronic, and their modular structure allowed phoneticians to apply the knowledge they had acquired of speech directly to the synthesis problem. Further development in formant synthesis progressed by the investigation of enhancements such as parallel formants, nasal branches, anti-formants and so on. Two of the leading figures in this stage were Denis Klatt [252], [254] and John Holmes [218], [219].

The development of LP synthesis was really a by-product of research into speech compression or coding. Linear prediction is of course the basis of LP coding (LPC), which was developed in the late 1960s and early 1970s [20], [233], [294], [301]. The compression was achieved by source–filter separation in which the LP coefficients were transmitted but the source was replaced by just a pitch value. One of the most famous developments in speech technology was the Texas Instruments Speak 'n Spell toy [485], which used LP coding to compress a considerable number of words, with the aim of speaking that word and getting a child to spell it on a keyboard so as to learn how to spell.[3] One of the first complete synthesisers to use LP was developed at Bell Labs [338], [339] and this started a line of TTS development that remained unbroken for many years.

13.5.2 Summary

There are three main techniques that make use of the classical acoustic theory of speech in the production model.

Formant synthesis

- Formant synthesis works by using individually controllable formant filters, which can be set to produce accurate estimations of the vocal-tract transfer function.
- An impulse train is used to generate voiced sounds and a noise source to generate obstruent sounds. These are then passed through the filters to produce speech.
- The parameters of the formant synthesiser are determined by a set of rules concerning the phone characteristics and phone context.
- In general formant synthesis produces intelligible but not natural-sounding speech.
- It can be shown that very natural speech can be generated so long as the parameters are set very accurately. Unfortunately, it is extremely hard to do this automatically.
- The inherent difficulty and complexity in designing formant rules by hand has led to this technique largely being abandoned for engineering purposes.

Classical linear prediction

- This adopts the all-pole vocal-tract model, and uses an impulse and noise-source model.

---

[3] As a child I tried this, and, while I marvelled at the technology then, and do so even more today, it had no noticeable effect on my awful spelling.

- In terms of production, it is very similar to formant synthesis with regard to the source and vowels. It differs in that all sounds are generated by an all-pole filter, whereas parallel filters are common in formant synthesis.
- Its main strength is that the vocal-tract parameters can be determined automatically from speech.

Articulatory synthesis

- These models generate speech by direct modelling of human articulator behaviour.
- They are by their very nature the most "natural" way of generating speech, and in principle speech generation in this way should involve control of a simple parameter space with only a few degrees of freedom.
- In practice, acquiring data to determining rules and models is very difficult.
- Mimicking the human system closely can be very complex and computationally intractable.
- Because of these difficulties, there is little engineering work in articulatory synthesis, but it is central in the other areas of speech production, articulator physiology and audio-visual or talking-head synthesis.

14 Synthesis by concatenation and signal-processing modification

We saw in Chapter 13 that, while vocal-tract methods can often generate intelligible speech, they seem fundamentally limited in terms of generating natural-sounding speech. We saw that, in the case of formant synthesis, the main limitation is not so much in generating the speech from the parametric representation, but rather in generating these parameters from the input specification which was created by the text-analysis process. The mapping between the specification and the parameters is highly complex, and seems beyond what we can express in explicit human-derived rules, no matter how "expert" the rule designer. We face the same problems with articulatory synthesis and in addition have to deal with the facts that acquiring data is fundamentally difficult and improving naturalness often necessitates a considerable increase in complexity in the synthesiser.

A partial solution to the complexities of specifiction-to-parameter mapping is found in the classical LP technique whereby we bypassed the issue of generating of the vocal-tract parameters explicitly and instead measured them from data. The source parameters, however, were still specified by an explicit model, which was identified as the main source of the unnaturalness.

In this chapter we introduce a set of techniques that attempt to get around these limitations. In a way, these can be viewed as extensions of the classical LP technique in that they use a data-driven approach: the increase in quality, however, largely arises from the abandonment of the over-simplistic impulse/noise source model. These techniques are often collectively called **second-generation** synthesis systems, in contrast to the **first-generation** systems of Chapter 13. While details vary, we can characterise second-generation systems as ones in which we directly use data to determine the parameters of the verbal component as with classical linear prediction. The difference is that the source waveform too is now generated in a data-driven fashion. The *input* to the source, however, is still controlled by an explicit model. So for instance we might have an explicit F0-generation model of the type described in Section 9.5, which generates an F0 value every 10 ms. The second-generation technique then realises these values in the synthesized speech by use of a data-driven technique, rather than the impulse/noise model (Section 13.3.2).

The differences among the second-generation techniques mainly arise from how explicitly they use a parameterisation of the signal. All use a data-driven approach, but some use an explicit speech model (for example using LP coefficients to model the vocal tract) whereas others perform little or no modelling at all, and just use "raw" waveforms as the data.

As we saw in Chapter 9, the acoustic manifestation of prosody is complex and affects many parts of the speech signal. In second-generation synthesis, this is greatly simplified and generally just reduces to modelling pitch and timing. This in many ways is a matter of expediency; while pitch and timing are clearly heavily influenced by the prosodic form of an utterance, they also happen to be the two aspects of the signal that are the easiest to modify. Other aspects such as voice quality (say breathiness at the end of a sentence) can't easily be accommodated within this framework and therefore are usually ignored.

The standard set-up in a second-generation system is to have a specification composed of a sequence of items such that each contains a phone description, an F0 value and a duration for that phone. The phone description is matched to the data, and various techniques are used to modify its pitch and duration. In general, the verbal or phonetic component of second-generation systems is quite trivial. Typically, just one recorded speech unit is available for each unique phonetic specification, and to generate the synthesis waveform we simply find these units in a database and concatenate them. Modifying the pitch and timing without introducing any unwanted side effects can, however, be quite tricky, and it is for this reason that the majority of interest in second-generation systems lies in the development of these techniques. The following sections therefore mostly concentrate on the signal-processing techniques used to change the pitch and timing of concatenated waveforms.

14.1 Speech units in second-generation systems

Historically, we can view second-generation systems as the further development of classical LP concatenative synthesis, where the goal of the new systems was to eliminate the problems caused by assuming the over-simplistic impulse/noise source model. A subtly different perspective can be formed by considering that in first-generation systems (those developed in the 1970s and early 1980s) the hard reality of memory limitations meant that alternatives to explicit source models were impossible; in fact storing the LP parameters themselves was often prohibitive. By the time second-generation systems were mature, memory had become so much cheaper that it was possible to store significant numbers of whole waveforms recorded at higher sampling rates (often 16 KHz). The quality of unanalysed speech that is simply played back is of course perfect, which leads to a different view on the nature of second-generation systems: rather than seeing them as a way of acquiring model parameters directly from data, alternatively we can see them as a technique whereby we record good-quality natural speech and use recombination techniques to generate phone and word sequences that weren't originally in the data. The only problem is that, although we can acquire enough real data to cover the range of phonetic effects, we can't do this and also cover all the prosodic effects. Hence we use signal processing to help us with this last stage. Given that the quality of the basic recombined speech can be very high, the goal of the signal-processing algorithms can then be stated as achieving the required prosody but doing so in a way that leaves all the other aspects of the signal as they were. From this perspective, we see why classical LP synthesis can sound so poor, since in this technique we are performing the quite brutal modification of replacing the natural source with a simple impulse.

14.1.1 Creating a diphone inventory

The most common type of unit in second-generation systems is the diphone, so a key task is careful determination of the full set of possible diphones. Our investigation into phonotactics (Section 7.4.1) showed that phonemes occurred in characteristic sequences, so at the beginning of a word we may find /s t r/ but not /s b r/. We can use such a phonotactic grammar to help us determine the full set of diphones; if a pair of phonemes in sequence is impossible according to the grammar we will not include it in our legal diphone set. It is worth noting that, while phoneme sequences within words can be quite constrained, because we need to consider sequences across word boundaries too, the number of possibilities is quite large. In fact, compared with the theoretical maximum (i.e. all possible pairs), there are only a few that are not legal. These all arise from constraints on what phones can occur word-finally, so, because /h/, /w/ and /y/ don't occur word-finally, /y-l/, /w-t/ and /h-ng/ are all out.

The full list of diphone types is called the **diphone inventory**, and, once that has been determined, we need to find units of such diphones in real speech. Since we are extracting only one unit for each diphone type, we need to make sure that this is in fact a good example for our purposes. We can state three criteria for this.

1. We need to extract diphones that will join together well. At any join, we have a left diphone and a right diphone, which share a phoneme. For each phoneme we will have about 40 left diphones and 40 right diphones, and the aim is to extract units of these 80 diphones such that the last frame of all the left diphones is similar acoustically to the first frame of all the right diphones.
2. We wish for the diphones to be typical of their type – since we have only one unit of each, it makes sense to avoid outliers.
3. Because we will have to change the duration and pitch of this unit, it makes sense to pick units that are least likely to incur distortion when these changes are made. Since the distortion incurred is proportional to the amount of change, it makes sense again to pick units with average pitch and duration values, because this minimises the average amount of change. One subtlety to this is that it is easier to make a long unit shorter than to make a short unit longer; this is because short units often lack the clearly defined target regions of the longer units and if these are lengthened the result can sound unnatural. If, on the other hand, a long unit is shortened, some perception of "over-articulation" may occur, but the result is generally acceptable.

14.1.2 Obtaining diphones from speech

Several techniques have been proposed for acquiring the diphone units on the basis of these three criteria. Firstly we can simply search for them in any corpus of speech from a single speaker. This can have the advantage that we need not necessarily collect new data for our synthesiser, but it can be difficult to find units that meet the criteria. An alternative is careful design and recording of a corpus of isolated words in which one or perhaps two diphones are embedded in each word. The advantage here is twofold. Firstly we can control the phonetic context of the diphone so that longer-range allophonic and coarticulation effects are absent. For example, if we wish to obtain a typical diphone /t-ae/, it is probably best not to record this in the context of a following /n/ phone, as in the word /t ae n/. The /n/ will colour the /ae/ vowel, and this will sound unnatural when the /t-ae/ diphone is used in non-nasal contexts. Secondly, the speaking and recording conditions can be carefully controlled and it is generally easier to elicit a diphone with the necessary prosodic characteristics under such circumstances. One approach along these lines is to search a lexicon and from this define a word list that covers all required diphones. A variation on this is to design a special set of invented words (called **nonsense words** or **logotomes**) that exactly match the criteria we require. In doing so, we can create a template (e.g. /ax X Y t ax/) for a diphone such that the context is nearly always the same for every unit, for example

/ax r uh t ax/
/ax l uh t ax/
/ax p uh t ax/
/ax b uh t ax/

One last major difficulty is that, while our phonotactic grammar may give us all the possible phoneme pairs which exist in a language, in reality a significant minority may never occur in any known word. In English this is particularly problematic with the /zh/ phoneme, which is nearly always followed by /er/ (e.g. MEASURE). For cases like this, imaginative use of phoneme pairs across word boundaries or the creation of nonsense words will be required.

Experience has shown that, with a suitably recorded and analysed diphone set, it is usually possible to concatenate the diphones without any interpolation or smoothing at the concatenation point. This is to be expected if the steady-state/transition model is correct (see Section 13.2.6). As we shall see in Chapter 16, however, this assumption is really possible only because the diphones have been well articulated, come from neutral contexts and have been recorded well. It is not safe to assume that other units can always be successfully concatenated in phone middles.

Once we have a sequence of diphone units that match the phonetic part of the specification, we need to modify the pitch and timing of each to match the prosodic part of the specification. The rest of the chapter focuses on techniques for performing this task.

14.2 Pitch-synchronous overlap and add (PSOLA)

Perhaps the mostly widely used second-generation signal-processing techniques are the family called **pitch-synchronous overlap and add** (shortened to **PSOLA** and pronounced /p ax s ow l ax/). These techniques are used to modify the pitch and timing of speech but do so without performing any explicit source-filter separation. The basis of all the PSOLA techniques is to isolate individual pitch periods in the original speech, perform modification, and then resynthesise to create the final waveform.

14.2.1 Time-domain PSOLA

Time-domain pitch-synchronous overlap and add or **TD-PSOLA** is widely regarded as the most-popular PSOLA technique and indeed may well be the most-popular algorithm overall for pitch and timing adjustment [194], [322], [473].

The technique works pitch-synchronously, which means that there is one analysis frame per pitch period. A prerequisite for this is that we need to be able to identify the epochs in the speech signal, and with PSOLA it is vital that this is done with very high accuracy. To perform this step, an algorithm of the the type discussed in Section 12.7 is used. The epoch positions are often taken to be at the instant of glottal closure for each period (Sections 12.6.2 and 12.7.2) but so long as the epoch lies in the same relative position for every frame PSOLA should work well. The signal is then separated into frames with a hanning window, which extends one pitch period before and one pitch period after the epoch, as shown in Figure 14.1. These windowed frames can then be recombined by placing their centres back on the original epoch positions, and adding the overlapping regions (hence the name, overlap and add). When this is done, the result is a speech waveform perceptually indistinguishable from the original. The waveform is not *exactly* the same, since the sinusoid multiplication carried out during analysis is not exactly reversed during synthesis, but the overlap–add procedure comes close enough that the difference is not noticeable.

Time-scale modification is achieved by elimination or duplication of frames, as shown in Figure 14.2. For a given set of frames, if we duplicate one of these frames and insert it back into the sequence and then perform overlap and add, we will create a speech waveform that is nearly identical to the original except that it is one frame longer. In general, listeners can't detect that this operation has been performed and the desired effect of a longer stretch of natural speech is perceived. Importantly, the listeners can't detect that in the new signal two consecutive frames are identical, rather than slowly evolving, which is what we see in real speech. By eliminating a frame we can achieve the converse effect, and again listeners normally do not detect that a frame is missing. A rule of thumb is often quoted, namely that these processes can be used to lengthen or shorten a section of speech by a factor of about two without any or much noticeable degradation. In reality, it is fairer to say that the more modification one performs the more likely it is that the listener will notice.

Pitch-scale modification is performed by recombining the frames on epochs that are set at different distances apart from the analysis epochs, as shown in Figure 14.3. All other things being equal, if we take for example a section of speech with an average pitch of 100 Hz, the epochs will lie 10 ms apart. From these epochs we perform the analysis and separate the speech into the pitch-synchronous frames. We can now create a new set of epochs that are closer together, say 9 ms apart. If we now recombine the frames by the overlap–add method, we find that we have created a signal that now has a pitch of $1.0/0.009 = 111$ Hz. Conversely, if we create a synthetic set of epochs that

14.2 Pitch-synchronous overlap and add (PSOLA)

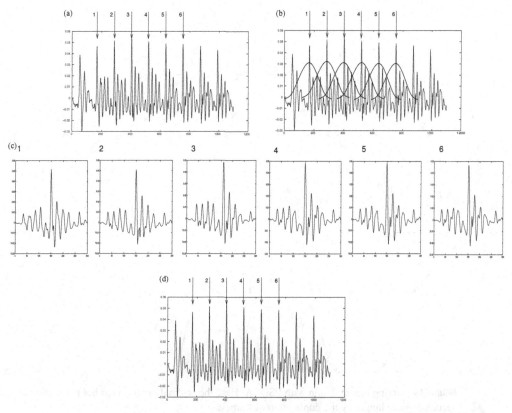

Figure 14.1 Basic operation of the PSOLA algorithm. (a) A section of voiced waveform, with epoch positions shown by the arrows. (b) For every epoch, a frame centred on the epoch is created. Here, a series of hanning windows is shown; note that the windows overlap. (c) The sequence of separate frames created by the hanning-window process. Each is centred on the epoch, which in this case is the point of maximum positive excursion. (d) A waveform resynthesised by overlapping and adding the separate frames, with positions governed by the original epoch positions. This results in a perceptually identical waveform to the original.

are further apart, and overlap and add the frames on those, we find that we generate a synthetic waveform of lower pitch. This lowering process partly explains why we need frames that are twice the local pitch period; this is to ensure that, up to a factor of 0.5, when we move the frames apart we always have some speech to add at the frame edges. Just as with timing modification in general, listeners cannot detect any unnaturalness in slight modifications.

14.2.2 Epoch manipulation

One of the key steps in both TD-PSOLA and FD-PSOLA is proper manipulation of the epochs. First, an epoch detector of the type described in Section 12.7.2 is used to find the instants of glottal closure. This results in the **analysis epoch sequence** $T^a = \langle t_1^a, t_2^a, \ldots, t_M^a \rangle$. From this, the local pitch period at any epoch can be found. A simple means of determining this is just to measure the distance between the previous and next

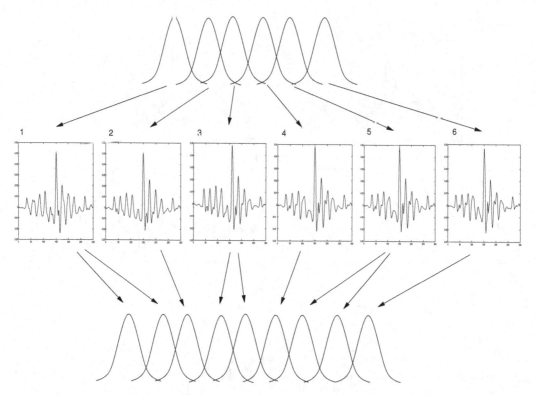

Figure 14.2 Timing manipulation with PSOLA. Here the original pitch is kept but the section of speech is made longer by the duplication of frames.

epochs and divide by two:

$$p_m^a = \frac{t_{m+1}^a - t_{m-1}^a}{2} \qquad (14.1)$$

Alternatively, more-sophisticated means such as taking a running average of the epoch distances can be employed. Given the sequence of epochs and local pitch periods, we can extract a sequence of analysis frames by successively windowing the original signal with a hanning window:

$$x_m^a[n] = w_m[n]x[n] \qquad (14.2)$$

The limits of n in the above are proportional to twice the local analysis pitch period.

Next, a set of synthesis epochs $T^s = \langle t_1^s, t_2^s, \ldots, t_N^s \rangle$ is created from the F0 and timing values given in the specification. If a synthetic F0 contour has already been created, the synthesis epochs can simply be calculated from this by placing t_0^s at 0, reading the first frequency value, taking the reciprocal of this to find the first pitch period, advancing this distance and setting t_1^s at this position. The F0 at this point is then found, the reciprocal taken to find the next pitch period and so on.

A mapping function $M[i]$ is then created. The job of the mapping function is to specify which analysis frames should be used with which synthesis epoch. We should note at this

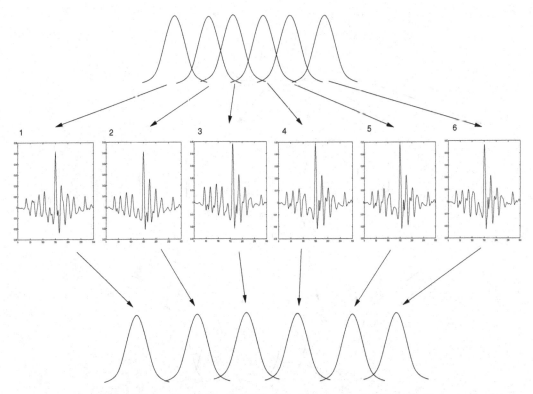

Figure 14.3 Pitch manipulation with PSOLA. A new set of synthesis epochs that are spaced further apart than the analysis set is created. When the frames are overlapped and added, the new pitch is now lower due to the wider frame spacing.

stage that the duration and pitch modifications interact. Figure 14.4 demonstrates this. Taking the case of duration modification first, we see in the example that we have five frames of 100 Hz speech spanning 40 ms, so we can produce a segment with the same pitch (100 Hz) but with a new duration. We do this by creating a sequence of synthesis epochs that have the required pitch but span the new duration. If we wish to increase the duration we add more epochs; if we wish to decrease the duration we use fewer epochs. The mapping function M is then used to specify which analysis frames should be used. In the case of increasing the duration, we may need to create (say) seven new synthesis frames from the five analysis frames. This effect is achieved by duplicating two of the analysis frames. Conversely, if we wish to decrease the duration, we skip frames.

Now consider the case of changing the pitch. If we have a simple example of a segment of speech having a constant pitch of 100 Hz and spanning five frames, we see that the distance between the first epoch t_1^a and the last t_5^a is 40 ms. Imagine now that we wish to change the pitch of this segment to 150 Hz. We can do this by creating a set of synthesis epochs T^s that are $1/150 = 0.0066$ s $= 6.6$ ms apart. If we use a straightforward linear-mapping function M such that $M[i] = i$ for all i, then the analysis frames are simply copied one by one onto the synthesis epochs. The final signal is then created by overlapping and adding of these frames. In doing this we find, however, that our five

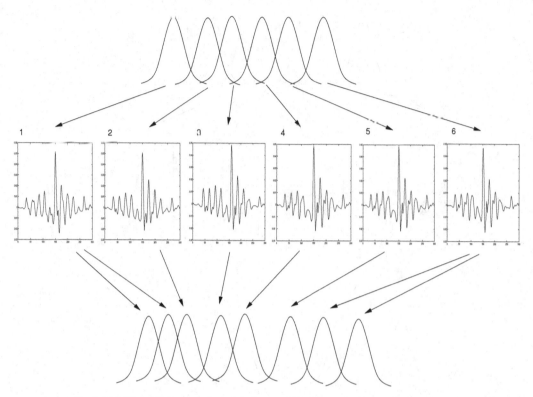

Figure 14.4 Simultaneous pitch and timing modification, whereby a new set of synthesis epochs and a single mapping that performs pitch and timing modification are used.

frames of speech, which originally spanned 40 ms, $(5 - 1) \times 10\,\text{ms} = 40\,\text{ms}$ peak to peak, now span only $(5 - 1) \times 6.6\,\text{ms} = 26\,\text{ms}$ peak to peak: because the frames are closer they take up less overall duration. We can compensate for this by using the frame-duplication/elimination technique used for duration modification; in this example we would duplicate two frames, which would give us a segment of $(7 - 1) \times 6.6\,\text{ms} = 40\,\text{ms}$. The mapping function specifies which frames to duplicate.

Hence duration modification can be performed without reference to pitch values, but pitch modification will necessarily change duration as a by-product of moving the frames closer together or further apart. In practice, duration and pitch modifications are done at once, by a segment-by-segment calculation of the mapping function. This ensures that each segment has the required pitch and duration as specified by the TTP component. Several options are available for the form of the mapping function. The simplest is to have a simple linear function, whereby if we wish to create 15 frames from 10, we duplicate every second frame. Likewise, if we wish to create 5 from 10, we eliminate every second frame. More sophisticated functions are also possible, such that for instance frames are duplicated and eliminated in the middles of phones in an attempt to preserve transition dynamics and have malleable steady-state portions as with formant synthesis.

14.2.3 How does PSOLA work?

We can now ask ourselves how TD-PSOLA works. Or, in other words, after all we have said about explicit source–filter separation, how is it that we have been able to change the characteristics of the source without performing any separation? This is certainly a surprising result – we appear to be able to modify the pitch of a signal simply by rearranging the positions where we place the frames – no local modification of the frame is required at all. While the explanation is now known (and given below), at the time this was certainly a surprising result. TD-PSOLA is massively simpler than the other approaches mentioned here, requires a fraction of the computational cost and often produces the best-sounding speech.[1]

A simple explanation can in fact be provided. In Figure 14.5(a), we have the time-domain signal for a single impulse fed into an IIR filter, which has been given realistic vocal-tract coefficients. We see clearly that the single impulse causes decaying oscillation and produces a characteristic time-domain pattern. Crucially, this pattern is a consequence of the filter only; all the impulse does is provide energy to the system. Now consider Figure 14.5(b), where we have a series of well-separated impulses exciting the same filter. We see the pattern of Figure 14.5(a) repeated; each impulse causes a time-domain pattern whose characteristics are governed by the filter only. In Figure 14.5(c), we see that it is easy to change the pitch of this signal by moving the impulses nearer together, which has no effect on the characteristic ringing of the filter. This shows that provided we can separate the effects of each impulse we can safely change the pitch without changing the filter. In effect TD-PSOLA is attempting to mimic this process on real speech. The fact that the technique is pitch synchronous means that we are in effect isolating individual excitations into the vocal tract. In real speech we have the problem that the ringing of the filter is not well isolated from the direct effects of the excitation. We get around this by use of the hanning windows, which attempt to capture as much of the filter effect from each pitch period without running into the effect of the excitation from the next pitch period. The hanning window does this by tapering near the window edges.

14.3 Residual-excited linear prediction (RELP)

When the TD-PSOLA algorithm is used with speech that has been epoch-marked accurately, the quality of synthesis is extremely high, and so long as the pitch and timing modifications aren't too big (say within 25% of the original) the quality of the speech can be "perfect" in the sense that listeners can't tell that this is not completely natural speech. In terms of speed, it is nearly inconceivable that any algorithm could be faster. Hence for many situations TD-PSOLA is regarded as a complete solution to the issue of pitch and timing modification. The algorithm is not, however, ideal for every situation,

---

[1] While the explanation given here is satisfactory, it is certainly "after the event" – had it been possible to deduce this from basic signal-processing principles the technique would have been discovered at least 20 years earlier.

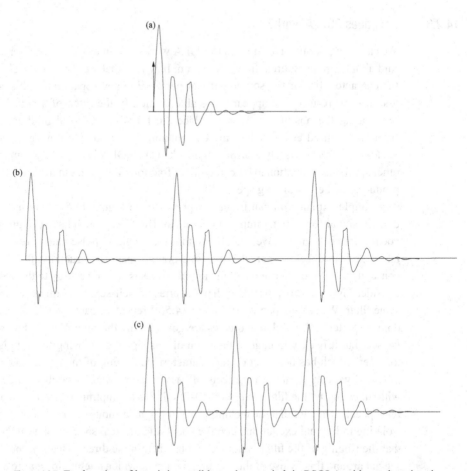

Figure 14.5 Explanation of how it is possible to change pitch in PSOLA without changing the spectral-envelope characteristics. (a) The time-domain pattern of a single impulse fed through an IIR filter. The shape of the waveform is entirely dependent on the filter and not dependent on the impulse. (b) A series of impulse responses, clearly separated in time. As before, the pattern of the response for each period is dependent only on the filter characteristics. The period of this waveform is determined separately, by the distance between the impulses. (c) The same series but the impulse responses are positioned so close together that they overlap. The only difference between the perception of this and that of case (b) is that here the pitch will be higher because the separation is so much less. The spectral envelope will be the same because the impulse response itself is hardly altered.

not because it fails in its goals (it does not) but because we have additional requirements. Firstly, TD-PSOLA can perform *only* pitch and timing modification; no spectral or other modification is possible. Secondly, the storage requirement is that of the original speech database. Whether or not this is a problem depends of course on the size of the database; for a basic diphone synthesiser with 1600 diphones, we might require 2M–5M storage for a sampling rate of 16 000 Hz. TD-PSOLA can also be used with much larger unit-selection databases, whose size may be several hundred megabytes. In such cases, some compression may be used, and, because this is often based on linear prediction (LP), we do in fact end up using LP in any case. We now therefore turn to techniques that use the LP model, but avoid the problems caused by impulse excitation.

To start with, we know that we can find, by inverse filtering or some other method, a residual signal that when fed back into the LP filter produces perfect speech. Our goal is therefore to see what we can learn or incorporate from this residual signal in order to produce the type of source signal we require. Modelling the noise part of the source is not an issue; we can feed the residual back into the filter and the output will sound perfect. If we need to lengthen or shorten the signal we can use the eliminate/duplicate model and the result sounds very good for duration modifications by less than a factor of two. The main challenge lies with voiced speech: while feeding the real residual back into the filter produces perfect speech, it will only of course produce speech with the original F0 and duration. Since we want to produce speech that has the F0 and duration of the specification, we can't simply use the residual as is.

14.3.1 Residual manipulation

Hunt [228] proposed a technique for modifying the original residual so as to produce a synthetic residual of desired F0 and duration. The key to this technique is to find and isolate each pitch period in the residual. Once this has been done, a special asymmetrical window function is applied to each pitch period, which isolates it from the rest of the residual. We can therefore use the elimination/duplication technique to change the timing. The F0 can be increased by moving the pitch periods closer together, and decreased by moving the pitch periods further apart. This rearrangement of the pitch periods is analogous to creating an impulse train from scratch; there we created a high F0 with impulses that were close together and a low F0 with impulses further apart. With the Hunt technique we are performing a similar operation, but adjusting the spacing of the primary impulse plus secondary impulses. Once the synthetic residual has been created, it is fed into the LP filter in exactly the same way as the impulse/noise residual.

The resultant speech is much more natural than speech generated with the impulse/noise model. If the pitch and duration are not modified at all the speech sounds indistinguishable from the original; and if slight (less than 25%) pitch and duration modifications are made hardly any degradation can be heard. The main downside is that the storage requirements are significantly higher because now the residual waveform for each frame must be stored as well as the coefficients. In addition, pitch-synchronous analysis must be performed, which is prone to error if not done properly.

14.3.2 Linear-prediction PSOLA

The above technique bears some similarities to the TD-PSOLA technique in that it uses a pitch-synchronous analysis to isolate individual pitch periods, after which modification and resynthesis are performed. In fact, in a technique called **linear-prediction pitch-synchronous overlap and add** or **LP-PSOLA**, we can use PSOLA more or less directly on the residual rather than on the waveform. As above, epoch detection is used to find the epochs. The residual is then separated into a number of symmetrical frames centred on the epoch. Pitch modification is performed by moving the residual frames closer together or further apart, and duration modification is performed by duplication or elimination of frames, in just the same way as in TD-PSOLA. The only difference is that these operations are performed on the residual, which is then fed into the LP filter to produce speech. This technique differs from the Hunt technique only in terms of the shape of the frames. Both techniques use window functions with their highest point at the epoch: in Hunt's technique the windows are asymmetrical, with the idea that they are capturing a single impulse; in LP-PSOLA the windows are symmetrical. In listening tests, the two techniques produced speech of virtually identical quality.

Other refinements on residual excited LP have been proposed. Hirai et al. [203] proposed a technique fairly similar to Hunt's, which performs a particular type of manipulation on the residual before using it to excite the filter. Authors of a number of studies have attempted to find an LP analysis technique that gives a parameterised source rather than simply a residual [135], [336], [337], [475].

There is no convincing evidence that these techniques produce better-quality modifications than those obtained using TD-PSOLA, but they do have the advantage that they offer a unifed framework for pitch/timing modification and compression. A typical system might be able to encode the LP parameters efficiently as reflection coefficients or line-spectral frequencies. In classicial LP the main compression gain comes from being able to encode the source signal as a single F0 value, which is then used to generate an impulse sequence at synthesis time. In a residual-excited model we cannot use this approach because we have to store the complete residual. It is, however, known that the dynamic range of this is less than for real speech, so savings can be made in this way.

In addition to these advantages we also have the possibility of modifying the spectral characteristics of the units. One use of this would be to ensure that the spectral transitions at diphone joins are completely smooth. While careful design and extraction of the units from the database should help ensure that we obtain smooth joins, they can't always be guaranteed, so some spectral manipulation can be useful.

14.4 Sinusoidal models

Recall that, in our introduction to signal processing, we saw that we could use the Fourier series to generate any periodic signal from a sum of sinusoids (Section 10.1.2):

$$x(t) = \sum_{l=1}^{L} A_l \cos(\omega_0 l + \phi_l) \qquad (14.3)$$

The techniques known as **sinusoidal models** use this as their basic building block and perform speech modification by finding the sinusoidal components for a waveform and carrying out modification by altering the parameters of the above equation, namely the amplitudes, phases and frequencies. This has some advantages over models such as TD-PSOLA in that it allows adjustments in the frequency domain. Frequency-domain adjustments are possible in the LP techniques, but the sinusoidal techniques facilitate this with far fewer assumptions about the nature of the signal and in particular don't assume a source and all-pole filter model.

14.4.1 Pure sinusoidal models

In principle we could perform Fourier analysis to find the model parameters, but, for reasons explained below, it is in fact advantageous to follow a different procedure that is geared towards our synthesis goals. For purposes of modifying pitch, it is useful to perform the analysis in a pitch-synchronous manner and in fact one of the main advantages of sinusoidal modelling is that the accuracy of this does not have to be as high as that for PSOLA [293], [420], [519].

For a frame of speech, basic sinusoidal analysis can be found by a means similar to LP (Section 12.4), whereby we use the model to create an artificial signal $\hat{s}(n)$:

$$\hat{s}(n) = \sum_{l=1}^{L} A_l \cos(\omega_0 l + \phi_l)$$

We then attempt to find the set of values for A_l, ω_0 and ϕ_l and choose the best artificial signal as the one that has the lowest error between it and the original $s(n)$. So in the expression

$$E = \sum_n w(n)^2 (s(n) - \hat{s}(n))^2 \qquad (14.4)$$

$$= \sum_n w(n)^2 \left(s(n) - \sum_{l=1}^{L} A_l \cos(\omega_0 l + \phi_l) \right)^2 \qquad (14.5)$$

we find the signal $\hat{s}(n)$ that minimises E. Unlike the case with LPC, there is no direct solution to this. Rather, we determine the model parameters by a complex linear regression [364], [421]. The reason why we use Equation (14.5) rather than Fourier analysis is that Equation (14.5) uses a window function that concentrates the modelling accuracy in the centre of the window. This has advantageous effects when the synthesis step is performed. Secondly, this analysis can be performed on relatively short frames of speech, rather than the fixed frame lengths required in standard Fourier analysis.

Given the parameters of the model, we can reconstruct a time-domain waveform for each frame by use of the synthesis equation (14.3). Figure 14.6 shows a real and resynthesised frame of speech. An entire waveform can be resynthesised by overlapping and adding the frames just as with the PSOLA method (in fact the use of overlap–add techniques was first developed for conjunction with sinusoidal models).

Modification is performed by separating the harmonics from the spectral envelope, but this is achieved in a way that doesn't perform explicit source–filter separation as with LP analysis. The spectral envelope can be found by a number of numerical techniques. For example, Kain [244] transforms the spectra into a power spectrum and then uses an inverse Fourier transform to find the time-domain autocorrelation function. The LP analysis is performed on this to give an all-pole representation of the spectral envelope.

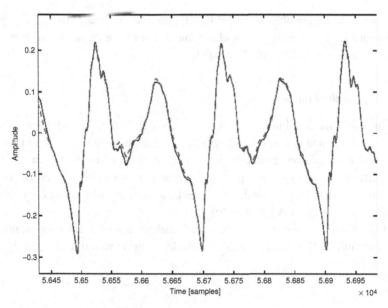

Figure 14.6 The original waveform (solid line) and reconstructed synthetic waveform (dashed line) of a short section of speech. (From Kain [244].)

This has a number of advantages over standard LP analysis in that the power spectrum can be weighted so as to emphasise the perceptually important parts of the spectrum. Other techniques use peak picking in the spectrum to determine the spectral envelope. Once the envelope has been found, the harmonics can be moved in the frequency domain and new amplitudes found from the envelope. The standard synthesis algorithm can be used to generate waveforms from this.

14.4.2 Harmonic/noise models

The technique just described works very well for perfectly periodic signals, but performance can degrade because the signal is rarely perfectly periodic. This is particularly true for signals that have a high sampling rate, for which the inherent roll-off characteristics of the voiced source mean that very little periodic source information is found at high frequencies. To see this, consider Figure 14.7, which shows a magnitude spectrum of a vowel from a signal sampled at 16 000 Hz. While the lower frequencies clearly exhibit the "spikes" that we associate with the harmonics of a periodic signal, at higher frequencies we see that this pattern breaks down, and the signal is significantly noisier. The non-periodic part of the signal arises from many sources, including breath passing through the glottis and turbulence in the vocal tract. It should be realised that normal voiced speech is in fact significantly noisier than we often assume. We can see this by comparing stable sung notes with speech; in the case of singing the noise is often greatly reduced. In fact a well-known exercise in classical singing is to sing a note at full volume in front of a candle flame. The idea is to sing without causing the candle to flicker, which it will do if there is any breath in the singing. The fact that this is incredibly hard to

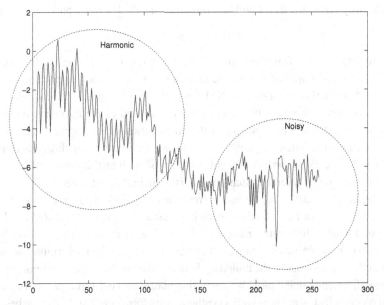

Figure 14.7 Log magnitude DFT at 16 000 Hz sampling rate. We can see that the lower part of the spectrum is more regular and clearly exhibits the spikes spaced at even intervals that we expect from periodic sounds. At higher frequencies the spectrum is considerably less regular as the roll-off from the voice source takes effect and other sound sources such as breath appear.

do demonstrates that there is nearly always a small but significant amount of breath in normal speech (and in rock singing, which is one reason why this sounds different from classical singing).

This observation has led to a number of models that include a stochastic component in addition to the harmonic one, giving

$$\hat{s}(t) = \hat{s}(t)_p + \hat{s}(t)_r \tag{14.6}$$

$$= \sum_{l=1}^{L} A_l \cos(\omega_0 l + \phi_l) + \hat{s}(t)_r \tag{14.7}$$

where the noise component $s(t)_r$ is assumed to be Gaussian noise. Several techniques based on this principle have been proposed, include the **multi-band-excitation (MBE)** model [185], [281] and a parallel hybrid approach [2]. The same idea has been applied to LP residuals in an attempt to separate the periodic and noise components of the source [118], [460], [484].

We will now consider the **harmonic/noise model (HNM)** of Stylianou [421] in detail, because this model was developed specifically for TTS. As the name suggests, the model is composed of a harmonic component (as above) and a noise component. This noise component is more sophisticated than in some models in that it explicitly models the fact that noise in real speech can have very specific temporal patterns. In stops, for example, the noise component is rapidly evolving over time, such that a model that enforces uniformity across the frame will lack important detail. The noise part is therefore given by

$$s(t)_r = e(t)[h(t, \tau) \otimes b(t)] \tag{14.8}$$

where $b(t)$ is white Gaussian noise, $h(t, \tau)$ is a spectral filter applied to the noise and $e(t)$ is a function applied to give the filtered noise the correct temporal pattern.

The first analysis stages in HNM concern classifying the frames into voiced and unvoiced portions, which then dictate the parameters of the harmonic and noise components and the relative strength of the contribution of each component to the frame. First we estimate pitch and with this perform a basic pitch-synchronous analysis. This does not have to be performed with the aim of finding the instant of glottal closure; rather a simple location of pitch periods is sufficient. Next a harmonic model, using a fundamental frequency ω_0 estimated from the pitch tracking, is fitted to each frame, and from this the error between the speech generated with the model and the real waveform is found. Highly harmonic frames will have a low error; noisy frames will have a high error. For those frames which we believe are voiced, we then determine the highest harmonic frequency, that is, the boundary between the low-frequency harmonic part and the higher-frequency noisy part. We carry this out by moving through the frequency range and continually testing how well a synthetic waveform generated from the best-matching model parameters fits the real waveform.

Once this cut-off has been found, we proceed to find a more-accurate estimation of the pitch using only the lower part of the signal. This is now possible since we are not attempting to pitch track noise. The amplitudes and phases are now found from Equation (14.5) as before, by minimising the error between the real and synthetic waveforms. This search can be performed in a number of ways; Stylianou presents a fast technique for minimising Equation (14.5) directly [420]. The noise component essentially has two parts: $h(t)$, which describes the impulse response of the noise (which describes the spectrum of the noise envelope when transformed to the frequency domain); and $e(t)$, which describes the sub-frame time evolution of the noise. For each frame, $h(t)$ is found in a manner similar to LP and $e(t)$ is estimated indirectly by calculating several values of $h(t)$ at shifted positions within the frame.

A final step is required to ensure that phase mismatches don't occur between frames. Since the pitch-synchronous analysis was performed without reference to the instant of glottal closure, they do not necessarily align. A time-domain technique is now used to adjust the relative positions of the waveforms within the frames to ensure that they all align.

At synthesis time, a set of synthesis epochs is generated and a mapping function from the analysis epochs is calculated in exactly the same way as for PSOLA. Timing modification is performed by this mapping function in the same way as for PSOLA. For pitch modification, we have to adjust the harmonics of the frames; this is done by resampling the spectrum of the frame and applying these values to each sinusoidal component. The noise component is created by passing the Gaussian noise $b(t)$ through the filter $h(t)$. This is performed several times per frame to ensure that the temporal characteristics are generated successfully. For voiced frames, the noise is high-pass filtered at the cut-off point of the harmonic component; in this way, no low-frequency noise is generated. The noise is then modulated in the time domain to ensure that it is synchronised with the harmonic component. This stage is essential to ensure that the perception is of a single sound rather that two separate unrelated sounds. Finally, synthesis is performed by overlap and add, as with PSOLA.

14.5 MBROLA

The only real drawback of TD-PSOLA in terms of quality is that it is very sensitive to errors in epoch placements. In fact, it is safe to say that, if the epochs are not marked with extremely high accuracy, then the speech quality from TD-PSOLA systems can sound very poor. The effect of inaccurate epochs is to make the synthetic speech sound **hoarse**, as if the speaker were straining their voice, or had an infection or some other ailment. This is not surprising, since it is known that the effect of hoarseness in natural speech arises because of irregular periodicity in the source.

Dutoit [148] investigated the issue of epoch accuracy and concluded that the problem is complicated, partly due to a lack of formal definition as to where exactly the epoch should lie (the instant of glottal closure is only one possibility). Since then, more-accurate epoch-detection algorithms have been developed (see Section 12.7.2) but there are cases (for example with poorer-quality recordings) for which perfect automatic epoch detection might not be possible.

The MBROLA technique was developed partly as a solution to this problem. It uses a synthesis technique similar to the harmonic/noise model, but uses an additional step to help deal with the problems associated with epochs. The MBROLA method operates pitch-synchronously as before, but unlike TD-PSOLA the exact position of the epoch in the frame need not be consistent from frame to frame. Hence a more-coarse-grained epoch detection suffices. During analysis, the frames are found as before, but during resynthesis the phases are adjusted so that every frame in the database has a matching phase. This step, which is easy to perform with a sinusoidal model, effectively adjusts all the epochs so as to make them lie in the same relative positions within the frames. Minor epoch errors are hence eliminated. Another way of looking at this is to view the analysis algorithm as analysing the speech and resynthesising it at a constant pitch for the whole database. TD-PSOLA can then be performed with complete accuracy.

14.6 Synthesis from cepstral coefficients

We now turn to techniques used to synthesize speech from cepstral representations and in particular the mel-frequency cepstral coefficients (MFCCs) commonly used in ASR systems. Synthesis from these is not actually a common second-generation technique, but it is timely to introduce this technique here because it is effectively performing the same job as pure second-generation techniques. In Chapter 15 we will give a full justification for wanting to synthesise from MFCCs, but the main reason is that they

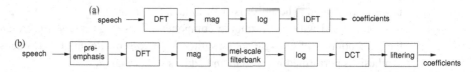

Figure 14.8 Steps in cepstral analysis. (a) Steps involved in standard cepstral analysis. (b) Steps involved in MFCC-style cepstral analysis.

are a representation that is highly amenable to robust statistical analysis because the coefficients are statistically independent of one another.

Figure 14.8(a) shows the operations involved in creating the normal, complex cepstrum. Most of these operations are invertible, though the fact that we use the magnitude spectrum only and discard the phase means that information has been lost at this stage. Figure 14.8(b) shows the steps involved in MFCC analysis and we see that the situation is somewhat more complicated. First, it is common to perform pre-emphasis so as to remove the inherent tilt in the spectrum. Next we perform the filter-bank operation that smoothes the spectrum and performs the mel-scaling. Finally, after the cepstrum has been created, we perform a "liftering" operation whereby we discard the higher cepstral coefficients, to leave typically only 12. The liftering operation and the filter-bank operation are not invertible because information is lost at these points.

Techniques intended to reverse these operations and generate speech have been developed [85], [229], [261], [419]. Here we follow one such technique described by Milner and Shao [313]; described in terms of the following steps.

1. Remove the pre-emphasis and the influence that the mel-scaling operation has on spectral tilt. This can be done by creating cepstral vectors for each process, and then simply subtracting these from the MFCC vector.
2. Perform an inverse-liftering operation by padding the mel-scale cepstrum with zeros. This then gives us a vector of the correct size.
3. Carry out an inverse cosine transform, which gives us a mel-scaled spectrum. This differs from the analysis mel-scale cepstrum because of the liftering, but in fact the differences between the envelopes of the original and reconstructed spectra have been shown to be minor, particularly with respect to the important formant locations.
4. Partially reverse the filter-bank operation. This is not trivial and it is impossible to recover even an approximation of the original spectrum because we threw away all the information about the harmonics in the original filter-bank operation. Instead, we attempt to find the spectral envelope, by a process of "up-sampling" the filter bank to a very-detailed spectrum, and then sampling this at the required intervals, so as to give a spectrum with 128 points.
5. From the magnitude spectrum, we calculate the power spectrum, and from this calculate a set of autocorrelation coefficients by performing an inverse DFT.
6. Calculate the LP coefficients from the autocorrelation coefficients by solving Equation (12.26).
7. Choose a value for F0, and use this to generate an impulse train, which then excites the LP coefficients. Use a noise source for unvoiced components.

This technique successfully reverses the MFCC coding operation. The main weakness is that, because we threw away the harmonic information in the filter-bank step, we have to resort to a classical LP-style technique of using an impulse to drive the LP filter. Improvements to this have been made, with the motivation of generating a more-natural source, while still keeping a model system in which the parameters are largely statistically independent. For example, the technique of Yoshimura *et al.* [509] uses various excitation parameters that allow mixing of noise and impulse, and allow a degree of aperiodicity in the positions of the impulses.

14.7 Concatenation issues

Having described a number of techniques for prosodic modifcation, we turn to the final issue in second-generation synthesis, that of how to join sections of waveform successfully, such that the joins cannot be heard so that the final speech sounds smoothly continuous and not obviously concatenated.

When discussing concatenation, it is useful to make a distinction between **micro-concatenation** and **macro-concatenation**. Micro-concatenation concerns the low-level issues of joining signals. For example, if we join two voiced waveforms such that there is a large jump between samples at the boundary, this may result in a "click" being heard. This is because the effect of the jump is the same as that of a high-frequency impulse, and, even though such an effect may last for only one or two samples, it is easily noticed by the listener. Macro-concatenation is concerned with higher-level, longer-range effects, such as ensuring that the general pattern of spectral evolution across a join is natural. In this section we concentrate on the issues of micro-concatenation, since the issue of macro-concatenation is discussed extensively in Section 16.5.1.

In general, the goal in concatenating speech is to do so in such a way that the listener cannot hear the join at all. The problem of clicks, caused by discontinuities in waveforms, can nearly always be solved by tapering the frames to be overlapped at their edges so that the samples near the end of the frame (and hence the join) are zero or close to zero. We can see that this works from PSOLA, where we successfully "join" frames of speech all the time in the overlapp–add operation. Furthermore, the timing-modification part of PSOLA, which eliminates or duplicates frames, is proof that frames not originally contiguous can successfully be joined. This principle can be extended to cases in which the frames to be joined are from different diphones; by simply overlapping and adding at diphone edges, successful micro-concatenation can be achieved.

One significant problem in many types of synthesis is that there may be **phase mismatches** between units. An example of this is shown in Figure 14.9, where we have a situation in which both units have accurate epochs, resulting in "correct" pitch-synchronous analysis, but the point of reference for the epochs is different for each unit. When concatenated, this will result in a single irregular period at the join, which is usually detectable. Many proposals have been made as to how to fix this problem [149]. Obviously, we can try to ensure consistent epoch detection in the first place. Good epoch detectors are now more common, but this was not always possible in the past. Other solutions include a

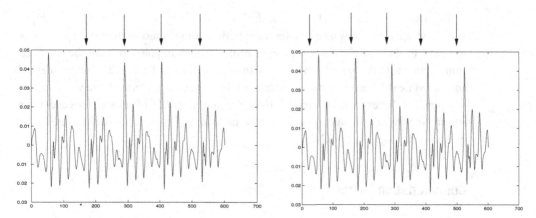

Figure 14.9 Phase problems can be caused by inconsistencies in epoch locations across join boundaries.

cross-correlation technique whereby the frames at the joins are shifted over one another until their cross-correlation is at a minimum [420]. Techniques such as MBROLA were in fact developed to handle such problems; recall that the solution there was effectively to resynthesise the whole database at a fixed pitch to ensure phase consistency. Bellegarda has proposed a technique similar to latent semantic analysis, whereby the DFT decomposition of the signal is avoided altogether, and joins are judged according to a measure of their separation in a transformed space [39].

A further issue in micro-concatenation concerns *where* in the unit we should make the join. So far, in describing our example of diphone synthesis, we have talked about placing the diphone boundary in the "middle" of the phone, without defining exactly what we mean by this. Options include taking the temporal middle (i.e. the half-way point through the phone), but we can also opt for a strategy that selects the acoustically most-stable region, since this is after all our motivation for using diphones in the first place. The best place to join units may of course depend on the particular units themselves and the technique of **optimal coupling** was developed to find a unique boundary location for every pair of diphones to be joined [105], [444]. In this, the diphones are stored as complete phone pairs, extending from the start of the first phone to the end of the second. When two phones are to be joined, the two halves of each diphone are "slid over" each other frame by frame, and the distance between acoustic representations (e.g. cepstra) of the phones is measured for that alignment. The idea is to match the general trajectories of the phones, not just the values at the concatenation point. The alignment with the lowest distance is chosen and the diphones are cut and then joined at these points.

While in second-generation synthesis signal processing is used mainly to modify pitch and timing, it can also be used in concatenation. If we are using a technique that gives us some sort of spectral representation, such as residual-excited LP or sinusoidal modelling, then we can smooth or interpolate the spectral parameters at the join. This is possible only in models with a spectral representation, and is one of the reasons why residual-excited LP and sinusoidal models are chosen over PSOLA.

14.8 Discussion

Second-generation techniques are characterised by a hybrid approach of using data to determine the behaviour of the verbal or phonetic part of synthesis and an explicit model plus signal processing to generate the correct prosody.

It should be clear that, apart from the cepstral technique, there is a considerable degree of similarity among the main signal-processing techniques described here. All operate pitch-synchronously and all avoid an explicit model of the source. It is in their detail that we can see differences. For example, the PSOLA technique clearly avoids a modelling approach altogether, and assumes only that the signal is composed of pitch periods. The sinusoidal models also make assumptions, but they are little more than general ones that are true of all periodic or quasi-periodic signals. While the LP techniques make the strongest assumptions, many of the negative consequences of these are lessened by the use of whole natural residuals. While all the techniques are pitch-synchronous, again we see subtle differences, such that the PSOLA techniques in particular are very sensitive to epoch-location errors, whereas the sinusoidal and MBROLA techniques are more robust with respect to these because they can control the phase details during synthesis. In general, the synthesis quality of all these techniques is very similar, such that the choice of one technique over another is usually governed by other factors, such as speed-versus-size tradeoffs and other practical factors.

14.8.1 Further reading

Several books cover the area of speech modification quite thoroughly. Kleijn and Paliwal [256] give a good introduction to many aspects of speech analysis, coding, modification and synthesis. Dutoit [148] stands out as being one of the few books to tackle the signal-processing issue from a TTS-only perspective, describing all the algorithms mentioned here in considerable detail. Many more-general-purpose books exist, which cover a number of topics in analysis, synthesis, coding and modifcation, including Quatieri [364], Rabiner and Schaffer [367] and Huang *et al.* [224].

14.8.2 Summary

- Second-generation synthesis systems are characterised by using a data-driven approach to generating the verbal content of the signal.
- This takes the form of a set of units such that we have one unit for each unique type. Diphones are the most-popular type of unit.
- The synthesis specification generated by the TTP system is in the form of a list of items, each with a verbal specification, one or more pitch values and a duration.
- The prosodic content is generated by explicit algorithms, and signal-processing techniques are used to modify the pitch and timing of the diphones to match that of the specification.

- Techniques to modify the pitch and timing have been developed. They include the following.

  PSOLA, which operates in the time domain. It separates the original speech into frames pitch-synchronously and performs modification by overlapping and adding these frames onto a new set of epochs, created to match the synthesis specification.

  Residual-excited linear prediction performs LP analysis, but uses the whole residual in resynthesis rather than an impulse. The residual is modified in a manner very similar to that of PSOLA.

  Sinusoidal models use a harmonic model and decompose each frame into a set of harmonics of an estimated fundamental frequency. The model parameters are the amplitudes and phases of the harmonics. With these, the value of the fundamental can be changed while keeping the same basic spectral envelope.

  Harmonic/noise models are similar to sinusoidal models, except that they have an additional noise component, which allows accurate modelling of noisy high-frequency portions of voiced speech and all parts of unvoiced speech.

  MBROLA is a PSOLA-like technique that uses sinusoidal modelling to decompose each frame and from this resynthesise the database at a constant pitch and phase, thus alleviating many problems arising from inaccurate epoch detection.

  MFCC synthesis is a technique that attempts to synthesise from a representation that we use because of its statistical modelling properties. A completely accurate synthesis from this is not possible, but it is possible to perform fairly accurate vocal-tract filter reconstruction. Basic techniques use an impulse/noise-excitation method, whereas more-advanced techniques attempt a complex parameterisation of the source.

- The quality of these techniques is considerably higher than that of classical, impulse-excited LP.

- All these methods produce results of roughly similar quality, meaning that the choice of which technique to use is mostly made on the basis of other criteria, such as speed and storage requirements.

15 Hidden-Markov-model synthesis

We saw in Chapter 13 that, despite the approximations in all the vocal-tract models concerned, the limiting factor in generating high-quality speech is not so much in converting the parameters into speech, but in knowing which parameters to use for a given synthesis specification. Determining these by hand-written rules can produce fairly intelligible speech, but the inherent complexities of speech seem to place an upper limit on the quality that can be achieved in this way. The various second-generation synthesis techniques explained in Chapter 14 solve the problem by simply measuring the values from real speech waveforms. Although this is successful to a certain extent, it is not a perfect solution. As we will see in Chapter 16, we can never collect enough data to cover all the effects we wish to synthesize, and often the coverage we have in the database is very uneven. Furthermore, the concatenative approach always limits us to recreating what we have recorded; in a sense all we are doing is reordering the original data.

An alternative is to use statistical, machine-learning techniques to infer the specification-to-parameter mapping from data. While this and the concatenative approach can both be described as **data-driven**, in the concatenative approach we are effectively *memorising* the data, whereas in the statistical approach we are attempting to *learn* the general properties of the data. Two advantages that arise from statistical models are that firstly we require orders of magnitude less memory to store the parameters of the model than to memorise the data, and secondly we can modify the model in various ways, for example to convert the original voice into a different voice.

While many possible approaches to statistical synthesis are possible, most work has focused on using **hidden Markov models (HMMs)**. These and the unit-selection techniques of the next chapter are termed **third-generation** techniques. This chapter gives a full introduction to these approaches and explains how they can be used in synthesis. In addition we also show how they can be used to label speech databases automatically, which finds use in many areas of speech technology, including unit-selection synthesis. Finally, we introduce some other statistical synthesis techniques.

15.1 The HMM formalism

HMMs themselves are quite general models and, although they were originally developed for speech recognition [36], they have been used for many tasks in speech and language technology, and are now in fact one of the fundamental techniques in biology (to such an extent that their use there now completely dwarfs speech-technology applications). General introductions to HMMs can be found in [224], [243], but here we focus only on the type of HMMs most relevant to modern speech recognition and synthesis. In doing so, we closely follow the HTK speech-recognition toolkit of Young et al. [510]. This is the most widely used HMM system in research labs today, and also forms the basis of some of the most-successful ASR engines of the past few years [170], [171], [192], [407]. An introduction/refresher course in probability can be found in Appendix A.

15.1.1 Observation probabilities

The input to an ASR system is a sequence of frames of speech, known as **observations** and denoted as

$$O = \langle o_1, o_2, \ldots, o_T \rangle$$

The frames are processed so as to remove phase and source information. The most common representation is based on mel-scale cepstral coefficients (MFCCs), which were introduced in Section 12.5.7. Hence each observation o_j is a vector of continuous values. For each phone we build a **probabilistic model**, which tells us the probability of observing a particular acoustic input.

So far in the book, all the probabilities we have considered have been of **discrete** events, for example, the probability that we will observe a word w_i in a particular context. In all the cases we have so far considered, this effectively means keeping a unique record of each event (e.g. the probability of observing word w_i in the context of word w_j); no "general" properties of the process have been modelled. We often find, however, that the probabilities with which events occur are not arbitrary; rather, we find that the probabilities with which events occur can be described by a parameterised function. The advantage of this is that we can effectively summarise the probabilistic behaviour of an event with a small number of parameters. For continuous variables, such functions are called **probability density functions (pdfs)**. The integral, or total area under the curve, always sums to exactly 1, in just the same way as the probabilities for all possible events in a discrete system sum to 1.

One important function is the **Gaussian distribution**, often called the **normal distribution** or **bell curve**, which is defined by two parameters. The **mean**, μ, describes its "average" value. The **variance** denoted σ^2 describes whether the distribution is narrow or dispersed. The square root of the variance is called the **standard deviation** and is denoted σ. The Gaussian distribution is shown in Figure 15.1 and defined by the equation

$$\mathcal{N}(o; \mu, \sigma) = \frac{1}{\sigma\sqrt{2\pi}} e^{\frac{-(o-\mu)^2}{2\sigma^2}} \quad (15.1)$$

The Gaussian has many interesting mathematical properties and because of these and the fact that a large range of natural phenomena seem to belong to this distribution (e.g. the range of peoples' heights in a population), it is often seen as the most "natural" or elementary probability distribution.

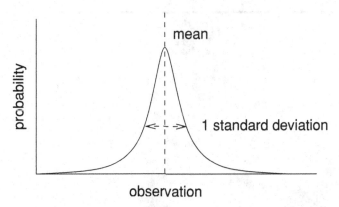

Figure 15.1 The Gaussian function.

When dealing with vector data, as in our case where the observations are acoustic frames, we need to use a **multivariate Gaussian**. This is the natural extension of the univariate one, but it is important to note that, while we have one mean value for each component in the vector, we have a **covariance matrix**, Σ, not a variance vector. This is because we wish to model not only the variance of each component, but also the *co*variance of components. The pdf of an N-dimensional Gaussian is given by

$$\mathcal{N}(o; \mu, \Sigma) = \frac{1}{\sqrt{(2\pi)^N |\Sigma|}} e^{-\frac{1}{2}(o-\mu)'\Sigma^{-1}(o-\mu)} \qquad (15.2)$$

where N is the dimensionality, μ is the vector of means and Σ is the covariance matrix.

With this, we can therefore build a system in which we have a model for every phone, each described by its own multivariate Gaussian. For an unknown utterance, if we know the phone boundaries, we can therefore test each phone model in turn and find which model gives the highest probability to the observed frames of speech, and from this find the sequence of phones that is most likely to have given rise to the observations in the utterance in question. A diagram of this is shown in Figure 15.2.

We can improve on the accuracy of these models in a number of ways. First, we note that the true density of a phone model is in fact rarely Gaussian. Rather than use other types of distribution, we adopt a general solution whereby we use a **mixture of Gaussians**, as shown in Figure 15.3 and given by

$$b(o_t) = \sum_{m=1}^{M} c_m \mathcal{N}(o_t; \mu_m, \Sigma_m) \qquad (15.3)$$

In this, we have M Gaussians, each with a mean vector μ_m and covariance matrix Σ_m. The parameters c_m are known as the **mixture weights** and are used to determine the relative importance of each Gaussian. As with any pdf, the area under the curve should sum to 1. We can draw a comparison between mixtures of Gaussians here and the Fourier series of Section 10.1.2. Here we can model a pdf of arbitrary complexity by summing weighted Gaussians; there we could build a periodic signal of arbitrary complexity by summing weighted sinusoids.

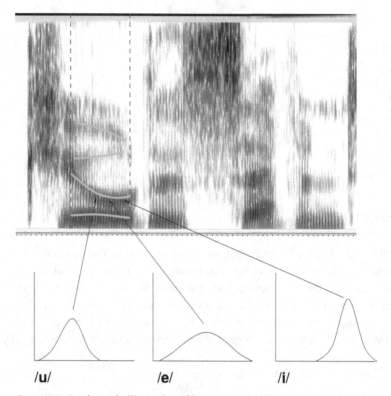

Figure 15.2 A schematic illustration of how we can build a Gaussian model for a phone. The figure shows three phones and a Gaussian showing the probability distribution for the second formant. We can use this probability distribution to tell us which phone was most likely to have generated the data.

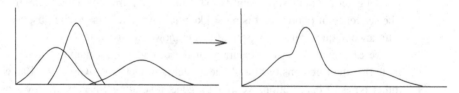

Figure 15.3 Mixtures of Gaussians. This shows how three weighted Gaussians, each with its own mean and variance, can be "mixed" (added) to form a non-Gaussian distribution. Note that the total probability, given by the area under the curve, must equal 1 and so the mixture weights are set to ensure this.

15.1.2 Delta coefficients

A second common addition to the observation function arises from our knowledge that the frames within a phone are not static; rather they evolve as a function of time. For this reason it is standard to include in the observations extra coefficients that describe not only the data, but also the rate of change of the coefficients and the rate of change of the rate of change. These extra coefficients are termed the **velocity** or **delta** coefficients and the **acceleration** or **delta delta** coefficients, respectively. As we shall see in Section 15.2, these coefficients not only encode this rate-of-change information, but also in quite a powerful way make up for some of the weakness in the modelling power of the HMMs.

A simple rate-of-change calculation is given by

$$d_t = \frac{c_{t+1} - c_{t-1}}{2}$$

where d_t is the delta coefficient and c_t is the original cepstral coefficient. Such calculations can, however, be quite unstable since a little noise in either of the coefficients can give rise to a large variation in d_t. A more-robust solution then is to calculate the rate of change over several frames. One way of doing this is with

$$d_t = \frac{\sum_{l=1}^{L} l(c_{t+l} - c_{t-l})}{2 \sum_{l=1}^{L} l^2} \tag{15.4}$$

where L can be thought of as the length of window over which the rate-of-change calculation is performed. This can also be written as

$$d_t = \sum_{l=-L}^{L} w_l(c_{t+l}) \tag{15.5}$$

where the weights w can be calculated from Equation (15.4).

15.1.3 Acoustic representations and covariance

A final point regarding the probabilistic modelling of observations concerns issues with the covariance matrix. If we use, say, 13 acoustic coefficients (often a standard number made from 12 cepstral coefficients and an energy coefficient), and then use delta and acceleration coefficients, we have a total of 39 coefficients in each observation frame and so the mean vector has 39 values. The covariance matrix however has $39^2 = 1521$ values. It is often difficult to find enough data to determine each of these values accurately, so a common solution is to ignore the covariance between coefficients, and simply model the variance of each coefficient alone. This is termed using a **diagonal covariance**, because only the diagonal values in the covariance matrix are calculated; the other values are set to zero. In general, using such a covariance matrix is not advised because the correlations (covariances) between coefficients often contain useful information. However, if we can show that the coefficients in the data vary more or less independently of one another, no modelling power is lost in this assumption.

It is partly for this reason that we find that mel-frequency cepstral coefficients (MFCCs) are used as the representation of choice in many ASR systems (in addition, they are deemed to have good discrimination properties and are somewhat insensitive to differences between speakers). It is important to note, however, that HMMs themselves are neutral with respect to the type of observation used, and in principle we could use any of the signal-processing derived representations of Chapter 12. Another way of producing data with independent coefficients is to perform a projection of the original data into a new space. Principal-component analysis (PCA) does this, as do a number of more-sophisticated techniques. These not only ensure that the coefficients in the new space are independent, but also can be used to reduced the dimensionality of the data being modelled.

15.1.4 States and transitions

As we noted above, the pattern of frames within a phone is not static. In addition to modelling the rate of change, it is also normal to split each phone model into a number of **states**, each of which represents a different part of the phone. In principle any number of states is allowable, but it is quite common to use three, which can informally be thought of as modelling the beginning, middle and end of each phone. We use **transition probabilities**, which give us the probability of moving from one state to the next (this "moving" process is most easily visualised if we think of the models as the generators of the acoustic data). In general, we can move from any state to any other state, so, for a model with P states, the transition probabilities can be stored in a $P \times P$ matrix. This matrix stores a set of discrete probabilities a_{ij}, which give the probability of moving from state i to state j.

One common HMM topology is to use three states, each with its own observation probability. Each state is linked to the next state and back to itself again. The last-mentioned transition is known as the **self-transition** probability and is basically the probability that we will generate the next observation from the state we are already in. A phone's state-transition probabilities govern the durational characteristics of the phone; if the self-transition probabilities are high, it is more likely that more observations will be generated by that phone, which means that the overall phone length will be longer. Such a model is shown in Figure 15.4.

15.1.5 Recognising with HMMs

Each HMM state has a pdf, so for any single observation o_t we can find the state whose pdf gives the highest probability for this. Note that, because of the way HMM recognisers are configured, we don't in fact have an explicit function that gives us the most-probable model given the data. Rather, we have a set of models, each of which has a pdf. Hence the way recognition works is that we assume that one of the models **generated** the data that we are observing,[1] and that the recognition job is that of finding out which model. It does this by calculating the probability that each model generated the observation, and picks the one with the highest probability as being correct. This is why HMMs are called **generative models**.

---

[1] Of course it didn't really; the speaker generated the data, but the idea is that we pretend that a speaker has an HMM generator inside his/her head and it is this process we are trying to model.

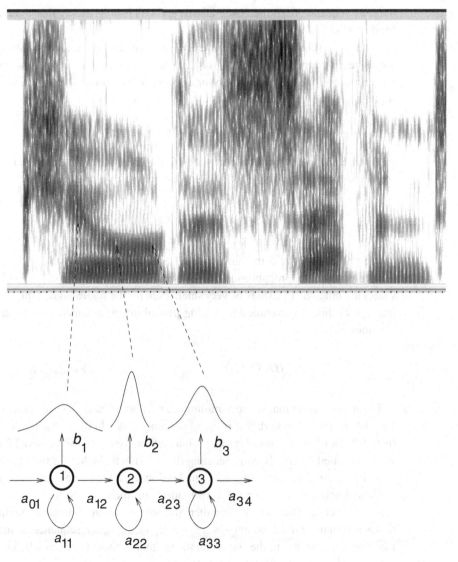

Figure 15.4 A schematic illustration of HMM with three states, each with a separate Gaussian, denoted b_1, b_2 and b_3. The transition probabilities are given by a_{ij}, where i is the start state and j is the destination state. If $i = j$ then this is the self-transition probability, that is, the probability that the system will stay in the same state. The transition probabilities exiting a state always sums to 1.

If we had only one observation, it would be easy to find the state which gives the highest probability. Instead, of course, we have a sequence of observations, which we assume has been generated by moving through a sequence of states. In principle *any one* of the possible state sequences could have generated these observations; it's just that some are more likely than others. Because of this, we cannot deterministically find the state sequence from the observations, and this is why we say that these are *hidden* Markov models.

For a state sequence Q and model M, we can find the total probability that this sequence generated the observations by calculating the probabilities of the observations and transitions of that sequence, so for a sequence that moves through states 1, 2 and 3, for example, we would have

$$P(O, Q|M) = a_{12}b_1(o_1)a_{23}b_2(o_2)a_{34}b_3(o_3)$$

In general, then, for a sequence of states $Q = \langle q_1, q_2, \ldots, q_t \rangle$, the probability for that sequence is given by

$$P(O, Q|M) = a_{q(0)q(1)} \prod_{t=1}^{T} b_{q(t)}(o_t) a_{q(t)q(t+1)} \qquad (15.6)$$

where $q(0)$ is a special entry state and $q(T+1)$ is a special exit state.

By definition, all probabilities are <1, so it should be clear that the final calculated values in Equation (15.6) may be very small, even for the single-highest-probability case. Because of this, it is common to use **log probabilities**, in which case Equation (15.6) becomes

$$\log(P(O, Q|M)) = a_{q(0)q(1)} + \sum_{t=1}^{T} b_{q(t)}(o_t) + a_{q(t)q(t+1)} \qquad (15.7)$$

To use these equations for recognition, we need to connect state sequences with what we wish eventually to find, that is, word sequences. We do this by using the lexicon, so that, if the word HELLO has a lexicon pronunciation /h eh l ou/, then a model for the whole word is created by simply concatenating the individual HMMs for the phones /h/, /eh/, /l/ and /ou/. Since the phone model is made of states, a sequence of concatenated phone models simply generates a new word model with more states: there is no qualitative difference between the two. We can then also join words by concatenation; the result of this is a sentence model, which again is simply made from a sequence of states. Hence the Markov properties of the states and the language model (explained below) provide a nice way of moving from states to sentences.

Given the input observation sequence $O = \langle o_1, o_2, \ldots, o_T \rangle$ and a word sequence $W = \langle w_1, w_2, \ldots, w_N \rangle$, we can use Equation (15.7) to find the probability that this word sequence generated those observations. The goal of the recogniser is to examine all possible word sequences and find the one $\hat{W}$ which has the highest probability according to our model:

$$\hat{W} = \underset{w}{\operatorname{argmax}} \left\{ P(W|O) \right\}$$

HMMs are generators, configured so that we have a model for each linguistic unit that generates sequences of frames. Hence the HMM for a single word w_i, which has been concatenated from single-phone models, is of the form

$$P(O|w_i)$$

So the HMM itself gives us the probability of observing a sequence of frames given the word, i.e. not the probability we require, which is that of observing the word given the frames. We convert one to the other using Bayes' rule (see Appendix A) so that

$$P(w_i|O) = \frac{P(w_i)P(O|w_i)}{P(O)}$$

and, if we consider this for the entire sentence, we have

$$P(W|O) = \frac{P(W)P(O|W)}{P(O)} \qquad (15.8)$$

$P(W)$ is called the **prior** and represents the probability that this particular sequence of words occurs independently of the data (after all, some words are simply more common than others). $P(O|W)$ is called the **likelihood** and is the probability given by our HMMs. $P(O)$ is the **evidence** and is the probability that this sequence of frames will be observed independently of everything else. Since this is fixed for each utterance we are recognising, it is common to ignore this term.

Our recognition algorithm then becomes

$$\hat{W} = \underset{w}{\mathrm{argmax}} \left\{ P(W)P(O|W) \right\}$$

15.1.6 Language models

In Equation (15.8) the term $P(W)$ gives us the prior probability that the sequence of words $W = \langle w_1, w_2, \ldots, w_N \rangle$ will occur. Unlike in the case of acoustic observations, we know of no natural distribution that models this. Partly this is due to the fact that the number of possible sentences (i.e. unique combinations of words) is extremely large. We therefore model sentence probabilities by a counting technique as follows.

The probability of a sentence $W = \langle w_1, w_2, w_3, \ldots, w_M \rangle$ is given by

$$P(W) = P(w_1, w_2, w_3, \ldots, w_M)$$

The chain rule in probability (see Appendix A) can be used to break this into a number of terms:

$$P(w_1, w_2, w_3, \ldots, w_M)$$
$$= P(w_1)P(w_2|w_1)P(w_3|w_1, w_2)\ldots P(w_M|w_1, \ldots, w_{M-1}) \qquad (15.9)$$
$$= \prod_{i=1}^{M} P(w_i|w_1, \ldots, w_{i-1}) \qquad (15.10)$$

The estimation of this becomes tractable if we shorten the number of words such that we approximate the term $P(w_i|w_1, \ldots, w_{i-1})$ as

$$\hat{P}(w_i|w_1, \ldots, w_{i-1}) \approx P(w_i|w_{i-n+1}, \ldots, w_{i-1}) \tag{15.11}$$

That is, we estimate the probability of seeing word w_i on a fixed-window **history** of the N previous words. This is known as an **N-gram** language model. Its basic probabilities are estimated from counting occurrences of the sequences in a corpus.

Many sequences are not observed, and in such cases we do not assume that these could *never* occur, but rather that they are missing from the training data because of problems to do with sparsity of data. To counter this, a number of **smoothing** and **back-off** techniques are employed, which ensure that some extra probability is given to rare and unseen sequences. It is worth noting that these techniques provide the answer to Chomsky's false assertion that, because most sentences will never have been observed by a person, this implies that human language cannot have a statistical basis [89]. In his argument (often called "poverty of the stimulus" rather than data sparsity), all unseen events would have a zero probability, meaning that a listener would be baffled on hearing a new sentence. No serious probabilistic model would adopt such a notion, and the ability to infer the probabilities of unseen events from the patterns of seen events is the basis of much of the entire field of probabilistic modelling.

15.1.7 The Viterbi algorithm

The key to recognition is to find the single path with the highest probability. To be sure of finding the single best (i.e. highest-probability) path, we have to check every sequence of states. The number of possible sequences is very large indeed, in fact far too large to calculate no matter how much computer power we have available. Instead we make use of the principle of dynamic programming and use the **Viterbi algorithm**, which allows us to find the best path in linear time [476].

The search can be visualised as a **trellis** of size N by T, where N is the number of states and T is the number of observations. The trellis is composed of T **time slots**, shown arranged horizontally in Figure 15.5. In each time slot, we have a set of N **nodes**, one for each state, and between a time slot t and the next, $t+1$, there exists a set of transitions. For simplicity we will assume that the trellis is fully connected, that is, all nodes in time slot t are connected to all nodes in time slot $t+1$. The search is performed by "moving" through the trellis, calculating probabilities as we go.

Consider the probabilities associated with a single node u_k at time slot $t+1$, and the nodes in the previous time slot t, as shown in Figure 15.6. For a given node u_i at time slot t, the log probability of *moving* from u_i to u_k is simply

$$\log \text{probability} = a_{ik} + b_i(o_t) \tag{15.12}$$

that is, the log probability of observation of the node u_i generating observation o_t plus the log probability of transition between the two nodes u_i and u_k. If we accumulate these

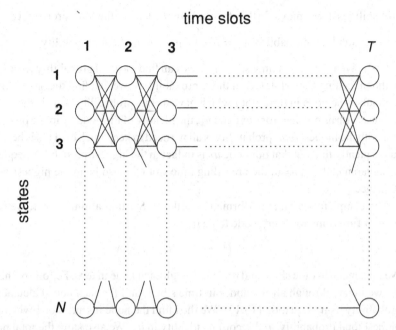

Figure 15.5 The Viterbi search shown as a trellis. The states are arranged vertically as nodes, which are duplicated once for each time slot. The connections between the nodes indicate transitions.

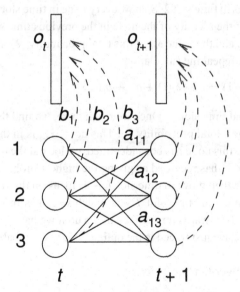

Figure 15.6 Calculation of probabilities for movement between time slot t and time slot $t+1$. The transition probabilities a_{12} etc. connect the nodes and show the probability of moving from one node to another. The observation probabilities b_1 etc. give the probability of that state generating the observation.

probabilities as we move further forwards in the lattice, the *total* probability is given by

$$\text{total log probability}_{t+1} = a_{ik} + b_i(o_t) + \text{total log probability}_t \tag{15.13}$$

For a given node u_k in time slot $t+1$, we can find the total probability for all the nodes in the preceding time slot. Given those probabilities, we can find the single highest and remember the node in time slot t which gave us this probability. This is the crucial step, because it turns out that, for our node u_k, the identity of the node $u_{\hat{x}}$ in the preceding time slot with the highest *total* probability is all we need to keep track of. This is because, if we subsequently find out that our node u_k is in fact in the highest-probability sequence, then it is guaranteed that $u_{\hat{x}}$ in the preceding time slot will also be in the highest-probability sequence.

The complete search is performed as follows. Starting at time slot 1, we initialise a set of δ **functions** for every node to be 0:

$$\delta_i(1) = 0 \qquad i \leq i \leq N \tag{15.14}$$

We then move to time slot 2 and iterate through each node in turn. For one of these nodes, u_k, we iterate through all the nodes in time slot 1, and for each node calculate the total probability using Equation (15.13). We then find the node in time slot 1 which gives the highest total probability, and record its identity in u_k. We also store the total probability associated with that node:

$$\delta_k(2) = \max_i \delta_i(1) + a_{ij} + b_i(o_1) \tag{15.15}$$

This is repeated this for all nodes in time slot 2. So now every node in time slot 2 has two search values associated with it: the identity of the node in the previous time slot which has the highest total probability, and the value δ_{kt} of this total probability. We then move forwards through the trellis and repeat the calculation:

$$\delta_k(t+1) = \max_i \delta_i(t) + a_{ij} + b_i(o_t) \tag{15.16}$$

The search terminates at the final time slot T. Once we reach there, we find the node in the final time slot with the highest total probability δ_T. This node, $\hat{x}_T$, is in the optimal (highest-probability) path. Once this node has been identified, we look at the node in the previous time slot $T-1$ which $\hat{x}_T$ has recorded as giving the highest-probability path. We shall call this node $\hat{x}_{T-1}$. We then move to time slot $T-1$, focus on the single node $\hat{x}_{T-1}$ and find the node in the previous time slot $T-2$ which $\hat{x}_{T-1}$ has recorded. We repeat this procedure of moving backwards through the trellis until we have reached the initial time slot. The path $\hat{X}$ we have just traced is the optimal, highest-probability path. This is shown in Figure 15.7.

The key ideas to the Viterbi algorithm are as follows.

1. For a single node x_k at any time slot, we need record only the best path leading to this node. If it turns out that this particular node x_k is in fact on the global best path, then the node in the preceding time slot that it has recorded is also on the best path.
2. We must reach the end of the time slots before finding the best path. At any given time slot t when moving forwards through the trellis, it is possible to find the node

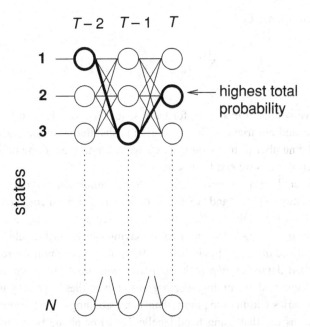

Figure 15.7 Back-tracing. Once we have calculated all the probabilities, we find the node in the last time slot with the highest total probability, shown as the thick circle. From this, we find the node in the previous time slot that was recorded as having the highest probability leading to the current slot (shown by the thick line). We then move to that slot and repeat the procedure.

x_{kt} with the highest total probability up to this point. By back-tracing we can then find the best path nodes in all the preceding time slots. But, no matter how high the probability associated with x_{kt}, there is no guarantee that this will end up on the best path when we do the full back-tracing from time slot T. Imagine for instance that there are simply no nodes in time slot $t+1$ that have high transition probabilities to node x_{kt}.

3. The search works in linear time. If the time taken to calculate a single transition probability is O_a and the time taken to calculate a single observation probability is O_b, then, for a database with N nodes, the total time for each time slot is $O_a(N^2) + O_b(N)$. For T time slots this is then $T[O_a(N^2) + O_b(N)]$.

15.1.8 Training HMMS

If the alignment of states to observations is known, then it is a simple matter to use the data to estimate the means and covariances directly. Assuming that we have a set of frames $O_j = o_{j1}, o_{j2}, \ldots, o_{jT}$ that are aligned with state j, we find the mean by calculating the simple average

$$\hat{\mu}_j = \frac{1}{T_j} \sum_{t=1}^{T_j} o_{jt} \tag{15.17}$$

and calculate the covariance by

$$\hat{\Sigma}_j = \frac{1}{T_j} \sum_{t=1}^{T_j} (o_{jt} - \mu_j)(o_{jt} - \mu_j)' \qquad (15.18)$$

So long as we have sufficient examples for each state, this provides a robust way to estimate the means and covariances. The transition probabilities can be calculated by simply counting the number of times for a state i we move to state j and dividing this by the total number of times we exit from state i.

In general, however, this state-to-observation alignment information is not given to us, so we can't use Equations (15.17) and (15.18) on their own. The most common situation in training is to have a list of the correct words for the sentence, $w_1, w_2, \ldots, w_M$, and the observations, $o_1, o_2, \ldots, o_T$ but no information saying which word should be aligned with which observation. In the past (say before 1990), it was common for researchers to used hand-labelled databases in which labellers would mark the word and phone boundaries by listening and examining spectrograms. Given these, the only unknowns were the state boundaries within each phone. This hand labelling is a laborious process, but thankfully it turns out that using hand-labelled word or phone boundaries is not in fact necessary, since we can instead use the HMMs themselves to give the state sequence.

We do this iteratively, so that if we have a set of models M_1 we can use these, and run the recogniser in **forced-alignment** mode, whereby instead of "recognising" the words in the utterance, we instead give it only the correct words in the correct sequence to choose from. The recognition will obviously be correct, but, since the Viterbi algorithm produces the state alignment as a by-product, we now have a state–observation alignment sequence. This can then be used to train a new set of models M_2 and this process can be repeated until the final set of models has been obtained. This technique, called **Viterbi training**, is a very quick and simple way of of training HMMs.

Viterbi training does, however, have a serious drawback in that it enforces hard decisions during training. If for instance our alignment picks completely wrong boundaries, it will then train on these, and may pick the same wrong boundaries next time; in effect the iteration gets in a rut from which it is hard to recover.

An alternative is to use the **Baum–Welch** algorithm.[2] Instead of using a single state sequence as in the Viterbi case, the Baum–Welch algorithm considers all possible alignments of states and observations, not simply the best one as is the case with the Viterbi algorithm. In these alignments, each state will be aligned with each observation, but in doing this we also make use of the probability that that state generated the observation. In other words, instead of assigning each observation vector to a specific state as in the above approximation, each observation is assigned to every state in proportion to the probability of that state generating the observation. Thus, if $L_j(t)$ denotes the probability of being in state j at time t, then Equations (15.17) and (15.18) can be modified to give

---

[2] The following explanation is taken from the HTK book [510].

the following weighted averages:

$$\hat{\mu}_j = \frac{\sum_{t=1}^{T} L_j(t) o_t}{\sum_{t=1}^{T} L_j(t)} \qquad (15.19)$$

and

$$\hat{\Sigma}_j = \frac{\sum_{t=1}^{T} L_j(t)(o_t - \mu_j)(o_t - \mu_j)'}{\sum_{t=1}^{T} L_j(t)} \qquad (15.20)$$

where the summations in the denominators are included to give the required normalisation.

Equations (15.19) and (15.20) are the Baum–Welch re-estimation formulas for the means and covariances of an HMM. A similar but slightly more-complex formula can be derived for the transition probabilities. Of course, to apply Equations (15.19) and (15.20), the probability of state occupation $L_j(t)$ must be calculated. This is done efficiently using the **forward–backward** algorithm. Let the **forward probability** $\alpha_j(t)$ for some model M with N states be defined as

$$\alpha_j(t) = P(o_1, \ldots, o_t, x(t) = j | M).$$

That is, $\alpha_j(t)$ is the joint probability of observing the first t speech vectors and being in state j at time t. This forward probability can be efficiently calculated by the following recursion:

$$\alpha_j(t) = \left[\sum_{i=2}^{N-1} \alpha_i(t-1) a_{ij} \right] b_j(o_t).$$

This recursion depends on the fact that the probability of being in state j at time t and seeing observation o_t can be deduced by summing the forward probabilities for all possible predecessor states i weighted by the transition probability a_{ij}. The slightly odd limits are caused by the fact that states 1 and N are non-emitting. The initial conditions for the above recursion are

$$\alpha_1(1) = 1$$
$$\alpha_j(1) = a_{1j} b_j(o_1)$$

for $1 < j < N$ and the final condition is given by

$$\alpha_N(T) = \sum_{i=2}^{N-1} \alpha_i(T) a_{iN}.$$

Notice here that, from the definition of $\alpha_j(t)$,

$$P(O|M) = \alpha_N(T).$$

Hence, the calculation of the forward probability also yields the total likelihood $P(O|M)$.

The **backward probability** $\beta_j(t)$ is defined as

$$\beta_j(t) = P(o_{t+1}, \ldots, o_T | x(t) = j, M).$$

As in the forward case, this backward probability can be computed efficiently using the following recursion:

$$\beta_i(t) = \sum_{j=2}^{N-1} a_{ij} b_j(o_{t+1}) \beta_j(t+1)$$

with initial condition given by

$$\beta_i(T) = a_{iN}$$

for $1 < i < N$ and final condition given by

$$\beta_1(1) = \sum_{j=2}^{N-1} a_{1j} b_j(o_1) \beta_j(1).$$

Notice that, in the definitions above, the forward probability is a joint probability whereas the backward probability is a conditional probability. This somewhat asymmetrical definition is deliberate since it allows the probability of state occupation to be determined by taking the product of the two probabilities. From the definitions,

$$\alpha_j(t) \beta_j(t) = P(O, x(t) = j | M).$$

Hence,

$$L_j(t) = P(x(t) = j | O, M) \qquad (15.21)$$

$$= \frac{P(O, x(t) = j | M)}{P(O | M)} \qquad (15.22)$$

$$= \frac{1}{P} \alpha_j(t) \beta_j(t) \qquad (15.23)$$

where $P = P(O|M)$.

All of the information needed to perform HMM parameter re-estimation using the Baum–Welch algorithm is now in place. The steps in this algorithm may be summarised as follows.

1. For every parameter vector/matrix requiring re-estimation, allocate storage for the numerator and denominator summations of the form illustrated by Equations (15.19) and (15.20). These storage locations are referred to as accumulators.
2. Calculate the forward and backward probabilities for all states j and times t.
3. For each state j and time t, use the probability $L_j(t)$ and the current observation vector o_t to update the accumulators for that state.
4. Use the final accumulator values to calculate new parameter values.
5. If the value of $P = P(O|M)$ for this iteration is not higher than the value at the previous iteration then stop, otherwise repeat the above steps using the new re-estimated parameter values.

All of the above assumes that the parameters for an HMM are re-estimated from a single observation sequence, that is, a single example of the spoken word. In practice, many examples are needed in order to get good parameter estimates. However, the use of multiple observation sequences adds no additional complexity to the algorithm. Steps 2 and 3 above are simply repeated for each distinct training sequence.

15.1.9 Context-sensitive modelling

As we saw in Section 7.3.2, the acoustic realisation of a phoneme is heavily dependent on the phonemes that surround it. We can make use of this in speech recognition by using separate models for separate phonetic contexts. So, instead of having, say, a single [p] model for all situations, we have a large number of specialised [p] models, each of which more precisely describes that phone in those contexts. This reduces the level of confusability in the recogniser and thereby increases recognition accuracy. The most common way to do this is to use so-called **triphone models**, which are not models made of three phones, but single-phone models in the context of the preceding and following phone. Hence, for a phoneme set of N phones, there will be slightly fewer than N^3 triphones (some combinations never occur).

Unfortunately, we rarely have enough data to have a sufficient number of examples of every triphone; either we have insufficient examples to train a model (**low occupancy**), or no examples at all (**zero occupancy**). The solution to this is to **cluster** the data; in effect we borrow parameters from well-trained models for use in the ones that suffer from data sparsity. There are several ways we could do this for the case of models with low occupancy, in that we can just find examples that are close to the known examples in acoustic space. The situation is more difficult for those with zero occupancy, since we don't have any acoustic examples at all.

We solve this by making use of some of the common properties of phones, and the most common way of doing this is to use the phones' **distinctive features** (see Section 7.4.3). In doing so, we are for instance positing that phones that share the same place of articulation may have more-similar acoustic realisations than do those which don't. The commonest way of performing this feature-based clustering is to use a **decision tree**: the clever thing about this is that, while we use the features to *suggest* commonality between triphones, we use the actual data to determine how close any particular feature combination actually is. In other words, the features serve as constraints or structure on the clustering, but don't determine what should be clustered with what.

The decision tree operates in a top-down manner. It operates on binary features, so as a pre-processing step we convert the original features into binary ones (so for instance the feature CONSONANT TYPE which can take values *stop, fricative* etc. gets converted into a new set of features encoding the same information but in a binary-valued system: often this is done by simply using simple question features, such as "is this a stop?", "is this a fricative?"). Initially, all the data points for the state of one phone are grouped together in a single cluster. This cluster is characterised by a measure of **impurity**, choices of which include variance, log likelihood and entropy. The process of growing a tree is as follows.

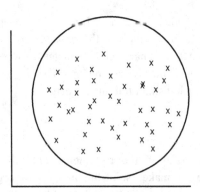

Figure 15.8 A schematic representation of two coefficients of the cepstrum for all the contexts of a single phone. The ellipse indicates a line two standard deviations from the mean.

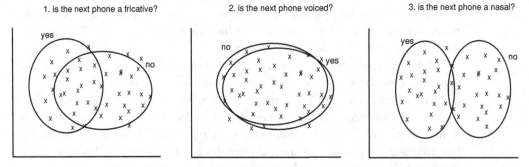

Figure 15.9 The effect of asking three different questions on the data. Each question is asked, the data are partitioned according to the answer, and the variance is calculated for each new cluster. Here we see that question 1 gives a reasonable separation of the data, whereas question 2 does not, meaning that this question is irrelevant, while question 3 gives the best separation.

1. Create an initial cluster containing all the data points.
2. For each feature
   (a) form two new clusters based on the value of the feature;
   (b) measure the combined variance of the two new clusters.
3. Find the feature which gives the biggest reduction in variance.
4. Delete this feature from the list of features to be examined.
5. Form two new clusters based on this feature.
6. Repeat steps 1–4 on each new cluster until the stopping criteria have been met.

Stopping criteria usually involve specifying a minimum decrease in impurity, and the stipulation that clusters should have a minimum occupancy (say ten data points). The process of examining and splitting clusters is shown in Figures 15.8–15.10. A trained decision tree is shown in Figure 15.11.

15.1 The HMM formalism

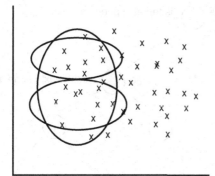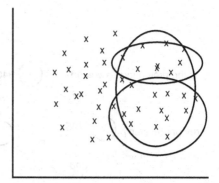

Figure 15.10 After choosing question 3 as the best split, the process continues and the set of questions is asked again on each cluster, resulting in two further splits. In all cases the variance (denoted by the width of the ellipse) is decreasing.

An important point about the decision tree grown in this way is that it provides a cluster for *every* feature combination, not just those encountered in the training data. To see this, consider the tree in Figure 15.11. One branch of this has the feature set

$$\begin{bmatrix} \text{NEXT:NASAL} & yes \\ \text{PREV:VOICED} & yes \\ \text{PREV:STOP} & no \\ \text{NEXT:VOICED} & no \\ \text{PREV:VOWEL} & no \end{bmatrix} \quad (15.24)$$

which will contain all the training examples matching this feature combination. This is only a subset of the features we use: the stopping criteria mean that, even though further feature differences may occur after this point, the differences are deemed slight or the model occupancy so low as to prevent further splitting. If we now have a model with features

$$\begin{bmatrix} \text{NEXT:NASAL} & yes \\ \text{PREV:VOICED} & yes \\ \text{PREV:STOP} & no \\ \text{NEXT:VOICED} & no \\ \text{PREV:VOWEL} & no \\ \text{PREV:ALVEOLAR} & yes \\ \text{NEXT:DENTAL} & no \\ \text{NEXT:VELAR} & no \end{bmatrix}$$

which is unobserved in the data, we see that the feature combination of (15.24) is a subset of this, which means that we will use those parameters for this model. While there is no

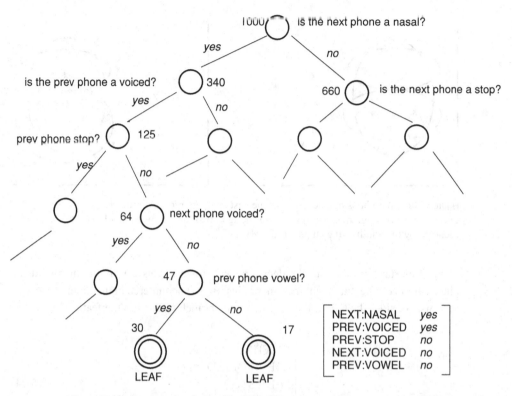

Figure 15.11 An example of part of a trained decision tree. The questions are asked at each node, after which the tree splits. The numbers indicate the numbers of data points at each node. The double circles indicate leaf nodes, where the splitting processes stops. For clarity, only part of the tree is shown. The feature structure which describes one leaf is shown.

path that exactly matches this, we find the subset of features that do (and this is always unique) and this gives us the cluster we require.

As we shall see in Sections 15.2.4 and 16.4.1, decision-tree clustering is a key component of a number of synthesis systems. In these, the decision is not seen so much as a clustering algorithm, but rather as a mapping or function from the discrete feature space onto the acoustic space. Since this is fully defined for every possible feature combination, it provides a general mechanism for generating acoustic representations from linguistic ones.

15.1.10 Are HMMs a good model of speech?

It is often asked whether HMMs are in fact a good model of speech [63], [102], [317]. This is not so straightforward a question to answer. On the surface, the answer is "no" because we can identify many ways in which HMM behaviour differs from that of real speech, and we shall list some of these differences below. However, in their defence, it is important to realise that many of the additions to the canonical HMM alleviate these weaknesses, so one must be careful to define exactly what sort of HMM is being discussed. Criticisms include the following.

1. Independence of observations. This is probably the most-cited weakness of HMMs. While an observation depends on its state, and the state depends on the previous states, one observation is not statistically dependent on the previous one. To see the effect of this, consider the probability of a sequence of observations at a single value. If we have a density with a mean of 0.5, we would then find that a sequence of observations of 0.3, 0.3, 0.3, 0.3, ... would have a certain probability, and, since the Gaussian is symmetrical, this would be the same probability as a sequence of 0.7, 0.7, 0.7, 0.7, ... But, since the observations are independent of one another, this would have the same probability as a sequence that "oscillates" such as 0.3, 0.7, 0.3, 0.7, ... Such a pattern would be extremely unlikely in real speech, a fact not reflected in the model.

2. Discrete states. This problem arises because, whereas real speech evolves more or less smoothly, the HMM switches characteristics suddenly when moving from one state to the next. Both this and the independent-observation problems are largely solved by the use of dynamic features. When these are included, then we can distinguish between the stable and oscillating sequences. In fact, as we shall see in Section 15.2, very natural trajectories can be generated by an HMM that uses dynamic features.

3. Linearity of the model. HMMs enforce a strictly linear model of speech, whereby a sentence is composed of a list of words, which are composed of a list of phones, a list of states and, finally, a list of frames. While there is little problem with describing an utterance as a list of words, or as a list of frames, we know that the strict linear model of phones within words and, worse, states within phones, differs from the more-complex interactions in real speech production (see Section 7.3.5). There we showed that it was wrong to think that we could identify phones or phonemes in a waveform; rather we should think of speech production as a process whereby phonemes are input and speech is output and we can only approximately identify where the output of a phone lies in the waveform. When we consider the alignment of states and frames in HMMs we see that we have in fact insisted that every frame belongs to exactly one state and one phone, which is a gross simplification of the speech-production process.

4. Generative nature. HMMs are generative models, meaning that they declaratively describe the phones, but are not a recognition algorithm in their own right. Furthermore, the validity of the maximum-likelihood training normally used has been questioned in terms of why we want to have a system that maximises the likelihood of the data rather than one that maximises recognition performance.

While many of these points are valid, there are often solutions that help alleviate any problems. As just explained, the use of dynamic features helps greatly with the problems of observation independence and discrete states. As we shall see, the linearity issue is potentially more of a problem in speech synthesis. Models such as neural networks that perform classification directly have been proposed [375] and have produced reasonable results. More recently, discriminative training has become the norm in ASR [360], [495], whereby HMMs as described are used, but their parameters are trained to maximise discrimination, not data likelihood.

An alternative argument about HMMs is simply that they have consistently been shown to outperform other techniques. The HMM recognition systems of today are very different from the systems of the early 1990s, so it is important to realise also that HMMs are more of a framework within which a number of ideas can be tried, rather than constituting just a single, unified approach.

15.2 Synthesis from hidden Markov models

While HMMs are nearly always described in terms of their use in recognition, there is nothing about the models themselves that performs recognition; this is achieved only by using the models in conjunction with a search algorithm such as Viterbi. The models themselves are just declarative representations of the acoustics of a phone and, as such, should be just as amenable for use in speech synthesis as in recognition.

If we were to use HMMs for synthesis, we know that the specification gives us the required phone sequence and possibly a duration for each phone. The phone sequence tells us which models to use in which order, but not which states to use, or which observations to generate from the state Gaussians. If we have a duration in the specification, this tells us how many observations we should generate from this model, but again not which states to generate from. The question then is, if we want to use the models for synthesis, how do we actually go about this?

A possible approach is just to sample randomly from each HMM. So, starting in the first state of the first phone, we simply use a random-number generator to generate a value, which we use with the state's Gaussian to generate an observation. We generate another random number and use this to decide which state we will use next. This process is repeated until we have generated the entire utterance. This approach is valid in a statistical sense in that, over a sufficiently large number of syntheses, the utterances generated will have the same statistical properties as the models. (That is, if we were to retrain on the generated data, we should end up with very similar models.) This approach does not, however, produce natural speech, the main reason being that such an approach causes the spectra to change rapidly and randomly from one frame to the next. Real speech, in contrast, evolves with some level of continuity.

A second approach is to use a maximum-likelihood solution, in which, instead of randomly sampling from the model in accordance with its mean and variance, we just generate the likeliest sequence of observations from the sequence of models. (Unlike in the first approach, retraining on this data will not give us the same models because the values for all states will be exactly the same and hence all the variances will be zero.) It should be obvious that in all cases each state will generate its mean observation. This avoids the problem of the previous approach, in which the observations were "jumping around" from one frame to the next, but now has the problem that the spectrum clearly "jumps" at each state boundary. Furthermore, the spectra are the same during each state, meaning that, in one dimension, the generated speech is a flat line followed by a discontinuity, followed by a different flat line. This again does not look or sound like natural speech. In effect we are ignoring all variance information – if we retrained on speech generated by this model, we would always see the same observation for each state and hence would calculate the same mean but a zero variance in all cases.

15.2.1 Finding the likeliest observations given the state sequence

The approach just described can be stated as one that generates the likeliest observations from the models in the normal sense. The problem with this is that the states always generate their mean values, which results in jumps at state boundaries. A solution to this problem was presented in a series of articles by Tokuda and colleagues [451], [452], [453], which showed how to generate a maximum-likelihood solution that took the natural dynamics of speech into account. The key point in this technique is to use the delta and acceleration coefficients as constraints on what observations can be generated.

Here we show the original system by Tokuda *et al.* [452], which uses delta coefficients as the constraint. It is simple to extend this to cases that use acceleration and higher-order constraints also. As before we represent the observations as

$$O = \langle o_1, o_2, \ldots, o_T \rangle$$

but we split each observation into its constituent parts of coefficients and delta coefficients:

$$o_t = \{c_t, \Delta c_t\}$$

All we require for actual synthesis is the coefficients $C = \langle c_1, c_2, \ldots, c_T \rangle$, so the problem is that of how to generate the highest-probability sequence of these that *also* obeys the constraints of the delta coefficients $\Delta C = \langle \Delta c_1, \Delta c_2, \ldots, \Delta c_T \rangle$. Since we have to find both the observations and the state sequence, we can't use the Viterbi algorithm as above but rather use an algorithm specific to this problem. First, we assume that we have a state sequence q. This probability of observing a sequence of acoustic vectors is therefore equal to

$$P(O|q, \lambda) = b_{q_1}(o_1) b_{q_2}(o_2), \ldots, b_{q_T}(o_T) \qquad (15.25)$$

Recall that the observation probability for a single N-dimensional Gaussian is given by

$$\mathcal{N}(o; \mu, \Sigma) = \frac{1}{\sqrt{(2\pi)^N |\Sigma|}} e^{-\frac{1}{2}(o-\mu)'\Sigma^{-1}(o-\mu)} \qquad (15.26)$$

The observation probability of a state j, which includes the coefficients, and delta coefficients is given by

$$b_j(o_t) = \mathcal{N}(c_t; \mu_j, \Sigma_j) \mathcal{N}(\Delta c_t; \Delta \mu_j, \Delta \Sigma_j). \qquad (15.27)$$

It is easier to work with log probabilities. Equation (15.25) becomes

$$\log P(O|q, \lambda) = b_{q_1}(o_1) + b_{q_2}(o_2) + \cdots + b_{q_T}(o_T) \qquad (15.28)$$

If we substitute Equation (15.26) into (15.27), we get the following expressed in log probabilities:

$$\log b_j(o_t) = -(1/2)\Big[2N\log(2\pi) + |\Sigma| + (c_t - \mu)'\Sigma^{-1}(c_t - \mu)$$
$$+ |\Delta\Sigma| + (\Delta c_t - \Delta\mu)'\Delta\Sigma^{-1}(\Delta c_t - \Delta\mu)\Big]$$

If we then put this into Equation (15.28), we can generate an expression that gives us the log probability for the state sequence. Since everything is in the log domain, all these terms add, giving

$$\log P(O|q,\lambda) = -(1/2)\Big[(c-\mu)'\Sigma^{-1}(c-\mu) + (\Delta c - \Delta\mu)'\Delta\Sigma^{-1}(\Delta c - \Delta\mu) + K\Big]$$

where

$$c = [c_1, c_2, \ldots, c_T]$$
$$\Delta c = [\Delta c_1, \Delta c_2, \ldots, \Delta c_T]$$
$$\mu = [\mu_1, \mu_2, \ldots, \mu_T]$$
$$\Delta\mu = [\Delta\mu_1, \Delta\mu_2, \ldots, \Delta\mu_T]$$

$$\Sigma = \begin{bmatrix} \Sigma_1 & 0 & \cdots & 0 \\ 0 & \Sigma_2 & 0 & \cdots & 0 \\ \vdots & & \ddots & & \vdots \\ 0 & & \cdots & & \Sigma_T \end{bmatrix}$$

$$\Delta\Sigma = \begin{bmatrix} \Delta\Sigma_1 & 0 & \cdots & 0 \\ 0 & \Delta\Sigma_2 & 0 & \cdots & 0 \\ \vdots & & \ddots & & \vdots \\ 0 & & \cdots & & \Delta\Sigma_T \end{bmatrix}$$

$$K = 2TN\log(2\pi) + \sum_{t=1}^{T}|\Sigma| + \sum_{t=1}^{T}|\Delta\Sigma|$$

Note that now we have vectors such as c that represent the entire sequence of coefficients for the sentence. We now introduce Equation (15.5) into our derivation. Recall that this is how we calculate the delta coefficients from the normal coefficients, and by replacing the Δc terms with weighted versions of c we now have our maximum-likelihood expression expressed purely in terms of the normal coefficients c:

$$-(1/2)\Big[(c-\mu)'\Sigma^{-1}(c-\mu) + (Wc - \Delta\mu)'\Delta\Sigma^{-1}(Wc - \Delta\mu) + K\Big] \quad (15.29)$$

Here W is a matrix that expresses the weights of Equation (15.5). It is given by

$$W = \begin{bmatrix} w(0)I & \cdots & w(L)I & & 0 \\ \vdots & w(0)I & & \ddots & \\ w(L)I & & \ddots & & w(L)I \\ & \ddots & & \ddots & \vdots \\ 0 & & w(0)I & \cdots & w(0)I \end{bmatrix} \quad (15.30)$$

To maximise Equation (15.29), we differentiate with respect to c, which gives

$$\Sigma^{-1} + W' \Delta\Sigma^{-1} Wc = \Sigma^{-1}\mu + W' \Delta\Sigma^{-1} \Delta\mu \quad (15.31)$$

which can be solved to find c by any matrix-solution technique.

15.2.2 Finding the likeliest observations and state sequence

This gives us the likeliest observation sequence for a given state sequence. The state sequence is partly determined by the synthesis specification in that we know the words and phones, but not the state sequence within each phone. Ideally, we would search every possible state sequence to find the observation sequence that maximise c in Equation (15.31), but this is too expensive, firstly because the number of possible state sequences is very large and secondly because the solution of Equation (15.31) for each possible state sequence is expensive.

Rather than solve Equation (15.31) every time we wish to consider a new state sequence, Tokuda *et al.* developed a technique that allows the calculation of $P(O|Q', \lambda)$ for a new state sequence Q', once $P(O|Q, \lambda)$ for a state sequence Q has been calculated. The only remaining problem then is how to search the space of possible state sequences effectively. There is no ideal solution to this, since finding the optimal sequence requires searching all the possibilities. Furthermore, because of the nature of the optimisation, we cannot use the Viterbi algorithm as in recognition mode. Rather we use a greedy-algorithm approach whereby we choose the best states locally and then calculate the global probability from this.

This result shows us how to generate from HMMs while also obeying the dynamic constraints of the delta coefficients. In further papers, Tokuda, his colleagues and others have extended this basic result to include any number of delta coefficients (e.g. acceleration coefficients, third-order deltas and so on) [453], explicit duration modelling [514], use of more-sophisticated trajectory modelling [133], [134], [454], [513], use of multiple Gaussian mixtures and use of **streams**, which will be explained below.

15.2.3 Acoustic representations

While it is true to say that an HMM itself is neutral with respect to the purpose to which it is put (synthesis recognition or other), it is also true to say that the manner in which it is normally configured in ASR systems is specific to the recognition task. For instance, we might use 12 MFCC coefficients with energy and their delta and acceleration since these are known to be good features for phone discrimination. We do not, however, include any F0, other source information or prosody. In fact even the transition probabilities, which in a sense model duration and can be seen as the one aspect of prosody that is present in an ASR HMM, are often ignored in modern ASR systems because they are not thought to include much information that discriminates phones.

For synthesis therefore, we are required to make some modifications to ASR-style HMMs to ensure that we obtain good-quality speech. MFCCs on their own are not sufficient to generate speech, since we need (as a minimum) to include F0 information as well. Given a vector of MFCCs and an F0 value, a number of techniques, detailed in Section 14.6, can be used to generate the waveform. We can include an F0 value directly in the observation vector o_t, but it is more common to make use of an additional aspect of the HMM formalism known as **streams**. Each stream has a mixture of Gaussian observation pdfs as before. The addition is that we can now have a number of these observation pdfs, which is particularly useful when we want to generate observations that are different in nature, as in our case of MFCCs and F0 values. A generalisation of the output probability specification of Equation (15.3) using streams is

$$b_j(o_t) = \prod_{s=1}^{S} \left[\sum_{m=1}^{M} c_{jms} \mathcal{N}(o_t; \mu_{jms}, \Sigma_{jms}) \right]^{\gamma_s}$$

The streams are considered statistically independent and their relative importance is scaled with the factors $\gamma_1, \gamma_2, \ldots$. Note that these weights are set by the system builder, not trained automatically.

In Figures 15.12–15.16 we show some of the synthesis properties of HMMs. The basic operation of the algorithm is most easily seen with a single-dimensional observation, and in Figures 15.12–15.14 we show an example of this where we have used an HMM to synthesize an F0 contour. One noted weakness of the HMM approach is that often the observations generated are over-smoothed; informally we can view this as a result of the HMMs being "overly cautious" when generating. The effect of this is shown in Figure 15.13. It is possible to alleviate this effect by simply boosting the relative importance of the mean over the dynamics; this is shown in Figure 15.14. In Figures 15.15 and 15.16 we show synthesis for a multi-dimensional representation of an evolving spectrum, which again shows the difference between simply synthesising from the state mean and synthesising when including the state dynamics.

A proper solution to this problem was proposed by Toda and Tokuda [450]. This technique of **global variance** (GV) ensures that the generated trajectory maximizes not only the output probability used for the conventional method but also that of the global variance of the generated trajectory. The latter probability works as a penalty for a reduction of the variance of the generated trajectory.

15.2 Synthesis from hidden Markov models

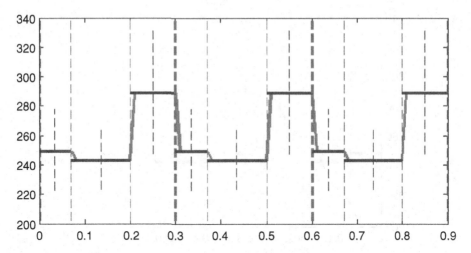

Figure 15.12 Synthesis of an F0 contour from an HMM, where we ignore the dynamic information. The thick vertical lines represent model boundaries and the thin lines state boundaries. The horizontal lines indicate the state means. It is clear that in all cases the algorithm generates the state means, which results in discontinuities at state boundaries.

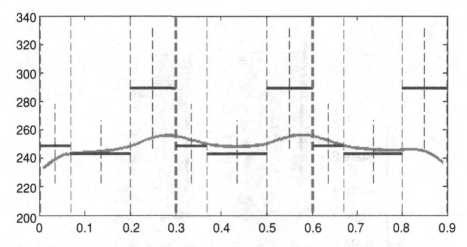

Figure 15.13 Synthesis of an F0 contour where we use Equation (15.31) and model the dynamic behaviour. The contour is now clearly continuous and the state boundaries are not observable from the contour alone. Only in some cases is the mean value for a state reached.

GV maximizes the following criterion which is based on a product of the two output probabilities with respect to the static feature sequence C,

$$L = \log\{p(O|Q,\lambda)^\omega \cdot p(v(C|\lambda_v))\} \qquad (15.32)$$

where $p(O|Q,\lambda)$ is the conventional output probability and $p(v(C|\lambda_v))$ is the output probability modeled by a single Gaussian distribution considering the GV static features $v(C)$. A set of model parameters λ_v consists of the mean vector μ_v and the covariance

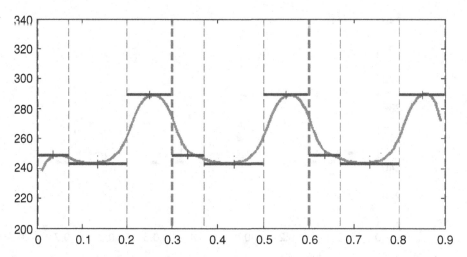

Figure 15.14 Synthesis of an F0 contour where we have boosted the relative strength of the state means, which has the effect that the contour now always reaches the state mean. Informally, we can say that the difference between this and Figure 15.13 is that here we are imitating the effect of the speaker speaking in a more enunciated fashion, whereas there the effect is of the speaker speaking more casually.

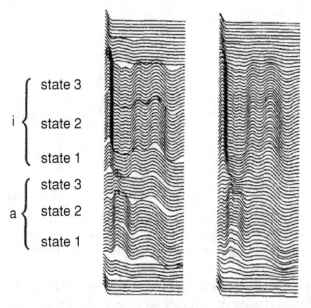

Figure 15.15 Comparison of evolution of generated spectra between synthesising from the mean (left) and using dynamic information (right). In the left figure, we can clearly see that the shape of the spectra jumps at state boundaries. Once we have included dynamic information the evolution is smooth across the utterance.

15.2 Synthesis from hidden Markov models

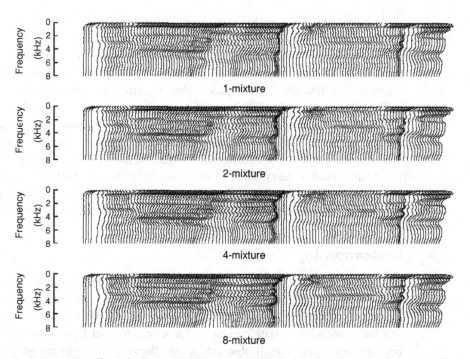

Figure 15.16 The effect of using more mixtures in synthesis. It is clear that as we use more mixtures we can more accurately account for spectral detail.

matrix Σ_v. The constant ω denotes the weight controlling a balance between the two probabilities.

To generate speech parameters, i.e. to determine C that maximizes L, a trajectory is first generated with the conventional parameter generation algorithm. Then the generated trajectory is converted so that its GV is equal to the mean of the Gaussian distribution. Using the converted trajectory as the initial value, the speech parameter trajectory that maximizes the objective function defined in 15.32 is iteratively optimized using the gradient method,

$$C^{(i+1)-th} = C^{(i)-th} + \alpha \cdot C^{(i)-th} \tag{15.33}$$

where α is the step size parameter.

15.2.4 Context-sensitive synthesis models

We use context-sensitive models in ASR because they have lower variance than that of the general phone models. In synthesis, we use context-sensitive models for a similar reason, so that a model in a particular context generates observations appropriate to that context only. In synthesis, however, we are interested in a broader range of factors, since we have to generate the prosodic variation as well as simply the verbal, phonetic variation.

Because of this, it is common to use a much broader notion of "context". In essence we decide what linguistic features we think are required for synthesis and build a separate model for every possible combination. The issue of features in synthesis is discussed fully in Section 16.2; for now we can just assume that, in addition to phonetic context, we want to include also features such as phrasing, stress and intonation.

The context-sensitive models are built in exactly the same way as described in Section 15.1.9, and the resultant decision tree provides a unique mapping from every possible feature combination to HMM model parameters. One significant point of note is that in synthesis the extra prosodic features mean that the number of possible unique feature combinations can be many orders of magnitude larger than in ASR. Given that we may be training on fewer data than in ASR, we see that the problems arising from sparsity of data can be considerably worse. We will return to this issue in Section 16.4.

15.2.5 Duration modelling

In a standard HMM, transition probabilities determine the durational characteristics of the model. If we take a standard topology in which a state can loop back to itself (with, say, probability 0.8) or move to the next state (with probability 0.2), we see that the distribution of state occupancy has an exponential form, such that it is most common to generate a single observation, with the next most common being two observations, then three and so on. In reality, we know that this does not follow the pattern of observed phone durations, which are much more accurately modelled by a Gaussian, or by a skewed Gaussian-like distribution that has low probability for both very short and very long sequences, and higher probability for sequences at or near the mean.

This disparity has been known for some time, and several attempts were made in ASR to correct this, the most notable proposal being the **hidden semi-Markov model (HSMM)** [282]. In this model, the transition probabilities are replaced by an explicit Gaussian duration model. It is now known that this increase in durational accuracy does not in fact improve speech recognition to any significant degree, most probably because duration itself is not a significant factor in discrimination. In synthesis, however, modelling the duration accurately is known to be important and for this reason there has been renewed interest in hidden semi-Markov models [503], [513].

15.2.6 Signal Processing in HMM synthesis

The output of the HMM synthesis process is a sequence of cepstral vectors and F0 values, and so the final task is to convert these into a speech waveform. This can be accomplished in a number of ways, see for example Section 14.6. In general though the approach is to use the generated cepstral output to create a spectral envelope, and use the generated F0 output to create an impulse train. The impulses are then fed into a filter with the coefficients derived from the cepstral parameters. While reasonably effective, this **vocoder** style approach is essentially the same as that used in first generation systems and so can suffer from the "buzz" or "metallic" sound characteristic of those systems (see Section 13.3.5). A major focus of current research is to improve on this.

While cepstral generation via inversion of the cepstral analysis technique is possible, it is more common in HMM synthesis to use the more direct technique of Mel Logarithmic Spectrum Approximation (MLSA) [229]. This technique is quicker (in that it doesn't require the expensive inverse DFT operations) and can be more accurate at modelling spectral envelopes. MLSA uses more sophisticated signal processing techniques than have so far been introduced and so is beyond our present scope. See however Imai's original paper for details of how this is performed [229].

Standard cepstral analysis can be used for a number of purposes, for example F0 extraction and spectral envelope determination. One of the main reasons that cepstral coefficients are used for spectral representations is that they are robust and well suited to statistical analysis because the coefficients are to a large extent statistically independent. In synthesis however, measuring the spectral envelope accurately is a critical to good quality and many techniques have been proposed for more accurate spectral estimation than classic linear prediction or cepstral analysis.

One of the most widely used techniques is the STRAIGHT set of analysis tools [?]. This system is operates as high quality speech analysis-modification-synthesis method implemented as a channel vocoder and has separate components for instantaneous-frequency-based F0 extraction and pitch-adaptive spectral smoothing. STRAIGHT attempts to obtain a more accurate spectral estimation *and* a use more sophisticated source model than simple impulses. A comparison of STRAIGHT and standard cepstral analysis is shown in Figure 15.17.

STRAIGHT extracts the F0 values with fixed-point analysis and carries out F0-adaptive spectral analysis combined with a surface reconstruction method in the time frequency region to remove signal periodicity. It also extracts aperiodicity measurements on the frequency domain. These are based on a ratio between the lower and upper smoothed spectral envelopes and represent the relative energy distribution of aperiodic components.

During synthesis, **mixed excitation** is used to create a more natural excitation signal than the impulse source used in standard vocoder style resynthesis. This operates as a weighted sum of a pulse train with phase manipulation and Gaussian noise. The weighting process is carried out in the frequency domain using the aperiodicity measurements. Using the smoothed spectrum and mixed excitation, STRAIGHT synthesizes a speech waveform with FFT-based processing.

It is not possible to use the STRAIGHT parametrization in the HMMs, since estimating statistically reliable acoustic models using high-dimensional observations is very difficult. To avoid this problem, some systems (e.g. [?]) have used mel-cepstral coefficients converted from the smoothed spectrum with a recursive algorithm [?]. For the same reason, the aperiodicity measurements must also be averaged, usually on five frequency sub-bands (0-1000, 1000-2000, 2000-4000, 4000-6000 and 6000-8000 Hz).

15.2.7 HMM synthesis systems

We now describe some complete HMM synthesis systems. In the HMM system described by Zen and Tokuda [513] five streams are used, one for the MFCCs and one each for

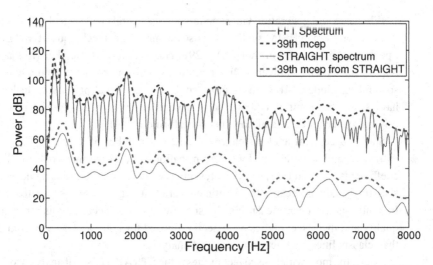

Figure 15.17 Spectral examples extracted from 39 order cepstrum calculated from a DFT and STRAIGHT.

log F0, delta log F0, delta delta log, F0 and voicing information. Since separate streams are used, these parts of the models can in fact be treated separately, to the extent that the decision-tree clustering can use different clusters for each stream. So, while the states of one set of models may be clustered together for modelling MFCCs, a completely different set of states would be clustered for F0. Durations within phones are modelled by hidden semi-Markov models, and a global variance technique [450] is used to compensate for the over-smoothing effect that HMM synthesis can create. The STRAIGHT technique [189] is used for signal analysis and synthesis (see Section 14.6).

Acero [4] describes a technique that uses HMMs as described above but with formants as the acoustic observations. This is particularly interesting in that this can be seen as a direct attempt to fix the problems of traditional formant synthesisers. Formants are indeed a good spectral representation for HMMs because we can assume, as with MFCCs, that each formant is statistically independent of the others. Acero notes that HMMs do in fact generate the sort of formant trajectories often observed in natural data. In particular, he notes that, when a model is used to generate only a few frames of speech, its generated observations rarely reach their mean value, but, when the model is used to generate longer sequences, the mean values are in fact reached. This concurs with the explanation of coarticulation and targets given in Section 7.3.4, but, significantly, because the rates of change of the individual formants can change independently in the HMM, the over-simplistic transition-and-target model of rule-based formant synthesis (see Section 13.2.6 and Figure 13.5) is avoided. In addition to simply replacing MFCCs with formants, Acero investigates the issue of the unnatural source excitation associated with formant synthesis [3].

Taylor [438] has performed some preliminary work on HMM topologies other than the three-state left-to-right models used above. He shows that a standard unit-selection system of the type described in Chapter 16 can in fact by modelled by a more-general

HMM that has many hundreds of states for each model. He shows that one can reduce the number of states by an arbitrary amount, allowing one to scale the size of a synthesis system in a principled manner.

15.3 Labelling databases with HMMs

In all types of data-driven synthesis, we not only require data, but also require the data to be **labelled** in some way. As a minimum, this normally means that we require the words and phones, but any feature that we require for model building or unit selection must be provided by some means. In this section, we concentrate on one such way of carrying out this labelling, which is based on the HMM principles introduced above.

The use of automatic labelling algorithms is normally justified in terms of saving time [5], [279], [384]. In the past, we had small databases, which could be labelled by hand, but, since the databases we use today are larger, and because sometimes we wish to label the database very quickly, we require an automatic system. This argument basically says that automatic labelling is therefore a matter of *convenience*; if we had an infinite supply of hand labellers we wouldn't have a problem. A second justification is that automatic systems can outperform human labellers in terms of accuracy and consistency, so convenience alone is not the only justification; automatic systems are in fact *better*: in our experience this certainly appears to be true.

15.3.1 Determining the word sequence

On its own the issue of identifying the word sequence $\langle w_1, w_2, \ldots, w_M \rangle$ for an utterance is a very difficult problem since it in fact amounts to the speech-recognition problem itself. Because of this it is nearly universally accepted that the word sequence is given in some form, and that the labelling problem then reduces to finding the word boundaries, the phones and the phone boundaries. The words can be provided in a number of ways. One way is to write a **script**, which the speaker reads aloud; from this it is possible to find the words. This is not always a perfect solution insofar as a speaker may deviate from the script. This can be dealt with either by carefully monitoring the speaker and making them repeat any sentences that deviate, or by using some sort of verification process to check that the words in the utterance correspond to those being used as the input to the labeller [15]. A further problem is that the script will be in the form of text, not words, so some processing to determine the words themselves is required.

Instead of, or in addition to, this technique it is possible to use a human labeller to listen to the speech. This labeller would either correct the words when the speaker deviates from the script, or, if no script is available, simply record what words were spoken. While this use of a human labeller means of course that the technique is not fully automatic, this labelling can be done quickly and accurately compared with determining the phone sequence or either of the boundary-type locations.

15.3.2 Determining the phone sequence

Given the word sequence, our next job is to determine the phone sequence. In considering this, we should recall the discussions of Sections 7.3.2, 7.3.6 and 8.1.3. There we stated that there was considerable choice in how we mapped from the words to a sound representation, with the main issue being whether to choose a representation that is close to the lexicon (e.g. phonemes) or one that is closer to the signal (e.g. phones with allophonic variation marked).

By far the main consideration in determining the phone sequence is that this should *match* the sort of sequences produced by the part of the TTS system which maps from text to phones. In both HMM and unit-selection synthesis, we are going to match the sequences in the specification with the sequences in the database labels, and obviously any disparity in labelling procedures or representations will lead to there being fewer correct matches. One way of doing this is to use the text-analysis system for both processes: while we of course always use the text-analysis system for synthesis, we can in fact also generate the phone string using the script as input. While the automatic nature of the text-analysis system may cause errors, the fact that two representations are consistent will ensure accurate matching. Sometimes we cannot determine the single best phone sequence for an utterance. Most often this occurs when we have a pronunciation variant such as EITHER, which can be pronounced as /iy dh er/ or /ay dh er/. We cannot tell which version is used from the text, so both are passed into the labelling algorithm.

Another important consideration is that, if we are going to use an HMM-based labeller, then it makes sense to use a sound-representation system that is amenable to HMM labelling. In general this means adopting a system that represents speech sounds as a linear list, and unfortunately precludes some of the more-sophisticated non-linear phonologies described in Section 7.4.3.

One approach best avoided (but unfortunately very common) is to base the phone sequence on labels created by an "expert" labeller. Often this entails an "expert" looking at a portion of the data, deciding the phones and their boundaries, and then using this to train, test or guide the design or the automatic labeller. This runs the risk of firstly being disconnected from the output that the text-analysis system will produce at run time and secondly being based on what are probably highly errorful labels provided by the human "expert" labeller.

15.3.3 Determining the phone boundaries

Given the word and phone sequence we can construct an HMM model network to recognise just those words. "Recognition" is obviously performed with perfect accuracy, but in carrying out the recognition search we also determine the most likely state sequence, which gives us the phone and word boundaries. Often this operation is called **forced alignment**.

Authors of a number of studies have investigated using a state-of-the-art general-purpose speaker-independent speech recogniser to perform the alignment. This works to a certain extent, but most of these studies have shown that the labelling accuracy is not high enough, with frequent errors occurring [279], [384], [399], [498], [517]. This then is not a complete solution, and the realisation that this is so has led to a number of alternatives. Some systems use HMMs as a first pass and then perform some fine tuning, others use HMMs, but not in the configuration of a general-purpose speech recogniser, and, finally, some systems take a non-HMM approach.

The two-pass approach, which uses a general-purpose recogniser followed by a fine-tuning second stage has been investigated. Studies advocating this approach include [262], [263], [279], [348], [355], [356], [390], and [399]. Kominek *et al.* refine the ASR output with a **dynamic-time-warping (DTW)** technique.[3] The DTW approach works by using an existing synthesiser to synthesise a new sentence that contains the same words as the sentence to be labelled. We therefore have two sentences with the same words (and, we hope, phones) and we use the boundaries of the synthesized sentence (which we know because we have generated it) to determine the boundaries of the sentence to be labelled. This is achieved by using the DTW algorithm to find the alignment which gives the smallest distance between the two sentences. From this, we can determine the best phone and word boundaries of the sentence being labelled. Sethy and Narayanan [399] and Wang *et al.* [481] use a technique whereby an HMM alignment is then fine tuned with a Gaussian mixture model. Lee and Kim [279] propose a technique that uses a neural network to refine the model boundaries. Sandri and Zovato describe techniques for refining very specific problems that occur at pause boundaries and in sections of speech with low energy [390]. Adell and Bonefonte performed perhaps the most exhaustive study, in which they compared HMMs, neural-networks, DTW, Gaussian mixture models and decision trees and showed that the neural-network and decision-tree approaches are the best. We should also note that techniques that use some of these approaches without the HMM step that have been proposed [348], [357], [357].

A different approach is to reason that it is not the HMM formulation itself which is the source of labelling inaccuracies, but rather the particular way in which it has been configured in a general-purpose recogniser. Recall that, while algorithmically a recogniser operates in a maximum-likelihood sense, in fact the choice of its acoustic representations, model topologies, context modelling and so on has been made on the basis of results from experiments that measure recognition performance, which in effect means that such systems are designed for word discrimination. Perhaps most importantly, the systems have been trained on a large number of speakers because they are built in order to recognise any speaker. Hence we should maybe not be surprised that general-purpose recognisers do not perform well at the labelling task when they have in fact been designed for quite different purposes.

This realisation has led to the study of alternative HMM configurations built specifically for the purpose of alignment. Matousek *et al.* [235] report a number of experiments using the HTK toolkit to label the data, whereby a small amount of hand-labelled data is used to provide initial models, which are then retrained on the full corpus. In Clark *et al.*

---

[3] See Jurafsky and Matrin [243] for an explanation of the DTW algorithm, its history and its relation to the Viterbi algorithm.

[99], [366], the system is trained only on the synthesis data, using a flat-start approach. In recognition mode these systems are then run on the same data as used in training. In normal recognition tasks this is of course anathema – the training data should never be used for testing – but here of course the problem is quite different in that we are building a separate aligner for each speaker and so the generalisation problem does not arise. Because we are training and aligning on the same data, we do not encounter problems to do with data sparsity; the data to be aligned appear in exactly the same proportion as the data we use for training.

Multiple pronunciations, caused by pronunciation variants, can easily be dealt with in an HMM framework, no matter whether a specially trained aligner or general-purpose recogniser is used. When a case of multiple pronunciation occurs, the recognition network simply splits, and allows two separate state paths from the start of the word to the end. During alignment, the decoder will pick the path with the highest probability, which can be taken to be the correct pronunciation.

15.3.4 Measuring the quality of the alignments

It is common when building any labelling system to evaluate its quality by comparing its performance against that of a human labeller. In fact this type of evaluation is so common and accepted as not normally to warrant comment (nearly all the techniques cited in the previous section were evaluated in this way). As we shall see in Chapter 17, this type of evaluation is inherently flawed, insofar as it assumes that in some way the human labeller is inherently better than the automatic algorithm. This is now known not to be the case; in fact humans in nearly all cases make frequent labelling mistakes. More importantly, their labelling is subject to somewhat arbitrary decisions in close cases, which can produce a high level of inconsistency in the labels. Comparing an automatic system with a human one is therefore often best avoided.

Apart from the issue of the errors that human labellers make, comparing the quality of the labelling in this way ignores the fact that this is simply an interim stage in the synthesis process. Automatic algorithms often have a consistency that means that, even if "errors" occur, they often cancel out at synthesis time. To take a simple example, if every single phone boundary were marked one frame too late, this would clearly look wrong, but would probably have little effect on the quality of the synthesis. Matousek et al. [235] point this out and show that even their poorest system can produce good-quality synthesis because of this "consistency of errors". So the only real evaluation of labelling quality is in terms of final synthesis quality.

It is genuinely difficult to reconcile the various opinions and results on this issue, to settle the issue of whether HMMs alone are of sufficiently high quality to be used in labelling. Some points we can make, though.

1. Human labelling should be avoided if possible.
2. State-of-the-art speech-recognition systems do not perform labelling adequately.
3. HMM systems set up and trained specifically for the purpose perform the task well.

4. A level of fine adjustment can be performed by a number of machine-learning algorithms.
  5. The quality of labelling should be measured in terms of the final synthesis, not by comparison with human labels.

15.4 Other data-driven synthesis techniques

While the HMM techniques described in this chapter constitute the leading approach, other data-driven techniques have been developed. All in a sense share the same basic philosophy, namely that it is inherently desirable to use a model to generate speech since this enables compact representations and manipulation of the model parameters, and all are attempts at solving the problems of specifying the model parameters by hand.

Hogberg [215] uses a decision-tree technique to generate formant patterns. This in a sense shows the general power of the decision-tree technique; just as with the context-sensitive clustering described in Section 15.1.9, Hogberg uses the decision tree to create a mapping between the discrete linguistic space and the space of formant values, and in doing so uses the decision tree to alleviate problems arising from data sparsity. Mori *et al.* [320] also attempt to learn formant synthesiser rules from data, by building a number of modules using an auto-regressive procedure, which is similar to the LP analysis of Section 12.4.

15.5 Discussion

HMM synthesis is an effective solution to the problem of how to map from the specification to the parameters. While most approaches aim to generate cepstral parameters, some generate formants and in this sense the HMM approach can be seen as a direct replacement for the provision of these rules by hand as described in Chapter 13. Issues still remain regarding the naturalness of the parameter-to-speech part of HMM synthesis, but as confidence in the ability to solve the specification-to-parameter part is gained, new techniques will be developed to solve these naturalness problems.

15.5.1 Further reading

General introductions to HMMs can be found in [224], [243]. The original papers on HMMs include Baum *et al.* [36], Viterbi [476], Baker [33] and Jelinek *et al.* [237]. These are mentioned for purposes of general interest; the ASR systems of today are quite different and, apart from the basic principle, not too much of that early work has survived in today's systems. The best practical guide to modern HMM systems is Cambridge University's HTK system [510]. This is a general-purpose practical toolkit that allows quick and easy building of elementary ASR systems, and serves as the basis for state-of-the-art systems for the research community. The HTK book is an excellent introduction to the theory and practice of HMM recognition.

HMM synthesis started with the now-classic paper by Tokuda *et al.* [452], which explained the basic principles of generating observations that obey the dynamic constraints. Papers explaining the basic principle include [305], [451], and [453]. From this, improvements have gradually been proposed, resulting in today's high-quality synthesis systems. The enhancements include provision of more-powerful observation modelling [508], duration modelling in HMMs [503], [514], trended HMMs [133], [134], trajectory HMMs [514] and HMM studies on emotion and voice transformation [232], [404], [429], [504].

15.5.2 Summary

HMM synthesis

- A trained HMM system describes a model of speech and so can be used to generate speech.
- A number of representations can be used as observations; a common set-up is to use MFCCs together with F0 and their delta values, and perhaps additional information about the source.
- The most obvious way to do this is via a maximum-likelihood approach, but this will simply generate the state mean vectors if used normally.
- Instead we use an approach that generates the maximum-likelihood sequence of vectors that also obey the state dynamics, as determined by the delta and acceleration coefficients, which are determined during training.
- A feature system is defined and a separate model is trained for each unique feature combination.
- Decision-tree clustering is used to merge parameters for low- and zero-occupancy states.
- Refinements to the basic HMM technique have been proposed, including more-realistic duration modelling and accounting for global variance.
- Advantages of HMM synthesis.
  1. HMM synthesis provides a means by which to train the specification-to-parameter module automatically, thus bypassing the problems associated with hand-written rules.
  2. The trained models can produce high-quality synthesis, and have the advantages of being compact and amenable to modification for voice transformation and other purposes.
- Disadvantages of HMM synthesis.
  1. The speech has to be generated by a parametric model, so, no matter how naturally the models generate parameters, the final quality is very much dependent on the parameter-to-speech technique used.
  2. Even with the dynamic constraints, the models generate somewhat "safe" observations and fail to generate some of the more-interesting and delicate phenomena in speech.

HMMs for labelling databases
- Since the state-to-frame alignment is a by-product of recognition with HMMs, they can be used to label and segment a speech database.
- This is done by running a recognition system in forced-alignment mode, such that the correct word sequence is given to the recogniser beforehand.
- It has been shown that general-purpose speaker-independent HMM recognisers do not produce alignments of acceptable quality.
- One approach therefore is to carry out a first pass with a speech recogniser and then adjust the boundaries with a fine-tuning technique.
- An alternative approach is to train an HMM system solely on the synthesis data, thus ensuring consistency.
- It is inadvisable to measure the accuracy of an alignment system by comparing it with hand-labelled data, because the quality of hand-labelled data is usually worse than that produced by the best automatic systems.

16 Unit-selection synthesis

We now turn to **unit-selection synthesis** which is the dominant synthesis technique in text-to-speech today. Unit selection is the natural extension of second-generation concatenative systems, and deals with the issues of how to manage large numbers of units, how to extend prosody beyond just F0 and timing control, and how to alleviate the distortions caused by signal processing.

16.1 From concatenative synthesis to unit selection

The main progression from first- to second-generation systems was a move away from fully explicit synthesis models. Of the first-generation techniques, classical LP synthesis differs from formant synthesis in that it uses data, rather than rules, to specify vocal-tract behaviour. Both first-generation techniques, however, still used explicit source models. The improved quality of second-generation techniques stems largely from abandoning explicit source models as well, regardless of whether TD-PSOLA (no model), RELP (use of real residuals) or a sinusoidal model (no strict source/filter model) is employed. The direction of progress is therefore clear: a movement away from explicit, hand-written rules, towards implicit, data-driven techniques.

By the early 1990s, a typical second-generation system was a concatenative diphone system in which the pitch and timing of the original waveforms were modified by a signal-processing technique to match the pitch and timing of the specification. In these second-generation systems, the assumption is that the specification from the text-analysis system comprises a list of items as before, where each item is specified with phonetic/phonemic identity information, a pitch and a timing value. Hence, these systems assume the following.

1. Within one type of diphone, all variation is accountable by invoking pitch and timing differences
2. The signal-processing algorithms are capable of performing all necessary pitch and timing modifications without incurring any unnaturalness.

In turns out that these assumptions are overly strong, and are limiting factors on the quality of the synthesis. While work on developing signal-processing algorithms still continues, even an algorithm that changed the pitch and timing perfectly would still not address the problems that arise from assumption 1. The problem here is that it is simply not true that

16.1 From concatenative synthesis to unit selection

all the variation within a diphone is accountable by invoking pitch and timing differences. As an example, we can see that low pitch can occur in an unaccented function word, in an accented syllable with a "low" intonation accent and in the last syllable in a sentence, and in all three cases the actual F0 values may be the same. However, all three sound quite different: the function word may have low energy and inexact articulation; the low accented syllable may have a high energy and more careful articulation; and the phrase-final syllable may have prominent voice-quality effects such as creakiness and breathiness because low energy from the lungs can create a significantly different source signal.

This overly simplistic second-generation model was adopted mainly for pragmatic, engineering reasons. Firstly, making this three-way distinction is fairly clean, since we can readily identify and measure the behaviour of each part. Secondly, pitch and timing are simply easier to modify with signal processing than other factors. Finally, in a higher-level sense, it is a reasonable approximation to say that phones come from the verbal component and pitch and timing from the prosodic component of language. In other words, there is a mixture of linguistic and physiological reasons for concentrating on these two factors, which were further promoted because they were the only things we could modify in any event.

16.1.1 Extending concatenative synthesis

The observations about weaknesses of second-generation synthesis led to the development of a range of techniques known collectively as unit selection. These use a richer variety of speech, with the aim of capturing more natural variation and relying less on signal processing. The idea is that for each basic linguistic type we have a number of **units**, which vary in terms of prosody and other characteristics. During synthesis, an algorithm **selects** one unit from the possible choices, in an attempt to find the best overall sequence of units which matches the specification.

We can identify a progression from the second-generation techniques to full-blown unit selection. With the realisation that having exactly one example (i.e. one unit) of each diphone was limiting the quality of the synthesis, the natural course of action was to store more than one unit. Again, the natural way to do this is to consider features beyond pitch and timing (e.g. stress or phrasing) and to have one unit for each of the extra features. So, for example, for each diphone, we could have a stressed and unstressed version and a phrase-final and non-phrase-final version. So, instead of the type of specification used for second-generation systems, namely

$$s_t = \begin{bmatrix} \text{STATE 1} \begin{bmatrix} \text{PHONEME} & n \\ \text{F0} & 121 \\ \text{DURATION} & 50 \end{bmatrix} \\ \text{STATE 2} \begin{bmatrix} \text{PHONEME} & t \\ \text{F0} & 123 \\ \text{DURATION} & 70 \end{bmatrix} \end{bmatrix}$$

we include additional linguistic features relating to stress, phrasing and so on:

$$s_t = \begin{bmatrix} \text{STATE 1} \begin{bmatrix} \text{PHONEME} & n \\ \text{F0} & 121 \\ \text{DURATION} & 50 \\ \text{STRESS} & \textit{true} \\ \text{PHRASE FINAL} & \textit{false} \end{bmatrix} \\ \text{STATE 2} \begin{bmatrix} \text{PHONEME} & t \\ \text{F0} & 123 \\ \text{DURATION} & 70 \\ \text{STRESS} & \textit{true} \\ \text{PHRASE FINAL} & \textit{false} \end{bmatrix} \end{bmatrix} \qquad (16.1)$$

One way of realising this is as a direct extension of the original diphone principle. Instead of recording and analysing one version of each diphone, we now record and analyse one version for each combination of specified features. In principle, we can keep on expanding this methodology, so that, for example, if we wish to have phrase-initial, -medial and -final units of each diphone, or a unit for every type or variation of pitch accent, we simply design and record the data we require.

As we use more features, we see that in practical terms the approach becomes increasingly difficult. This is because we now have to collect significantly more data and do so in just such a way as to collect exactly one of each feature value. Speakers cannot of course utter specific diphones in isolation; rather they must do so in carrier words or phrases. This has the consequence that the speaker is uttering speech in the carrier phrases that is not part of the required list of effects. If we adhere strictly to this paradigm, we should throw this extra speech away, but this seems wasteful. The unit-selection approach offers a solution to both these problems, which enables us to use the carrier speech and also lessen the problems arising from designing and recording a database that creates a unit for every feature value.

In unit selection, the idea is that we obtain a database and perform the analysis such that potentially the entire database can be used as units in synthesis. Systems vary regarding the degree to which the content of database is designed. In some systems, an approach similar to that just outlined above is taken, in which the words, phrases or sentences to be spoken are carefully designed so as to illicit a specific range of required feature values. Any extra speech is also added to the database as a beneficial side effect. At the other extreme, we can take any database (designed or not), analyse it and take all the units we find within it as our final unit database. The difference is really one of degree, since in both cases we will end up with an arbitrary number of each of the features we want; and, depending on how rich the feature set we use is, we may end up with many cases of missing units, that is, feature combinations that we may require at synthesis time but for which there are no examples in the database. This means that we need some technique for choosing amongst those units which match the specification and for dealing with cases in which an exact matching of features is not possible.

A further issue concerns how we concatenate units in unit selection. Recall that in second-generation synthesis the diphones were specifically designed to join together well, in that they were all taken from relatively neutral phonetic contexts such that, when the two diphones were joined, the left side of the first diphone could be relied upon to join well with the right side of the second diphone. The whole point of extending the range of units on offer is to increase variability, but this has the side effect of increasing the variability at the unit edges. This results in a situation in which we cannot rely on the units always joining well, so steps must to be taken to ensure that only unit combinations that will result in good joins are used.

Unit selection is made possible by the provision of a significantly larger database than with second-generation techniques, and in fact it is clearly pointless having a sophisticated selection system if the choice of units is very limited. With a large database we often find that long contiguous sections of speech are chosen, and this is one of the main factors responsible for the very high quality of the best utterances. Often in unit selection no signal-processing modification is performed, and we refer to this approach as **pure unit selection**. In fact, an alternative view of unit selection is that it is a **resequencing algorithm** [84], which simply cuts up speech and rearranges it. Thinking of unit selection in this way can be helpful because it leads us to the **principle of least modification**. This states that the naturalness of the original database is of course perfect, and that any modification we perform, whether cutting, joining or using signal processing, runs the risk of making the original data sound worse. Hence our aim should be to meet the specification by rearranging the original data in as few ways as possible so as to try to preserve the "perfect" quality of the original.

16.1.2 The Hunt and Black algorithm

Various proposals were put forth in answer to these problems of managing larger databases of units, enabling selection within a class of units, coping with missing units and ensuring good joins [234], [327], [387], [502]. However, in 1996 Andrew Hunt and Alan Black [227] proposed a general solution to the problem, which was the culmination of many years of unit-selection work at ATR labs (Figure 16.1). In this (now-classic) paper, Hunt and Black put forward both a general framework for unit selection and specific algorithms for calculating the various components required by the framework.[1]

For easy comparison with second-generation techniques, we will assume that we also use diphones in unit selection, but, as we will see in Section 16.2.1, a wide variety of other types is possible. As before, the specification is a list of diphone items $S = \langle s_1, s_2, \ldots, s_N \rangle$, each described by a feature structure. The database is a set of diphone

---

[1] There are really several related ideas in this work, and we need to be careful in distinguishing these since it is more the general framework proposed by Hunt and Black which has become standard, rather than the specific way in which the system was configured. In our explanation, then, we will propose a somewhat more-general approach than that put forth in the original paper; but in spirit this approach is a direct descendant.

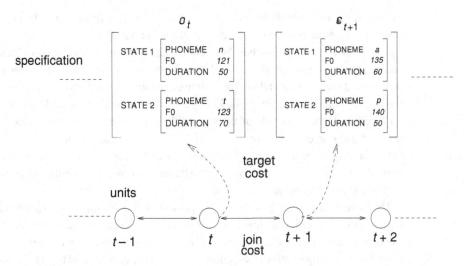

Figure 16.1 A diagram of the Hunt and Black algorithm, showing one particular sequence of units and how the target cost measures a distance between a unit and the specification, and how the join cost measures a distance between the two adjacent units.

units, $\mathbf{U} = \{u_1, u_2, \ldots, u_M\}$, each of which is also described by a feature structure. The **feature system** is the set of features and values used to describe both the specification and the units, and this is chosen by the system builder in such a way as to satisfy a number of requirements. The purpose of the unit-selection algorithm is to find the best sequence of units $\hat{U}$ from the database $\mathbf{U}$ that satisfies the specification S.

In the Hunt and Black framework, unit selection is defined as a **search** through every possible sequence of units to find the best possible sequence of units. There are several options as to how we define "best", but in the original Hunt and Black formulation this is defined as the lowest **cost**, as calculated from two local components. First we have the **target cost**, $T(u_i, s_t)$, which is a cost or **distance** between the specification s_t and a unit in the database u_i. This cost is calculated from specified values in the feature structure of each unit. Second, we have the **join cost**, $J(u_t, u_{t+1})$, which is a measure of how well two units join (low values mean good joins). This is calculated for a pair of units in the database, again from specified values in the units' feature structures. The total combined cost for a sentence is given by

$$C(U, S) = \sum_{t=1}^{N} T(u_t, s_t) + \sum_{t=1}^{N-1} J(u_t, u_{t+1}) \qquad (16.2)$$

and the goal of the search is to find the single sequence of units $\hat{U}$ which minimises this cost:

$$\hat{U} = \operatorname*{argmin}_{u} \left\{ \sum_{t=1}^{N} T(u_t, s_t) + \sum_{t=1}^{N-1} J(u_t, u_{t+1}) \right\} \qquad (16.3)$$

Section 16.6 will explain how to perform this search, but for now we will simply note that it can be performed as a Viterbi-style search (introduced in Section 15.1.7).

The power of the Hunt and Black formulation as given in Equation (16.2) is that it is a fully general technique for unit selection. We can generalise the idea of target and join costs in terms of target and join **functions**, which don't necessarily have to calculate a cost. The target function is so called because it gives a measure of how well a unit in the database matches the "target" given by the specification. The join function again can accommodate a wide variety of formulations, all of which can encompass the notion of how well two units join. Finally, the formulation of the algorithm as a search through the whole space of units allows us to ensure that the algorithm has found the optimal set of units for the definitions of target and join functions that we have given.

Of course, the very general nature of this algorithm means that there is enormous scope in how we specify the details. In the next sections, we will discuss the issues of what features to use, how to formulate the target function, the join function, the issue of the choice of base type and, finally, search issues.

16.2 Features

16.2.1 Base types

The first "feature" we will examine is the **base type**, that is the type of units we will use in the synthesiser. The base type chosen in second-generation systems was often the diphone, since diphones often produced good joins. In unit selection, the greater variability in the units means that we can't always rely on diphones joining well, so the reasons for using diphones are somewhat less convincing. Indeed, from a survey of the literature, we see that almost every possible kind of base type has been used. In the following list we describe each type by its most common name,[2] cite some systems that use this base type, and give some indication of the number of each type, where we assume that we have N unique phones and M unique syllables in our pronunciation system.

frames Individual frames of speech, which can be combined in any order [204].
states Parts of phones, often determined by the alignment of HMM states [138, 140].
half-phones These are units that are "half" the size of a phone. Thus, they are either units that extend from the phone boundary to a mid point (which can be defined in a number of ways), or units that extend from this mid point to the end of the phone. There are $2N$ different half-phone types [315].
diphones These units extend from the mid point of one phone to the mid point of the next phone. There are just fewer than N^2 diphones, since not all combinations occur in practice (e.g. /h-ng/) [99], [108], [273], [383].
phones Phones or phonemes as normally defined. There are N of these [54, 227, 388].
demi-syllables The syllable equivalent of half-phones, that is, units that either extend from a syllable boundary to the mid point of a syllable (the middle of the vowel) or

---

[2] One unfortunate aspect of base types is that there is no consistent naming scheme for types, and therefore the names of the base types are somewhat haphazard.

extend from this mid point to the end of the syllable. There are $2M$ demi-syllables [349].

di-syllables Units that extend from the middle of one syllable to the middle of the next. There are M^2 di-syllables [86], [277].

syllables Syllables as normally defined [236], [388], [511].

words Words as normally defined [359], [416], [479].

phrases Phrases as normally defined [139].

The reasons why a developer chooses one base type over another are varied, and the choice is often simply down to personal preference. Many systems use the "family" of that which have joins in the middle of phones (half-phones, diphones, demi-syllables) because these are thought to produce better joins. Sometimes the phonology of the language is the main consideration. European languages are often considered phoneme-based, therefore phones, diphones and half-phones are normal choices. Chinese, by contrast, is often considered syllable-based, and many systems in that language use the syllable or variant as their main unit [93], [190], [499].

In addition to **homogeneous** systems, which use a single type of unit, we have **heterogeneous** systems, which use a mixture of two or more types of unit. One reason for using different types of units is for dealing with cases when we have a primary unit of one type that requires some units of another type for joining purposes; a good example of this is the phrase-splicing system of Donovan [139], which concatenates whole canned phrases with smaller units for names. The term **non-uniform unit synthesis** was popular in the early development of unit selection since it was seen that the explicit use of long sequences of contiguous speech was the key to improving naturalness [502]. For the remainder of this chapter, we shall use continue to use diphones for explanatory reasons, but it should be noted that in general the unit-selection framework works for all base types. We consider again the issue of base type in Section 16.6.1 on searching, and show in fact that, so long as certain conditions are met, the choice of base type does not have as big an impact on quality as do other factors in the set-up of the system.

16.2.2 Linguistic and acoustic features

Both the target and the join function operate on a feature structure of the units and specification, and conveniently we can use the feature-structure formalism for both. The Hunt and Black framework doesn't limit the type or nature of the features or their values in any way, so in principle we are able to use any features we want.

Each unit has a waveform representation, i.e. the waveform found from the recordings of the speaker. This in a sense *is* the fundamental representation of the unit. In addition we can have one or more linguistic representations, and systems differ in how these representations are created. One common technique is to use a **script**, which the speaker reads during the recording session. This script may be raw text, annotated text, or text in a form that is close to the underlying words (i.e. all ambiguity has been removed or marked). The script can then serve as a fundamental representation from which other features can be derived. Another possibility is to use **hand labelling**, whereby a human labeller listens to or views the recordings and makes annotations, for example to show where phrase boundaries occur.

We can view both the waveform and text representations as **original features** and from these we can now generate a number of **derived features**. From the waveform, we can derive any of the acoustic representations introduced in Chapter12 by using signal-processing algorithms. From the text we can use automatic processes to generate other linguistic representations, but, significantly, we can just treat this text as input to the TTS system and use the text-analysis system to generate all the linguistic representations normally used in live synthesis. This has the advantage that the feature structures of the units will have the same type of information as the feature structures of the specification, which will obviously help us in assessing their similarity.

Unlike the units, the specification items have no acoustic representation (after all, it is this we are trying to generate) and so are impoverished with respect to the units in this regard. So in choosing features for the join function, we to some extent have an easier problem in that we have two entities of the same type with equal access to the same information. In the case of the target function, we have one unit with both linguistic and acoustic features (the unit) and one with only linguistic features (the target).

All specification items and units then have a **target feature structure** that is used in the target function. In addition, units have a **join feature structure**, or more precisely, a **left join feature structure** and a **right join feature structure**, one for each side of the unit. These feature structures can be completely separate, but it is common for features to be shared (e.g. F0 is often used in both the target and the join feature structure). As with all the feature structures we consider, each is an instance drawn from a defined feature system, comprising defined features and their values, which can be categorical, binary, discrete or continuous. In the following, we use the term **feature combination** to refer to a unique set of values for a feature structure. If any of the features take a continuous value there will be an infinite number of feature combinations, but for simplicity and without loss of generality we shall assume that all features in fact take discrete values, meaning that there is a finite set of feature combinations within both the target and the join feature system.

16.2.3 Choice of features

In choosing our features, we are free to use any features that can be obtained from the utterance structure. Of course, the whole point of building the utterance structure has been to provide the features for use at this point, so it is putting the cart before the horse to suggest that we are fortunate that we can pick any features we have previously derived. Rather, we should think of the process as one of deciding what features we really want in unit selection, finding what features we can derive from the text input, and then formulating the rest of the system so as to provide these. It is important to stress that the rest of the system really does serve to provide features for this stage; all too often the process of TTS design is seen as one whereby the features are somehow set in stone (this is particularly common with systems that provide just pitch and timing as features), or one in which the features are first produced and then just fed into the unit selection with little thought as to what is really required. In using features from the utterance we have a choice as to whether we simply use the values as they are or perform further processing.

As we will see in Section 16.3, each feature has a distance function associated with it and, if the features are used directly, these distance functions can become quite complicated. Hence it is sometimes appropriate to perform further processing on the features to convert them into a form that simplifies the distance function. As a simple example, let us assume that we want to measure F0 differences logarithmically; to do this we can either use the absolute F0 value and perform the log conversion in the function, or convert the F0 value into log form and store this in the specification and units, so that the distance function can be simply a difference. Often this pre-processing of features is beneficial since it can be performed offline and saves computation effort in the distance-function calculations at run time.

In general, we should see that, in terms of the target function, the richer the features we use, the more exact we can be in our requirements. So in a simple system we may just have a feature for stress, whereas in a more-detailed system we would have features for stress, intonation, phrasing, emotion, speaking style and so on. It should be clear, however, that we have a fundamental tradeoff between precision and data sparsity. If we have only a small number of features we will quite often get exactly the feature combination we want, whereas with a large, detailed feature system we will rarely get exact matches. The number and type of features don't affect the unit-selection algorithm itself, but we should note that for detailed feature combinations the target function has to be "cleverer", in that, since it will rarely find an exact match, it will have to find close matches. This is not a simple process, as we will see in Section 16.3. Target-function complexities aside, there is still considerable scope and discretion in what features we can and should use, and we will now investigate this issue.

16.2.4 Types of features

One of the fundamental problems in unit selection is that the specification items lack the acoustic description(s) that would make matching them with units a fairly easy process. We can approach this problem in two ways. Firstly, we can just ignore the fact that the units have acoustic features and match on the linguistic features alone. Alternatively, we can try to perform a **partial synthesis**, whereby we attempt to generate some or all of the acoustic features and then match these with the acoustic features derived by signal processing from the waveforms.

To demonstrate the issue of choice of feature system, let us consider how to represent intonation in the specification. In second-generation systems and in some unit-selection systems, this is represented solely by the F0 contour (sampled in some appropriate way). However, as we saw in Sections 9.4 and 9.5, when F0 contours are generated, they are generated from other features that have been calculated previously. Since we cannot actually generate new information, the F0 contour is clearly just a transformation of the information contained in the original features into a different form. Hence there is no a-priori reason why we can't just use the features we use to calculate the F0 contour directly and forgo any actual F0 calculation.

How then should we choose one approach over the other? This comes down to a tradeoff between **dimensionality reduction** and **accuracy**. If we consider the possible input features to an F0-generation algorithm, we might have stress, accentuation, phrasing and phonetic identity. These multiple input features give rise to a F0 value, and we can see the virtue of calculating the F0 value because it very succinctly captures and effectively summarises the values of the other features. As we explained above, having a large number of features incurs a considerable cost in the complexity of the design of the target function, so on its own this dimensionality reduction is highly desirable. However, there are potential drawbacks to representing the prosody in this way. Firstly, any algorithm attempting this transformation process is bound to make errors, so the generated F0 will be a less-accurate representation than the original features. Because of this we have therefore to weigh up the benefits of dimensionality reduction against the disadvantages of inaccurate generation. This really is an empirical matter and is dependent on the particular F0 algorithms used. Two further points influence our decision regarding our choice of features. If we generate an F0 contour from the higher-level features and then use that alone, we are in fact stating that the *only* influence of these features is on the F0 contour. This is a very strong claim and, as we explained in Chapter 9, the traditional view of prosody as being expressed only in pitch, timing and phrasing is too restricted. These high-level features are also likely to influence voice quality and spectral effects and if these are left out of the set of specification features then their influence cannot be used in the synthesiser. This can lead to a situation in which the high-level features are included *in addition* to the F0 contour. This adds even more complexity to the target-function design insofar as any statistically formulated target function has to consider the statistical dependences between the features. Finally, it can be argued that, since features such as F0 are real-valued, there is a natural distance metric between the specification and units and this simplifies the design of the target function. To a certain extent this is true, but when we consider the complexities of the perceptual comparison of F0 values (whereby for instance listeners are much more sensitive to peak and timing F0 differences than to F0 differences in unstressed syllables) we see that this may provide only partial help.

When we consider the issue of describing units with features we see that to some extent the reverse problem is apparent. With units we have the waveform and so of course measure low-level features such as F0 and timing very accurately. It is obtaining high-level features that is harder. To see this, consider the issue of finding values for stress. This can be done by examining the acoustics and determining stress on that basis, by using the stress values of the unit's word from the lexicon, or by using a combination of the two methods. Regardless, there is still some significant chance that an error will be made.

The choice of features is therefore dependent on many factors and we cannot a priori say that one approach is better than the others. However, we should note that that there has been a continual move in TTS away from exact specification of low-level features towards using higher-level features, see for example [65], [92], [103], [436], [447]. Given this, an approach whereby we use high-level features and leave the influence that they have on the speech up to a trainable target function seems to be more in keeping with the principles of modern systems.

16.3 The independent-feature target-function formulation

16.3.1 The purpose of the target function

We will now turn to the issue of how to design a target function. The job of the target function is to assess the suitability of a unit. This can be formulated as a function that when given a specification item and unit returns a **distance** or **cost**. While there are other possible formulations, which for instance simply rank the candidates (no "spacing" between each candidate and the next is given) or we have a generative probabilistic model with which we calculate the probability that specification s_t generates unit u_i, we will limit ourselves here to the distance. In the most general sense this function will return a ranked list of all the units in the database, each with a score, cost or distance. In practice, though, we usually eliminate from consideration any unit that does not match the base type of the specification. In other words, if our specification is a /n-iy/ unit, we consider only those units in the database which have this base type. We term the set of all units that match the base type the **full set of candidates**. The size of this set varies considerably, from sometimes only one or two units for rare cases (e.g. /zh-uw/) to large numbers (in the thousands or tens of thousands) for common units such as /s-ax/ and /ax-ng/); an average number, however, might be 500. Since the total number of candidates can be quite large, sometimes only a subset of this full set is used in the search, in which case we call this the set of **search candidates**.

When examining the full set of candidates, if we are lucky we may find units that are exact matches to the specification, and these will therefore have a target cost of zero. Usually, though, we find insufficient units with exact matches or no units with exact matches. As we increase the number of features we wish to consider, low- and zero-match situations become the norm, so that often we never have exactly the units we want. In such cases we have to consider non-matching units, and this is where the target cost really does its work. We might ask ourselves how units with different feature combinations from the specification will sound: surely, because they have different features, the listener will hear this and decide that the speech simply sounds wrong? It turns out that the situation is not as bleak as just portrayed for two reasons. Firstly, the acoustic space within which units from a particular feature structure lie often overlaps with the acoustic space of other units. This means that, although two units may have different feature combinations, they do in fact have the same acoustic representation. This many-to-one mapping is of course what makes speech recognition difficult; it is well known that acoustic spaces associated with one phone overlap with those of other phones, and this extends to different feature definitions within phones as well. This leads us to one aspect of the definition of the target cost, namely that it should be a measure of *how similar two units sound*. If two units sound the same to a listener, then it is safe to use one in place of the other. Significantly, in such cases the listener will not realise that a unit with the "wrong" feature combination is being used. A very good demonstration of the use of ambiguity is given by Aylett [25], who uses a normal unit-selection database to synthesise a type of speech that was completely absent from the training data.

Often, however, we will have neither exact feature matches nor units that have different features but the same acoustics. We therefore have to use units that definitely do sound different from the specification. This leads us to the second aspect of target cost, which is a measure of how much **disruption** is caused when one feature combination is used instead of another. In these cases, we are not trying to "bluff" the listener that what they are hearing is true to the specification, but rather to choose a unit that is different but "acceptable". This is a subtle notion and somewhat difficult to define, but to see this consider our initial decision to restrict ourselves to using just the matching base types. If we wished to synthesize /n-iy/ and used a unit /m-aa/ the results would be quite disastrous and the feeling of disruption – that the speech simply sounds wrong – very strong. Hence matching the phonemes is clearly very important. After this, simple experimentation shows that it is also important to match features such as stress and phrasing and certain context features. After this, we then find that often finding a unit with exactly the right context isn't so important, as is the case for instance in finding a unit that has slightly lower or higher F0 than we ideally require. In a sense, this idea of minimal disruption is what inspires the design of second-generation diphone databases. There we have to use one unit for all circumstances. We know that it is impossible to pick a unit that will cover all eventualities, so we instead pick one that is safe, and that causes least disruption when used in the "wrong" context.

We can combine the principles of ambiguity and disruption minimisation into a single general principle of **perceptual substitutability**. The idea here is to define a target cost that, on comparing specification and unit, gives a measure of how well one feature combination substitutes for another.

16.3.2 Defining a perceptual space

Upon consideration, we see that the nature of the target function is very general. Given the feature combinations of the unit and specification item, the function could potentially calculate a separate distance for every single possible difference in the feature combinations. Even for a modest number of features (say 20) this can quickly run into millions of different combinations, all of which require a distance calculation.

Two main, quite different, approaches to the target cost function have been proposed. The first, described in this section, is called the **independent feature formulation (IFF)** and was proposed in the original Hunt and Black paper. Since then, many variations of this have been put forward, but all can be described in the same general form. The second, described in Section 16.4, is termed the **acoustic-space formulation (ASF)** and shares many similarities with the HMM systems of Chapter 15. The fact that there are only two main target-function formulations is probably more due to historical accidents of system development than because these are particularly salient solutions to the problem. As we shall see, we can consider a more-general formulation that encompasses both these and many other potential formulations.

To examine these two main formulations, and to demonstrate other possibilities, we shall adopt a general formulation based on **perceptual spaces**. In this, we first define a perceptual space in which each feature combination lies. This space is defined so that we can use a simple distance metric (e.g. Euclidean) to measure the perceptual distances between feature combinations. Secondly, we define a mapping that specifies where each feature combination lies in this space.

16.3.3 Perceptual spaces defined by independent features

In the IFF, a distance for each feature is calculated independently and weighted, and then the total target cost is calculated as a function of these. In the original Hunt and Black paper, this is given as

$$T(s_t, u_i) = \sum_{p=1}^{P} w_p\left(T_p(s_t[p], u_j[p])\right) \qquad (16.4)$$

where

s_t is the tth specification, described by a feature combination of size P.
u_i is a unit in the database, described by a feature combination of size P.
$s_t(p)$ is the value for feature p in the specification s_t. Likewise with $u_j[p]$.
$T_p(x, y)$ is a function that returns the distance between x and y for the values of feature p.
$T(s_t, u_j)$ is the total cost between the unit and the specification.
w_p is the weight for function T_p.

Hence the total target cost T is a **Manhattan distance**, calculated by summing P sub-costs, one for each feature p in the unit and item feature structures. Another mechanism is to allow each sub-cost to be calculated using any subset of the feature structure, so the numbers of sub-costs and features are not necessarily equivalent. This, however, can be generalised to the above case by pre-processing the features.

An alternative formulation is to use the **Euclidean distance** [449]:

$$T(s_t, u_j) = \sqrt{\sum_{p=1}^{P} w_p\left(T_p(s_t[p], u_j[p])\right)^2} \qquad (16.5)$$

The sub-costs define separate distance functions for each of the features. For the categorical features, this may simply be a binary decision saying whether they match or not. For the continuous features such as F0, this may be some standard difference such as absolute or log distance. Some feature differences are seen as more important than others: this is reflected in the choice of values for the weights w_p for each feature distance.

We can visualise this using our perceptual-space formulation as follows. If we take a simple feature system in which every feature is binary-valued, we see that the IFF

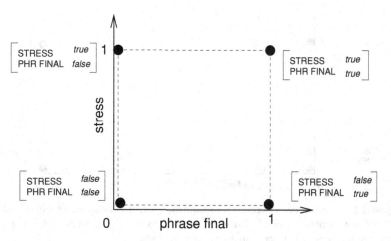

Figure 16.2 A diagram of perceptual space for two ($P = 2$) dimensions. The four unique combinations of the two binary features lie at the four corners of the square.

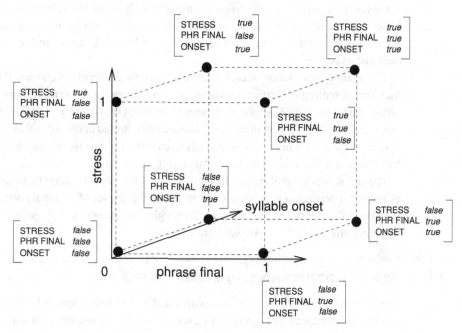

Figure 16.3 A diagram of perceptual space for three ($P = 3$) dimensions. The eight unique combinations of the three binary features lie at the eight corners of the cube.

defines a perceptual space of dimensionality P, where each axis represents exactly one feature. The features themselves map onto the corners of a P-dimensional hypercube in this space. This is shown in Figure 16.2 for $P = 2$ and in Figure 16.3 for $P = 3$. In these figures, we can clearly see that the more two feature combinations differ the greater the distance between them. Continuous features have a single axis too, and the metric chosen to measure their difference is used to scale the axis (so, if the metric is linear, then

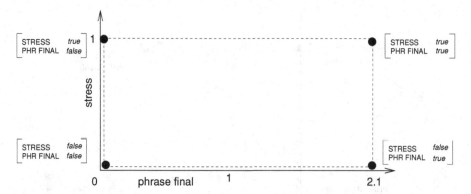

Figure 16.4 The effect of the weights is to *scale* the axes; in this case the weight for stress is set at 1.0 and that for phrasing at 2.1. The effect of this is to make differences in phrasing give rise to bigger differences in distance than do differences in stress.

the axis just represents the feature values in their original form). For N-ary categorical variables, we can use a single axis if the differences between the values can be ordered. Otherwise, we simply encode the category values into a number of binary features (so that a category with eight values would have three binary features), and use an axis for each one as before.

The relative contribution of each feature is given by the set of weights w_p. These have the effect of **scaling** the feature axes in the perceptual space, as shown in Figure 16.4. The difference between the Manhattan distance (Equation (16.4)) and the Euclidean distance (16.5) is that in the Manhattan case the distance is composed of distances that run parallel to each of the axes, whereas in the Euclidean case the shortest line is drawn between the two points and its length gives the distance (Figure 16.5).

The IFF solves the problem of the complexity of the target function by its assumption of feature independence. If we take an example feature set of 20 binary variables, we see that in the IFF this requires setting 20 weights, whereas in a fully general system it would require $2^{20} > 1\,000\,000$ separate weights.

16.3.4 Setting the target weights using acoustic distances

Various ways of setting the weights automatically have been proposed. Since the Hunt and Black algorithm is not probabilistic, we cannot directly use the most-standard probabilistic training technique of maximum likelihood. We can, however, investigate similar types of techniques that try to set the parameters so as to generate data that are closest to a set of training data.

A maximum-likelihood approach would be to find the set of weights $W = \{w_1, w_2, \ldots, w_p\}$ that generates utterances that are closest to the training data. To do this we require a distance metric that measures the distance between a synthesised utterance and the natural utterance. As we shall see in Section 16.7.2, defining such a distance is a difficult problem because it is in effect the same problem as defining the target function itself; a good distance would be one at which utterances that are perceptually similar

16.3 The independent-feature target-function formulation

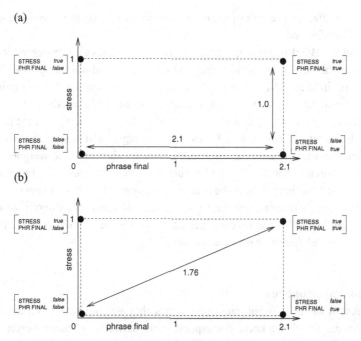

Figure 16.5 (a) Illustration of the Manhattan distance between two feature combinations. The distance is calculated separately for each feature and the totals added, coming in this case to $2.1 + 1.0 = 3.1$. (b) Illustration of the Euclidean distance between the same two features. In this case, the shortest line between the two points is found and its distance measured. In this case, the distance is $(2.1)^2 + 1.0^2 = 1.76$.

would receive a low score. For the time being, however, let us assume that we do in fact have a distance function D that can tell us how similar two utterances are. A further complication arises because the unit-selection system isn't a generative model in the normal probabilistic sense. Rather it is a hybrid model that uses a function to select units (which is not problematic) but then just concatenates the actual units from the database, rather than generating the units from a parameterised model. This part is problematic for a maximum-likelihood type of approach. The problem arises because, if we try to synthesise an utterance in the database, the unit-selection system should find those actual units in the utterance and use them. Exact matches for the whole specification will be found and all the target and join costs will be zero. The result is that the synthesized sentence in every case is identical to the database sentence.[3]

The solution to this then is to use a **held-out** test set, from which we leave out say 10% of the training data, and train on the remainder. When we now attempt to synthesize the

---

[3] In fact, synthesising the training data is a very useful diagnostic tool in unit selection. In all cases, if we synthesize a sentence in the training data, the speech synthesised should be exactly that which was recorded. If the unit selection picks different units then this is often indicative that something has gone wrong. A common cause for this is that the TTS system has produced different pronunciations, or that the prosody generated is very different from the natural prosody.

test set, these utterances aren't of course present in the unit database and so meaningful synthesis is performed.

The Hunt and Black paper proposes this technique and uses a linear-regression method to set the weights. In one experiment they use a separate set of weights for each base type. For each unit in the held-out test set, they find the N closest units in the training data using a distance metric D, which in their case is just the Euclidean distance between the cepstra of the original and synthesized utterances. These are taken to be the units which we wish the target function to find for us. A linear-regression algorithm is then used to train a function that predicts the distances from the sub-costs. The weights W of this linear regression are then taken to be the sub-cost weights themselves. In principle any function-learning technique could be used in this manner; the idea is to design a function that predicts the distances from the sub-costs and then use this directly or indirectly to set the weights. Other approaches include the use of neural networks [131], genetic algorithms [9] and discriminative techniques [385].

Perceptual approaches

The weakness of the left-out-data approach is that we are heavily reliant on a distance function, which we know corresponds only partially to human perception. Various approaches have therefore been taken to create target functions that more directly mimic human perception. One way of improving the distance metric is to use acoustic representations based on models of human perception; for example Tsuzaki [456] uses an auditory modelling approach. However, as we shall see in Section 16.7.2, even a distance function that perfectly models human auditory perception is only part of the story; the categorical boundaries in perception between one discrete feature value and another mean that no such measure on its own can sufficiently mimic human perception in the broader sense.

A more "global" perceptual technique is to use real human judgments of similarities between units to assist in the design of the target function. The most-comprehensive approach to this was proposed by Lee [292], who has listeners **rank** sentences and then trains a model to predict the rankings. The parameters of this model are then taken as the weights. Although the actual training technique is different (the **downhill simplex** method as opposed to linear regression), the framework of this is somewhat similar to the original Hunt and Black technique. The difference is that now the system is trained to predict real human perception, not a distance based on acoustic measurements. Another approach is that advocated by Wouters and Macon [496], who played listeners sets of pairs of words with one phoneme different and asked them to rank the differences. In principle this technique could be extended to any difference in values for a feature, and the resulting data would be a truly accurate reflection of human judgment of sound substitutability. Toda et al. [448] developed a technique that uses subjective evaluations of individual sub-costs to train the weights.

The only drawback with perceptual techniques is that they can be time-consuming to implement, since human subjects are involved in the training process. It can be argued, however, that the weights in the IFF are in fact largely independent of the particular database being used; so, once the weights have been found, they can be used again, or as time goes on more and more experiments can be performed, which will add to the total quantity of data available for training.

Tuning by hand

A final approach to "training" the weights is the strategy of not attempting any automatic training at all but simply setting the weights "by hand" [65], [99], [108]. Typically a system designer listens to some synthesis output and then, using a mixture of previous knowledge and results from this listening, sets each weight individually. Such an approach is of course anathema to all that we believe in the modern world of machine learning, statistics and science, and it can be claimed that this is a retrograde step back to the setting of rules by hand in formant synthesis.

We can, however, justify this approach to some extent. Firstly there is significant evidence that it produces the best results (and of course the real reason why we believe in statistical rather than knowledge-based approaches is actually because they give better results). It should also be stressed that in a typical system there may be only 10 or 20 weights, and this is a manageable amount for a person. By comparison with setting rules by hand in formant synthesis, all we are doing here is biasing a data-driven system in a particular direction, and the setting of a few parameters is a long way from setting *every* parameter by hand.

A final point is that nearly all machine-learning approaches adopt an **equal-error-rate** approach to learning, in which one classification error is considered as bad as another. In real life of course, we know this to be false (consider the relative importance of recognising the word "not" in the sentence "I am definitely not guilty"), but this simplification is adopted in many machine-learning tasks, with the idea that, biases reflecting different degrees of error can be set independently. The advantage of setting the weights by hand is that, by using global judgments about how good the speech sounds, we are implicitly defining which features it is most important to get right. As we saw in Section 16.3.1, we can rank the importance of features independently of any acoustic distances. This is probably most evident in word identity; no matter what effects the other features have, any feature that causes mis-recognition of a word is likely to be regarded as particularly important to get right. When humans tune a system "by hand" they are implicitly making judgments of this type and incorporating them into the weights, and it is possible that it is this that is responsible for the high quality of systems in which we set the weights by hand.

16.3.5 Limitations of the independent-feature formulation

The main design feature of the IFF is that it assumes that the features operate independently. This ensures that the number of parameters (weights) to be determined is very low, and hence that they are easy to learn from data or set by hand. The formulation doesn't in general suffer from data-sparsity problems, since there are nearly always enough examples of each feature present in the data.

The assumption that features operate independently is, however, very strong, and it is not difficult to find instances where this falls apart. The starkest demonstration of this is that two different feature combinations will always have a non-zero distance: this clearly contradicts what we know from above, namely that ambiguity is rife in speech and that there are frequent cases of different feature combinations mapping onto the same acoustics (and hence the same sound). So long as a weight is not zero (which would imply that that feature is irrelevant in every single case) the IFF ensures that we cannot make use of the power of ambiguity.

To show a real example of this, imagine our simple two-feature system as shown in Figure 16.2. From this, we see that the feature combinations

$$\begin{bmatrix} \text{STRESS} & \textit{false} \\ \text{PHRASING} & \textit{true} \end{bmatrix} \text{ and } \begin{bmatrix} \text{STRESS} & \textit{true} \\ \text{PHRASING} & \textit{false} \end{bmatrix}$$

and

$$\begin{bmatrix} \text{STRESS} & \textit{false} \\ \text{PHRASING} & \textit{false} \end{bmatrix} \text{ and } \begin{bmatrix} \text{STRESS} & \textit{true} \\ \text{PHRASING} & \textit{true} \end{bmatrix}$$

give the largest distance (regardless of whether Manhattan or Euclidean distances are used).

In fact, it is well known that stressed and phrase-final versions of a phone are longer than the unstressed and non-phrase-final versions. Imagine for the sake of demonstration that both these features double the length of the phone, then we would expect the following duration values for each feature combination:

$$\begin{bmatrix} \text{STRESS} & \textit{false} \\ \text{PHRASING} & \textit{false} \end{bmatrix} = D$$

$$\begin{bmatrix} \text{STRESS} & \textit{true} \\ \text{PHRASING} & \textit{false} \end{bmatrix} = 2D$$

$$\begin{bmatrix} \text{STRESS} & \textit{false} \\ \text{PHRASING} & \textit{true} \end{bmatrix} = 2D$$

$$\begin{bmatrix} \text{STRESS} & \textit{true} \\ \text{PHRASING} & \textit{true} \end{bmatrix} = 4D$$

This clearly shows that

$$\begin{bmatrix} \text{STRESS} & \textit{false} \\ \text{PHRASING} & \textit{true} \end{bmatrix} \text{ and } \begin{bmatrix} \text{STRESS} & \textit{true} \\ \text{PHRASING} & \textit{false} \end{bmatrix}$$

both have the same duration value, which means that one is well suited to be used in place of the other. However, as we just saw, the IFF gives the largest distance to these, and, furthermore, gives the *same* distance to two feature combinations that are entirely unsuitable.

These problems arise from the very severe constraints the IFF puts on the nature of features, by insisting that all feature combinations with the same value for one feature lie on a single line. It fails to take into account any interaction between features, and therefore is unable to make use of ambiguity and other complex aspects of the feature-to-acoustic mapping.

16.4 The acoustic-space target-function formulation

The **acoustic-space formulation (ASF)** is a quite different way of defining the target function, so different in fact that often it is regarded as a completely different way of performing unit selection. This approach attempts a different solution to the problem of the specification items lacking an acoustic description. The formulation uses a **partial-synthesis function** to synthesize an acoustic representation from the linguistic features. Once this acoustic representation has been obtained, a search is used to find units that are acoustically similar to it. In general, the partial-synthesis function does not go "all the way" and generate an actual waveform; this would amount to a solution of the synthesis problem itself. Rather, an approximate representation is found, and synthesis is performed by using real units that are close to this.

We can also describe the ASF using the idea of perceptual spaces. The key idea of the ASF is that an existing, predefined, acoustic space is taken to be the perceptual space. Thus, rather than defining an abstract perceptual space, we use one that we can measure directly from speech data. Often the cepstral space is used, but in principle any space derivable from the waveform is possible. In the acoustic space, the distance between points is Euclidean, so that the only issue is how to place feature combinations within this space. In the case of feature combinations for which we have plenty of units, we can define a *distribution* for this feature combination, which we find from the units. This can be done in a number of ways, but an obvious way to do it is to use the observations to build a hidden Markov model (HMM). The simplest HMM would have one state, whose observations would be modelled with a single multivariate Gaussian [54]. The means and variances of this Gaussian can be calculated directly from the units in the usual way (i.e. using Equations (15.17) and (15.18)). Since the speech representing the feature combination will have dynamic properties, it is also natural to consider using more than one state in the HMM; common choices are to use three states, as is standard in ASR [138], [140]. If a diphone base-type is used, we should also consider using one

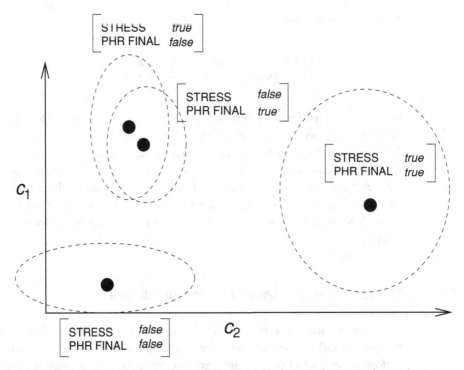

Figure 16.6 A diagram of four feature combinations lying in acoustic space, where only two dimensions of the high-dimensional acoustic space are shown for clarity. Note that, unlike in Figure 16.4, the positions of the feature combinations are not determined by the feature values, but rather by the acoustic definitions of each feature combination. Hence, these can lie at any arbitrary point in the space. In this case, we see that two feature combinations with quite different values lie close to each other, a situation that would not be possible in the IFF. The dotted ellipses indicate the variances of the feature combinations, which are used in some algorithms to measure distances.

state for each half of the unit. From these parameters, we can easily measure the target cost by measuring the distances, in acoustic space, between the model and any unit. The only remaining problem now is that of how to calculate the distance if one or both of the feature combinations are unobserved.

16.4.1 Decision-tree clustering

The key part of the ASF is the design of a partial-synthesis function that can take any feature combination and map it onto the chosen acoustic space. The most common way of doing this is to use the decision-tree method, in more or less the same way as in HMM approaches (see Section 15.1.9). Since context accounts for a significant level of variance within a phone model, using separate phone models for each possible context greatly reduces the overall variance of the models. The problem faced, however, is that, while many of the required models have few or no observations in the training data, their parameters still have to be estimated. The similarity to our problem can now be seen: if

16.4 The acoustic-space target-function formulation

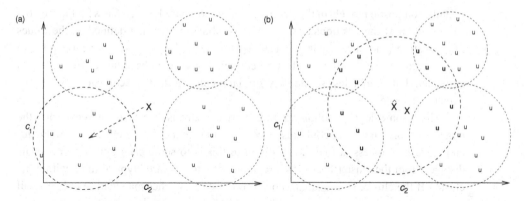

Figure 16.7 (a) Decision-tree behaviour of ASF for unobserved feature combinations. In this figure, we have a feature combination whose true position is given by the point X. In the decision-tree formulation, we find the cluster which has the most feature values which match the unobserved feature combination. Once found, we use this cluster "as is" (in effect we set a pointer from the unobserved feature combination to the chosen cluster). All the units in this cluster are used, even though some are a considerable distance from the true value and are further away than units in other clusters. (b) General projection-function behaviour of ASF for unobserved feature combinations. In this case, we use a general function to generate a value $\hat{X}$, which is an estimate of where the unobserved feature combination should lie in acoustic space. Given this, we then find the closest N units, which are shown in bold inside the ellipse. This technique has the advantage that it finds the units which are closest to the predicted value, not simply a whole cluster for a different feature combination.

we describe HMM context in terms of features, we see that the problem in ASR is one in which a set of model parameters is required for each unique feature combination, many of which have not been included in the training data, whereas in unit selection the problem is that we require a distance between each unique feature combination and the next, again many of which have not been seen in any training data. In the ASF of unit selection, the decision tree is trained in just the same way as with HMMs, the only difference being that "prosodic" features such as stress, phrasing and so on are included. While the method just described is now more or less standard, it is important to note that the decision-tree method itself is based on the **context-oriented-clustering** technique developed for speech synthesis by Nakajima and colleagues [326, 327]. This was later extended and incorporated into the ASR-style decision-tree method by Donovan and Woodland [140]. Other systems of this type include [54], [115], [197], [220], [355] and [381]

The usefulness of the tree is that it provides natural metrics of similarity between *all* feature combinations, not just those observed. The clever bit is that the tree will always lead us to a cluster even for feature combinations completely missing from the data. As with observed feature combinations, we can just measure the distance between their means to give us the value for our target function.

The main benefit of the ASF is that it does not assume that the features operate independently. It therefore avoids two of the weaknesses of the IFF. Firstly we can define complex feature dependences. Secondly, we can use ambiguity in speech to our advantage; it is the acoustic similarity of two different feature structures which allows us to build composite models in the first place. If we compare Figure 16.6 with Figure 16.4 we see that now the positions of the feature combinations are not defined by the values of the features, but rather by the acoustic space and the feature-to-acoustic mapping. In Figure 16.6 we see that the two feature combinations which have no feature values in agreement but nonetheless sound very similar are now in fact close to each other in the perceptual space.

We have some degree of choice as to how we actually use the decision tree to form the target function. Firstly, we could just accept the units in the chosen cluster as candidates and leave the process at that. An addition to this is to score each unit in terms of its distance from the cluster mean [54], in an attempt to "reward" typical units within the cluster. If the cluster size is small, however, these approaches may lead to only a small number of units being used as candidates. Alternatives include going back up the tree and accepting leaves that are neighbours to the current one. Again, these can be given some extra cost to show that they are further away from the specification than the original cluster was. Finally, it is possible to expand any cluster without reference to the tree. To do this, we decide how many candidates we require and then simply find the nearest N units to the mean of the cluster. This can be done with or without using the variances of the cluster.

16.4.2 General partial-synthesis functions

There are two main weaknesses to the decision-tree technique. Firstly, the the assumption that an acoustic space such as the cepstral space is a good approximation of the true perceptual space is a very strong claim; we examine the full extent of this problem in Section 16.7.2. The second weakness is a consequence of using decision trees and is not a weakness of the ASF itself. The basic problem is that the decision tree divides the observations into fixed clusters by taking into account only the particular feature combinations which are observed: no thought is given to the general shape of the space or the properties of the missing feature combinations. The decision tree effectively assigns each missing feature combination to an observed cluster; in computer terms we can think of this as being a "pointer" from the missing feature combination to an observed one. With this, however, it is possible to find a better set of units for the unobserved feature combination by building a new cluster that doesn't necessarily correspond to any existing one. This is shown schematically in Figure 16.7(a).

An alternative to the decision-tree approach is to train a function that estimates an acoustic point directly for a given feature combination. The difference between this and the decision-tree approach is that, while the decision-tree function gives an acoustic point for every feature combination, all the feature combinations which share the feature values in a branch of a tree are given the *same* point. In the general approach, all feature combinations can in principle generate unique points in acoustic space, so there is no assumption of equivalence. Given this generated point, it is possible to find the closest N points to this and therefore form a new cluster. This is shown schematically in Figure 16.7(b). One solution to this is to use a neural network to learn this function [437]. It can be argued that, despite some often-noted disadvantages to these algorithms, neural networks are quite suited to this task insofar as they can easily accommodate binary input and generate continuous vector output. The strength of the neural network is that it can learn the idiosyncrasies of certain feature combinations while generalising over others. The use of a neural network is only one possible solution to this; in principle any function-learning algorithm could be used for this task. While this helps solve the problem of the fixed clusters, it is still, however, reliant on the assumption that the acoustic space is a good approximation to the true perceptual space.

16.5 Join functions

16.5.1 Basic issues in joining units

The purpose of the join function is to tell us how well two units will join together when concatenated. In most approaches this function returns a cost, such that we usually talk about **join costs**. Other formulations are possible, however, including the **join classifier**, which returns true or false, and the **join probability**, which returns the probability that two units will be found in sequence.

Before considering the details of this, it is worth making some basic observations about concatenation in general. We discussed the issue of micro-concatenation in Section 14.7, which explained various simple techniques for joining waveforms without clicks and so on. We also use these techniques in unit selection, but now, because the variance at unit edges is considerably greater, we can't use the concatenation method of second-generation systems which solved this problem by using only very neutral, stable units. Instead, because the units have considerable variability, we now have to consider also the issue of **macro-concatenation**, often simply called the join problem.

Knowing whether two units will join together well is a complex matter. It is frequently the case, however, that we do in fact find a "perfect" join (that is, one that is completely inaudible). We stress this because it is often surprising to a newcomer to TTS that such manipulations can be performed so successfully. Of course, many times when we join arbitrary sections of speech the results are bad; the point is that there are many cases in which speech from completely different sentences and contexts can be joined with no audible join whatsoever. We strongly suggest that anyone involved in unit selection should spend some time just playing around with joining waveforms in this way. Furthermore, when we examine some objective representations of the speech, such as the spectrogram, we can clearly observe the join *visually*, but not audibly. We mention these facts to show that in many cases considerable separation and recombination of speech can be achieved with no noticeable effects; this is of course a fortunate outcome for our general concatenation framework.

The join-feature issue is considerably more straightforward than that of target features. Firstly, we have access both to linguistic features and to the true acoustic features; secondly, because all units have had their features measured and derived in the same way, we don't have to worry about issues of comparing generated acoustic values with ones determined by signal processing. In addition to these "theoretical" advantages,

Table 16.1 How detectability of joins varies with phonetic class (from Syrdal [428])

Broad phonetic class	Percentage of joins detected
Liquid	54.7
Glide	54.7
Glottal fricative	54.3
Nasal	47.4
Unvoiced weak fricative	35.3
Voiced weak fricative	23.3
Unvoiced strong fricative	22.2
Voiced strong fricative	13.6

experience has shown that joining units can be easier in practice: in the best systems there are often virtually no bad joins.

The basic principle of the join cost is to use the features from the two units to decide whether the join will be good or not. While this forms the basis of most calculations, it is common to employ an additional "trick" whereby we use a separate route and always assign a zero join cost to any two units that were neighbouring in the original data. Since these two units were "joined" originally, we know that they form a perfect join and receive a cost of 0, regardless of what the features or join cost proper tells us. This is a powerful addition, and it biases the search towards choosing sequences of units that were originally concatenated. Thus, while nearly all the joins we consider in the search are chosen by the join cost proper, when we consider the joins of the units which are actually chosen, we find that many have been calculated by this other method.

In the following sections we describe techniques for deciding how well two units join, and limit ourselves to doing this solely on the basis of properties of the units. In addition to this, we can also use the techniques mentioned in Section 14.7, where we discussed further techniques for smoothing and interpolation of the units on either side of the join.

16.5.2 Phone-class join costs

A significant factor is the type of phone that is being joined. In the formulations which join in the middle of phone (e.g. diphone, half-phone) we are always joining units that share the same phone across the boundary (e.g. we join a /s-iy/ to an /iy-n/). The identity of this phone has a significant impact on the quality of the join insofar as it appears that some phones are more sensitive to being chopped and recombined, whereas others are more robust.

Syrdal and colleagues have conducted extensive formal studies of these effects [426, 428] and showed that there is a basic hierarchy of phone classes, ranked by how often listeners can detect a join. The results of this are shown in Table 16.1 and show that, regardless of the actual units involved, some phone types simply join better than others. Informally, we can think of this as a type of "prior" on the join cost, which is used to bias

Table 16.2 Correlation of perceptual ratings of "goodness" of join with various acoustic measures and distance functions for a single vowel /ey/ (from Vepa et al. [472])

	Euclidean	Absolute	Mahalanobis
MFCC	0.6	0.64	0.66
MFCC + Δ	0.55	0.55	0.50
LSF	0.63	0.64	0.64
LSF + Δ	0.63	0.64	0.58
Formant	0.59	0.58	0.55
Formant + Δ	0.46	0.46	0.62

the search to pick those joins which fall within the low-detection classes (i.e. fricatives). This **phone-class join cost** should play an important part in any well-designed join cost function; such an approach is taken by Bulyko and Ostendorf [74], who use a phone-class cost (termed **splicing cost** in their paper) in addition to other costs measured between actual units.

16.5.3 Acoustic-distance join costs

It is clear that the acoustic differences between the units being joined have a significant impact on how successful the join will be, and this observation is the basis of the **acoustic-distance join cost**. Perhaps the simplest way of implementing this is to compare the last frame in the left diphone with the first frame in the right diphone. For instance, in Hunt and Black [227], two cepstral vectors are compared.

This on its own is known to produce a number of bad joins, which has prompted an area of research in which researchers have examined various acoustic measures and distance metrics. Authors of several studies investigated these with the aim of seeing which agreed best with human perception. Two main approaches were used. In Klabbers and Veldhuis [250, 251] and Syrdal [426, 428] the perceptual studies focused on asking listeners whether they could detect a join. In a second approach, used in Wouters and Macon [495, 496] and Vepa et al. [469], [470], [471], listeners were asked to rate the quality of the join on a scale. A typical set of results for the second approach is given in Table 16.2.

Among the acoustic representations investigated were the following.

1. Cepstral coefficients section (Section 12.3) [496]
2. LP cepstral coefficients (Section 12.5.6) [496]
3. Mel-scaled cepstral coefficients (MFCCs) (Section 12.5.7) [251], [471]
4. Line-spectral frequencies (Section 12.5.5) [471], [495]
5. Perceptual LP (Section 12.5.8)
6. Formants and formant-like measures [471]
7. LP log area ratios (Section 12.5.4) [495]

Among the metrics used to calculate the distances were the following.

1. Manhattan distance,

$$D = \sum_{i=1}^{N} \mathrm{abs}(x_i - y_i)$$

2. Euclidean distance,

$$D = \sqrt{\sum_{i=1}^{N} (x_i - y_i)^2}$$

3. Mahalanobis distance. The Euclidean distance treats all components of the vector equally, whereas the Mahalanobis distance computes a distance in which each component is scaled by the inverse of its variance. It can be thought of as a measure that normalises the space before computing the distance:

$$D = \sqrt{\sum_{i=1}^{N} \left(\frac{x_i - y_i}{\sigma_i}\right)^2}$$

4. Kullback–Leibler-style distances. The Kullback–Leibler divergence is a measure of "distance" between two probability distributions. From this, we can derive an expression that calculates the distance between two spectral vectors [251], [471]:

$$D = \sum_{i=1}^{N} (x_i - y_i) \log\left(\frac{x_i}{y_i}\right)$$

16.5.4 Combining categorical and and acoustic join costs

The above studies shed interesting light on the relationship between acoustic measures and their perception, but also show that there seems to be an upper limit to how far this approach can go. From Table 16.2, we see that the *best* correlation between an acoustic cost and perceptual judgment is only 0.66, which is far from the type of correlation that we would be happy to accept as a scientific rule. Given the number of studies and that nearly all the well-known acoustic measures (MFCCs, LSFs, formants etc.) and all the distance metrics (Euclidean, Mahalanobis, Kullback–Leibler) have been studied, we can be fairly sure that this area has been investigated thoroughly and that no combination of features and distance metric is likely to improve significantly on the results in Table 16.2.

While it is possible that the distance metrics are at fault, a more likely explanation is that the acoustic features alone don't contain the information required. If we think about joins as a purely low-level phenomenon, this may be surprising, but in fact by concentrating solely on acoustics we may be enforcing low costs on units that are quite different from a purely linguistic perspective. Recall from Section 7.3.2 that often the acoustic differences between linguistic categories are slight, but, because the listener always perceives these categorically, a slight difference across a category boundary may cause a big difference in perception. Hence small acoustic differences within categories may be fine, whereas small acoustic differences across categories may cause problems. The obvious thing to do is therefore to include categorical/linguistic information in the join cost. By doing this, we are saying that units that share the same stress value, or the same general phonetic context, are likely to produce good joins. Some systems have their join costs based *entirely* on linguistic features [14], [65], [441], [442], and often the quality of the joins in these systems is as good as those of systems that use purely acoustic measures. A commoner approach is to use a mixture of acoustic and categorical features in the join cost [74], [108], [210], [427]. As far as we are aware, only one serious study by [427] has looked at join functions that use categorical features and in this it was found that the use of these features in addition to acoustic features made significant improvements.

One possible reason for this is that acoustic distances are limited in that they usually only consider frames near the join, whereas we know that context effects can be felt over a distance of two or three phones. Since the categorical features often operate at the phone level or above, they compensate for the short-term nature of the acoustic features. As Coorman et al. [108] state, it is well known that vowels followed by an [r] are heavily influenced by the [r] and this fact alone is useful in designing the join cost (we could for instance have a sub-cost stating that units to be joined should either both have an [r] context or both not have an [r] context). Syrdal [427] investigates the listener responses of a mixed system.

16.5.5 Probabilistic and sequence join costs

Another way of improving on the basic acoustic-distance join cost is the **probabilistic sequence join function**, which takes more frames into account than just those near the join. Vepa and King [469] used a Kalman filter to model the dynamics of frame evolution, and then converted this into a cost, measured in terms of how far potential joins deviated from this model. A full probabilistic formulation, which avoids the idea of cost altogether, was developed by Taylor [438].

The Taylor technique uses a model that estimates the probability of finding a sequence of frames, with the idea that, if we find a high-probability frame sequence across a join, this indicates a natural sequences of frames, which in turn implies a natural join. The model uses an *n*-gram approach of the type used for HMM language models (see Section 15.1.6). *N*-grams work on discrete entities, so frames have to be quantised as a pre-processing step. This is done via a codebook of the type used in discrete-probability HMMs [367], [510], which uses a bottom-up clustering to quantise the data, followed by a mapping, which assigns any frame a number corresponding to its nearest cluster. Given this, we can calculate the probability $P(O)$ of observing a sequence of frames:

$$P(O) = P(o_1, o_2, o_3, \ldots, o_M)$$

The idea is to measure the probability of this for the sequence of frames across every pair of candidate units across a join, and use the result as a measure of join quality in the search. $P(O)$ is too difficult to estimate in full, so we make the *n*-gram assumption and estimate it on a shorter sequence of frames (say two before and two after the join). The powerful thing about this approach is that the model can be trained on all the data in the speech database, not just examples near unit boundaries. This often results in many millions or tens of millions of training examples, which means that robust estimates of $P(O)$ can be obtained.

This approach has a number of advantages. Firstly, it gets around a core weakness of the acoustic-distance technique. There, frames that are identical have a zero cost, and this in general is the only way in which a zero join cost can be achieved. In real speech, however, we see that this is hardly ever observed: the speech signal is changing from frame to frame. The key point is that in natural speech, while the signal evolves from frame to frame, all these frames "join" perfectly: in other words, if we separated these frames and recombined them, the joins would be perfect, but in none of these cases would the acoustic distance between the frames be zero. Thus we have an enormous number of cases in which the perceptual cost is clearly zero but the acoustic cost as measured by acoustic distances is not zero. This is shown up in the trick whereby we set the join cost of units that were originally side by side to be zero (Section 16.5.1). While this is a very useful addition to the overall system, it shows the inherent weakness of an acoustic-distance join cost, since naturally neighbouring units rarely get a zero cost as calculated by that method. Secondly, we know that dynamic information is often important in speech. While we can use dynamic features in an acoustic-distance join cost, the probabilistic sequence join function models this more directly. Finally, the probabilistic sequence join function gives us a probability rather than a cost. Vepa and King [469] convert this back into a cost and use it in the standard way, but Taylor [438] shows that it can be kept as a probability, which is one way of formulating a unit-selection system based on probabilities rather than costs.

Bellegarda [39] describes a join-cost algorithm that uses sequences of frames, but in a quite different way. Bellegarda's approach defines a function for each type of join, which in a diphone system would be one for each phoneme. This technique is particularly interesting in that it works on waveforms, rather than relying on any derived acoustic representation. This is so that the technique can solve not only the macro-concatenation problems normally associated with join costs, but also micro-concatenation problems such as phase mismatches. The algorithm works by forming a matrix for each type of join, such that the rows represent the frames and the columns the instances of units. This matrix captures all the acoustic information relevant to that type of join. Because the data are stored as raw waveforms direct comparison is inadvisable (for all the usual reasons), so a real-valued transform is performed by means of a **singular-value decomposition (SVD)**. The join cost is then measured using the cosine of the angle between two vectors in the matrix. The reason for using an SVD is that it transforms the waveforms into representations in which the most salient parts are most prominent, but does so in a way that does not throw away phase information as would be the case with any measure based on Fourier analysis. Bellegarda reports that this technique significantly outperforms standard acoustic measures such as MFCCs in listening experiments.

16.5.6 Join classifiers

Another angle on the problem of why acoustic measures and perception correlate so badly is to consider whether the notion of a join *cost* is really the correct formulation in

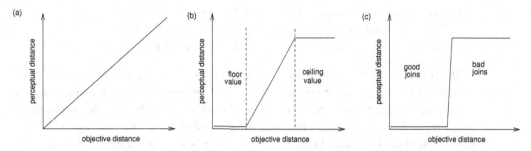

Figure 16.8 (a) The normal formulation whereby the relationship between objective distance and perceptual distance is linear. (b) The use of floor and ceiling values, which say that beyond a threshold all distances are perceptually equivalent. Between these values the relationship is linear. (c) The situation in which the linear region is negligibly small, meaning that the join cost function effectively becomes a join classifier categorising every join as either "good" or "bad".

the first place. Coorman *et al.* [108] show that using costs directly often contradicts basic properties of human perception. They argue that it is good policy to ignore any objective cost differences below a certain value because humans will not perceive such small differences. Additionally, they argue that, above a certain value, the costs are so high as to make any distinction between one high cost and another meaningless: once the join sounds "terrible" there is no point in making the distinction between this and an even-worse pair of units whose join is "truly appalling" etc. We can term these thresholds the **floor** value and the **ceiling** value; their effect on the cost function can be seen in Figure 16.8 and this formulation can apply to both target and join costs. For the case of join costs it shows that there are basically three regions: the first, where all units join perfectly, the middle, where units join well to a greater or lesser degree, and the third region, where all units join badly.

A more extreme view is to consider the middle region very small, such that we take the view that units either join together well (we can't hear the join) or don't (we can hear the join). Then, the join cost becomes a function that returns a binary value. Another way of stating this is that the join function is not returning a cost at all, but is in fact a **classifier**, which simply returns true if two units will join and false if they don't. To the best of our knowledge Pantazis *et al.* [345] is the only published study that has examined this approach in detail. The authors of that study asked listeners to state whether they could hear a join or not, and used this to build a classifier that made a decision on the basis of acoustic features. Pantazis *et al.* used a harmonic model as the acoustic representation (see Section 14.4.2) and Fisher's linear discriminant as the classifier [45], [160], but in essence any features or classifier would be amenable to this approach.

As with all approaches that directly use perceptual evidence, the amount of data available for training is often very limited due to the high cost of collection. It is possible, however, to consider classifiers that do not rely on the making of human judgments. This **sequence join classifier** has a similar philosophy to the probabilistic sequence join function. We use the fact that our entire speech corpus is composed of large amounts of naturally occurring perfect "joins", since every single example of two frames found side by side is an example of such. Therefore we have three situations.

1. Two frames that occurred naturally and therefore form a good join.
2. Two frames that did not occur naturally but form a good join.
3. Two frames that did not occur naturally and form a bad join.

The goal of this approach is to use the data of case 1 (which we have plenty of) to build a classifier that separates cases 2 and 3. In naturally occurring speech, we have of course only data of type 1, but there are in fact machine-learning techniques that are trainable on only positive data, or large amounts of positive data and a small amount of negative data, which can be obtained as before [298]. To the best of our knowledge this approach has not been tried, but should be worthy of investigation because the very large amount of training data should allow a robust classifier to be built.

16.6 Searching

Recall that in the Hunt and Black algorithm (Equations (16.2) and (16.3)) the total cost of one sequence of units from the database $U = \langle u_1, u_2, \ldots, u_N \rangle$ with the specification $S = \langle s_1, s_2, \ldots, s_N \rangle$ is given by

$$C(U, S) = \sum_{t=1}^{N} T(u_t, s_t) + \sum_{t=1}^{N-1} J(u_t, u_{t+1}) \tag{16.6}$$

This calculates the cost for just one possible sequence of units. Our goal is of course to find the single sequence of units in the database which gives the lowest cost:

$$\hat{U} = \operatorname{argmin}\left\{ \sum_{t=1}^{N} T(u_t, s_t) + \sum_{t=1}^{N-1} J(u_t, u_{t+1}) \right\} \tag{16.7}$$

and the issue of unit-selection searching concerns how to find this one best sequence of units from all the possible sequences.

To give some idea of the scale of this, consider a typical case where we have, say, 20 words in the input sentence, which gives rise to 100 diphone items in the specification. If we have 100 units for each diphone, this means that there are $100^{100} = 10^{200}$ unique possible sequences of diphones. If we calculated Equation (16.2) for every sequence this would still take an astonishing length of time regardless of computer speed.[4] Luckily, we can employ the well-known **dynamic-programming** algorithm, a speciality of which is the **Viterbi** algorithm described in Section 15.1.5 This algorithm operates in time $M^2 N$, where M is the number of units and N is the length of the specification sequence. This algorithm takes linear time in N, which is obviously an improvement on the exponential M^N run time of the exhaustive search.

As described in Section 15.1.5 the Viterbi algorithm seeks the highest-probability path through an HMM network. The only real difference between that and a unit-selection

---

[4] For example, if we could perform 10 000 calculations per second, it would still take about a billion times the known lifetime of the Universe to calculate all the costs. (This is assuming that the Universe is 16 billion years old. For creationist readers, the number is even larger!)

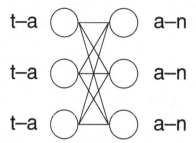

Figure 16.9 A diphone search, where every unit which has edge matches is linked in the network.

search is that here we are trying to find the *lowest*-cost path. This is in fact a trivial difference, so the Viterbi algorithm as previously described can be used directly to find the lowest-cost sequence of units in unit selection.

16.6.1 Base types and searching

On the surface, it seems that the issue of what base type we should pick for our system (e.g. phones, diphones, half-phones etc.) is a fundamental design choice. Here we will show that in fact, so long as certain conditions are met, the choice of base type does not greatly affect the quality of the final speech, but is really an issue of size and speed in the search (Figure 16.9).

Consider again the case of a diphone unit-selection system. In a well-designed system with a good database, we should have at least one example of nearly every diphone. However, even in the best systems it is sometimes the case that we have missing diphones in our unit database, and we need to be able to synthesize these at run time. A potential solution is to use the same approach to missing data as we used in the target-function formulation. Recall that the problem there was that, while we had plenty of examples of units for the required base type, we did not have all of the possible feature combinations within this. Since this problem seems similar, one solution would therefore be to use decision trees; adopting this for missing base types would mean for instance that if we were missing the diphone /t-i/ we would simply use the units from a similar diphone, say /t-iy/. Experience, however, has shown that this approach is rarely successful; while listeners may be "fooled" by using one feature combination instead of another, using an entirely different phone combination is simply going too far; the listeners often hear this substitution and may even misrecognise the word because of this.

An alternative solution, which we will adopt here, is called **unit back-off**. In this, we *manufacture* new units from pieces of existing units. Thus, instead of using a /t-iy/ unit instead of a /t-ih/ unit, we instead use only the first part of /t-iy/, and join this to the second half of a /d-ih/ unit to create the required /t-ih/ diphone. There is no standard way of doing this, but, since the number of missing diphones is often small, it is not too difficult to specify how to build the missing diphones by hand. As in the example just

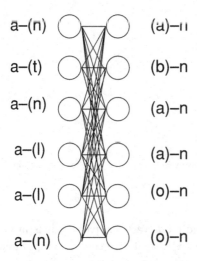

Figure 16.10 An ergodic half-phone search. Here a-(n) means a half-phone unit of the second half of [a] in the context of a following [n] and n-(a) means a half-phone unit of [n] in the context of a preceding [a].

given, we can use criteria such as place of articulation to guide us, and so advocate using a /d/ instead of a /t/.

In effect, for our missing diphones, we are using a half-phone solution. We maintain phone identity, but use diphones where we have them and half-phones elsewhere. Now consider a half-phone system in which we can assume that there are no missing half-phones (which is a fairly safe assumption because we have only $2N$ base types for half-phones but just fewer than N^2 base types for phones, where N is the number of phonemes). In the construction of the diphone lattice, it is standard to allow only diphone units whose base type matches the specification. In other words, in the lattice all edges match in the phone identity to be considered. So, for instance, if we have a set of /t-ih/ candidates in slot t, we will consider only diphones starting with /ih/ in slot $t + 1$. In a half-phone lattice, we have more flexibility, in that, while it makes sense that the phone middle of each half-phone must match across the boundary, we can in principle join any half-phone at the other edge. This is shown in Figure 16.10. This would be the most-general formulation, but if we adopt this we see that the number of possible joins to be calculated at each time slot is now very high.

If we have N phonemes in our system, we will have $2N$ half-phone base types and N^2 diphones. Assuming that we have on average 500 units for each diphone base type, we would have on average $250N$ units for each half-phone base type. Thus, if we consider every unit that matches the specification, we now have $250N$ units at each time slot in the search for half-phones, compared with 500 for diphones, so if $N = 40$ this means that we have 10 000 units, i.e. 20 times more units at each slot. Given that the join function is calculated over every combination of units at a join, this means that instead of $500^2 = 250\,000$ join costs to calculate we now have $(250N)^2 = 100\,000\,000$ joins, 400 times more. These calculations are of course for the same-sized database, which

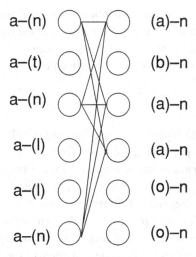

Figure 16.11 A partially connected half-phone search. Here only those half-phones which match on context are linked; this network is equivalent to the diphone network of case (a).

means that just changing the base type means that we now have to compute 400 times more join costs.

To speed up the search when dealing with half-phones, it is possible to restrict the type of units for which we calculate the join cost. The simplest way to do this is to restrict the numbers of joins in base types with plenty of units, so as to consider joining only those units which have a matching context at their edge. So, for example, if we have an /-i/ half-phone unit that is followed by a /t/ in the original data, we consider joining this only to those /t-/ units which are proceed by an /i/ in the original data. We do this for all base types that have more than a threshold N number of units. For below this, we consider more and possibly all possible joins. This is shown in Figure 16.11.

We should now see that in effect the diphone formulation with the half-phone back-off and the half-phone formulation with restricted join possibilities are in fact identical in terms of what speech is passed to the search. This equivalence between base types, back-off strategies and search topologies can be shown for all base types that share the same type of join (e.g. this is true for all those that allow joins in phone middles). This therefore shows that the choice of base type really isn't as fundamental as it seems; the choices of which features to use and how to formulate the target and join functions are much more significant in determining final quality.

Using non-uniform units complicates the search somewhat since now we don't have a lattice of time slots where each time slot aligns one-to-one with the specification. We can, however, still perform a fully optimal search. First we define a "grammar", which specifies which units can follow which others (this grammar is no different from the rules which state which diphones or half-phones can be joined). Next we expand each large unit into a number of nodes equivalent to the smallest base type in the system. This then gives a lattice with time slots and nodes as before, which can be searched in the same manner as above. This further demonstrates that the impact of unit choice is neutralised by the search; the resulting lattice is in many ways identical to one constructed for a uniform unit system built around the smallest base type. In effect, all the non-uniform system is doing is restricting the possible joins from all those that agree a basic phone match to ones specified by the particular choice of non-uniform units.

16.6.2 Pruning

While the Viterbi search is clearly vastly superior to any exhaustive search, in its pure form it is often still too slow for many practical applications. To speed up the search, we can use one or more methods of **pruning**, all of which reduce the number of unit sequences under consideration. Pruning is a bit of an art: in the full Viterbi search we are *guaranteed* to find the lowest-cost sequence of units, and if we eliminate any of the possible sequences from consideration, we run the risk of eliminating the best path. With a little skill, knowledge and judgement, it is often possible to configure a search that considers far fewer sequences, but still has the correct sequence within it. Alternatively, it is often the case that, while the single best path might not be found, a path with a very close score to that is found instead. It is not always a disaster if the best path is missed; this very much depends on whether or not there are plenty of paths close to the score of the best. This in turn mostly depends on whether there are plenty of units that are close matches to the specification; if not, we may have only a few close-matching units and bad results will ensue if these are not considered in the search. It should be stressed, however, that there is no silver bullet to pruning; if we remove any unit sequences from consideration there is always a risk that the best unit sequence will not be found, and, in cases with rare base types and feature combinations, the consequences of this can be severe.

In considering issues of speed, recall that the Viterbi search operates by considering all possible joins at each time slot. If we assume that there are the same numbers of units, N, on the right and left sides of a join this gives us N^2 join-function calculations. We will use O_J to denote the time taken for one join function, so the total time for one set of join functions is $O_J(N^2)$. Each time slot requires a target-function calculation for each unit, and, if we use O_T to denote the time for a single target-function calculation, the total time per time slot is $O_T(N)$. So we see immediately that the number of join calculations is N larger than the number of target calculations. For larger values of N, the target calculation will have a significant impact on time only if $O_T \gg O_J$.

16.6.3 Pre-selection

The first pruning technique we will consider is that of **pre-selection**, whereby we aim to reduce N, that is, the number of candidates being considered during the search. In general, the target function will return a score or cost for all the units of the base type, and thereby create a ranked list. It is easy then to place a cut-off at either the number of units passed to the search or the absolute score. It is normally seen as better practice to define the cut-off in terms of numbers of units, because this ensures an evenness in the search, and also incidentally allows us to ensure that N really is the same for every time slot, thus allowing us to predict accurately the time taken for the search. This method still requires a time of $O_T(N)$ for each time slot, which may be acceptable. If it is not, then a two-pass strategy can be adopted (see below). To give an example, assume for simplicity that $O = O_T = O_J$, that N_T is the total number of units available, and that this equals 500. We see that a full search will take $O(N_T) + O(N_T^2) = 500O + 250\,000O = 250\,500O$ (which of course demonstrates quite clearly that the join-cost calculation dominates in the full search). Let us now define N_C as the number of the chosen sub-sets-of units to be passed, and set this to be 50. We still have to calculate all the target costs (N_T), but after doing so pass only the best 50 into the search. This means that we now have a time of $O(N_T) + O(N_C^2) = 500O + 50^2 O = 3000O$. Here we see that, while the join-function time still dominates, it does not do so to the degree of the full search. If we set N_C lower still, say 10, we see that now the search takes $500O + 100O = 600O$, wherein finally the target-function time dominates. This of course is under the assumption that the calculation times for each join and target function are equivalent; this is seldom true in practice, but it is hard to make generalisations about which takes more time because this is very system-specific.

16.6.4 Beam pruning

In cases with $O_T \gg O_J$, the above technique does not help us much since it always requires a full set of target costs to be calculated. An alternative is to perform a different type of searching known as **beam pruning**, which dynamically prunes the candidates during the search.

In beam pruning we move from time slot to time slot as before, but, instead of enumerating the full set of paths, we enumerate only the N_b best paths. At a given time slot, after we have calculated the join costs, for each node we have the best path up to that point and the total score of the best path that leads to that point. Now we sort these by total score, and discard all but the top N_b paths. We then move to the next time slot by calculating all the join costs between these N_b nodes and the full set of N nodes in the next slot. This process is repeated until the search is complete. Beam pruning reduces the number of target-cost calculations from TN to TN_b and the number of join-cost calculations from TN^2 to TNN_b.

16.6.5 Multi-pass searching

The risk of employing the pre-selection technique is that we reduce the number of candidates being considered in the search without taking the join function into account; the operation considers target functions alone. Thus, while the reduction of the list of candidates may have eliminated those candidates with the highest target costs, the ones left may end up having the worst join costs. A more-sophisticated version of this is to employ a **multi-pass search**. First we do a search using all the candidates but using approximate and quick calculations of both costs. We then select the candidates from the best paths and pass them on to a search that uses the more-sophisticated (and slower) cost calculations. We are not even limited to doing two passes; in principle we can perform as many as we need, with each being more detailed than the ones before.

16.7 Discussion

At the time of writing, unit selection is judged the highest-quality synthesis technique. The *intelligibility* obtained with unit selection compares well with results attained using other techniques, and is sometimes better, sometimes worse. The *naturalness* with unit selection is generally considered *much* better and this is why the technique wins overall. Unit selection is not perfect, however, and a frequent criticism is that the quality can be inconsistent. This is to a degree inherent in the technique: occasionally completely unmodified originally contiguous sections of speech are generated, which of course will have the quality of pure recorded waveforms. On the other hand, sometimes there simply aren't any units that are good matches to the specification or join well. It is clear that in these cases synthesis will sound worse than the stretches of contiguous speech.

It is vital to note that the quality of the final speech in unit selection is heavily dependent on the quality of the database used; much more so than with other techniques. This makes assessments of individual algorithms quite difficult unless they are using the same data. The point here is that it is very difficult to conclude that, say, linguistic join costs are really better than acoustic join costs just because a system that uses the former sounds better than one that uses the latter. Only when such systems are trained and tested on the same data can such conclusions be drawn. This is not to say, however, that using a good-quality database is "cheating"; in fact it can be argued that obtaining a high-quality, large, speech database is the single most-vital factor in making a unit selection system sound good.

That the size of the database is a critical factor in determining overall quality shouldn't be too surprising: we know success largely rests on how many close or exact matches we find in target costs, so the more data we have the more likely we are to find such matches. That said, it is difficult to come up with hard and fast rules about how much data we need. While bigger is generally better, **coverage**, which is a measure of how diverse the units are in terms of their feature combinations, is a vital issue also. For instance, should we design the database so as to have a "flat" coverage whereby we have at least one example of every feature combination, or should we design it to have a "natural" coverage whereby the feature combinations are distributed according to how often we will require them? These and other database issues will be dealt with in more detail in Chapter 17.

It is useful to look at the tradeoff between the relative contributions of the data and unit-selection algorithms in another way. We can imagine that for any given specification there is a single best set of units in the data. That is, if we had all the time in the world[5] and could listen to all the possibilities, we would always find one sequence of units that sounded the best. The design of the unit-selection algorithm, then, is concerned with finding this sequence automatically. The design of the database, on the other hand, is a question of ensuring that there is at least one good sequence of units for every specification. It is very hard to prove this in a formal manner, but informally it appears to be the case that, with a state-of-the-art system, most errors, or poor-sounding sequences

---

[5] Or many billions of times this, to be frank.

of units, arise from algorithm rather than data problems. That is, if the search produces a bad sequence of units, it is often possible to find other units that do in fact sound good. This is actually quite encouraging, since it is logistically easier to work on algorithmic rather than database improvements.

Whether unit selection remains the dominant technique is hard to say. While the systems continually improve, and at a rapid rate, it is also worth noting that there are some fundamental limits. Most importantly, while we recognise that the size of the database is a crucial factor in creating high-quality speech, there is a limit to how much one can record from a single speaker. Often it is simply not practical to record for weeks on end, and therefore improvement in quality by database improvement alone is inherently limited.

16.7.1 Unit selection and signal processing

Once we have found the units by our search, the actual synthesis process is often trivial. In pure unit selection, we have stored our units as waveforms, so application of a simple overlap-and-add technique at the waveform boundaries is sufficient. If we have stored the units in a more-parametric representation (e.g. as LP coefficients and residuals), these can be synthesized in the same way as with second-generation diphone synthesis. Some systems perform spectral smoothing between units, but again this can be done with the same techniques as with second-generation systems.

Some systems have a somewhat hybrid nature in that they perform unit selection as above, but in addition perform prosodic modification by signal processing. In cases where, for instance, the specification requires a unit with a particular F0 value and the database has no unit of this value, it is possible to use signal processing to adjust the unit to the required value. If the degree of modification is too high, poor quality will result. In particular, if the modified speech is present in the same utterance as unmodified speech then a listener may be able to sense the shift in quality. If the degree of modification is slight, however, it is possible to use signal processing with no adverse affect. A further possibility is to anticipate signal processing in the search. We do this by giving lower priority (i.e. weights) to the features we know we can compensate for with signal processing.

16.7.2 Features, costs and perception

Here we make some observations about the nature of features and how they are employed in both the target and the join functions. Firstly, it is important to note that the term "perceptual" is often used in a very general manner, which can lead to confusion over what we are actually talking about. The main point is not to confuse any idea of *auditory* perception with *linguistic* perception. It is well known that the ear cleverly processes the signals it receives and has a non-linear response over the spectrum. While this can, and perhaps should, be modelled, it is important to note that this will go only so far, since "on top" of this sits an independent linguistic perceptual space, which behaves in a quite different manner.

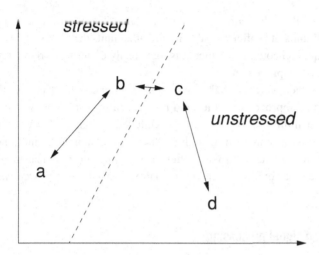

Figure 16.12 The effect of category boundaries on perception. The acoustic distance between units a and b and between c and d is relatively high, and that between units b and c is relatively low. However, because units b and c belong to different linguistic categories they will be perceived to have the largest perceptual distance.

To see this, consider the fact that within a base type there are many feature combinations, and, while there is an *acoustic* continuum within the base type, as one moves from one point to another there may be sudden changes in the values of the categorical features. To take an example, consider Figure 16.12 where we consider the stress feature. If we have a feature system in which units are either stressed or unstressed, then, while the acoustic values that determine stress may be continuous, listeners will judge any given unit as being stressed or unstressed. Even if the acoustic difference between two units is small, if one is judged stressed and the other unstressed, then a bad join may ensue if these are concatenated, because the listener perceives a mismatch in the identity of the feature combination being used.

Another angle on the problem of why acoustic measures and perception correlate so badly is that the cost function as normally portrayed contradicts basic properties of human perception. As we saw in Section 16.5.6, it is good policy to ignore any objective cost differences below a certain value because humans will not perceive such small differences. Additionally, above a certain value, the costs are so high as to make any distinction between one high cost and another meaningless.

16.7.3 Example unit-selection systems

Here we review a few well-known unit-selection systems to give a feeling of how they use the various components described in this chapter to effect full synthesis.

The ATR family and CHATR

Advanced Telecommunications Research (ATR) in Kyoto has played a particularly central role in the development of unit selection. It was there that Yoshinori Sagisaka proposed the idea of unit selection [387] and then demonstrated the significant increase in naturalness that could be brought about by using larger units in synthesis [502]. A second major development was the proposal by Naoto Iwahashi *et al.* [234] that one could use a dynamic programming framework for unit selection. From 1993, Alan Black and Paul Taylor developed the CHATR system [56], which served as a general-purpose synthesis system and development platform, which was to form the backbone of much ATR synthesis research. In 1996, Andrew Hunt and Alan Black [227] published the unit-selection framework used in this chapter, incorporating this into CHATR and giving demonstrations in both English and Japanese. Partly because of the formulation itself and partly because CHATR could speak English, this work attracted considerable attention; this was the first time, after all, that anyone had succeeded in building a complete unit-selection system in English. Additionally, the system did not use signal processing for prosody modification and thus can lay claim to being the first "pure" unit-selection system.[6] The CHATR implementation was not without faults, and was seen as notoriously inconsistent, but the quality of the best speech it generated was so good as to make people believe that a significant quality improvement over second-generation systems was possible.

Partly because of its institutional nature, where visiting researchers would come for a few months to a few years and then leave to work for another organisation, ATR spawned a whole "family tree" of synthesis systems, including the Festival work at Edinburgh University [49] and the AT&T Next Gen system [42], [104], which in turn spawned many other commercial systems. In fact, it is probably true to say that the only major unit-selection systems that are not in some way associated with ATR are those of Laureate and Donovan (see below).

Laureate The Laureate system, developed by Andrew Breen and colleagues at BT labs in England, was developed independently of and simultaneously with the original ATR systems, but was published for the first time only in 1998 [65]. It has some similarity with the ATR systems, in that it uses costs similar to the target and join costs, and also uses a dynamic-programming search to find the best sequence of units. Laureate has many other interesting features, though; for example it nearly exclusively used categorical features for its target and join costs. Laureate also used signal processing in much the same way as second-generation systems.

AT&T NextGen The NextGen system developed by AT&T labs directly used components from CHATR, and so is included in the ATR family [42], [104]. The AT&T group, however, made significant improvements to all aspects of the system, for example in search issues [291] and the perceptual design of the cost functions

---

[6] In fact, I remember quite clearly listening to an early demo of the unit selection from CHATR (I had left ATR at this stage). It sounded very impressive and natural, and I remember saying to Alan Black, "Well I admit this sounds good, but you haven't used any signal processing yet, and when you do it will degrade the speech quality significantly". Alan then told me that he had no intention of using any signal processing, which was a shocking idea to me. Of course, from my own statement it is clear that not doing this was the obvious thing to do because the side effects of the signal processing were the main source of the unnaturalness.

[426], [428]. The most-significant impact of NextGen, however, is that it can lay claim to being the first unit-selection system that actually *worked*. While CHATR and other systems often produce good speech, there were many bad examples, and to some extent the "jury was still out" regarding whether unit selection could be made sufficiently reliable to be used as a replacement for second-generation systems. To me, one of the most significant moments in speech synthesis was at the Jenolan Caves Speech Synthesis Workshop in 1998. There the theme of the workshop was evaluation, and the idea was that everyone would bring their systems and test them on the same sentences and listening conditions. In listening to the various systems, it soon became clear that the AT&T system was a clear winner and had a naturalness that was significantly better than that of any other system. This to me was the crossing of the Rubicon, after which it became virtually impossible to argue that second-generation systems were superior to unit-selection systems.

Cambridge University and IBM The HMM system developed by Rob Donovan, initially at Cambridge University [140] and then at IBM, is notable as one of the systems independent of the ATR family. It was based on Cambridge University's HTK ASR system, and used decision trees to segment and cluster state-sized units [138], [150], [196]. Particularly interesting recent developments have concerned expressiveness and emotion in text-to-speech [151], [195].

RealSpeak The RealSpeak system [108] was another significant landmark in that it was the first commercial unit-selection system to be deployed. RealSpeak achieved yet further heights in terms of naturalness, and it showed that unit selection could be deployed in real-time commercial systems, proof that practical constraints such as database size and search speed could be solved. Because of commercial acquisitions, today's RealSpeak has brought together ideas from many other systems, including the SpeechWorks system (itself a successor of the NextGen system), the Nuance system (a successor of Laureate) and rVoice, described below.

rVoice rVoice was the TTS system developed by Rhetorical Systems, a spin-out from Edinburgh University's Centre for Speech Technology Research. rVoice was the leading system in terms of naturalness during the period 2001–2004. Noteworthy aspects of rVoice include it being the first system to offer custom voices, meaning that Rhetorical could add any voice into the system and consistently produce high-quality results. The rVoice unit-selection system was developed by Matthew Aylett, Justin Fackrell, Peter Rutten, David Talkin and Paul Taylor.

16.7.4 Further reading

Despite unit selection being the dominant technique, there are few good review articles on it. The best sources of information are the papers associated with the above-mentioned systems.

16.7.5 Summary

Unit-selection framework
- Unit-selection synthesis operates by selecting units from a large speech database according to how well they match a specification and how well they join together.
- The specification and the units are completely described by a feature structure, which can be any mixture of linguistic and acoustic features.
- Two functions, normally defined as costs, are used. The target function/cost gives a measure of similarity between specification and unit, and the join function/cost gives a measure of how well two units join.
- A Viterbi-style search is performed to find the sequence of units with the lowest total cost.

Target functions
- The independent feature formulation (IFF) calculates a weighted sum of sub-costs, such that each sub-cost concentrates on one or more features.
- The acoustic-space formulation (ASF) uses the acoustic distance between feature combinations to determine distance. A function is used to project from a feature combination onto a point in acoustic space, and this function is normally trained from example data. A number of techniques can be employed to estimate this point in acoustic space for feature combinations unobserved during training.
- Both these formulations have weaknesses, and the problem of how to calculate and train target functions remains unsolved.

Join functions
- The join function tells us how well two units will join.
- Quite often units join perfectly, with no audible join whatsoever.
- Some types of phones class (e.g. fricatives) are easier to join than others (e.g. vowels) and a good join function should account for this.
- An acoustic join cost that measures the acoustic distance across a join gives a reasonably good indication of how well two units will join.
- A linguistic join cost that looks at the differences between categorical/linguistic features across a join can be defined. Acoustic and linguistic join costs can be combined in a number of ways.
- An alternative view is to see join functions as classifiers that give us a binary decision as to whether two units will join well.
- A further view is to define the join function as something that gives us a probability of seeing the frames across a join, with the idea that high-probability sequences will sound good and low-probability ones will sound bad.

Features and costs
- The issue of which features to use in both the target and the join cost is a complicated one. The following factors are involved.

- High-level, linguistic features are often easy to determine, but lack natural distances and can lead to an explosion in feature combinations.
- Low-level, acoustic features are usually easy to determine for units, but very hard to determine for the specification.
- Measuring costs in acoustic space often does not agree with perception insofar as a small acoustic distance that crosses an important linguistic boundary may result in a large perceptual distance.

17 Further issues

This chapter contains a number of final topics, which have been left until last because they span many of the topics raised in the previous chapters.

17.1 Databases

Data-driven techniques have come to dominate nearly every aspect of text-to-speech in recent years. In addition to being affected by the algorithms themselves, the overall performance of a system is increasingly dominated by the quality of the databases that are used for training. In this section, we therefore examine the issues in database design, collection, labelling and use.

All algorithms are to some extent data-driven; even hand-written rules use some "data", either explicitly or in a mental representation wherein the developer can imagine examples and how they should be dealt with. The difference between hand-written rules and data-driven techniques lies not in whether one uses data or not, but concerns how the data are used. Most data-driven techniques have an automatic training algorithm such that they can be trained on the data without the need for human intervention.

17.1.1 Unit-selection databases

Unit selection is arguably the most data-driven technique because little or no processing is performed on the data, rather it is simply analysed, cut up and recombined in different sequences. As with other database techniques, the issue of coverage is vital, but in addition we have further issues concerning the actual recordings.

There is no firm agreement on how big a unit-selection system needs to be, but it is clear that, all other things being equal, the larger the better. As far as we know, no formal studies have been performed assessing the effect of database size on final unit-selection speech quality. Informally, it seems that about 1 hour of speech is the minimum and many systems make use of 5 hours or more. Below 1 hour, data-sparsity issues start to dominate, and, since little can be done about data sparsity in the unit-selection framework, the problems can be quite serious.

In general, unit-selection databases are acquired by having a single speaker read a specially prepared **script**. Some approaches advocate that the script should be carefully designed in order to have the most efficient coverage of phonetic contexts. Others advocate that the script should rather be normal text materials (for example from a news-service website) since these are more representative of the language as a whole. Hybrid approaches are also possible; here a special set of **core** sentences is designed, and supplemented with more-natural materials from other sources. If the TTS system is to be used for a particular application, it makes sense that the non-core materials should come from that application. So, for instance, in a weather-reading application, the script might contain a core section to cover all phonetic effects and then a series of weather reports to ensure that common phrases such as EARLY MORNING MIST, HIGHS MID-60S and RAIN OVER SCOTLAND occur.

In terms of the actual recording, we face two main issues: the choice of speaker and the recording conditions. Speaker choice is perhaps one of the most vital areas, and all commercial TTS operations spend considerable time choosing a "good" speaker. The speaker must be able to read the script accurately, without straining his or her voice, and keep their voice in a regular style throughout. In addition, it makes sense to pick a speaker who has a "good" voice. This is of course a very subjective matter, but there should be no doubt that some people have voices that listeners greatly prefer over others, otherwise radio announcers, hosts, DJs and readers of talking books would not be specifically chosen. No objective means of selecting a good speaker is as yet known, but a good idea is to play recordings of potential speakers to a target listening group and ask them which they prefer. Since a good unit-selection system will sound very similar to the speaker it is based on, selecting a speaker whom a group of listeners like is a good step towards having the final system sound good.

Insofar as recording conditions are concerned, it is highly desirable that these be of as high a quality as possible because background noise, reverberations and other effects will all be heard in the synthesised speech. In addition, it is essential to ensure that consistency in recordings is maintained. The speaker should speak in the same tone of voice throughout, with a constant distance kept between the microphone and the speaker's mouth. That said, it is important not to put too much pressure on the speaker to speak in one way or another; the best results are obtained when the speaker is relaxed and speaking in their natural voice.

17.1.2 Text materials

It is not always necessary to use material that has actually been spoken in TTS; for many of the text modules text data alone will suffice. This is highly beneficial since it is much quicker and cheaper to collect text data than speech data. Many public text corpora are available, but today by far the best source of text material is on the internet. Many billions of words of data exist there, and, so long as copyright restrictions are adhered to, this can provide a plentiful supply of high-quality data.

17.1.3 Prosody databases

Great care must be taken when recording prosodic databases since the recording conditions can influence the speaker to the extent that the prosody is unrepresentative of real speech. In particular it is inadvisable to instruct the speaker to speak in different types of pitch accent and so on because this will inevitably lead to artificial exaggeration of the intonation. It can even be argued that prosody recorded in a studio will nearly always be artificial because the speaker is not engaged in a normal discourse [81].

17.1.4 Labelling

Regardless of whether the data are to be used for homograph disambiguation, F0 synthesis, unit selection or any other purpose, the data nearly always have to be **labelled**. The labels differ from component to component but often they are intended to represent the truth or gold standard in terms of the linguistic content of the data, and this can be used in training and testing the modules. The main issues in labelling concern the following questions.

1. What sort of linguistic description is needed?
2. What is the inventory of labels that we should select from?
3. How is the labelling to be performed?

The type of linguistic description is closely bound to the purpose of the component it is being used for. As an example, for a homograph-disambiguation module we would first list all the homographs we know (e.g. BASS-FISH, BASS-MUSIC), identify all the tokens belonging to each group, e.g. bass, and then label each token of this type that we find in a corpus as being either BASS-FISH or BASS-MUSIC. Since the nature of the problem is clearly expressed in the identity of the token, we can even search for cases of bass in raw text data and so specifically collect enough examples for this particular case.

This may all seem rather obvious, but it should be noted that this approach differs from more-traditional approaches in which databases were labelled with general linguistic labels (e.g. POS tags), and used for a variety of purposes besides training TTS components. By labelling just the type of information required, by contrast, we ensure that the database is more relevant to the purpose to which it is put.

In many cases the labelling is quite easy to perform. In the above example of homographs, the relevant tokens are easy to identify and the disambiguation is easy to do. However, this is not always the case. If we consider intonation labelling, described in Sections 6.10.1 and 9.3, we see that the process is much more difficult. First, there is no commonly agreed theory of intonation and hence there is no commonly agreed set of labels. The choice of which intonation model to use is of course based on which model is used in the intonation component, but, since this may be decided on the availability of data, the two issues are intimately connected. As reported in Section 6.10.1, intonation labelling is one of the most difficult tasks, with inter-labeller agreement figures being as low as 70%. If a database is labelled in this way, we face the issue of having a high degree of noise in the labels, which will hamper training and testing.

17.1.5 What exactly is hand labelling?

An issue that is seldom addressed in labelling is that of just what a labeller is doing when he or she hand labels some data. One sees the terms "hand labelling" and "expert labeller" quite frequently in this context, but what do they mean?

Here we take the position that there are really two types of labelling: **intuitive** and **analytical**. Intuitive labelling is where the human labeller makes a judgment using their own language ability, but without using any explicit reasoning based on this. The homograph example above is a good case of this. If we present the sentence he plays bass guitar to someone and ask them which word bass corresponds to, they will quickly and certainly say that it is BASS-MUSIC. No knowledge of linguistics or of any of the issues involved is required in order to make this judgement. Now consider a case in which someone is asked which ToBI pitch accent is used on a word in a spoken sentence. It is impossible to perform this task without specialist training; even most experienced linguists or speech engineers would have trouble with this unless they had direct experience in ToBI labelling itself.

Our position is that intuitive labelling is nearly always useful and reliable, whereas analytical labelling is usually neither. With intuitive labelling, we are in effect using a person's in-built, natural linguistic ability to make a judgment. Since most people are perfect speakers of their language, tapping into this knowledge gives access to a very powerful knowledge source. The tasks in TTS that come under this sort of labelling include work with homographs, sentence boundaries, semiotic classification and verbalisation.

Analytical labelling normally relies on the labeller applying a set of procedures or labelling rules to the data. Consider a case in which a labeller is asked to decide whether a prosodic phrase break should occur between two words in a spoken sentence. To do this, the labeller must first be instructed as to what a prosodic phrase break is. Next they need to have an explanation of how to find one, and this may consist of clues such as "listen for lengthening at the ends of words" and "examine pauses and see whether these constitute phrase breaks". In many cases the labeller will apply these rules easily, but in a significant number of cases the labeller will be in doubt, and maybe further rules will need to be applied. This is where the problem lies. From considerable experience in labelling databases of all kinds and with all types of labels it is clear that there are many problematic cases in which the "correct" label is not immediately obvious. This can lead to labellers making spot judgments, which can often lead to considerable inconsistency in labelling, both between labellers and between different sessions of work by the same labeller.

One of the main sources of difficulty in labelling is that the labelling model which provides the labels to be chosen is in some sense lacking. As we saw in Section 9.3, there is an enormous range of intonation models and theories, and, while we reserve judgment on which one is better than another, it is clear that they can't *all* be right. A more-accurate statement is that they all have good aspects but none is complete or an accurate model of the reality of intonation. Hence, in every labelling situation, in many cases the labeller is trying to force a square peg into a round hole, and difficulties will always ensue.

The problem results in very poor labelling agreement. As previously stated, "good" agreement in intonation labelling is considered to be 70% [406], which not only shows how poor the agreement is but also how low the expectations in the community are, that this can be considered good. Furthermore, experience has shown that the more complicated or difficult the labelling system is, the longer this takes to perform, and the result might be not only an inaccurately labelled database but also one that was time-consuming and expensive to collect. This situation strongly contrasts with intuitive labelling, whereby labellers usually make quick, effortless judgments with a high degree of accuracy.

17.1.6 Automatic labelling

We believe that there are two possible answers to these problems in analytical labelling. The reason why intuitive labelling works so well is that we are tapping a fantastic source of linguistic knowledge, namely the labeller's own language faculty. With analytical labelling, however, while some intuitive aspects may be present, we are really asking the labeller to follow a set of rules. Labelling schemes can be improved to some extent by specifying more and more detailed rules, but this has the effect of making the labelling scheme more complicated. If, however, the answer to improving labelling is specifying more rules, why use humans at all? Instead why not use something that excels at following rules, namely a computer? Using a computer for analytical labelling certainly has difficulties, but a major advantage is that consistency is guaranteed. In many processes that make use of the resulting labels we would rather have a consistent set of labels with some inaccuracies than a set of labels with an unknown degree of consistency. In addition, automatic analysis is of course much quicker and, when training on multiple databases from many speakers, this speed increase becomes invaluable. The pros and cons of automatic labelling differ from problem to problem, and there may still be cases for which analytical hand labelling is better, but in our experience computer labelling is generally to be preferred. We should note, however, that this statement applies only to analytical labelling; the ability of humans to perform intuitive labelling still greatly out-performs automatic techniques.

17.1.7 Avoiding explicit labels

A second solution to the problem of inadequate human labelling is to ask whether we need the labels at all. Older systems used a large number of modules, each of which had specific linguistic inputs and outputs. More-modern systems are in general more integrated, such that HMM and unit-selection systems often subsume tasks such as phonetic assimilation, F0 generation and timing generation that existed separately in previous systems. This is not to say that these modern systems have no concept of (say) phonetic assimilation, but rather that these processes are carried out *implicitly* in the system, so no explicit rules of the type /n/ → /m/ in the context of a labial are needed. In statistical terminology, we say that these processes are hidden, and that the levels of representation in which these inputs and outputs occur are hidden variables. Hence a good system can learn these implicitly and there is no need for explicit representation and hence no need for explicit labels in training. Such an approach is taken, for example, in the SFC system of Bailly and Holm [31], where the F0 is generated directly from functional input (e.g. syntax and semantics), bypassing the need for intonation labels for pitch accents. Some concept of pitch accent is still present in the model, but this is implicit and does not need labelled training data.

In conclusion, labelling should be performed in line with a module's requirements. Intuitive labelling is a useful and accurate method of labelling training data for modules. Analytical labelling is often best performed by computer, and in many cases can be avoided altogether by the use of hidden variables in a module. Analytical labelling by expert human labellers is best avoided if at all possible.

17.2 Evaluation

Throughout the book, we have made statements to the effect that statistical text analysis outperforms rule-based methods, or that unit selection is more natural than formant synthesis. But how do we know this? In one way or another, we have **evaluated** our systems and come to these conclusions. How we go about this is the topic of this section.

There is no single answer to the question of how or what to evaluate. Much synthesis research has proceeded without any formal evaluation at all; the researchers simply examined or listened to the output of a system and made an on-the-spot judgment. Beyond this, certain very specific tests have been proposed, and these are still employed by many in the field. The result is that testing in TTS is not a simple or widely agreed-on area. Throughout the book, we have adopted an engineering mentality in our approach to all issues; and we will continue to do so in this section. As well as keeping a certain homogeneity, the field of engineering is very rich in the area of testing in that real systems have to undergo rigorous testing before they are deployed. It therefore follows that, when interested in testing TTS systems designed for practical purposes, we can follow the testing and evaluation procedures used in mainstream engineering.

The first thing to do is not to ask what testing we should perform, but rather to specify a set of **user requirements** that we expect or hope our system will meet. These will often be specific to an application, so, in a telephony system that reads addresses, we might state that the system should be at least as good as a human at reading addresses. If the address-reading application is simply to confirm a user's address we might be happy with significantly lower performance: so long as the system reads an address that sounds something like the user's address this may suffice. A screen reader for the blind would be expected to read a very wide range of information accurately, and it is quite likely in such a situation that the user would prefer the content to be read very accurately but with an unnatural voice, rather than vice vera. Hence a good engineer will always state the user requirements first.

Once the user requirements have been stated, we can then move on to the issue of how to test whether a system meets these. Such tests are called **system tests** because they test the overall performance of the system. For the above examples, a test might include measuring how many addresses drawn from a random sample are understandable by a listener. Often the results of these tests can be surprisingly poor, so another key part of a test would be to get a human to read the same addresses. We may then find that the human struggles also, and the comparison between human and machine may give a much better indication of the usefulness of a TTS system in an application.

Testing the system as a whole need not always be the best approach; TTS developers may be working on a specific component and may want to know how well it is performing and whether it is performing better than a previous version. This sort of testing is called **unit testing** because it tests a particular "unit" or component in the system. Sometimes system testing is called **black-box** testing because the evaluation progresses without any knowledge of how the system operates, and by contrast unit testing is sometimes called **glass-box** testing since we are "seeing inside" the system ("box") when doing the tests.

17.2.1 System testing: intelligibility and naturalness

It is widely agreed that the two main goals of a TTS system are to sound intelligible and natural. This much is agreed, but defining exactly what we mean by intelligible and natural can be somewhat trickier.

Considering the issue of intelligibility first, we have to define again what exactly our goals are. For example, in many cases we may find that so long as the message is *understood* it need not be necessary for the listener to recognise every single word. It is often the case in normal conversation that we get the "gist" of a message and miss some parts, and this may be acceptable. In such cases **comprehension tests** can be used. Often comprehension is judged by asking listeners to answer questions on speech they have just heard [425]. Other approaches attempt to determine the comprehension in some other way, e.g. Swift *et al.* [425] describe a technique in which eye-movement response times are measured.

17.2.2 Word-recognition tests

A quite different approach focuses on **word recognition** and tests of this type are perhaps the most common sort of intelligibility test. Here listeners are played words either in isolation or in sentences and asked which word(s) they have heard. One group of tests plays listeners similar words and asks them to make distinctions. The most commonly used test is perhaps the **modified rhyme test** (MRT) [221]. In this test, listeners are played a set of 50 groups, 6 words each. The words in each group are similar, differing in only one consonant, and the users are asked to record which word they have heard on a multiple-choice sheet. A typical set of words might be BAD BACK BAN BASS BAT BATH. Various tests have been developed for testing word intelligibility at the sentence level. The **Harvard sentences** [341] are sentences such as THESE DAYS A CHICKEN LEG IS A RARE DISH, and have been designed to provide a test that follows the natural distributions of phonemes in English. Care must be taken with sentence design in these tests because listeners can quite often guess what word is coming next, which turns the test from a recognition one into a verification one. To counter this, **semantically unpredictable sentences** are often used. Here the words in the syntax obey the "rules" of normal syntax, but the sentences don't make any particular sense [181], [358], [414]. While some syntactic predictability is therefore present, it is very difficult to guess correctly the word following the current one. Examples of these include the **Haskins sentences**, such as THE WRONG SHOT LED THE FARM [353].

It is arguable whether such tests have much value as real system tests since the types of words or sentences being used have little validity in terms of a real-world application. The tests are certainly useful for diagnostic purposes and hence should really be considered as unit tests for testing, for instance, a measure of raw consonant ability in a system. For real applications their use is not recommended because the tests are not good samples of the type of material that a system might actually be used for: highly confusable words only rarely occur in contexts where they might be confused, and in general normal sentences are highly semantically predictable. Despite this, the use of the MRT and semantically unpredictable sentences is unfortunately quite commonplace in system evaluations. To see why these tests are inappropriate, consider the hypothetical situation that a developer might build a system that could read *only* confusable isolated content words well; this system could not synthesise sentences or function words. Such a system might score very well on the MRT, but would clearly be useless as a real TTS system.

17.2.3 Naturalness tests

Most approaches to measuring naturalness adopt an assessment technique whereby listeners are played some speech and simply asked to rate what they hear. Often this is in the form of a **mean opinion score**, where listeners indicate their assessments on a scale of 1 to 5, where 1 is bad and 5 is excellent. These tests are fairly simple to perform and give a reasonably accurate picture of a system's performance. The main difficulty lies in how to interpret the results; specifically, in determining what the listeners are actually assessing. In principle, we would like the listeners to state how *natural* the system sounds, but it seems that in general listeners are actually recording how much they *like* the system. This can be seen in tests where real speech is included along side synthetic speech [308]. Often the likeability and naturalness of a system are highly correlated, but problems can occur when we consider that different scores are achieved when real speech is being used in the test. If naturalness were being assessed all real speech would receive a perfect 5 rating; but this is not the case: "pleasant"-sounding voices outscore unpleasant or jarring voices.

A further problem can occur in that it is sometimes difficult to compare results from tests carried out at different times and under different conditions. Just because a system scores 3.5 in one test it is not necessarily better than another system that scored 3.2 in a different test. MOS tests are probably best seen as ranking tests, such that if, say, five systems are compared in a single test we can rank them with confidence, but should be wary of comparing their absolute scores with those from other tests. Some investigators include some natural speech in MOS tests, since this should always achieve a score of 5 and hence "anchor" the scale. A common technique that bypasses the problem of absolute quality judgements is to perform **comparison tests**. This is often done when a developer wishes to assess the quality contribution from a new component or technique. The same sentence is played through the old and new synthesisers and a listener is asked which is preferred (without of course knowing which system is which) [462].

17.2.4 Test data

It is vital that the **test set** used in testing is not used in training. If it is, any automatic algorithm will **over-fit the data**, meaning that it will show very good performance on these data, but, since it has not generalised, will give poor performance elsewhere. In addition, strictly speaking a developer should not even *look* at the test data, since doing so could bias any design decisions. A commonly adopted solution is to have a **development test set**, which is used many times during the development of an algorithm, and an **evaluation test set**, which is saved and left unexamined, to be used as a one-off test at the end of a development period. Ideally the size of the test sets should be set on the basis of statistical significance such that results on one test set will be very close to results on another.

17.2.5 Unit or component testing

This type of testing is used by TTS developers to help them improve specific aspects or components of the system they are working on [516]. Each system will of course have its own set of components and hence we can't provide a single exhaustive testing framework that will encompass all systems. We can, however, identify several important areas in which testing can take place.

When considering designing a test for a component, the two most important questions are is the output discrete or continuous, and is there a single correct answer? The easiest components to assess are those with discrete outputs in which there is a single correct answer. Such components are usually classifiers and include homograph disambiguation, semiotic classification, POS tagging and LTS rules. The methodology for assessing these is to design a test set in which the correct answers have been marked by hand. The components are given the text of the test sentences as input and their output is compared with the truth of the hand-marked versions. From this a percentage accuracy can be computed.

The situation is more difficult in cases in which there is more than one correct answer. This arises in verbalisation, where for example there are many possible ways of converting 3:15 into words, for example THREE FIFTEEN or A QUARTER PAST THREE. One way of handling this is to mark all the possible outputs for a test sentence. This can prove impractical, so an alternative is to have the output assessed "live" by a developer or test subject to see whether the output is acceptable.

The problem is even more complicated when the output is continuous, as is the case in F0-synthesis algorithms and when assessing the speech-synthesis modules themselves. For F0 synthesis, an equivalent of the above tests is often used, whereby a set of real-speech recordings is used for the test corpus. The F0 contours of these, measured by a pitch-detection algorithm, are compared with the F0 contours produced by the system. Popular metrics include the Euclidean distance and correlation measures [98]. The problem here is that the Euclidean distance between two contours only vaguely correlates with human perception. For instance, slight differences in alignment may result in quite different perceptions by the listener, whereas large differences in F0 values between pitch accents may be irrelevant [269]. Further difficulties arise because nearly all synthesis components have more than one valid answer. As with verbalisation, one solution is to have multiple versions of the correct answer, but again too many sentences may be required to make this a comprehensive solution.

It is when assessing speech itself that we are presented with the most-difficult problems. This aspect is not only continuous and has many correct answers, but, worse than with F0 comparison, there isn't even a simple distance metric (such as the Euclidean distance) that can help in such cases. Some attempts at objective testing have been tried, for instance by measuring the distances between cepstra of the synthetic and reference sentences, but this again suffers from the problem that such a comparison only crudely compares with human perception. When assessing speech output, listening tests are nearly always required. Even this can be problematic, since it is very hard to get normal listeners (i.e. not linguists or speech engineers) to listen to just one aspect of the output. Alvarez and Huckvale [12] conducted an extensive study of several systems along several dimensions (e.g. listening effort, word distinguishability) and found a broad correlation, indicating that some idea of overall quality was always present.

One solution to assessing components separately was devised by Bunnel et al. [75], who developed a technique of transplanting prosody from one utterance to another, and, by comparing scores on all possible combinations, could determine individual contributions to overall quality separately. Hirai et al. [204] describe a technique whereby listening tests are performed on fully automatic and hand-corrected versions of sentences, which helps to show which errors listeners judge most harshly. Sonntag and Portele [408] describe a technique whereby sentences are delexicalised, meaning that the listener could listen only to the prosody. Several techniques were then used to elicit and compare judgments from listeners.

One final point to consider when looking at unit testing is the danger of local optimisation. This can occur when the individual components each give a good score, but when combined do not work together effectively, resulting in a poor overall performance. Wherever possible, unit tests should be compared with system tests to ensure that an increase in score in a unit test does in fact lead to an overall improvement in the system as a whole.

17.2.6 Competitive evaluations

The speech-recognition research community has long adopted a policy of **competitive evaluation** whereby different research groups run their systems on the same test data and report results. In this way, performance differences between systems can easily be seen, and the competitive nature of the evaluation programme has been credited with driving progress forwards.

At the time of writing, the Blizzard Challenge [40], [48], [50] is becoming a recognised standard in TTS testing. Blizzard involves testing a number of TTS systems on the same materials and at the same time. This ensures that the testing conditions are as even as possible. While this development is new, if it can match the success of evaluation competitions in other areas of speech and language processing it should have enormous positive benefits on the rate of progress in TTS.

17.3 Audio-visual speech synthesis

In many real-world situations, we communicate face-to-face. While speech, encoding language, *seems* to be the medium by which communication is primarily performed, face-to-face situations provide a considerable amount of what is frequently called **non-verbal communication**, and it is clear that people attribute significant value to this. In principle, everything in business could be conducted over the telephone, but in fact busy people will travel considerable distances just to engage in face-to-face communication.

It is natural then that many have considered the issue of how to generate facial and other non-verbal aspects of communication automatically, so that, in addition to the synthetic speech, a more-comprehensive communication process can be created. The field of **audiovisual synthesis**, also known as **talking heads**, is concerned with the artificial generation by means of computer graphics of natural-looking faces that speak in a realistic fashion. The problem involves creating realistic-looking faces, animating them in realistic ways, and then *synchronizing* their movement and behaviour with a speech signal, to generate the illusion that the face is doing the actual talking.

Developments in this field have mirrored those in speech synthesis to a considerable degree and Bailly et al. [30] give a review of the developments in this field.[1] Many early systems comprised **mesh** models in which the face and the vocal organs are described as a series of polygons whose vertices are repositioned to create movement. In essence articulation is performed by moving the points in the three-dimensional mesh, and subsequently a face can be drawn by filling in the polygons in the mesh and by other basic graphical techniques such as smoothing. The models can be rotated and viewed from any angle and give a recognisable, but somewhat artificial-looking, talking head [41], [101], [346], [347].

These models offer a basic set of movements that far exceeds what a real face can achieve, so a research goal has been to find the natural patterns of movements which constitute an effective set of control parameters. A basic problem concerns *where* in the overall process the control should be manifested. While we may be able to build a realistic lip model that exhibits the correct degrees of freedom for the lips, the lips themselves are the outputs of other more-fundamental controls such as speaking, smiling and other actions governed by other facial movements. An attractive solution, then, is to use more-fundamental controls based on articulatory degrees of freedom.

An alternative approach is to create models that mimic the actual biomechanical properties of the face, rather than simply rely on deformations of the grid [154], [354], [490], [491], [492]. This is the parallel approach to articulatory synthesis described in Section 13.4, and the pros and cons are just the same. While this in a sense is the "proper" and ultimate solution, the enormous complexities of the muscle movements involved make it a complex process. Furthermore, as with articulatory synthesis, there is no single solution as to how complex each muscle model should be; approaches range from simple models to close mimicry. At present the computational requirements and

---

[1] This section draws heavily on Bailly and Holm's paper [31].

complexity of this approach rule it out for engineering purposes. It continues to be an interesting field for scientific purposes, though, as is articulatory synthesis itself.

These approaches are attractive from a modelling point of view, and offer a powerful means of controlling all the movements of the face. On their own, though, they don't generate highly natural faces; it is clear that a computer image is being viewed. Unlike with speech, however, highly artificial, stylised faces are in many applications completely acceptable, for example in cartoons and other types of animation. The annoyance factor which nearly everyone feels when listening to highly unnatural speech is absent when viewing highly artificial faces. In many applications, however, more-realistic face generation is required. To achieve this with the above techniques, we can use a process of **texture mapping**. Here we first obtain a three-dimensional image of a real face (from photographs at various angles), and identify a series of points on this that correspond to the points on the mesh. The image can then be warped using the movements of the mesh [371].

Again following parallels in speech synthesis,[2] data-driven techniques have also been developed. For example, the **video rewrite** approach [66] collects a series of video images of a real face talking. The face is then divided up into sections and a unit-selection-style approach of picking units and concatenating them together is used. A number of techniques specific to graphics can be used to cope with artefacts arising from the concatenation. A second approach is to use statistical analysis to find the principal components of facial movement, resulting in an **eigenface** model (cf. Section 17.4.2 on eigenvoices). Once identified the eigenfaces can be linearly combined to generate a full scale of facial patterns.

17.3.1 Speech control

Turning to the issue of making the face talk, we see that any idea of co-production of the speech signal and facial images is discarded, and instead the speech signal is taken as primary and used to drive the control of the face. **Visemes** are the minimal set of distinct facial movements related to speech. In general there are fewer of these than there are phonemes, since the dimensions of distinctive features for speech which are not visible can be ignored. The most important of these is voicing, so that /b/ and /p/ are represented by a single viseme, as are /k/ and /g/. The distinctiveness of facial movements related to speech is what makes lip reading possible, and, although the confusability is inherently higher than for speech, the basic redundancy in language normally makes lip reading by skilled practitioners possible. Generally the synchronization between speech and facial movements is performed at the phone level, where each phone and its duration are given as input to the process controlling the facial movements. Greater precision in synchronization (e.g. with stop consonants) would potentially be possible if sub-phone units were used.

Just as with speech, however, we do not have a discrete model in which one viseme has one facial shape, which is held constant until the next viseme. Rather the face is in constant and smooth motion. Furthermore, content plays an important part, just as

---

[2] Note, we are not making any claim that TTS was *first* in any of these developments.

with speech. One approach to production of realistic facial articulation is Cohen and Massaro's **co-production** model [101], which uses targets and interpolation techniques. A different approach is to collect a much larger set of units in context and use patterns in the data to model the context effects.

17.4 Synthesis of emotional and expressive speech

Until recently, most synthetic speech was delivered in what can be described as a neutral style, somewhat akin to the tone of voice that a television newsreader would use. Recently, though, there has been considerable interest in going beyond this and building systems capable of generating emotional or expressive speech. The interest in this is partly due to a realisation that synthesis systems need to generate a wider range of affects if they are to become fully usable in certain applications. In addition, there is a general feeling that many of the problems of neutral speech synthesis, while not quite solved, are in the "end game", so researchers, as beings who like a challenge, are moving on to some of these more-advanced topics.

17.4.1 Describing emotion

The scientific study of emotion can be seen as starting with Darwin [124], who postulated that human emotions were innate, biological mechanisms that had existed before the development of language. A key study was that performed by Ekman [152], who conducted experiments across cultures and showed that there is a considerable universality in human emotion. Ekman used what are now called the **big six** emotional categories of **disgust, anger, fear, joy, sadness** and **surprise** in his experiments, and these have been adopted widely as a popular taxonomy of emotions. Quite how emotions should be described is a difficult topic; while we can all describe the meanings given by the terms "hesitancy", "aggression", "humiliation" and so on, it is quite a different matter to come up with a solid scientific model of the full range of emotions. Practical research has to proceed on some basis, however, and the commonest approaches are to adopt the six emotions described above or a subset of them, usually happiness, anger and sadness. An alternative is to use the system developed by Schlosberg [393], who described emotion with the three continuous variables of **activation, evaluation** and **power**. The debate about defining a fundamental taxonomy of emotions will not be settled anytime soon. Good reviews from a speech perspective of the full range of theories of emotion and their interaction with language are given in Cowie *et al.* [113] and Tatham and Morton [431].

17.4.2 Synthesising emotion with prosody control

The techniques for synthesizing emotion have followed closely the general developments in speech-synthesis algorithms. "First-generation" techniques include the work by Murray [324] and Cahn [79], [80], who used formant synthesisers, and thus were able to vary every parameter of the synthesiser as desired. The experimental paradigm used in these systems was to synthesise multiple versions of the same sentence each with a different emotion, and then perform a listening test in which a subject had to make a forced choice as to which emotion he or she is listening to. The results of these systems are quite good, with most emotions being recognised with a fair degree of accuracy. One of the problems with this approach, though, is that formant synthesis lends itself to being able to generate quite distinct linguistic effects, but in doing so can create speech in which the "distance" between two categories is artificially large. In other words, as with other aspects of formant synthesis, intelligibility (that is distinctiveness of emotions) is achieved only at the expense of naturalness.

Many "second-generation" techniques have been developed [73], [520]. The approach here is data-driven in nature, such that typically a database containing labelled emotional speech is analysed to determine the characteristics of each emotion. Often these are taken to be the traditional prosodic dimensions of F0 and timing. Once these patterns are known, normal speech can be converted into the emotional speech by using one of the signal-processing techniques described in Chapter 14 to change F0 and timing. In forced tests of the type described above, these systems perform quite well, but again naturalness suffers to some extent because only the F0 and timing are being changed; other important factors such as voice quality are often ignored.

As techniques become in general more data-driven, the issue of how to collect emotional data comes to the fore. There are basically two approaches. One can collect natural, perhaps spontaneous speech from a conversation and then label each sentence with regard to its emotional content. This produces the most-natural data, but has the difficulty that the speech then has to be labelled, difficulties which should not be lightly dismissed as we saw in Section 17.1.4. Databases collected in this way are described by Campbell [81] and Douglas-Cowie et al. [112]. An alternative strategy is to have actors produce sentences intentionally spoken with particular emotions. Since the type of emotion is an "input" to this process, no labelling is required. This approach has a different problem in that the resultant speech might not be particularly natural; an actor's version of a particular emotion may be too "hammed up".

17.4.3 Synthesising emotion with voice transformation

A further approach, which may perhaps still be called second generation, is to use more-general signal-processing techniques that change other aspects of the signal. The principle is still the same; a general speech database is used to build a neutral-sounding synthesizer, which can then be adapted to sound as if speaking with a particular emotion. One way of achieving this is to borrow from the field of **voice transformation**. The goal here is to take speech from one speaker and transform it so that it sounds like another. The words are unchanged but every other part of the signal can be adapted to sound like the new, target, speaker. It is beyond our scope to give an account of this interesting field; the main point is that some of the techniques developed for converting speech from one speaker to another can also be used to convert from one *emotion* to another [120], [222], [246], [430], [457], [458].

17.4.4 Unit selection and HMM techniques

There is nothing to stop the techniques just described being used for unit-selection synthesis [47]. The problem is not that the techniques would not create speech that reflects the desired emotion, but rather that the whole approach of using signal processing to modify the signal may lead to degradation that would lower the overall quality of the unit-selection speech. The "pure" unit-selection technique would of course be to record all the normal material but a number of times, once for each emotion. This is, however, nearly universally acknowledged to be impractical insofar as every additional emotion will mean that several hours of extra speech will have to be recorded. Alternative solutions have therefore been put forth. Van Santen *et al.* [465] describe a hybrid scheme in which a general unit-selection database is mixed with units recorded with specific emotion states. Aylett [25] describes a system in which a normal unit-selection system (i.e. one using a normal database without signal processing) is used to synthesise particularly emphatic speech. The result is quite convincing and works on the principle that in the natural distribution of units there will be ones that *when combined* can produce effects quite different from the original speech. In principle this technique could be extended to other emotions.

HMM synthesis arguably provides the best technique for synthesizing emotional speech [222], [230], [457]. This requires far fewer data than unit selection, so in principle it can be used in a situation where representative data from each required emotion are collected and used for system training. Beyond this, however, HMMs have a number of other advantages, primarily arising from the ability of the model parameters to be modified. This implies that a small amount of target data can be used to modify globally a normal HMM synthesizer to convey the emotion of the target data. Furthermore, spectral and temporal parameters can be modified as well as F0. Finally, HMMs themselves may offer a solution to the emotion-taxonomy problem. Shichiri *et al.* [404] describe a system whereby they use a principal-component technique to determine the **eigenvoices** of a set of trained models. These effectively describe the main dimensions of variation within a voice. Once found, their values can be altered and a number of different voices or emotions can be created. This is a purely data-driven technique. The result is quite curious and the variations sound *almost* like variations attributable to recognised emotions such as happiness and sadness.

We are only in the very early stages of research into emotion in speech in general and speech synthesis in particular. Thus much current interest involves database collection and analysis [22], [34], [81], [82], [112], [130] and ways to describe and control emotion [150]. The study of emotion and expression in speech synthesis will certainly grow into a rich and exciting major topic of research.

17.5 Summary

Databases
- Most modern synthesis techniques require a database.
- The quality of unit-selection databases is a critical factor in overall performance, with special attention being required for choice of speaker and recording conditions.

- There are two types of human labelling, intuitive and analytical.
- Intuitive labelling can be performed by non-specialists and draws on the labellers' own language faculty. It is generally very useful and accurate.
- Analytical labelling, also known as expert labelling, is performed by a labeller following rules and using a linguistic theory. In general it is error-prone and best avoided.
- Analytical labelling can usually be performed better by a computer or can be avoided altogether by use of hidden variables.

Evaluation
- A good testing regime is one within which user requirements are specified ahead of time and tests are subsequently designed to see whether the system meets these requirements.
- Intelligibility testing is often performed with the modified-rhyme test or semantically unpredictable sentences. Although these give a measure of word recognition, care should be taken when using results from these tests to indicate performance in applications.
- Naturalness is normally measured by obtaining mean opinion scores (MOS).
- Testing of individual components in TTS is often difficult due to there being continuous outputs, differences between objective measures and perception and components having more than one correct answer.

Audio-visual synthesis
- This can be performed in a number of ways, ranging from use of explicit models that afford a large degree of control to data-driven techniques that use video clips as basic units.

Emotion
- There is no widely agreed set of emotions, but common categories include the six emotions of disgust, anger, fear, joy, sadness and surprise, and the three continuous dimensions of activation, evaluation and power.
- "First-generation" synthesis systems have specific rules for each type of emotion.
- "Second-generation" techniques transplant prosody patterns of particular emotions onto neutral speech.
- Voice-conversion techniques can be used to transform speech from one emotion to another.
- HMM synthesis offers a powerful way to learn and manipulate emotion.

18 Conclusion

We finish with some general thoughts about the field of speech technology and linguistics and a discussion of future directions.

18.1 Speech technology and linguistics

A newcomer to TTS might expect that the relationship between speech technology and linguistics would parallel that between more-traditional types of engineering and physics. For example, in mechanical engineering, machines and engines are built on the basis of principles of dynamics, forces, energy and so on developed in classical physics. It should be clear to a reader with more experience in speech technology that this state of affairs does not hold between the engineering issues that we address in this book and the theoretical field of linguistics. How has this state of affairs come about and what is the relationship between the two fields?

It is widely acknowledged that researchers in the fields of speech technology and linguistics do not in general work together. This topic is often raised at conferences and is the subject of many a discussion panel or special session. Arguments are put forward to explain the lack of unity, all politely agree that we can learn more from each other, and then both communities go away and do their own thing just as before, such that the gap is even wider by the time the next conference comes around.

The first stated reason for this gap is the "aeroplanes don't flap their wings" argument. The implication of this statement is that, even if we had a complete knowledge of how human language worked, it would not help us greatly because we are trying to develop these processes in machines, which have a fundamentally different architecture. This argument certainly has some validity, but it does not explain the rift, since we can point to many cases where some knowledge of human language has proved useful and we can identify other areas where a breakthrough in fundamental knowledge would help greatly. A second stated reason is that "our goals are different", and this again is partly true. A quick glance at a linguistics journal will show many papers on languages other than the main European and Asian languages. Furthermore, we find that some areas of linguistic study (e.g. syntax) receive an enormous amount of attention, while others, such as prosodic phrasing, are rarely studied at all. This focus on particular areas is not of course unique to our fields; there are many areas of physics that are keenly investigated but have no immediate engineering purpose. A final reason that is often given is that there is a "cultural divide". This certainly exists; I am saddened by the fact that most linguists have little or no understanding of how an HMM speech-recognition system works, and that most engineers cannot read spectrograms or perform simple phonetic transcriptions. While it is unrealistic to expect people in one field to understand the advanced concepts in another, the fact that many courses regularly teach phonetics and HMMs to students shows that the concepts cannot be too difficult to grasp.

These reasons are certainly valid, but in my opinion don't give the full picture. Considering the issue of differences in culture, it is important to realise that this was not always the case. In the 1970s, most speech-technology groups were mixtures of people from linguistics, electrical engineering, artificial intelligence and so on. The gap in culture started to develop then and has been spreading ever since. The reason is quite clear, namely that the theories and models the linguists proposed proved not to work robustly enough for speech-technology purposes. A growing frustration was felt by the engineers, who gradually became tired of implementing yet another linguistic theory only to find that it did not work in a real system. This state of affairs was expressed by Fred Jelinek, who famously said that "whenever I fire a linguist our system performance improves" [238]. The question, though, is *why* did the knowledge of linguists prove so useless (and indeed damaging insofar as it diverted resources from other approaches)?

Linguistics itself is not a homogeneous field and there are many different schools and points of view. For discussion purposes, we can divide the field very crudely into a number of camps. The most-famous camp of all is the Chomskian camp, started of course by Noam Chomsky, which advocates a very particular approach. Here data are not used in any explicit sense, quantitive experiments are not performed and little stress is put on explicit formal description of the theories advocated.[1] While Chomskian linguistics was dominant from the 1960s onwards, it is now somewhat isolated as a field and rarely adopted as a model of language outside the immediate Chomskian camp. We can identify another area of linguistics, which we shall call experimental linguistics, which really should be the "physics of language" that would serve as the scientific bedrock of speech engineering. Within experimental linguistics we include the fields of experimental phonetics, psycholinguistics and many other areas of the study of discourse and dialogue. Here traditional scientific methods of experimentation are rigorously used and hence this field cannot be dismissed in the way the Chomskian field can. While accepting that to a certain extent the focus of this field is often on different issues, we do find that many studies are conducted in just the areas that would be of benefit in speech synthesis. So why, then, are these studies ignored?

In my opinion, the answer lies in the **curse of dimensionality** introduced in Section 5.2.1. Often experimental linguistics (and particularly phonetics) is conducted in the framework of a controlled experiment, where some data is collected with the

---

[1] Note that in his early work (pre-1970s) Chomsky put considerable emphasis on formally describing his theories, for example *The Sound Pattern of English* [91] is full of explicit rules and formulas. From this point on, however, the formal side more or less disappeared and recent work is not described in terms familiar to those from a traditional scientific background.

idea of holding everything constant apart from the phenomenon under investigation. For example, if we were conducting a study on the shapes of pitch accents, speech would be carefully collected which was the same in most ways, but just varied with regard to these pitch accents. Often patterns emerge and from these linguistic rules are proposed as a theory. Often such work can't be faulted in its experimental design, procedures and methodology and the results show consistent patterns in how speakers producte pitch accents. So why then do such studies not form the basis of pitch accent generation in TTS? What then is the problem in applying this knowledge?

The problem lies in the **generalisation** or **scalability** of the results. While the results are probably quite valid for the conditions in which they were measured, we simply can't extrapolate these results to other situations. For example, the phonetic contexts in which the pitch accents appeared are all of a particular type: what then do we do if we (as will definitely be the case in TTS) want to synthesize a pitch accent for a different segmental context? The study woul typically involve only speakers of one regional accent, but what about the others? What about differences in speaking rates? What happens when the accented syllables are at different positions in the words or phrases or sentences? What about different languages and do differences occur in conversational, informal or disfluent speech? As we can see, the number of unknowns is vast, and while the studies carried out in this area suffice to shed light on those areas; this amounts to only a tiny fraction in the types of situations that can occur. Worse still, as we explained in Section 5.2.1, it is generally not the case that features operate independently; so if we for instance adopted a tactic which is common in experimental linguistics where we hold all variables (features) constant and then change just one, we might think that by doing this for every feature we could build a complete model. But this would rarely suffice; the features interact and hence studying the effect of each feature separately will not lead to complete knowledge of the problem. These of course are exactly the same problems which occur in machine learning; rarely do simple rules or simple classifiers (like naive Bayes) work very well for exactly this problem of feature interaction. Successful machine learning techniques then are those which acknowledge problems of data sparsity, the interaction of features and the curse of dimensionality. While no silver bullet yet exists in machine learning, at least the problem is clearly identified and handled with the more sophisticated learning algorithms.

In my view, then, this is the main reason why linguistic knowledge of the traditional kind does not and *will not* be applicable in speech technology. I do not have an answer to this problem, but would at least argue for an acknowledgment of this fundamental difficulty of generalisation and scalability as being the source of the problem. I certainly think that the field of experimental linguistcs could benefit from engaging with the knowledge acquired in machine learning. Specific machine-learning techniques do not have to be adopted, but the procedures and understanding used for dealing with complex, messy and sparse data would surely help build more-scalable models of human language.

I will finish, though, by saying that a solid knowledge of the basics of linguistics has been invaluable to me in my own TTS research, and that, in my view, a solid knowledge of acoustic phonetics is one of the key skills required to build a good unit-selection system. I would therefore still recommend that anyone wishing to work in TTS should obtain a basic understanding of key areas in linguistics.

18.2 Future directions

In the words of Sam Goldwyn, "Never make predictions, especially about the future" [488]. In general, when we look back on predictions of the future made in the past, the comparisons between our life now and the prediction are laughable. So, considering the futility of making predictions, it would be easily forgivable if I were to forgo this section, but, perhaps mainly to give future readers just such a laugh, I will now state how I see the future of text-to-speech evolving.

Text analysis A sensible starting point for any prediction would be to extrapolate current trends. To this extent, it is reasonable to assume that TTS will become entirely data-driven. I think it is incontestable that the front end, or text-processing component, will become entirely statistical. In recent years the advances in statistical NLP have been enormous, having been fuelled by the use of search engines and the need to translate documents. I think many of these techniques are directly applicable to TTS and will be adopted.

Integrated systems I think that systems will in general become more integrated, to the extent that there may be only a single integrated text-analysis component and a single integrated speech-synthesis component. It is of course possible that a single system could do both, but from my perspective I can't see how this could be done. This will have the greatest consequence in the field of prosody. I don't think that the traditional approach to prosody in TTS, whereby F0 contours and timings are generated by algorithm, has much future. This is because prosody affects so much more in the signal, and for truly realistic prosody to be generated all aspects of voice quality need to be modelled. Since this is so intertwined with the phonetic aspects of synthesis it makes sense to generate the phonetics, voice quality and prosody parts of synthesis in an integrated fashion. This to some extent has already happened in many unit-selection and HMM systems. If explicit F0 modelling is still used, it will be done by statistical algorithms such as dynamic-system models and HMMs; deterministic models such as Tilt and Fujisaki will be used less.

Synthesis algorithms A harder prediction concerns speech-synthesis algorithms. A few years ago, unit selection seemed the dominant technique, but recently HMM synthesis has caught up, and, depending on how one evaluates the systems, HMM synthesis may even have the edge at present. The main thing holding HMM synthesis back is the fact that the signal-processing algorithms that are required to generate the speech from coefficient representations still have buzziness and other artefacts. Signal processing in TTS is now receiving much attention after having been somewhat neglected at the height of unit selection. If these problems can be overcome, I believe that HMM or other statistical synthesis techniques will become dominant, because they allow so much more control over the synthesis than the inherently hit-and-miss nature of unit selection.

Quality improvements In terms of overall quality and performance, I think that the problems of text analysis can be fully solved with today's technology; all that is needed are good-quality databases and a concerted drive to increase performance. I also believe that the best synthesis systems are close to being fully acceptable for certain styles of speaking, in particular the neutral styles required for news reading. In more-limited domains, e.g. weather reports, systems are on the brink of passing Turing tests to the extent that listeners can't tell (or, more significantly, don't care) that they are listening to a TTS system. The overall problem is far from solved, however; the amount of data needed in order to train a unit-selection system is unattractive for quickly adding a new voice, and the range of speech styles that can be elicited is limited.

Concept-to-speech In Section 3.4.2 we briefly mentioned **concept-to-speech** systems, which can directly take a structured linguistic input, bypassing all problems of text analysis. Concept-to-speech systems have to a large extent failed to take off, and it is not entirely clear why this should be so, given the potential advantages. One reason may be that the sophistication of dialogue systems and other systems that use language generation is not yet at a level that could, for instance, make good use of discourse prosody. An alternative explanation is that while text may be inherently ambiguous it is at least familiar and can even be described as a "standard", resulting in it still being an attractive interface to a system. So, regarding the future, the question isn't so much whether concept-to-speech systems will play a prominent role in future speech-synthesis systems, but rather why they haven't already done so.

Relationship with linguistics Going back to the issue of the interaction between the fields of speech-technology and linguistics, we find many who believe that speech-technology problems will be solved only by using the results of basic research in linguistics. It should be no surprise to the reader by now that I completely disagree with such an idea. To put my view clearly, *the field of linguistics, as currently practiced, will never produce any research that will facilitate a major leap forwards in speech technology*. This is for the reasons explained above, namely that much of linguistics is not experimental and so has no real evidence to back up its claims, while that part of linguistics which is experimental operates within a paradigm that makes its results inherently difficult to scale because of the fundamental problems of the curse of dimensionality.

In fact, my prediction is that just the opposite will occur, such that the field of linguistics will sooner or later cease to exist in its current form, and the rigorous experimental, data-driven, formal, mathematical and statistical techniques developed in speech and language technology will be adopted and start to provide the answers to many of the long-running problems in linguistics. This is already happening to some extent, in that research using probability is making inroads into mainstream linguistics (see [59] for instance). The progress is slow, though, and one gets the feeling that the researchers using probability in linguistics do so somewhat apologetically. It is to be hoped that this transition period will not last too long, and that the techniques used in speech technology will quickly find their way into mainstream linguistics. If this does in fact happen, the benefits would be enormous for all, since, for the first time, those in speech technology could start to use linguistic knowledge directly. Furthermore, the use of these techniques in linguistics should solve many of the problems in that field and lead to a more-accurate and richer knowledge of how human language works.

Applications   The most-difficult prediction concerns a topic that hasn't been mentioned much in the book, namely applications for TTS. After nearly 40 years of TTS research and system building, it is fair to say that no "killer application" for TTS has yet emerged. This is fundamental to all aspects of research insofar as it is ultimately the commercial use of TTS that funds and maintains interest in the field. In the late 1990s many thought that telephony dialogue systems were the killer application for speech technology, and many companies achieved considerable success in this area. At the time of writing, though, the market for this has been flat for several years, and shows no sign of improvement. I am doubtful that dialogue systems will be a killer for TTS or other speech technologies.

The use of the web will gradually replace nearly all automated telephony services and the types of systems currently being used will be discontinued. If given a choice, people simply do not want to converse with computers. The main problem with dialogue systems as they currently stand is that they do not perform any real *understanding*. Hence the dialogues have to be heavily constrained and managed or else low performance will result. What would make dialogue systems genuinely popular would be where people could spontaneously and naturally speak and be *understood*, in much the same way as they would talk to a real person. This of course requires speech-understanding technology, which is many years away (perhaps decades). TTS is simply a component in dialogue systems, but its adoption is inherently tied up in the success of the overall systems.

I have no real insight or ideas as to where the killer application for TTS lies, or even whether there will be one. This should not be interpreted as overly negative, however; the history of technology is full of sudden shifts and of technologies suddenly being found useful in areas their inventors could never have imagined. Another perspective can be found if we look at the use of speech, not just speech synthesis, in technology. Here we see that the use of pre-recorded messages is widespread, from car navigation systems to computer games. If we in the speech-synthesis community can make TTS as good-quality and as easy to use as recorded spoken messages, then the technology will surely be adopted due to its increased flexibility. As Moore [318] points out, one of the key problems of all speech technology today is that each component performs its function in a processing "vacuum", and that the model of communication described in Chapter 2 is ignored. Successful interaction by natural language can occur only when the components are sensitive to their roles in the communication process.

18.3 Conclusion

This then concludes *Text-to-Speech Synthesis*. Despite the continual exhortations to adhere to engineering principles through the book, I will finish by saying that building TTS systems should always be regarded as a fun activity. In academic life, I have seen that virtually everyone regards their own field as more important or more interesting than any other. I will make no such claims about TTS, but I will claim that there is something about making computers talk that is simply *fun* and that in this it may have an edge over some other activities. After doing this for 18 years I still find it a great moment when a new system speaks for the first time and I never fail to laugh at the sometimes absurd output of even the best systems. The demonstrations of speech synthesis I give at conferences nearly always attract a smile and often laughter, both with me and at me. I would encourage any reader who has not yet built a system to have a go, since doing so sheds immediate light on many ideas of how language and human communication work. In doing so a new convert might just agree that doing research and building systems in TTS is just about one of the most enjoyable things you can do with your working time.

Appendix A Probability

This appendix[1] gives a brief guide to the probability theory needed at various stages in the book. The following is too brief to be intended as a first exposure to probability, but rather is here to act as a reference. Good introductory books on probability include Bishop [45], [46] and Duda, Hart and Stork [142].

A.1 Discrete probabilities

Discrete events are the simplest to interpret. For example, what is the probability of

- it raining tomorrow?
- a **6** being thrown on a die?

Probability can be thought of as the chance of a particular event occurring. We limit the range of our **probability measure** to lie in the range 0 to 1, where

- lower numbers indicate that the event is *less* likely to occur, 0 indicates it will *never* occur;
- higher numbers indicate that the event is *more* likely to occur, 1 indicates that the event will *definitely* occur.

We like to think that we have a good grasp of both estimating and using probability. For simple cases such as "will it rain tomorrow?" we can do reasonably well. However, as situations get more complicated things are not always so clear. The aim of probability theory is to give us a mathematically sound way of inferring information using probabilities.

A.1.1 Discrete random variables

Let some event have have M possible outcomes. We are interested in the **probability** of each of these outcomes occurring. Let the set of possible outcomes be

$$\mathcal{X} = \{v_1, v_2, \ldots, v_M\}$$

---

[1] Nearly all this appendix is taken from the lecture notes of Mark Gales given for the MPhil in Computer Speech, Text and Internet Technology at the University of Cambridge.

The probability of a particular event occurring is

$$p_i = P(x = v_i)$$

where x is a discrete **random variable**. For a single die there are six possible events ($M = 6$):

$$\mathcal{X} = \{1, 2, 3, 4, 5, 6\}$$

For a fair die (for example),

$$p_1 = \frac{1}{6}, \quad p_2 = \frac{1}{6}$$

and so on.

A.1.2 Probability mass functions

It is more convenient to express the set of probabilities

$$\{p_1, p_2, \ldots, p_M\}$$

as a **probability mass function (PMF)**, $P(x)$. Attributes of a probability mass function (PMF) are

$$P(x) \geq 0; \quad \sum_{x \in \mathcal{X}} P(x) = 1$$

- The first constraint means that probabilities must always be positive (what would a negative probability mean?).
- The second constraint states that one of the set of possible outcomes must occur.

From these constraints it is simple to obtain

$$0 \leq P(x) \leq 1$$

A.1.3 Expected values

There are useful **statistics** that may be obtained from PMFs. The **expected value** (think of it as an average) is often extracted,

$$\mathcal{E}\{x\} = \sum_{x \in \mathcal{X}} x P(x) = \sum_{i=1}^{M} v_i p_i$$

This is also known as the **mean**:

$$\mu = \mathcal{E}\{x\} = \sum_{x \in \mathcal{X}} x P(x)$$

The mean value from a single roll of a die is

$$\mu = 1\frac{1}{6} + 2\frac{1}{6} + 3\frac{1}{6} + 4\frac{1}{6} + 5\frac{1}{6} + 6\frac{1}{6} = 3.5$$

The expected values over functions and linear combinations of the random variable are

$$\mathcal{E}\{f(x)\} = \sum_{x \in \mathcal{X}} f(x) P(x), \quad \mathcal{E}\{\alpha_1 f_1(x) + \alpha_2 f_2(x)\} = \mathcal{E}\{\alpha_1 f_1(x)\} + \mathcal{E}\{\alpha_2 f_2(x)\}$$

A.1.4 Moments of a PMF

The nth moment of a PMF is defined as

$$\mathcal{E}\{x^n\} = \sum_{x \in \mathcal{X}} x^n P(x)$$

The second moment and the **variance** are often used. The variance is defined as

$$\text{Var}\{x\} = \sigma^2 = \mathcal{E}\{(x - \mu)^2\} = \mathcal{E}\{x^2\} - (\mathcal{E}\{x\})^2$$

i.e. it is simply the difference between the second moment and the first moment squared. An attribute of the variance is

$$\sigma^2 \geq 0$$

Again taking the example of the die, the second moment is

$$\mathcal{E}\{x^2\} = 1^2 \frac{1}{6} + 2^2 \frac{1}{6} + 3^2 \frac{1}{6} + 4^2 \frac{1}{6} + 5^2 \frac{1}{6} + 6^2 \frac{1}{6} = 15.1667$$

Therefore the variance is

$$\sigma^2 = 15.1667 - 3.5 \times 3.5 = 2.9167$$

A.2 Pairs of discrete random variables

For many problems there are situations involving more than a single random variable. Consider the case of two discrete random variables, x and y. Here y may take any of the values of the set $\mathcal{Y}$. Now, instead of having PMFs of a single variable, the **joint PMF**, $P(x, y)$, is required. This may be viewed as the probability of x taking a particular value and y taking a particular value. This joint PMF must satisfy

$$P(x, y) \geq 0, \quad \sum_{x \in \mathcal{X}} \sum_{y \in \mathcal{Y}} P(x, y) = 1$$

Take a simple example of the weather – whether it rains or not on two particular days. Let x be the random variable associated with it raining on day 1, y being that for day 2. The joint PMF may be described by the table

$P(x, y)$	rain	sun
rain	0.4	0.2
sun	0.3	0.1
total		1.0

A.2.1 Marginal distributions

Given the joint distribution it is possible to obtain the probability of a single event. From the joint PMF **marginal** distributions can be obtained as

$$P_x(x) = \sum_{y \in \mathcal{Y}} P(x, y), \qquad P_y(y) = \sum_{x \in \mathcal{X}} P(x, y)$$

For ease of notation $P_x(x)$ is written as $P(x)$ where the context makes it clear. Take the rain and sun example,

$P(x, y)$	rain	sun	
rain	0.4	0.2	0.6
sun	0.3	0.1	0.4
total	0.7	0.3	1.0

Hence

$$P(x = \text{rain}) = 0.6$$
$$P(x = \text{sun}) = 0.4$$

and similarly for the marginal distribution for y.

A.2.2 Independence

An important concept in probability is **independence**. Two variables are statistically independent if

$$P(x, y) = P(x)P(y)$$

This is very important since it is necessary to know only the individual PMFs to obtain the joint PMF. Take the "sun and rain" example. Is whether it rains or not on the second day independent of what happens on the first day? Take the example of raining on both days and assume independence:

$$P(x = \text{rain}, y = \text{rain}) = P(x = \text{rain})P(y = \text{rain})$$
$$= 0.6 \times 0.7 = 0.42$$
$$\neq 0.4$$

So from the joint PMF the two random variables are *not* independent of one another.

A.2.3 Expected values

The expected value of two variables is also of interest:

$$\mathcal{E}\{f(x, y)\} = \sum_{x \in \mathcal{X}} \sum_{y \in \mathcal{Y}} f(x, y) P(x, y)$$

This follows directly from the single-variable case.

A.2.4 Moments of a joint distribution

Using vector notation, where

$$\mathbf{x} = \begin{bmatrix} x \\ y \end{bmatrix}$$

the first moment is defined as

$$\mu = \mathcal{E}\{\mathbf{x}\} = \begin{bmatrix} \mathcal{E}\{x\} \\ \mathcal{E}\{y\} \end{bmatrix} = \sum_{\mathbf{x} \in \{\mathcal{X}\mathcal{Y}\}} \mathbf{x} P(\mathbf{x})$$

A.2.5 Higher-order moments and covariance

In the same fashion as single-variable distributions, higher-order moments can be considerd. The one of most interest is the second-order moment

$$\mathcal{E}\{\mathbf{x}\mathbf{x}'\} = \sum_{\mathbf{x} \in \{\mathcal{X}\mathcal{Y}\}} P(\mathbf{x})\mathbf{x}\mathbf{x}'$$

This can be used to find the **covariance matrix** as

$$\Sigma = \mathcal{E}\{(\mathbf{x} - \mu)(\mathbf{x} - \mu)'\}$$
$$= \begin{bmatrix} \mathcal{E}\{(x - \mu_x)^2\} & \mathcal{E}\{(x - \mu_x)(y - \mu_y)\} \\ \mathcal{E}\{(x - \mu_x)(y - \mu_y)\} & \mathcal{E}\{(y - \mu_y)^2\} \end{bmatrix}$$

The covariance matrix may also be expressed as

$$\Sigma = \mathcal{E}\{\mathbf{x}\mathbf{x}'\} - \mu\mu'$$

Covariance matrices are *always* symmetrical.

A.2.6 Correlation

For two random variables x and y the covariance matrix may be written as

$$\Sigma = \begin{bmatrix} \sigma_{xx} & \sigma_{xy} \\ \sigma_{xy} & \sigma_{yy} \end{bmatrix} = \begin{bmatrix} \sigma_x^2 & \sigma_{xy} \\ \sigma_{xy} & \sigma_y^2 \end{bmatrix}$$

where

$$\sigma_{xy} = \mathcal{E}\{(x - \mu_x)(y - \mu_y)\}$$

and $\sigma_x^2 = \sigma_{xx}$. The correlation coefficient, ρ, is defined as

$$\rho = \frac{\sigma_{xy}}{\sigma_x \sigma_y}$$

This takes the values

$$-1 \leq \rho \leq 1$$

In general when $\rho = 0$ the two random variables are said to be **uncorrelated**. Note that independent random variables are always uncorrelated (you should be able to show this).

A.2.7 Conditional probability

The probability of an event occurring **given** that some event has already happened is called the **conditional probability**:

$$P(y|x) = \frac{P(x,y)}{P(x)}$$

This is simple to illustrate with an example. From the "sun and rain" example; "What is the probability that it will rain on the second day *given* that it rained on the first day?" From the above equation and using the joint and marginal distributions,

$$P(y = \text{rain}|x = \text{rain}) = \frac{P(x = \text{rain}, y = \text{rain})}{P(x = \text{rain})}$$
$$= \frac{0.4}{0.6} = 0.6667$$

It is worth noting that when the x and y are independent

$$P(y|x) = \frac{P(x,y)}{P(x)} = P(y)$$

A.2.8 Bayes' rule

One important rule that will be used often in the book is **Bayes' rule**. This is a very useful way of manipulating probabilities:

$$P(x|y) = \frac{P(x,y)}{P(y)}$$
$$= \frac{P(y|x)P(x)}{P(y)}$$
$$= \frac{P(y|x)P(x)}{\sum_{x \in \mathcal{X}} P(x,y)}$$
$$= \frac{P(y|x)P(x)}{\sum_{x \in \mathcal{X}} P(y|x)P(x)}$$

The conditional probability $P(x|y)$ can be expressed in terms of

- $P(y|x)$, the conditional probability of y given x;
- $P(x)$, the probability of x.

A.2.9 Sum of random variables

What happens if two random variables are added together (Figure A.1)?

- **Mean**: the mean of the sum of two random variables is

$$\mathcal{E}\{x+y\} = \mathcal{E}\{x\} + \mathcal{E}\{y\} = \mu_x + \mu_y$$

Appendix A

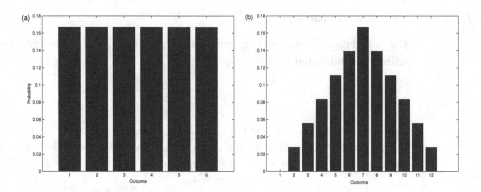

Figure A1 Outcomes with (a) a single die and (b) the sum of two dice.

- **Variance**: the variance of the sum of two independent random variables is

$$\mathcal{E}\{(x+y-\mu_x-\mu_y)^2\} = \mathcal{E}\{(x-\mu_x)^2\} + \mathcal{E}\{(y-\mu_y)^2\} + 2\mathcal{E}\{(x-\mu_x)(y-\mu_y)\}$$
$$= \sigma_x^2 + \sigma_y^2$$

A.2.10 The chain rule

The chain rule computes joint probabilities from conditional probabilities:

$$P(x,y,z) = P(x|y,z)P(y|z)P(z)$$

To see why this holds, consider what happens when we expand out the conditional probabilities with their definitions from Bayes' rule:

$$P(x,y,z) = \frac{P(x,y,z)}{P(y,z)} \frac{P(y,z)}{P(z)} P(z)$$

Each of the terms in the numerator cancels out with the previous term in the denominator, leaving us with the simple expression that $P(x, y, z)$ equals itself.

A.2.11 Entropy

It would be useful to have a measure of how "random" a distribution is. Entropy, H, is defined as

$$H = -\sum_{x \in \mathcal{X}} P(x) \log_2(P(x))$$
$$= \mathcal{E}\left\{\log_2\left(\frac{1}{P(x)}\right)\right\}$$

- $\log_2(\ )$ is log base 2, not base 10 or natural log.
- By definition $0 \log_2(0) = 0$.

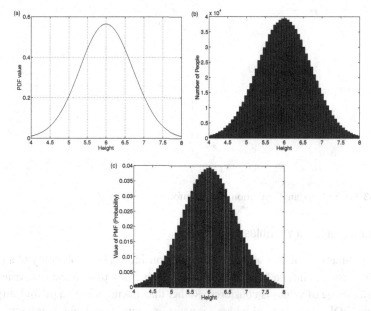

Figure A2 Plots of (a) a "true" distribution, (b) the raw counts and (c) the probability mass function.

For discrete distributions the entropy is usually measured in **bits**. One bit corresponds to the uncertainty that can be resolved with a simple yes/no answer. For any set of M possible symbols

- the PMF which has the **maximum** entropy is the uniform distribution

$$P(x) = \frac{1}{M}$$

- the PMF which has the **minimum** entropy is the distribution for which only a single probability is non-zero (and so must be unity).

A.3 Continuous random variables

Not everything in life is discrete. Many examples occur that are not, for example

- the correlation between height and weight
- precisely when the last student arrives at a lecture
- speech recognition ...

Continuous random variables are not as easy to interpret as discrete random variables. We usually convert them into discrete random variables, for example we quote our height to the nearest centimetre, or inch (Figure A.2).

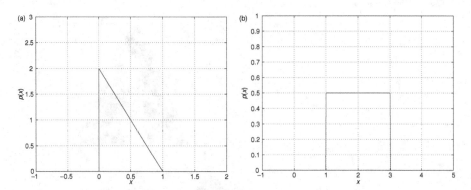

Figure A3 (a) A Gaussian distribution. (b) A uniform distribution.

A.3.1 Continuous random variables

Since continuous random variables are being considered, the probability of a *particular* event occurring is infinitely small. Hence the probability of an event occurring within a particular range of values is considred. Rather than having a PMF a **probability density function** (PDF), $p(x)$, is used, where x is now a continuous variable. This has the property

$$P(x \in (a, b)) = \int_a^b p(x)dx$$

In words this says that the probability that the random variable x lies within the range a to b is the *integral* of the probability density function between a and b. Consider a very small range, $(a, a + \Delta x)$, then it is possible to write

$$p(a) \approx \frac{P(x \in (a, a + \Delta x))}{\Delta x}$$

From this it is easy to see that

$$p(x) \geq 0, \qquad \int_{-\infty}^{\infty} p(x)dx = 1$$

As a simple example, consider the case

$$p(x) = \begin{cases} 2(1-x), & 0 \leq x \leq 1 \\ 0, & \text{otherwise} \end{cases}$$

What is $P(x \in (0.25, 0.75))$ for this distribution?

A.3.2 Expected values

In a similar fashion to the discrete case the expected values of continuous random variables can be obtained as

$$\mathcal{E}\{f(x)\} = \int_{-\infty}^{\infty} f(x)p(x)dx$$

This naturally leads to the moments of continuous variables. For example the mean, μ, is defined as

$$\mu = \mathcal{E}\{x\} = \int_{-\infty}^{\infty} x p(x) dx$$

For the simple example above,

$$\mu = \int_{-\infty}^{\infty} x p(x) dx = \int_{0}^{1} x(2(1-x)) dx = \frac{1}{3}$$

Similarly, the higher-order moments and variance may be calculated (if you have any doubts, work this out). The same rules apply for continuous variables as discrete variables for linear combinations.

A.3.3 The Gaussian distribution

The most commonly used distribution (for reasons explained later) is the **Gaussian** (or normal) distribution (Figure A.3). This has the form

$$p(x) = \frac{1}{\sqrt{2\pi\sigma^2}} \exp\left(-\frac{(x-\mu)^2}{2\sigma^2}\right)$$

Looking at the first few moments of a Gaussian distribution,

$$\mathcal{E}\{x^0\} = \mathcal{E}\{1\} = 1, \quad \mathcal{E}\{x^1\} = \mu, \quad \mathcal{E}\{x^2\} = \sigma^2 + \mu^2$$

A.3.4 The uniform distribution

Another distribution is the **uniform distribution** (Figure A.3). This is defined as

$$p(x) = \begin{cases} 1/(b-a), & a \leq x \leq b \\ 0, & \text{otherwise} \end{cases}$$

Again it is simple to compute the moments of the uniform distribution. For the example above

$$\mathcal{E}\{x\} = \int_{-\infty}^{\infty} x p(x) dx = \int_{1}^{3} \frac{1}{2} x dx = (9-1)/4 = 2$$

A.3.5 Cumulative density functions

In addition to PDFs we sometimes use cumulative density functions (CDFs). These are defined as

$$F(a) = P(x \leq a) = \int_{-\infty}^{a} p(x) dx$$

but, from earlier,

$$p(a) \approx \frac{P(x \in (a, a+\Delta x))}{\Delta x} = \frac{F(a+\Delta x) - F(a)}{\Delta x}$$

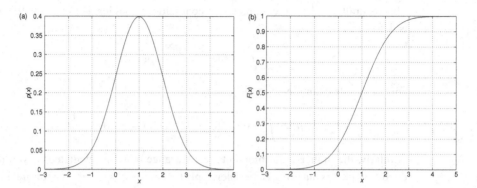

Figure A4 (a) The Gaussian distribution and (b) its CDF.

So, in the limit $\Delta x \to 0$,

$$p(a) = \left.\frac{\mathrm{d}F(x)}{\mathrm{d}x}\right|_a$$

In words, the value of the PDF at point a is the gradient of the CDF at a. Figure A.4 shows the Gaussian distribution and its associated CDF. Note: not all distributions are Gaussian, for example Figure A.5.

A.4 Pairs of continuous random variables

All the aspects of discrete distributions have continuous equivalents.

- **Marginal distributions**: by analogy

$$p_x(x) = \int_{-\infty}^{\infty} p(x, y) \mathrm{d}y$$

- **Conditional probability**:

$$p(y|x) = \frac{p(x, y)}{p(x)}$$

$$= \frac{p(x, y)}{\int_{-\infty}^{\infty} p(x, y) \mathrm{d}y}$$

- **Independence**: for two random variables to be independent of one another

$$p(x, y) = p(x)p(y)$$

The same definition for the correlation coefficient is used. It is interesting to look at uncorrelated versus independent data.

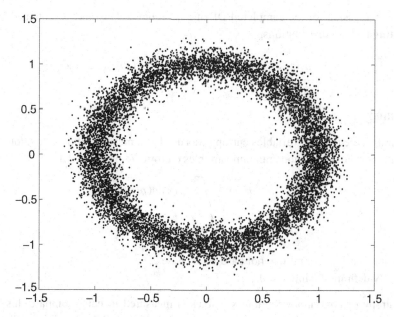

Figure A5 A non-Gaussian distribution.

A.4.1 Independent versus uncorrelated

Consider data generated by the function

$$\mathbf{x} = \begin{bmatrix} r \cos(\theta) \\ r \sin(\theta) \end{bmatrix}$$

θ is uniformly distributed in the range $(0, 2\pi)$ and r is Gaussian distributed, with $\mu = 1.0$ and $\sigma = 0.1$. From these data

$$\Sigma = \begin{bmatrix} 0.50 & 0.00 \\ 0.00 & 0.50 \end{bmatrix}$$

The data are uncorrelated – **are they independent?**

A.4.2 The sum of two random variables

In some situations (consider adding some noise to a clean speech signal) the distribution of the sum of two independent random variables, x and y, is required Let

$$z = x + y$$

$p_z(z)$ is required given that $p_y(y)$ and $p_x(x)$ are known (in the same fashion as for PMFs the subscripts on the PDFs have been added to make the distribution clear). From previous results it is easy to obtain

- the mean (even if x and y are not independent)
- the variance

The expression for the complete PDF of z is required. This may be obtained by **convoluting** the two distributions,

$$p_z(z) = p_x(x) \otimes p_y(y) = \int_{-\infty}^{\infty} p_x(x) p_y(z-x) dx$$

A.4.3 Entropy

For discrete random variables entropy was used as a measure of how "random" a distribution is. For continuous random variables entropy, H, is defined as

$$H = -\int_{-\infty}^{\infty} p(x) \ln(p(x)) dx$$

$$= \mathcal{E}\left\{ \ln\left(\frac{1}{p(x)}\right) \right\}$$

1. ln( ) is the natural log, not base 10, or base 2.
2. By definition $0 \ln(0) = 0$.

Entropy of continuous random variables is measured in nats (compare this with bits for the discrete distributions). It is worth noting that of all the continuous distributions having mean μ and variance σ^2 the distribution with the greatest entropy is the Gaussian distribution. (Try to compute the entropy of a Gaussian distribution.)

A.4.4 Kullback–Leibler distance

It will be useful to measure how "close" two distributions are. One measure is the Kullback–Leibler distance (or relative entropy). This is defined as

$$\mathcal{D}_{KL}(p(x), q(x)) = \int_{-\infty}^{\infty} q(x) \ln\left(\frac{q(x)}{p(x)}\right) dx$$

This has the attributes that

$$\mathcal{D}_{KL}(p(x), q(x)) \geq 0$$

and

$$\mathcal{D}_{KL}(p(x), q(x)) = 0$$

if, and only if, $p(x) = q(x)$. Note that Kullback–Liebler distance is not necessarily symmetrical and does not necessarily satisfy the triangular inequality.

Appendix B Phone definitions

Table B.1 The modified TIMIT phoneme inventory for General American, with example words

Symbol	Example word	Transcription	Symbol	Example word	Transcription
b	BEE	B iy	l	LAY	L ey
d	DAY	D ey	r	RAY	R ey
g	GAY	G ey	w	WAY	W ey
p	PEA	P iy	y	YACHT	Y aa t
t	TEA	T iy	iy	BEET	b IY t
k	KEY	K iy	ih	BIT	b IH t
dx	MUDDY	m ah DX iy	eh	BET	b EH t
jh	JOKE	JH ow k	ey	BAIT	b EY t
ch	CHOKE	CH ow k	ae	BAT	b AE t
s	SEA	S iy	aa	BOT	b AA t
sh	SHE	SH iy	aw	BOUT	b AW t
z	ZONE	Z ow n	ay	BITE	b AY t
zh	AZURE	ae ZH er	ah	BUTT	b AH t
f	FIN	F ih n	ao	BOUGHT	b AO t
th	THIN	TH ih n	oy	BOY	b OY
v	VAN	V ae n	ow	BOAT	b OW t
dh	THEN	DH e n	uh	BOOK	b UH k
hh	HAY	HH ey	uw	BOOT	b UW t
m	MOM	M aa M	er	BIRD	b ER d
n	NOON	N uw N	ax	ABOUT	AX b aw t
ng	SING	s ih NG	axr	BUTTER	b ah d axr

Appendix B

Table B.2 The British English MRPA phoneme inventory, with example words

Symbol	Word-initial example	Word-final example	Symbol	Word-initial example	Word-final example
p	PIP	RIP	j	YIP	–
t	TIP	WRIT	ii	HEAT	PEEL
k	KIP	KICK	ih	HIT	PILL
b	BIP	RIB	ei	HATE	PATE
d	DIP	RID	e	HET	PET
g	GAP	RIG	aa	HEART	PART
ch	CHIP	RICH	a	HAT	PAT
dz	GYP	RIDGE	uu	HOOT	POOT
m	MIP	RIM	u	FULL	PUT
n	NIP	RUN	ou	MOAT	BOAT
ng	–	RING	uh	HUT	PUTT
f	FOP	RIFF	oo	HAUGHT	CAUGHT
th	THOUGHT	KITH	o	HOT	POT
s	SIP	REECE	ai	HEIGHT	BITE
sh	SHIP	WISH	au	LOUT	BOUT
v	VOTE	SIV	oi	NOISE	BOIL
dh	THOUGH	WRITHE	i@	HEAR	PIER
z	ZIP	FIZZ	e@	HAIR	PEAR
zh	–	VISION	u@	SURE	PURE
h	HIP	–	@@	HER	PURR
l	LIP	PULL	@	CHINA	ABOUT
r	RIP	CAR	w	WIP	–

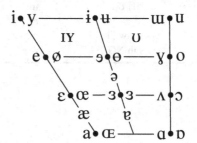

Figure B.1 The IPA vowel chart. The position of each vowel indicates the position of the tongue used to produce that vowel. Front vowels are to the left, back vowels to the right. Where an unrounded/rounded pair occurs, the unrounded version is on the left.

Table B.3 The IPA consonant chart; where there is an unvoiced/voiced contrast, the unvoiced symbol is shown on the left

	Bilabial	Labiodental	Dental	Alveolar	Post-alveolar	Retroflex	Palatal	Velar	Uvular	Pharyngeal	Glottal
Plosive	p b			t d		ʈ ɖ	c ɟ	k ɡ	q ɢ		ʔ
Nasal	m	ɱ		n		ɳ	ɲ	ŋ	N		
Trill	ʙ			r					ʀ		
Tap/flap				ɾ		ɽ					
Fricative	ɸ β	f v	θ ð	s z	ʃ ʒ	ʂ ʐ	ç ʝ	x ɣ	χ ʁ	ħ ʕ	h ɦ
Lateral fricative				ɬ ɮ							
Approximant		ʋ		ɹ		ɻ	j	ɰ			
Lateral approximant				l		ɭ	ʎ	L			

References

[1] ABNEY, S. Chunks and dependencies: Bringing processing evidence to bear on syntax. In *Computational Linguistics and the Foundations of Linguistic Theory*, J. Cole, G. Green, and J. Morgan, Eds. CSLI (1995), pp. 145–164.

[2] ABRANTES, A. J., MARQUES, J. S., AND TRANCOSO, I. M. Hybrid sinusoidal modeling of speech without voicing decision. In *Proceedings of Eurospeech 1991* (1991).

[3] ACERO, A. Source filter models for time-scale pitch-scale modification of speech. In *Proceedings of the International Conference on Speech and Language Processing 1998* (1998).

[4] ACERO, A. Formant analysis and synthesis using hidden Markov models. In *Proceedings of Eurospeech 1999* (1999).

[5] ADELL, J., AND BONAFONTE, A. Toward phone segmentation for concatenative speech synthesis. In *Proceedings of the 5th ISCA Speech Synthesis Workshop* (2004).

[6] ADRIEN, K. *Andean Worlds: Indigenous History, Culture and Consciousness*. New Mexico: University of New Mexico Press (2001).

[7] AGUERO, P., WIMMER, K., AND BONAFONTE, A. Joint extraction and prediction of Fujisaki's intonation model parameters. *Proceedings of International Conference on Spoken Language Processing* (2004).

[8] AINSWORTH, W. A system for converting English text into speech. *IEEE Transactions on Audio and Electroacoustics* **21** (1973), 288–290.

[9] ALIAS, F., AND LLORA, X. Evolutionary weight tuning based on diphone pairs for unit selection speech synthesis. In *Proceedings of Eurospeech 2003* (2003).

[10] ALLEN, J., HUNNICUT, S., AND KLATT, D. *From Text to Speech: The MITalk System*. Cambridge: Cambridge University Press (1987).

[11] ALLWOOD, J. On the analysis of communicative action. *Gothenburg Papers in Theoretical Linguistics* **38** (1978).

[12] ALVAREZ, Y. V., AND HUCKVALE, M. The reliability of the ITU-T p.85 standard for the evaluation of text-to-speech systems. In *Proceedings of the International Conference on Speech and Language Processing 2002* (2002).

[13] ALWAN, A., NARAYANAN, S., AND HAKER, K. Towards articulatory-acoustic models for liquid consonants based on MRI and EPG data. Part II: The rhotics. *Journal of the Acoustical Society of America* **101**, 2 (1997), 1078–1089.

[14] LEWIS, E., AND TATHAM, M. Word and syllable concatenation in text-to-speech synthesis. In *Proceedings of the European Conference on Speech 1999* (1999).

[15] AMDAL, I., AND SVENDSEN, T. Unit selection synthesis database development using utterance verification. In *Proceedings of Eurospeech 2005* (2005).

[16] ANDERSON, M. D., PIERREHUMBERT, J. B., AND LIBERMAN, M. Y. Synthesis by rule of English intonation patterns. In *Proceedings of the International Conference on Acoustics Speech and Signal Processing 1984* (1984).

[17] ANDERSON, S. R. *Phonology in the Twentieth Century*. Chicago, IL: University of Chicago Press (1985).

[18] ASANO, H., NAKAJIMA, H., MIZUNO, H., AND OKU, M. Long vowel detection for letter-to-sound conversion for Japanese sourced words transliterated into the alphabet. In *Proceedings of Interspeech 2004* (2004).

[19] ASCHER, M., AND ASCHER, R. *Code of the Quipu: A Study in Media, Mathematics, and Culture*. Michigan: University of Michigan Press (1980).

[20] ATAL, B. S., AND HANAUER, L. Speech analysis and synthesis by linear prediction of the speech wave. *Journal of the Acoustical Society of America* **50** (1971), 637–655.

[21] ATTERER, M., AND LADD, D. R. On the phonetics and phonology of "segmental anchoring" of F0: Evidence from German. *Journal of Phonetics* **32** (2004), 177–197.

[22] AUBERGE, V., AUDIBERT, N., AND RILLIARD, A. Why and how to control the authentic emotional speech corpora. In *Proceedings of Eurospeech 2003* (2003).

[23] AYLETT, M. P. The dissociation of deaccenting, givenness and syntactic role in spontaneous speech. In *Proceedings of the XIVth International Congress of Phonetic Science 1999* (1999), pp. 1753–1756.

[24] AYLETT, M. P. Stochastic suprasegmentals: Relationships between redundancy, prosodic structure and care of articulation in spontaneous speech. In *Proceedings of the International Conference on Speech and Language Processing 2000* (2000).

[25] AYLETT, M. P. Synthesising hyperarticulation in unit selection TTS. In *Proceedings of Eurospeech 2005* (2005).

[26] BACHENKO, J., AND FITZPATRICK, E. A computational grammar of discourse-neutral prosodic phrasing in English. *Computational Linguistics* **16**, 3 (1990), 155–170.

[27] BAGSHAW, P. Phonemic transcription by analogy in text-to-speech synthesis: Novel word pronunciation and lexicon compression. *Computer Speech & Language* **12**, 2 (1998), 119–142.

[28] BAGSHAW, P. C., HILLER, S. M., AND JACK, M. A. Enhanced pitch tracking and the processing of F0 contours for computer aided intonation teaching. In *Proceedings of Eurospeech 1993* (1993), pp. 1003–1006.

[29] BAILLY, G. No future for comprehensive models of intonation? In *Computing Prosody: Computational Models for Processing Spontaneous Speech*, Y. Sagisaka, N. Campbell and N. Higuchi, Eds. Berlin: Springer-Verlag (1997), pp. 157–164.

[30] BAILLY, G., BERAR, M., ELISEI, F., AND ODISIO, M. Audiovisual speech synthesis. *International Journal of Speech Technology* **6** (2003), 331–346.

[31] BAILLY, G., AND HOLM, B. SFC: A trainable prosodic model. *Speech Communication* **46**, 3–4 (2005), 348–364.

[32] BAILLY, G., AND TRAN, A. Compost: A rule compiler for speech synthesis. In *Proceedings of Eurospeech 1989* (1989), pp. 136–139.

[33] BAKER, J. K. The DRAGON system – an overview. *IEEE Transactions on Acoustics, Speech, and Signal Processing* **23**, 1 (1975), 24–29.

[34] BANZINER, T., MOREL, M., AND SCHERER, K. Is there an emotion signature in intonational patterns? and can it be used in synthesis? In *Proceedings of Eurospeech 2003* (2003).

[35] BARBOSA, P., AND BAILLY, G. Characterization of rhythmic patterns for text-to-speech synthesis. *Speech Communication* **15**, 1 (1994), 127–137.

[36] BAUM, L. E., PETERIE, T., SOULED, G., AND WEISS, N. A maximization technique occurring in the statistical analysis of probabilistic functions of Markov chain. *The Annals of Mathematical Statistics* **41**, 1 (1970), 249–336.

[37] BELL, P., BURROWS, T., AND TAYLOR, P. Adaptation of prosodic phrasing models. In *Proceedings of Speech Prosody 2006* (2006).

[38] BELLEGARDA, J. Unsupervised, language-independent grapheme-to-phoneme conversion by latent analogy. In *Acoustics, Speech, and Signal Processing, 2003. Proceedings (ICASSP'03). 2003 IEEE International Conference* (2003).

[39] BELLEGARDA, J. R. A novel discontinuity metrica for unit selection text-to-speech synthesis. In *5th ISCA Workshop on Speech Synthesis* (2004).

[40] BENNETT, C. L. Large scale evaluation of corpus-based synthesizers: Results and lessons from the Blizzard Challenge 2005. In *Proceedings of Interspeech 2006* (2005).

[41] BESKOW, J. Rule-based visual speech synthesis. In *Proceedings of Eurospeech 1995* (1995).

[42] BEUTNAGEL, M., CONKIE, A., AND SYRDAL, A. Diphone synthesis using unit selection. In *Proceedings of the Third ISCA Workshop on Speech Synthesis* (1998).

[43] BIRD, S. *Computational Phonology: A Constraint-Based Approach*. Cambridge: Cambridge University Press, 1995.

[44] BISANI, M., AND NEY, H. Investigations on joint-multigram models for grapheme-to-phoneme conversion. *Proceedings of the International Conference on Spoken Language Processing 1* (2002), pp. 105–108.

[45] BISHOP, C. M. *Neural Networks for Pattern Recognition*. Oxford: Oxford University Press (1995).

[46] BISHOP, C. M. *Pattern Recognition and Machine Learning*. Berlin: Springer-Verlag (2006).

[47] BLACK, A. Unit selection and emotional speech. In *Proceedings of Eurospeech 2003* (2003).

[48] BLACK, A., AND BENNETT, C. L. Blizzard Challenge 2006. In *Proceedings of Interspeech 2006* (2006).

[49] BLACK, A., AND TAYLOR, P. Automatically clustering similar units for unit selection in speech synthesis. In *Proceedings of Eurospeech 1997* (1997), vol. 2, pp. 601–604.

[50] BLACK, A., AND TOKUDA, K. Blizzard Challenge 2006: Evaluating corpus-based speech synthesis on common datasets. In *Proceedings of Interspeech 2005* (2005).

[51] BLACK, A. W., AND HUNT, A. J. Generation F0 contours from ToBI labels using linear regression. In *Computer Speech and Language* (1996).

[52] BLACK, A. W., AND LENZO, K. A. Limited domain synthesis. In *Proceedings of the International Conference on Speech and Language Processing 2000* (2000).

[53] BLACK, A. W., AND TAYLOR, P. CHATR: A generic speech synthesis system. In *COLING 1994* (1994), pp. 983–986.

[54] BLACK, A. W., AND TAYLOR, P. Assigning phrase breaks from part-to-speech sequences. In *Proceedings of Eurospeech 1997* (1997).

[55] BLACK, A. W., TAYLOR, P., AND CALEY, R. The Festival Speech Synthesis System. Manual and source code avaliable at http://www.cstr.ed.ac.uk/projects/festival.html, 1996–2006.

[56] BLACK, A. W., AND TAYLOR, P. A. A framework for generating prosody from high level linguistics descriptions. In *Spring Meeting, Acoustical Society of Japan* (1994).

[57] BLOOMFIELD, L. *Language*. New York: Henry Holt (1933).

[58] BLUMSTEIN, S., AND CUTLER, A. Speech perception: Phonetic aspects. In *International Encyclopedia of Language*, W. J. Frawley, Ed., vol. 4. Oxford: Oxford University Press (2003).

[59] BOD, R., HAY, J., AND JANNEDY, S. *Probabilistic Linguistics*. Cambridge, MA: MIT Press (1999).

[60] BOERSMA, P. *Functional Phonology Formalizing the Interactions between Articulatory and Perceptual Drives*. PhD thesis, University of Amsterdam (1998).

[61] BOSWELL, J. *The Life of Samuel Johnson*. London: Everyman's Library (1791).

[62] BOULA DE MAREÜIL, P., YVON, F., D'ALESSANDRO, C. ET AL. Evaluation of grapheme-to-phoneme conversion for text-to-speech synthesis in French. *Proceedings of First International Conference on Language Resources & Evaluation* (1998), pp. 641–645.

[63] BOULARD, H., HERMANSKY, H., AND MORGAN, N. Towards increasing speech recognition error rates. *Speech Communication* **18** (1996), 205–255.

[64] BOYER, C. B. *History of Mathematics*. Princeton, MA: Princeton University Press (1985).

[65] BREEN, A. P., AND JACKSON, P. A phonologically motivated method of selecting non-uniform units. In *International Conference on Speech and Language Processing* (1998).

[66] BREGLER, C., COVELL, M., AND SLANEY, M. Video rewrite: Visual speech synthesis from video. In *Proceedings of Eurospeech 1997* (1997).

[67] BRILL, E. Transformation-based error-driven learning and natural language processing: A case study in part of speech tagging. *Computational Linguistics* **21**, 4 (1995) 543–565.

[68] BROOKES, D. M., AND LOKE, H. P. Modelling energy flow in the vocal tract with applications to glottal closure and opening detection. In *Proceedings of the International Conference on Acoustics, Speech, and Signal Processing, 1999* (1999).

[69] BROWMAN, C. P., AND GOLDSTEIN, L. Towards an articulatory phonology. In *Phonology Yearbook* **3** (1986), pp. 219–252.

[70] BROWMAN, C. P., AND GOLDSTEIN, L. Articulatory phonology: an overview. *Phonetica* **49** (1992), 155–180.

[71] BU, S., YAMAMOTO, M., AND ITAHASHI, S. An automatic extraction method of F0 generation model parameters. *IEICE Transactions on Information and Systems* **89**, 1 (2006), 305.

[72] BUHMANN, J., VEREECKEN, H., FACKRELL, J., MARTENS, J. P., AND COILE, B. V. Data driven intonation modelling of 6 languages. In *Proceedings of the International Conference on Spoken Language Processing 2000* (2000).

[73] BULT, M., BUSSO, C., TILDIM, S. ET AL. Investigating the role of phoneme-level modifications in emotional speech resynthesis. In *Proceedings of the International Conference on Speech and Language Processing 2002* (2002).

[74] BULYKO, I., AND OSTENDORF, M. Unit selection for speech synthesis using splicing costs with weighted finite state transducers. In *Proceedings of Eurospeech 2001* (2001).

[75] BUNNELL, H. T., HOSKINS, S. R., AND YARRINGTON, D. Prosodic vs segmental contributions to naturalness in a diphone synthesizer. In *Proceedings of the International Conference on Speech and Language Processing 1998* (1998).

[76] BUSSER, B., DAELEMANS, W., AND VAN DEN BOSCH, A. Machine learning of word pronunciation: the case against abstraction. In *Proceedings of Eurospeech 1999* (1999).

[77] BUSSER, B., DAELEMANS, W., AND VAN DEN BOSCH, A. Predicting phrase breaks with memory-based learning. In *4th ISCA Tutorial and Research Workshop on Speech Synthesis* (2001).

[78] FILLMORE, C. J., KAY, P., AND O'CONNOR, C. Regularity and idiomaticity in grammatical constructions: The case of let alone. *Language* **64** (1988), 501–538.

[79] CAHN, J. The generation of affect in synthesized speech. *Journal of the American Voice I/O Society* **8** (1990), 1–19.

[80] CAHN, J. A computational memory and processing model for prosody. In *International Conference on Speech and Language Processing* (1998).

[81] CAMPBELL, N. Towards synthesizing expressive speech; designing and collecting expressive speech data. In *Proceedings of Eurospeech 2003* (2003).

[82] CAMPBELL, N. Conventional speech synthesis and the need for some laughter. *IEEE Transactions on Audio, Speech and Language Processing* **14**, 4 (2006), 1171–1178.

[83] CAMPBELL, W. N. Syllable-based segmental duration. In *Talking Machines: Theories, Models and Designs*, C. B. G. Bailly and T. R. Sawallis, Eds. Amsterdam: Elsevier Science Publishers (1992), pp. 211–224.

[84] CAMPBELL, W. N. A high-definition speech re-sequencing system. In *Proceedings of the Third ASA/ASJ Joint Meeting* (1996), pp. 373–376.

[85] CHAZAN, D., HOORY, R., COHEN, G., AND ZIBULSK, M. Speech reconstruction from mel frequency cepstral coefficients and pitch. In *Proceedings of the International Conference on Acoustics, Speech, and Signal Processing 2000* (2000).

[86] CHEN, S. F. Conditional and joint models for grapheme-to-phoneme conversion. In *Proceedings of Eurospeech 2003* (2003).

[87] CHEN, W., LIN, F., AND ZHANG, J. L. B. Training prosodic phrasing rules for Chinese TTS systems. In *Proceedings of Eurospeech 2001* (2001).

[88] CHOMSKY, N. *Syntactic Structures*. The Hague: Mouton (1957).

[89] CHOMSKY, N. *Aspects of the Theory of Syntax*. Cambridge, MA: MIT Press (1965).

[90] CHOMSKY, N. *Knowledge of Language: Its Nature, Origin and Use*. New York: Praeger (1986).

[91] CHOMSKY, N., AND HALLE, M. *The Sound Pattern of English*. London: Harper and Row (1968).

[92] CHU, M., PENG, H., YANG, H. Y., AND CHANG, E. Selecting non-uniform units from a very large corpus for concatenative speech synthesizer. In *Proceedings of the International Conference on Acoustics, Speech, and Signal Processing 2001* (2001).

[93] CHU, M., PENG, H., ZHAO, Y., NIU, Z., AND CHAN, E. Microsoft Mulan – a bilingual TTS system. In *Proceedings of the International Conference on Acoustics, Speech, and Signal Processing 2003* (2003).

[94] CHURCH, K. W., AND GALE, W. A. A comparison of the enhanced good-Turing and deleted estimation methods for estimating probabilities of English bigrams. *Computer Speech and Language* **5** (1991), 19–54.

[95] CLARK, H. H. *Using Language*. Cambridge: Cambridge University Press (1996).

[96] CLARK, H. H., AND CLARK, E. V. *Psychology and Language: An Introduction to Psycholinguistics*. London: Harcourt Brace Jovanovich (1977).

[97] CLARK, J., AND YALLOP, C. *An Introduction to Phonetics and Phonology*. Oxford: Basil Blackwell (1990).

[98] CLARK, R., AND DUSTERHOFF, K. Objective methods for evaluating synthetic intonation. In *Proceedings of Eurospeech 1999* (1999).

[99] CLARK, R. A. J., RICHMOND, K., AND KING, S. Festival 2 – build your own general purpose unit selection speech synthesiser. In *5th ISCA Workshop on Speech Synthesis* (2004).

[100] CLEMENTS, G. N. The geometry of phonological features. *Phonology Yearbook 2* (1985), pp. 225–252.

[101] COHEN, M. M., AND MASSARO, D. W. Modeling coarticulation in synthetic visual speech. In *Models and Techniques in Computer Animation* (1993), pp. 141–155.

[102] COLE, R. Roadmaps, journeys and destinations: Speculations on the future of speech technology research. In *Proceedings of Eurospeech 2003* (2003).

[103] COLOTTE, V., AND BEAUFORT, R. Linguistic features weighting for a text-to-speech system without prosody model. In *Proceedings of Eurospeech 2005* (2005).

[104] CONKIE, A. A robust unit selection system for speech synthesis. In *137th Meeting of the Acoustical Society of America* (1999).

[105] CONKIE, A. D., AND ISARD, S. Optimal coupling of diphones. In *Proceedings of Eurospeech 1995* (1995).

[106] COOPER, F. S., LIBERMAN, A. M., AND BORST, J. M. The interconversion of audible and visible patterns as a basis for research in the perception of speech. *Proceedings of the National Academy of Science* **37**, 5 (1951), 318–325.

[107] COOPER, W. E., AND SORENSEN, J. M. *Fundamental Frequency in Sentence Production*. Berlin: Springer-Verlag (1981).

[108] COORMAN, G., FACKRELL, J., RUTTEN, P., AND COILE, B. V. Segment selection in the LH RealSpeak Laboratory TTS system. In *Proceedings of the International Conference on Spoken Language Processing 2000* (2000).

[109] CORDOBA, R., VALLEJO, J. A., MONTERO, J. M. ET AL. Automatic modeling of duration in a Spanish text-to-speech system using neural networks. In *Proceedings of Eurospeech 1999* (1999).

[110] CORNELIUS, R. Theoretical approaches to emotion. In *ICSA Workshop on Speech and Emotion* (2000).

[111] CORTES, C., AND VAPNIK, V. Support-vector networks. *Machine Learning* **20**, 3 (1995), 273–297.

[112] COWIE, R., DEVILLERS, L., MARTIN, J.-C. ET AL. Multimodal databases of everyday emotion: Facing up to complexity. In *Proceedings of Eurospeech, Interspeech 2005* (2005).

[113] COWIE, R., DOUGLAS-COWIE, E., TSAPATSOULIS, N. ET AL. Emotion recognition in human–computer interaction. *IEEE Signal Processing Magazine* (2001), 32–80.

[114] CRISTIANINI, N., AND SHAWE-TAYLOR, J. *Nello Cristianini and John Shawe-Taylor. An Introduction to Support Vector Machines and Other Kernel-based Learning Methods*. Cambridge: Cambridge University Press (2000).

[115] CRONK, A., AND MACON, M. Optimized stopping criteria for tree-based unit selection in concatenative synthesis. In *Proceedings of the International Conference on Speech and Language Processing 1998* (1998).

[116] CROWE, A., AND JACK, M. A. A globally optimising format tracker using generalised centroids. *Electronics Letters* **23** (1987), 1019–1020.

[117] CRYSTAL, D. *Prosodic Systems and Intonation in English*. Cambridge: Cambridge University Press (1969).

[118] D'ALESSANDRO, C., AND MERTENS, P. Automatic pitch contour stylization using a model of tonal perception. In *Computer Speech and Language* (1995).

[119] DAELEMANS, W., VAN DEN BOSCH, A., AND ZAVREL, J. Forgetting exceptions is harmful in language learning. *Machine Learning* **34**, 1 (1999), 11–41.

[120] D'ALESSANDRO, C., AND DOVAL, B. Voice quality modification for emotional speech synthesis. In *Proceedings of Eurospeech 2003* (2003).

[121] D'ALESSANDRO, C., AND MERTENS, P. Automatic pitch contour stylization using a model of tonal perception. *Computer Speech & Language* **9**, 3 (1995), 257–288.

[122] DAMPER, R., AND EASTMOND, J. Pronunciation by analogy: Impact of implementational choices on performance. *Language and Speech* **40**, 1 (1997), 1–23.

[123] DAMPER, R., MARCHAND, Y., ADAMSON, M., AND GUSTAFSON, K. Evaluating the pronunciation component of text-to-speech systems for English: A performance comparison of different approaches. *Computer Speech and Language* **13**, 2 (1999), 155–176.

[124] DARWIN, C. *The Expression of the Emotions in Man and Animals*. London: John Murray (1872).

[125] DAVID D. LEWIS, W. A. G. A sequential algorithm for training text classifiers. In *17th ACM International Conference on Research and Development in Information Retrieval* (1994).

[126] DAVITZ, J. Auditory correlates of vocal expression of emotional feeling. In *The Communication of Emotional Meaning*. New York: McGraw-Hill (1964).

[127] DEDINA, M., AND NUSBAUM, H. Pronounce: A program for pronunciation by analogy. *Computer Speech & Language (Print)* **5**, 1 (1991), 55–64.

[128] DELLER, J. R., AND PROAKIS, J. *Discrete-Time Processing of Speech Signals*. New York: John Wiley and Sons (2000).

[129] DENG, L., BAZZI, I., AND ACERO, A. Tracking vocal track resonances using an analytical nonlinear predictor and a target guided temporal constraint. In *Proceedings of Eurospeech 2003* (2003).

[130] DEVILLERS, L., AND VASILESCU, I. Prosodic cues for emotion characterization in real-life spoken dialogs. In *Proceedings of Eurospeech 2003* (2003).

[131] DIAZ, F. C., ALBA, J. L., AND BANGA, E. R. A neural network approach for the design of the target cost function in unit-selection speech synthesis. In *Proceedings of Eurospeech 2005* (2005).

[132] DILLEY, L., LADD, D., AND SCHEPMAN, A. Alignment of L and H in bitonal pitch accents: Testing two hypotheses. *Journal of Phonetics* **33**, 1 (2005), 115–119.

[133] DINES, J., AND SRIDHARAN, S. Trainable speech with trended hidden Markov models. In *Proceedings of the International Conference on Acoustics, Speech, and Signal Processing 2001* (2001).

[134] DINES, J., SRIDHARAN, S., AND MOODY, M. Application of the trended hidden Markov model to speech synthesis. In *Proceedings of Eurospeech 2001* (2001).

[135] DING, W., AND CAMPBELL, N. Optimising unit selection with voice source and formants in the Chatr speech synthesis system. In *Proceedings of Eurospeech 1997* (1997).

[136] DIVAY, M., AND VITALE, A. J. A computational grammar of discourse-neutral prosodic phrasing in English. *Computational Linguistics* **23**, 4 (1997), 495–523.

[137] DONEGAN, P. J., AND STAMPE, D. The study of natural phonology. In *Current Approaches to Phonological Theory*, D. Dinnsen, Ed. Indiana: Indiana University Press (1979), pp. 126–173.

[138] DONOVAN, R. E., AND EIDE, E. The IBM trainable speech synthesis system. In *Proceedings of the International Conference on Speech and Language Processing 1998* (1998).

[139] DONOVAN, R. E., FRANZ, M., SORENSEN, J. S., AND ROUKOS, S. Phrase splicing and variable substitution using the IBM trainable speeech synthesis system. In *Proceedings of the International Conference on Acoustics, Speech, and Signal Processing 1999* (1999), pp. 373–376.

[140] DONOVAN, R. E., AND WOODLAND, P. C. Automatic speech synthesiser parameter estimation using HMMS. In *Proceedings of the International Conference on Acoustics, Speech, and Signal Processing 1995* (1995).

[141] DROPPO, J., AND ACERO, A. Maximum a posteriori pitch tracking. In *Proceedings of the IEEE International Conference on Acoustics, Speech, and Signal Processing 1998* (1998).

[142] DUDA, R. O., HART, P. E., AND STORK, D. G. *Pattern Classification*. New York: John Wiley and Sons (2000).

[143] DUDLEY, H. Remaking speech. *Journal of the Acoustical Society of America* **11** (1939), 169–177.

[144] DURAND, J. *Dependency and Non-Linear Phonology*. Croom Helm (1986).

[145] DURAND, J., AND ANDERSON, J. *Explorations in Dependency Phonology*. Foris (1987).

[146] DUSTERHOFF, K., AND BLACK, A. Generating f0 contours for speech synthesis using the tilt intonation theory. In *Proceedings of Eurospeech 1997* (1997).

[147] DUSTERHOFF, K. E., BLACK, A. W., AND TAYLOR, P. Using decision trees within the tilt intonation model to predict F0 contours. In *Proceedings of Eurospeech 1999* (1999).

[148] DUTOIT, T. *An Introduction to Text to Speech Synthesis*. Dordrecht: Kluwer Academic Publishers (1997).

[149] DUTOIT, T., AND LEICH, H. Text-to-speech synthesis based on an MBE re-synthesis of the segments database. *Speech Communication* **13** (1993), 435–440.

[150] EIDE, E., AARON, A., BAKIS, R. ET AL. Recent improvements to the IBM trainable speech synthesis system. In *Proceedings of the International Conference on Acoustics, Speech, and Signal Processing 2003* (2003).

[151] EIDE, E., AARON, A., BAKIS, HAMZA, W., AND PICHENY, M. J. A corpus-based approach to AHEM expressive speech synthesis. In *Proceedings of the 5th ISCA Workshop on Speech Synthesis* (2005).

[152] EKMAN, P. Universals and cultural differences in facial expressions of emotion. In *Nebraska Symposium on Motivation*. University of Nebraska Press (1972).

[153] EKMAN, P. *Basic Emotions The Handbook of Cognition and Emotion*. New York: John Wiley (1999).

[154] EKMAN, P., AND FRIESEN, W. V. *Unmasking the Face*. London: Consulting Psychologists Press (1975).

[155] ELOVITZ, H. S., JOHNSON, R., MCHUGH, A., AND SHORE, J. Letter-to-sound rules for automatic translation of English text to phonetics. *IEEE Transactions on Acoustics, Speech, and Signal Processing* **24** (1976), 446–459.

[156] GROSJEAN, L. G., AND LANE, H. The patterns of silence: Performance structures in sentence production. *Cognitive Psychology* **11** (1979), 58–81.

[157] FACKRELL, J. W. A., VEREECKEN, H., MARTENS, J. P., AND COILE, B. V. Multilingual prosody modelling using cascades of regression trees and neural networks. In *Proceedings of Eurospeech 1999* (1999).

[158] FANT, G. *Acoustic Theory of Speech Production*. The Hague: Mouton (1960).

[159] FANT, G. Glottal flow: models and interaction. *Journal of Phonetics* **14** (1986), 393–399.

[160] FISHER, R. A. the use of multiple measures in taxonomic problems. *Annals of Eugenics* **7** (1936), 179–188.

[161] FITT, S., AND ISARD, S. The generation of regional pronunciations of English for speech synthesis. In *Proceedings of Eurospeech 1997* (1997).

[162] FITT, S., AND ISARD, S. Representing the environments for phonological processes in an accent-independent lexicon for synthesis of English. In *Proceedings of the International Conference on Speech and Language Processing 1998* (1998).

[163] FITT, S., AND ISARD, S. The treatment of vowels preceding 'r' in a keyword lexicon of English. In *Proceedings of ICPhS 99* (1999).

[164] FLANAGAN, J. L. *Speech Analysis, Synthesis and Perception*. Berlin: Springer-Verlag (1972).

[165] FUJISAKI, H., AND HIROSE, K. Analysis of voice fundamental frequency contours for declarative sentences of Japanese. *Journal of the Acoustical Society of Japan* **5**, 4 (1984), 233–241.

[166] FUJISAKI, H., AND KAWAI, H. Modeling the dynamic characteristics of voice fundamental frequency with applications to analysis and synthesis of intonation. In *Working Group on Intonation, 13th International Congress of Linguists* (1982).

[167] FUJISAKI, H., AND KAWAI, H. Realization of linguistic information in the voice fundamental frequency contour of the spoken Japanese. In *Proceedings of the International Conference on Acoustics, Speech, and Signal Processing 1988* (1988).

[168] FURUI, S. *Digital Speech Processing, Synthesis and Recognition*. New York: Marcel Dekker (2001).

[169] GALE, W. A., CHURCH, K. W., AND YAROWSKY, D. Using bilingual materials to develop word sense disambiguation methods. In *International Conference on Theoretical and Methodological Issues in Machine Translation* (1992), pp. 101–112.

[170] GALES, M. J. F., JIA, B., LIU, X. ET AL. Development of the CU-HTK 2004 Mandarin conversational telephone speech transcription system. In *Proceedings of the International Conference on Acoustics, Speech, and Signal Processing 2005* (2005).

[171] GALES, M. J. F., KIM, D. Y., WOODLAND, P. C. ET AL. Progress in the CU-HTK Broadcast News transcription system. *IEEE Transactions on Audio, Speech, and Language Processing* **14**, 5 (2006), 1513–1525.

[172] GALESCU, L., AND ALLEN, J. Name pronunciation with a joint N-gram model for bi-directional grapheme-to-phoneme conversion. *Proceedings of ICSLP* (2002), pp. 109–112.

[173] GAROFOLO, J. S., LAMEL, L. F., FISHER, W. M. ET AL. The DARPA-TIMIT acoustic-phonetic continuous speech corpus. Technical report, US Department of Commerce, Gaithersburg, MD (CD-ROM, 1990).

[174] GIEGERICH, H. J. *English Phonology: An Introduction*. Cambridge: Cambridge University Press (1992).

[175] GLUSHKO, R. The organization and activation of orthographic knowledge in reading aloud. *Journal of Experimental Psychology: Human Perception and Performance* **5**, 4 (1979), 674–691.

[176] GOLD, B., AND MORGAN, N. *Speech and Audio Signal Processing: Processing and Perception of Speech and Music*. New York: John Wiley and Sons (1999).

[177] GOLDBERG, A. *Constructions: A Construction Grammar Approach to Argument Structure*. Chicago, IL: University of Chicago Press (1995).

[178] GOLDFARB, C. F. *The SGML Handbook*. Oxford: Clarendon Press (1990).

[179] GOLDSMITH, J. An overview of autosegmental phonology. *Linguistic Analysis* **2**, 1 (1976), 23–68.

[180] GOLDSMITH, J. *Autosegmental and Metrical Phonology*. Oxford: Blackwell (1990).

[181] GOLDSTEIN, M. Classification of methods used for the assessment of text-to-speech systems according to the demands placed on the listener. *Speech Communication* **16** (1995), 225–244.

[182] GOUBANOVA, O., AND KING, S. Predicting consonant duration with Bayesian belief networks. In *Proceedings of Interspeech 2005* (2005).

[183] GRANSTRÖM, B. The use of speech synthesis in exploring different speaking styles. *Speech Communication* **11**, 4–5 (1992), 347–355.

[184] GRICE, H. P. Logic and conversation. In *Syntax and Semantics: Speech Acts*, P. Cole and J. Morgan, Eds. New York: Academic Press (1975), vol. 3, pp. 41–58.

[185] GRIFFIN, D., AND LIM, J. Multiband excitation vocoder. *IEEE Transactions on Acoustics, Speech, and Signal Processing*, **36**, 8 (1988).

[186] GROSJEAN, F., AND GEE, J. P. Prosodic structure and spoken word recognition. *Cognition*, 156 (1987).

[187] GROVER, C., FACKRELL, J., VEREECKEN, H., MARTENS, J., AND VAN COILE, B. Designing prosodic databases for automatic modelling in 6 languages. In *Proceedings of ICSLP 1998* (1998).

[188] GUSTAFSON, J., LINDBERG, N., AND LUNDEBERG, M. The August spoken dialogue system. In *Proceedings of Eurospeech 1999* (1999).

[189] KAWAHARA, I. M.-K., AND DE CHEVEIGNE, A. Restructuring speech representations using a pitch-adaptive time-frequency smoothing and an instantaneous-frequency-based F0 extraction: Possible role of a repetitive structure in sounds. *Speech Communication* **27** (1999), 187–207.

[190] MENG, C. K. K., SIU, T. Y. F., AND CHING, P. C. Cu vocal: Corpus-based syllable concatenation for chinese speech synthesis across domains and dialects. In *Proceedings of the International Conference on Speech and Language Processing 2002* (2002).

[191] HAIN, H.-U. A hybrid approach for grapheme-to-phoneme conversion based on a combination of partial string matching and a neural network. In *Proceedings of the International Conference on Speech and Language Processing* (2000).

[192] HAIN, T., WOODLAND, P. C., EVERMANN, G. ET AL. Automatic transcription of conversational telephone speech – development of the CU-HTK 2002 system. *IEEE Transactions on Audio, Speech, and Language Processing* (2005).

[193] HALLIDAY, M. A. *Intonation and Grammar in British English*. Mouton (1967).

[194] HAMON, C., MOULINES, E., AND CHARPENTIER, F. A diphone synthesis system based on time-domain modifications of speech. In *Proceedings of International Conference on Acoustics, Speech, and Signal Processing 1989* (1989).

[195] HAMZA, W., BAKIS, R., EIDE, E. M., PICHENY, M. A., AND PITRELLI, J. F. The IBM expressive speech synthesis system. In *Proceedings of the International Conference on Spoken Language Processing 2004* (2004).

[196] HAMZA, W., BAKIS, R., SHUANG, Z. W., AND ZEN, H. On building a concatenative speech synthesis system from the blizzard challenge speech databases. In *Proceedings of Interspeech 2005* (2005).

[197] HAMZA, W., RASHWAN, M., AND AFIFY, M. A quantitative method for modelling context in concatenative synthesis using large speech database. In *Proceedings of the International Conference on Acoustics Speech and Signal Processing 2001* (2001).

[198] HAN, K., AND CHEN, G. Letter-to-sound for small-footprint multilingual TTS engine. In *Proceedings of Interspeech 2004* (2004).

[199] HARRIS, R. *Signs of Writing*. London: Routledge (1996).

[200] HENRICHSEN, P. J. Transformation-based learning of Danish stress assignment. In *Proceedings of Eurospeech 2001* (2001).

[201] HERTZ, S. R. *Papers in Laboratory Phonology I: Between the Grammar and the Physics of Speech*. Cambridge: Cambridge University Press (1990).

[202] HESS, W. *Pitch Determination of Speech Signals*. Berlin: Springer-Verlag (1983).

[203] HIRAI, T., TENPAKU, A., AND SHIKANO, K. Manipulating speech pitch periods according to optimal insertion/deletion position in residual signal for intonation control in speech synthesis. In *Proceedings of the International Conference on Speech and Language Processing 2000* (2000).

[204] HIRAI, T., AND TENPAKU, S. Using 5 ms segments in concatenative speech synthesis. In *5th ISCA Workshop on Speech Synthesis* (2005).

[205] HIROSE, K., ETO, M., AND MINEMATSU, N. Improved corpus-based synthesis of fundamental frequency contours using generation process model. In *Proceedings of the International Conference on Acoustics, Speech, and Signal Processing 2002* (2002).

[206] HIROSE, K., ETO, M., MINEMATSU, N., AND SAKURAI, A. Corpus-based synthesis of fundamental frequency contours based on a generation process model. In *Proceedings of Eurospeech 2001* (2001).

[207] HIRSCHBERG, J. Pitch accent in context: Predicting intonational prominence from text. *Artificial Intelligence* **63** (1993), 305–340.

[208] HIRSCHBERG, J. Communication and prosody: Functional aspects of prosody. *Speech Communication* **36** (2002), 31–43.

[209] HIRSCHBERG, J., AND RAMBOW, O. Learning prosodic features using a tree representation. In *Proceedings of Eurospeech 2001* (2001).

[210] HIRSCHFELD, D. Comparing static and dynamic features for segmental cost function calculation in concatenative speech synthesis. In *Proceedings of the International Conference on Spoken Language Processing 2000* (2000).

[211] HIRST, D. Automatic analysis of prosody for multilingual speech corpora. In *Improvements in Speech Synthesis*. Chichester: Wiley (2001).

[212] HIRST, D., AND DI CRISTO, A. *Intonation Systems: A Survey of Twenty Languages*. Cambridge: Cambridge University Press (1998).

[213] HIRST, D., DI CRISTO, A., AND ESPESSER, R. Levels of representation and levels of analysis for the description of intonation systems. *Prosody: Theory and Experiment*. Dordrecht: Kluwer Academic Publishers (2000).

[214] HOCKETT, C. F. The origin of speech. *Scientific American* **203** (1960), 88–96.

[215] HOGBERG, J. Data driven formant synthesis. In *Proceedings of Eurospeech 1997* (1997).

[216] HOLM, B., AND BAILLY, G. Generating prosody by superposing multi-parametric overlapping contours. *Proceedings of the International Conference on Speech and Language Processing* (2000), pp. 203–206.

[217] HOLM, B., AND BAILLY, G. Learning the hidden structure of intonation: Implementing various functions of prosody. In *Speech Prosody* (2002), 399–402.

[218] HOLMES, J. N. The influence of the glottal waveform on the naturalness of speech from a parallel formant synthesizer. *IEEE Transactions on Audio Electroacoustics* **21** (1980), 298–305.

[219] HOLMES, J. N., MATTINGLY, I. G., AND SHEARME, J. N. Speech synthesis by rule. *Language and Speech* **7** (1964), 127–143.

[220] HON, H., ACERO, A., HUANG, X., LIU, J., AND PLUMPE, M. Automatic generation of synthesis units for trainable text-to-speech systems. In *Proceedings of the IEEE International Conference on Acoustics, Speech, and Signal Processing 1998* (1998).

[221] HOUSE, A., WILLIAMS, C., HECKER, M., AND KRYTER, K. Articulation-testing methods: Consonantal differentiation with a closed-response set. *The Journal of the Acoustical Society of America* **37** (1965), 158.

[222] HSIA, C.-C., WE, C. H., AND LIU, T.-H. Duration embedded bi-HMM for expressive voice conversion. In *Proceedings of Interspeech 2005* (2005).

[223] HUANG, J., ABREGO, G., AND OLORENSHAW, L. System and method for performing a grapheme-to-phoneme conversion. In *Proceedings of Interspeech 2006* (2006).

[224] HUANG, X., ACERO, A., AND HON, H.-W. *Spoken Language Processing: A Guide to Theory, Algorithm and System Development*. Englewood Cliffs, NJ: Prentice-Hall (2001).

[225] HUCKVALE, M. Speech synthesis, speech simulation and speech science. In *Proceedings of the International Conference on Speech and Language Processing 2002* (2002), pp. 1261–1264.

[226] HUNNICUT, S. Phonological rules for a text-to-speech system. *Americal Journal of Computational Linguistics* **57** (1976), 1–72.

[227] HUNT, A. J., AND BLACK, A. W. Unit selection in a concatenative speech synthesis system using a large speech database. In *Proceedings of the International Conference on Speech and Language Processing 1996* (1996), pp. 373–376.

[228] HUNT, M., ZWIERYNSKI, D., AND CARR, R. Issues in high quality IPC analysis and synthesis. In *Proceedings of Eurospeech 1989* (1989), pp. 348–351.

[229] IMAI, S. Cepstral analysis synthesis on the mel frequency scale. In *Proceedings of the International Conference on Acoustics Speech and Signal Processing 1983* (1983), pp. 93–96.

[230] INANOGLU, Z., AND CANEEL, R. Emotive alert: HMM-based emotion detection in voicemail messages. In *Proceedings of the 10th International Conference on Intelligent User Interfaces* (2005), pp. 251–253.

[231] ISARD, S. D., AND PEARSON, M. A repertoire of British English contours for speech synthesis. In *Proceedings of the 7th FASE Symposium, Speech 1988* (1988).

[232] ISOGAI, J., YAMAGISHI, J., AND KOBAYASHI, T. Model adaptation and adaptive training using ESAT algorithm for HMM-based speech synthesis. In *Proceedings of Eurospeech 2005* (2005).

[233] ITAKURA, F., AND SAITO, S. Analysis synthesis telephony based on the maximum likelihood method. In *Reports of the 6th International Conference on Acoustics* (1968).

[234] IWAHASHI, N., KAIKI, N., AND SAGISAKA, Y. Speech segment selection for concatenative synthesis based on spectral distortion minimization. *Transactions of the Institute of Electronics, Information and Communication Engineers* **E76A** (1993), 1942–1948.

[235] MATOUSEK, D. T., AND PSUTKA, J. Automatic segmentation for Czech concatenative speech synthesis using statistical approach with boundary-specific correction. In *Proceedings of Eurospeech 2003* (2003).

[236] MATOUSEK, Z. H., AND TIHELKA, D. Hybrid syllable/triphone speech synthesis. In *Proceedings of Eurospeech 2005* (2005).

[237] JELINEK, F. Continuous speech recognition by statistical methods. *Proceedings of the IEEE* **64** (1976), 532–556.

[238] JELINEK, F. Talk. In *Workshop on Evaluation of NLP Systems, Wayne, Pennsylvania* (1988).

[239] JELINEK, F. *Statistical Methods for Speech Recognition*. Cambridge, MA: MIT Press (1998).

[240] JENSEN, K. J., AND RIIS, S. Self-organizing letter code-book for text-to-phoneme neural network model. In *Proceedings of the International Conference on Speech and Language Processing* (2000).

[241] JILKA, M., MOHLER, G., AND DOGIL, G. Rules for the generation of ToBI-based American English intonation. In *Speech Communication* **28** (1999), 83–108.

[242] JONES, D. *An Outline of English Phonetics*, 8th edn. Cambridge: Heffer & Sons (1957).

[243] JURAFSKY, D., AND MARTIN, J. H. *Speech and Language Processing: An Introduction to Natural Language Processing, Computational Linguistics, and Speech Recognition*. Englewood Cliffs, NJ: Prentice-Hall (2000).

[244] KAIN, A., AND MACON, M. Design and evaluation of a voice conversion algorithm based on spectral envelope mapping and residual prediction. In *Proceedings of the International Conference on Acoustics Speech and Signal Processing 2001* (2001).

[245] KATAMBA, F. *An Introduction to Phonology*. London: Longman (1989).

[246] KAWANAMI, H., IWAMI, Y., TODA, T., SARUWATARAI, H., AND SHIKANO, K. GMM base voice conversion applied to emotional speech synthesis. In *Proceedings of Eurospeech 2003* (2003).

[247] KAY, J., AND MARCEL, A. One process, not two, in reading aloud: Lexical analogies do the work of non-lexical rules. *Quarterly Journal of Experimental Psychology* **33a** (1981), 397–413.

[248] KAYE, J., LOWENSTAMM, J., AND VERGNAUD, J. R. The internal structure of phonological elements: A theory of charm and government. *Phonology Yearbook 2* (1985), pp. 305–328.

[249] KIENAPPEL, A. K., AND KNESER, R. Designing very compact decision trees for grapheme-to-phoneme transcription. In *Proceedings of Eurospeech 2001* (2001).

[250] KLABBERS, E., AND VELDHUIS, R. On the reduction of concatenation artefacts in diphone synthesis. In *Proceedings of the International Conference on Speech and Language Processing 1998* (1998).

[251] KLABBERS, E., AND VELDHUIS, R. Reducing audible spectral discontinuities. *IEEE Transactions on Speech and Audio Processing* **9**, 1 (2001), 39–51.

[252] KLATT, D. H. Acoustic theory of terminal analog speech synthesis. In *Proceedings of the International Conference on Acoustics, Speech, and Signal Processing 1972* (1972), vol. 1, pp. 131–135.

[253] KLATT, D. H. Interaction between two factors that influence vowel duration. *Journal of the Acoustical Society of America* **5** (1973), 1102–1104.

[254] KLATT, D. H. Software for a cascade/parallel formant synthesizer. *Journal of the Acoustical Society of America* **67** (1980), 971–995.

[255] KLATT, D. H. Review of text-to-speech conversion for English. *Journal of the Acoustical Society of America* **82**, 3 (1987), 793–850.

[256] KLEIJN, W. B., AND PALIWAL, K. K. *Speech Coding and Synthesis*. Amsterdam: Elsevier (1995).

[257] KOEHN, P., ABNEY, S., HIRSCHBERG, J., AND COLLINS, M. Improving intonational phrasing with syntactic information. In *Proceedings of the International Conference on Acoustics, Speech, and Signal Processing 2000* (2000).

[258] KOHLER, K. J. A model of German intonation. In *Studies in German Intonation*, K. J. Kohler, Ed. Kiel: Universität Kiel (1991).

[259] KOHLER, K. J. The perception of accents: Peak height versus peak position. In *Studies in German Intonation*, K. J. Kohler, Ed. Kiel: Universität Kiel (1991), pp. 72–96.

[260] KOHLER, K. J. Terminal intonation patterns in single-accent utterances of German: Phonetics, phonology and semantics. In *Studies in German Intonation*, K. J. Kohler, Ed. Kiel: Universität Kiel (1991), pp. 53–71.

[261] KOISHIDA, K., TOKUDA, K., AND IMAI, S. CELP coding based on mel cepstral analysis. In *Proceedings of the International Conference on Acoustics, Speech, and Signal Processing 1995* (1995).

[262] KOMINEK, J., BENNETT, C., AND BLACK, A. M. Evaluating and correcting phoneme segmentation for unit selection synthesis. In *Proceedings of Eurospeech 2003* (2003).

[263] KOMINEK, J., AND BLACK, A. W. A family-of-models approach to HMM-based segmentation for unit selection speech synthesis. In *Proceedings of the International Conference on Spoken Language Processing 2004* (2004).

[264] KRISHNA, N. S., TALUKDAR, P. P., BALI, K., AND RAMAKRISHNAM, A. G. Duration modeling for Hindi text-to-speech synthesis system. In *Proceedings of the International Conference on Speech and Language Processing 2004* (2004).

[265] KUBRICK, S. *2001: A Space Odyssey* (1968).

[266] LADD, D., MENNEN, I., AND SCHEPMAN, A. Phonological conditioning of peak alignment in rising pitch accents in Dutch. *The Journal of the Acoustical Society of America* **107** (2000), 2685.

[267] LADD, D. R. Declination reset and the hierarchical organization of utterances. *Journal of the Acoustical Society of America* **84**, 2 (1988), 530–544.

[268] LADD, D. R. Compound prosodic domains. Edinburgh University Linguistics Department Occasional Paper (1992).

[269] LADD, D. R. *Intonational Phonology*. Cambridge: Cambridge University Press (1996).

[270] LADD, D. R., AND SILVERMAN, K. E. A. Vowel intrinsic pitch in connected speech. *Phonetica* **41** (1984), 31–40.

[271] LADEFOGED, P. *A Course in Phonetics*. London: Thompson (2003).

[272] LADEFOGED, P. *An Introduction to Phonetic Fieldwork and Instrumental Techniques*. Oxford: Blackwell Publishing (2003).

[273] LAMBERT, T., AND BREEN, A. A database design for a TTS synthesis system using lexical diphones. In *Proceedings of the International Conference on Speech and Language Processing 2004* (2004).

[274] LASS, R. *An Introduction to Basic Concepts*. Cambridge: Cambridge University Press (1984).

[275] LAVER, J. *Principles of Phonetics*. Cambridge: Cambridge University Press (1995).

[276] LAVER, J., ALEXANDER, M., BENNET, C. ET AL. Speech segmentation criteria for the ATR/CSTR database. Technical report, Centre for Speech Technology Research, University of Edinburgh (1988).

[277] LAW, K. M., AND LEE, T. Using cross-syllable units for Cantonese speech synthesis. In *Proceedings of the International Conference on Spoken Language Processing 2000* (2000).

[278] LAWRENCE, W. The synthesis of speech from signals which have a low information rate. In *Communication Theory*, W. Jackson, Ed. London: Butterworth & Co., Ltd (1953), pp. 460–469.

[279] LEE, K.-S., AND KIM, J. S. Context-adaptive phone boundary refining for a TTS database. In *Proceedings of the International Conference on Acoustics Speech and Signal Processing 2003* (2003).

[280] LEE, Y. J., LEE, S., KIM, J. J., AND KO, H. J. A computational algorithm for F0 contour generation in Korean developed with prosodically labeled databases using K-ToBI system. In *Proceedings of the International Conference on Spoken Language Processing 1998* (1998).

[281] LEICH, H., DEKETELAERE, S., DBMAN, I., DOTHEY, M., AND WERY, B. A new quantization technique for LSP parameters and its application to low bit rate multi-band excited vocoders. In *EUSIPCO* (1992).

[282] LEVINSON, S. Continuously variable duration hidden Markov models for automatic speech recognition. *Computer Speech and Language* **1** (1986), 29–45.

[283] LI, X., MALKIN, J., AND BILMES, J. A graphical model approach to pitch tracking. In *Proceedings of the International Conference on Spoken Language Processing 2004* (2004).

[284] LIBERMAN, M. *The Intonational System of English*. PhD thesis, MIT (1975). Published by Indiana University Linguistics Club.

[285] LIBERMAN, M., AND PIERREHUMBERT, J. Intonational invariance under changes in pitch range and length. In *Language Sound Structure*, M. Aronoff and R. T. Oehrle, Eds. Cambridge, MA: MIT Press (1984), pp. 157–233.

[286] LIBERMAN, M. Y., AND PRINCE, A. On stress and linguistic rhythm. *Linguistic Inquiry* **8** (1977), 249–336.

[287] LIEBERMAN, P. *Intonation, Perception and Language*. Cambridge MA: MIT Press (1967).

[288] LILJENCRANTS, J. *Reflection-type Line Analog Synthesis*. PhD thesis, Royal Institution of Technology, Stockholm (1985).

[289] LLITJOS, A., AND BLACK, A. Knowledge of language origin improves pronunciation accuracy of proper names. In *Proceedings of Eurospeech 2001* (2001).

[290] LYONS, J. *Introduction to Theoretical Linguistics*. Cambridge: Cambridge University Press (1968).

[291] BEUTNAGEL, M. M., AND RILEY, M. Rapid unit selection from a large speech corpus for concatenative speech synthesis. In *Proceedings of Eurospeech 1999* (1999).

[292] LEE, D. P. L., AND OLIVE, J. P. A text-to-speech platform for variable length optimal unit searching using perceptual cost functions. In *Proceedings of the Fourth ISCA Workshop on Speech Synthesis* (2001).

[293] MACON, M. W. *Speech Synthesis Based on Sinusoidal Modeling*. PhD thesis, Georgia Tech (1996).

[294] MAKHOUL, J. Spectral analysis of speech by linear prediction. *IEEE Transactions on Audio and Electroacoustics* **3** (1973), 140–148.

[295] MAKHOUL, J. Linear prediction: A tutorial review. *Proceedings of the IEEE* **63**, 4 (1975), 561–580.

[296] MALFRERE, F., DUTOIT, T., AND MERTENS, P. Automatic prosody generation using suprasegmental unit selection. In *International Conference on Speech and Language Processing* (1998).

[297] MALKIN, J., LI, X., AND BILMES, J. A graphical model for formant tracking. In *Proceedings of the IEEE International Conference on Acoustics, Speech, and Signal Processing 2005* (2005).

[298] MANEVITZ, L. M., AND YOUSEF, M. One-class SVMs for document classification. *Journal of Machine Learning Research* **2** (2001), 139–154.

[299] MANNING, C. D., AND SCHUTZE, H. *Foundations of Statistical Natural Language Processing*. Cambridge, MA: MIT Press (1999).

[300] MARCHAND, Y., AND DAMPER, R. A multistrategy approach to improving pronunciation by analogy. *Computational Linguistics* **26**, 2 (2000), 195–219.

[301] MARKEL, J. D. The SIFT algorithm for fundamental frequency estimation. *IEEE Transactions on Audio and Electroacoustics* **20** (1972), 367–377.

[302] MARKEL, J. D., AND GRAY, A. H. *Linear Prediction of Speech*. Berlin: Springer-Verlag (1976).

[303] MARSI, E., BUSSER, B., DAELEMANS, W. ET AL. Combining information sources for memory-based pitch accent placement. In *Proceedings of the International Conference on Speech and Language Processing 2002* (2002).

[304] MASSARO, D. W. Categorical perception: Important phenomena or lasting myth. In *Proceedings of the International Conference on Speech and Language Processing 1998* (1998), pp. 2275–2278.

[305] MASUKO, T., TOKUDA, K., KOBAYASHI, T., AND IMAI, S. Speech synthesis using HMMs with dynamic features. In *Proceedings of the International Conference on Acoustics Speech and Signal Processing 1996* (1996).

[306] MCCARTHY, J. Feature geometry and dependency: A review. *Phonetica* **43** (1988), 84–108.

[307] MCILROY, D. M. Synthetic English speech by rule. Technical report, *Bell System Technical Journal* (1973).

[308] MCINNES, F. R., ATTWATER, D. J., EDGINGTON, M. D., SCHMIDT, M. S., AND JACK, M. A. User attitudes to concatenated natural speech and text-to-speech synthesis in an automated information service. In *Proceedings of Eurospeech 1999* (1999).

[309] MEDAN, Y., YAIR, E., AND CHAZAN, D. Super resolution pitch determination of speech signals. *IEEE Transactions on Signal Processing* **39** (1991), 40–48.

[310] MERON, J. Prosodic unit selection using an imitation speech database. In *Proceedings of the Fourth ISCA ITRW on Speech Synthesis* (2001).

[311] MERON, J. Applying fallback to prosodic unit selection from a small imitation database. In *Proceedings of the International Conference on Speech and Language Processing 2002* (2002).

[312] MICHAELIS, L. A., AND LAMBRECHT, K. Toward a construction-based model of language function: The case of nominal extraposition. *Language* **72** (1996), 215–247.

[313] MILNER, R. Bigraphs as a model for mobile interaction. In *Proceedings of the International Conference on Graph Transformation 2002* (2002).

[314] MIXDORFF, H. A novel approach to the fully automatic extraction of Fujisaki-model parameters. In *International Conference on Acoustics, Speech, and Signal Processing 2000* (2000).

[315] MÖHLER, G., AND CONKIE, A. Parametric modeling of intonation using vector quantization. In *Proceedings of the Third ESCA/IEEE Workshop on Speech Synthesis* (1998), pp. 311–314.

[316] MOORE, B. C. J., AND GLASBERG, B. R. Suggested formulae for calculating auditory-filter bandwidths and excitation patterns. *Journal of the Acoustical Society of America* **74** (1983), 750–753.

[317] MOORE, R. K. Twenty things we still don't know about speech. In *Proceedings of Eurospeech 1999* (1995).

[318] MOORE, R. K. Research challenges in the automation of spoken language interaction. In *ISCA Tutorial and Research Workshop on Applied Spoken Language Interaction in Distributed Environments* (2005).

[319] MORAIS, E., AND VIOLARO, F. Exploratory analysis of linguistic data based on genetic algorithm for robust modeling of the segmental duration of speech. In *Proceedings of Interspeech 2005* (2005).

[320] MORI, H., OHTSUKA, T., AND KASUYA, H. A data-driven approach to source-formant type text-to-speech system. In *Proceedings of the International Conference on Speech and Language Processing 2000* (2000).

[321] MORLEC, Y., BAILLY, G., AND AUBERGÉ, V. Generating prosodic attitudes in French: data, model and evaluation. *Speech Communication* **33**, 4 (2001), 357–371.

[322] MOULINES, E., AND VERHELST, W. Time-domain and frequency-domain techniques for prosodic modification of speech. In *Speech Coding and Synthesis*, W. B. Kleijn and K. K. Paliwal, Eds. Amsterdam: Elsevier Science B.V. (1995), pp. 519–555.

[323] MULLER, A. F., AND HOFFMANN, R. Accent label prediction by time delay neural networks using gating clusters. In *Proceedings of Eurospeech 2001* (2001).

[324] MURRAY, I. R., AND ARNOTT, J. L. Toward the simulation of emotion in synthetic speech: A review of the literature on human vocal emotion. *Journal of the Acoustical Society of America* **93**, 2 (1993), 1097–1108.

[325] NAGANO, T., SHINSUKE, M., AND NISHIMURA, M. A stochastic approach to phoneme and accent estimation. In *Proceedings of Interspeech 2005* (2005).

[326] NAKAJIMA, S. English speech synthesis based on multi-layered context oriented clustering. In *Proceedings of the Third European Conference on Speech Communication and Technology, Eurospeech 1993* (1993).

[327] NAKAJIMA, S., AND HAMADA, H. Automatic generation of synthesis units based on context oriented clustering. In *Proceedings of the International Conference on Acoustics, Speech, and Signal Processing 1988* (1988).

[328] NAKATANI, C. H. *The Computational Processing of Intonational Prominence: A Functional Prosody Respective*. PhD thesis, Harvard University (1997).

[329] NAKATANI, C. H. Prominence variation beyond given/new. In *Proceedings of Eurospeech 1997* (1997).

[330] NAKATANI, C. H. Coupling dialogue and prosody computation in spoken dialogue generation. In *Proceedings of the International Conference on Spoken Language Processing 2000* (2000).

[331] NARAYANAN, S., ALWAN, A., AND HAKER, K. An articulatory study of fricative consonants using MRI. *Journal of the Acoustical Society of America* **98**, 3 (1995), 1325–1347.

[332] NARAYANAN, S., ALWAN, A., AND HAKER, K. Towards articulatory-acoustic models for liquid consonants based on MRI and EPG data. Part I: The laterals. *Journal of the Acoustical Society of America* **101**, 2 (1997), 1064–1077.

[333] O'CONNOR, J. D., AND ARNOLD, G. F. *Intonation of Colloquial English*. Longman, 1973.

[334] OGDEN, R., HAWKINS, S., HOUSE, J. ET AL. Prosynth: An integrated prosodic approach to device-independent, natural-sounding speech synthesis. *Computer Speech and Language* **14** (2000), 177–210.

[335] ÖHMAN, S. Word and sentence intonation: A quantitative model. *STL-Quarterly Progress and Staatus Report* **8**, 2–3 (1967), 20–54.

[336] OHTSUKA, T., AND KASUYA, H. An improved speech analysis–synthesis algorithm based on the autoregressive with exogenous input speech production model. In *Proceedings of the International Conference on Speech and Language Processing 2000* (2000).

[337] OHTSUKA, T., AND KASUYA, H. Aperiodicity control in ARX-based speech analysis–synthesis method. In *Proceedings of Eurospeech 2001* (2001).

[338] OLIVE, J. P. Rule synthesis of speech from diadic units. In *Proceedings of the International Conference on Acoustics, Speech, and Signal Processing 1977* (1977).

[339] OLIVE, J. P., AND SPICKENAGLE, N. Speech resynthesis from phoneme-related parameters. *Journal of the Acoustical Society of America* **59** (1976), 993–996.

[340] OLIVER, D., AND CLARK, R. Modelling pitch accent types for Polish speech synthesis. In *Proceedings of Interspeech 2005* (1995).

[341] IEEE SUBCOMMITTEE ON SUBJECTIVE MEASUREMENTS, I. S. IEEE recommended practices for speech quality measurements. *IEEE Transactions on Audio and Electroacoustics* **17** (1969), 227–246.

[342] OPPENHEIM, A. L., AND SCHAFER, R. W. *Digital Signal Processing*. Englewood Cliffs, NJ: Prentice-Hall (1975).

[343] PAGEL, V., LENZO, K., AND BLACK, A. Letter to sound rules for accented lexicon compression. In *Proceedings of the ICSLP 1998* (1998).

[344] PALMER, H. *English Intonation with Systematic Exercises*. Cambridge: Cambridge University Press (1922).

[345] PANTAZIS, Y., STYLIANOU, Y., AND KLABBERS, E. Discontinuity detection in concatenated speech synthesis based on nonlinear speech analysis. In *Proceedings of Eurospeech, Interspeech 2005* (2005).

[346] PARKE, F. I. Computer generated animation of faces. In *ACM National Conference* (1972).

[347] PARKE, F. I. A parametrized model for facial animations. *IEEE Transactions on Computer Graphics and Animations* **2**, 9 (1982), 61–70.

[348] PAULO, S., AND OLIVEIRA, L. C. DTW-based phonetic alignment using multiple acoustic features. In *Proceedings of Eurospeech 2003* (2003).

[349] PEARSON, S., KIBRE, N., AND NIEDZIELSKI, N. A synthesis method based on concatenation of demisyllables and a residual excited vocal tract model. In *Proceedings of the International Conference on Speech and Language Processing 1998* (1998).

[350] PEIRCE, C. S. *The Essential Peirce, Selected Philosophical Writings*, Indiana: Indiana University Press (1998), vol. 2.

[351] PIERREHUMBERT, J. B. *The Phonology and Phonetics of English Intonation*. PhD thesis, MIT (1980). Published by Indiana University Linguistics Club.

[352] PIKE, K. L. *The Intonation of American English*. Michigan: University of Michigan (1945).

[353] PISONI, D., AND HUNNICUTT, S. Perceptual evaluation of MITalk: The MIT unrestricted text-to-speech system. In *Proceedings of the International Conference on Acoustics, Speech, and Signal Processing 1980* (1980).

[354] PLATT, S. M., AND BADLER, N. I. Animating facial expression. *Computer Graphics* **15**, 3 (2001), 245–252.

[355] PLUMPE, M., AND MEREDITH, S. Which is more important in a concatenative speech synthesis system – pitch duration or spectral discontinuity? In *Third ESCA/IEEE Workshop on Speech Synthesis* (1998).

[356] PLUMPE, M. D., QUATIERI, T. F., AND REYNOLDS, D. A. Modeling of the glottal flow derivative waveform with application to speaker identification. *IEEE Transactions on Speech and Audio Processing* **1**, 5 (1999), 569–586.

[357] POLLET, V., AND COORMAN, G. Statistical corpus-based speech segmentation. In *Proceedings of Interspeech 2004* (2004).

[358] POLS, L. Voice quality of synthetic speech: Representation and evaluation. In *Proceedings of the International Conference on Speech and Language Processing 1994* (1994).

[359] PORTELE, T., STOBER, K. H., MEYER, H., AND HESS, W. Generation of multiple synthesis inventories by a bootstrapping procedure. In *Proceedings of the International Conference on Speech and Language Processing 1996* (1996).

[360] POVEY, D., AND WOODLAND, P. Minimum phone error and I-smoothing for improved discriminative training. In *Proceedings of the International Conference on Acoustics, Speech, and Signal Processing 2002* (2002).

[361] POWER, R. J. D. The organisation of purposeful dialogues. *Linguistics* **17** (1979), 107–152.

[362] PRINCE, A., AND SMOLENSKY, P. *Optimality Theory: Constraint Interaction in Generative Grammar*. Oxford: Blackwell (2004).

[363] QANG, M. Q., AND HIRSCHBERG, J. Automatic classification of intonational phrase boundaries. *Computer Speech and Language* **6** (1992), 175–196.

[364] QUATIERI, T. F. *Speech Signal Processing*. Englewood Cliffs, NJ: Prentice-Hall (2002).

[365] QUENÉ, H. The derivation of prosody for text-to-speech from prosodic sentence structure. *Computer Speech & Language* **6**, 1 (1992), 77–98.

[366] CLARK, K. R., AND KING, S. Multisyn voices from ARCTIC data for Blizzard Challenge. In *Proceedings of Interspeech 2005* (2005).

[367] RABINER, L., AND JUANG, B.-H. *Fundamentals of Speech Recognition*. Englewood Cliffs, NJ: Prentice-Hall (1993).

[368] RABINER, L. R., AND SCHAFER, R. W. *Digital Processing of Speech Signals*. Englewood Cliffs, NJ: Prentice-Hall (1978).

[369] READ, I., AND COX, S. Stochastic and syntactic techniques for predicting phrase breaks. In *Proceedings of Eurospeech 2005* (2005).

[370] REICHEL, U. Improving data driven part-of-speech tagging by morphologic knowledge induction. In *Proceedings of Advances in Speech Technology* (2005).

[371] REVERET, L., BAILLY, G., AND BADIN, P. Mother: A new generation of talking heads providing flexible articulatory control for video-realistic speech animation. In *Proceedings of the International Conference on Speech and Language Processing 2000* (2000).

[372] RILEY, M. Tree-based modelling of segmental duration. In *Talking Machines: Theories, Models and Designs*, C. B. G Bailly and T. R. Sawallis, Eds. Amsterdam: Elsevier Science Publishers (1992), pp. 265–273.

[373] RIVEST, R. L. Learning decision lists. *Machine Learning* **2** (1987), 229–246.

[374] ROACH, P. Conversion between prosodic transcription systems: "Standard British" and ToBI. *Speech Communication* **15**, 1–2 (1994), 91–99.

[375] ROBINSON, T., AND FALLSIDE, F. A recurrent error propagation network speech recognition system. *Computer Speech and Language* **5**, 3 (1991).

[376] ROSENBERG, A. E. Effect of glottal pulse shape on the quality of natural vowels. *Journal of the Acoustical Society of America* **49** (1970), 583–590.

[377] ROSENBERG, C., AND SEJNOWSKI, T. NETtalk: A parallel network that learns to read aloud. EE & CS Technical Report no JHU-EECS-86/01. Johns Hopkins University, Baltimore, MD (1986).

[378] ROSS, K., AND OSTENDORF, M. Prediction of abstract prosodic labels for speech synthesis. In *Computer Speech and Language* **10**, 3 (1996), 155–185.

[379] ROSSI, P., PALMIERI, F., AND CUTUGNO, F. A method for automatic extraction of Fujisaki-model parameters. In *Proceedings of Speech Prosody 2002* (2002), pp. 615–618.

[380] ROUCO, A. AND RECASENS, D. Reliability of EMA data acquired simultaneously with EPG. In *Proceedings of the ACCOR Workshop on Articulatory Databases* (1995).

[381] ROUIBIA, S., AND ROSEC, O. Unit selection for speech synthesis based on a new acoustic target cost. In *Proceedings of Interspeech 2005* (2005).

[382] RUBIN, P., BAER, T., AND MERMELSTEIN, P. An articulatory synthesizer for perceptual research. *Journal of the Acoustical Society of America* **70** (1981), 32–328.

[383] RUTTEN, P., AYLETT, M., FACKRELL, J., AND TAYLOR, P. A statistically motivated database pruning technique for unit selection synthesis. In *Proceedings of the ICSLP* (2002), pp. 125–128.

[384] COX, R. B., AND JACKSON, P. Techniques for accurate automatic annotation of speech waveforms. In *Proceedings of the International Conference on Speech and Language Processing 1998* (1998).

[385] PARK, C. K. K., AND KIM, N. S. Discriminative weight training for unit-selection based speech synthesis. In *Proceedings of Eurospeech 2003* (2003).

[386] SAGEY, E. *The Representation of Features and Relations in Non-linear Phonology*. PhD thesis, MIT (1986).

[387] SAGISAKA, Y. Speech synthesis by rule using an optimal selection of non-uniform synthesis units. In *Proceedings of the International Conference on Acoustics, Speech, and Signal Processing 1988* (1988).

[388] SAITO, T., HASHIMOTO, Y., AND SAKAMOTO, M. High-quality speech synthesis using context-dependent syllabic units. In *Proceedings of the International Conference on Acoustics, Speech, and Signal Processing 1996* (1996).

[389] SAN-SEGUNDO, R., MONTERO, J. M., CORDOBA, R., AND GUTIERREZ-ARRIOLA, J. Stress assignment in Spanish proper names. In *Proceedings of the International Conference on Speech and Language Processing 2000* (2000).

[390] SANDRI, S., AND ZOVATO, E. Two features to check phonetic transcriptions in text to speech systems. In *Proceedings of Eurospeech 2001* (2001).

[391] SAUSSURE, F. DE. Cours de linguistique générale. In *Cours de linguistique générale*, C. Bally and A. Sechehaye, Eds. Dordrecht: Kluwer Academic Publishers (1916).

[392] SAUSSURE, F. DE. *Saussure's Second Course of Lectures on General Linguistics (1908–09)*. Amsterdam: Elsevier (1997).

[393] SCHLOSSBERG, H. Three dimensions of emotion. *Psychological Review* **61**, 2 (1954), 81–88.

[394] SCHWEITZER, A. Speech parameter generation algorithm considering global variance for HMM-based speech synthesis. In *Proceedings of Eurospeech, Interspeech 2005* (2005).

[395] SEARLE, J. R. *Speech Acts*. Cambridge: Cambridge University Press (1969).

[396] SEJNOWSKI, T., AND ROSENBERG, C. Parallel networks that learn to pronounce English text. *Complex Systems* **1**, 1 (1987), 145–168.

[397] SEJNOWSKI, T., AND ROSENBERG, C. *NETtalk: A Parallel Network that Learns to Read Aloud*. Cambridge, MA: MIT Press (1988).

[398] SELKIRK, E. O. *Phonology and Syntax*. Cambridge, MA: MIT Press (1984).

[399] SETHY, A., AND NARAYANAN, S. Refined speech segmentation for concatenative speech synthesis. In *Proceedings of the International Conference on Speech and Language Processing 2002* (2002).

[400] SHANNON, C. E. A mathematical model of communication. Technical report, *Bell System Technical Journal* (1948).

[401] SHANNON, C. E., AND WEAVER, W. *A Mathematical Model of Communication*. Urbana, IL: University of Illinois Press (1949).

[402] SHENG, Z., JIANHUA, T., AND LIANHONG, C. Prosodic phrasing with inductive learning. In *Proceedings of the International Conference on Spoken Language Processing, Interspeech 2002* (2002).

[403] SHI, Q., AND FISCHER, V. A comparison of statistical methods and features for the prediction of prosodic structures. In *Proceedings of the International Conference on Speech and Language Processing 2004* (2004).

[404] SHICHIRI, K., SAWABE, A., YOSHIMURA, T. ET AL. Eigenvoices for HMM-based speech synthesis. In *Proceedings of the 8th International Conference on Spoken Language Processing* (2002).

[405] SILVERMAN, K. *The Structure and Processing of Fundamental Frequency Contours*. PhD thesis, University of Cambridge (1987).

[406] SILVERMAN, K., BECKMAN, M., PIERREHUMBERT, J. ET AL. ToBI: A standard scheme for labelling prosody. In *Proceedings of the International Conference on Speech and Language Processing 1992* (1992).

[407] SINHA, R., GALES, M. J. F., KIM, D. Y. ET AL. The CU-HTK Mandarin Broadcast News transcription system. In *Proceedings of the International Conference on Acoustics, Speech, and Signal Processing 2006* (2006).

[408] SONNTAG, G. P., AND PORTELE, T. Comparative evaluation of synthetic prosody with the PURR method. In *Proceedings of the International Conference on Speech and Language Processing 1998* (1998).

[409] SPIEGEL, M. F. Proper name pronunciations for speech technology applications. In *Proceedings of the 2002 IEEE Workshop on Speech Synthesis* (2002).

[410] SPROAT, R. English noun-phrase accent prediction for text-to-speech. In *Computer Speech and Language* (1994), vol. 8, pp. 79–94.

[411] SPROAT, R. *Multilingual Text-to-Speech Synthesis: The Bell Labs Approach*. Dordrecht: Kluwer Academic Publishers (1997).

[412] SPROAT, R. Corpus-based methods and hand-built methods. In *Proceedings of the International Conference on Spoken Language Processing 2000* (2000).

[413] SPROAT, R. PMtools: A pronunciation modeling toolkit. In *Proceedings of the Fourth ISCA Tutorial and Research Workshop on Speech Synthesis* (2001).

[414] STEVENS, C., LEES, N., AND VONWILLER, J. Experimental tools to evaluate intelligibility of text-to-speech (TTS) synthesis: Effects of voice gender and signal quality. In *Proceedings of Eurospeech 2003* (2003).

[415] STEVENS, S., VOLKMAN, J., AND NEWMAN, E. A scale for the measurement of the psychological magnitude of pitch. *Journal of the Acoustical Society of America* **8** (1937), 185–190.

[416] STOBER, K., PORTELE, T., WAGNER, P., AND HESS, W. Synthesis by word concatenation. In *Proceedings of Eurospeech 1999* (1999).

[417] STRIK, H., AND BOVES, L. On the relation between voice source parameters and prosodic features in connected speech. *Speech Communication* **11**, 2 (1992), 167–174.

[418] STROM, V. From text to prosody without ToBI. In *Proceedings of the International Conference on Speech and Language Processing 2002* (2002).

[419] STYLIANOU, Y. Concatenative speech synthesis using a harmonic plus noise model. In *Proceedings of the Third ESCA Speech Synthesis Workshop* (1998).

[420] STYLIANOU, Y., AND SYRDAL, A. K. Perceptual and objective detection of discontinuities in concatenative speech synthesis. In *Proceedings of the International Conference on Acoustics, Speech, and Signal Processing 2001* (2001).

[421] STYLIANOU, Y., AND SYRDAL, A. K. Perceptual and objective detection of discontinuities in concatenative speech synthesis. In *International Conference on Acoustics, Speech, and Signal Processing* (2001).

[422] SULLIVAN, K., AND DAMPER, R. Novel-word pronunciation: A cross-language study. *Speech Communication* **13**, 3–4 (1993), 441–452.

[423] SUN, X., AND APPLEBAUM, T. H. Intonational phrase break prediction using decision tree and n-gram model. In *Proceedings of Eurospeech 2001* (2001).

[424] SWERTS, M., AND VELDHUIS, R. The effect of speech melody on voice quality. *Speech Communication* **33**, 4 (2001), 297–303.

[425] SYDESERFF, H. A., CALEY, R. J., ISARD, S. D. ET AL. Evaluation of speech synthesis techniques in a comprehension task. *Speech Communication* **11**, 2–3 (1992), 189–194.

[426] SYRDAL, A. K. Prosodic effects on listener detection of vowel concatenation. In *Proceeding of Eurospeech 2001* (2001).

[427] SYRDAL, A. K., AND CONKIE, A. D. Data-driven perceptually based join costs. In *International Conference on Spoken Language Processing 2004* (2004).

[428] SYRDAL, A. K., AND CONKIE, A. D. Perceptually-based data-driven join costs: Comparing join types. In *Proceedings of Eurospeech, Interspeech 2005* (2005).

[429] TACHIBANA, M., YAMAGISHI, J., MASUKO, T., AND KOBAYASHI, T. Speech synthesis with various emotional expressions and speaking styles by style interpolation and morphing. *IEICE Transactions on Information and Systems 2005* **E88-D11** (2004), 2484–2491.

[430] TAKAHASHI, T., TAKESHI, F., NISHI, M. ET AL. Voice and emotional expression transformation based on statistics of vowel parameters in an emotional speech database. In *Proceedings of Interspeech 2005* (2005).

[431] TALKIN, D. A robust algorithm for pitch tracking RAPT. In *Speech Coding and Synthesis*, W. B. Kleijn and K. K. Paliwal, Eds. Amsterdam: Elsevier (1995), pp. 495–518.

[432] TATHAM, M., AND MORTON, K. *Expression in Speech: Analysis and Synthesis*. Oxford: Oxford University Press (2004).

[433] TAYLOR, P. Hidden Markov models for grapheme to phoneme conversion. In *Proceedings of Interspeech 2005* (2005).

[434] TAYLOR, P. A. *A Phonetic Model of English Intonation*. PhD thesis, University of Edinburgh (1992). Published by Indiana University Linguistics Club.

[435] TAYLOR, P. A. The rise/fall/connection model of intonation. *Speech Communication* **15** (1995), 169–186.

[436] TAYLOR, P. A. Analysis and synthesis of intonation using the tilt model. *Journal of the Acoustical Society of America* **107**, 4 (2000), 1697–1714.

[437] TAYLOR, P. A. The target cost formulation in unit selection speech synthesis. In *Proceedings of the International Conference on Speech and Language Processing, Interspeech 2006* (2006).

[438] TAYLOR, P. A. Unifying unit selection and hidden Markov model speech synthesis. In *Proceedings of the International Conference on Speech and Language Processing, Interspeech 2006* (2006).

[439] TAYLOR, P. A., AND BLACK, A. W. Synthesizing conversational intonation from a linguistically rich input. In *Proceedings of the Second ESCA/IEEE Workshop on Speech Synthesis* (1994), pp. 175–178.

[440] TAYLOR, P. A., AND BLACK, A. W. Speech synthesis by phonological structure matching. In *Proceedings of Eurospeech 1999* (1999), pp. 623–626.

[441] TAYLOR, P. A., BLACK, A. W., AND CALEY, R. J. Heterogeneous relation graphs as a mechanism for representing linguistic information. *Speech Communication* special issue on annotation, 1–2 (2000), 153–174.

[442] TAYLOR, P. A., KING, S., ISARD, S. D., AND WRIGHT, H. Intonation and dialogue context as constraints for speech recognition. *Language and Speech* **41**, 3–4 (1998), 491–512.

[443] TAYLOR, P. A., KING, S., ISARD, S. D., WRIGHT, H., AND KOWTKO, J. Using intonation to constrain language models in speech recognition. In *Proceedings of Eurospeech 1997* (1997).

[444] TAYLOR, P. A., NAIRN, I. A., SUTHERLAND, A. M., AND JACK, M. A. A real time speech synthesis system. In *Proceedings of Eurospeech 1991* (1991).

[445] 'T HART, J., AND COHEN, A. Intonation by rule: A perceptual quest. *Journal of Phonetics* **1** (1973), 309–327.

[446] 'T HART, J., AND COLLIER, R. Integrating different levels of intonation analysis. *Journal of Phonetics* **3** (1975), 235–255.

[447] TIHELKA, D. Symbolic prosody driven unit selection for highly natural synthetic speech. In *Proceedings of Eurospeech, Interspeech 2005* (2005).

[448] TODA, T., KAWAI, H., AND TSUZAKI, M. Optimizing sub-cost functions for segment selection based on perceptual evaluations in concatenative speech synthesis. In *Proceedings of the International Conference on Acoustics, Speech, and Signal Processing 2004* (2004).

[449] TODA, T., KAWAI, H., TSUZAKI, M., AND SHIKANO, K. Segment selection considering local degradation of naturalness in concatenative speech synthesis. In *Proceedings of the International Conference on Acoustics, Speech, and Signal Processing 2003* (2003).

[450] TODA, T., AND TOKUDA, K. Speech parameter generation algorithm considering global variance for HMM-based speech synthesis. In *Proceedings of Eurospeech, Interspeech 2005* (2005).

[451] TOKUDA, K., KOBAYASHI, T., AND IMAI, S. Speech parameter generation from HMM using dynamic features. In *Proceedings of the International Conference on Acoustics, Speech, and Signal Processing 1995* (1995).

[452] TOKUDA, K., MASUKO, T., AND YAMADA, T. An algorithm for speech parameter generation from continuous mixture HMMs with dynamic features. In *Proceedings of Eurospeech 1995* (1995).

[453] TOKUDA, K., YOSHIMURA, T., MASUKO, T., KOBAYASHI, T., AND KITAMURA, T. Speech parameter generation algorithms for HMM-based speech synthesis. In *Proceedings of the International Conference on Acoustics, Speech, and Signal Processing 2000* (2000).

[454] TOKUDA, K., ZEN, H., AND KITAMURA, T. Trajectory modeling based on HMMs with the explicit relationship between static and dynamic features. In *Proceedings of Eurospeech 2003* (2003).

[455] TOUREMIRE, D. S. Automatic transcription of intonation using an identified prosodic alphabet. In *Proceedings of the International Conference on Speech and Language Processing 1998* (1998).

[456] TSUZAKI, M. Feature extraction by auditory modeling for unit selection in concatenative speech synthesis. In *Proceedings of Eurospeech 2001* (2001).

[457] TSZUKI, R., ZEN, H., TOKUDA, K. ET AL. Constructing emotional speech synthesizers with limited speech database. In *Proceedings of Interspeech 2004* (2004).

[458] TURK, O., SCHRODER, M., BOZKURT, B., AND ARSLAN, L. Voice quality interpolation for emotional speech synthesis. In *Proceedings of Interspeech 2005* (2005).

[459] UEBLER, U. Grapheme-to-phoneme conversion using pseudo-morphological units. In *Proceedings of Interspeech 2002* (2002).

[460] DARSINOS, D. G., AND KOKKINAKIS, G. A method for fully automatic analysis and modelling of voice source characteristics. In *Proceedings of Eurospeech 1995* (1995).

[461] VAN DEEMTER, K. Towards a blackboard model of accenting. *Computer Speech and Language* **12**, 3 (1998), 143–164.

[462] VAN HERWIJNEN, O., AND TERKEN, J. Evaluation of pros-3 for the assignment of prosodic structure, compared to assignment by human experts. In *Proceedings of Eurospeech 2001* (2001).

[463] VAN SANTEN, J. Assignment of segmental duration in text-to-speech synthesis. *Computer Speech and Language* **8** (1994), 95–128.

[464] VAN SANTEN, J. Quantitative modeling of pitch accent alignment. In *International Conference on Speech Prosody* (2002), pp. 107–112.

[465] VAN SANTEN, J., BLACK, L., COHEN, G. ET AL. Applications and computer generated expressive speech for communication disorders. In *Proceedings of Eurospeech 2003* (2003).

[466] VAN SANTEN, J., KAIN, A., KLABBERS, E., AND MISHRA, T. Synthesis of prosody using multi-level unit sequences. *Speech Communication* **46** (2005), 365–375.

[467] VATIKIOTIS-BATESON, E., AND YEHIA, H. Unified physiological model of audible–visible speech production. In *Proceedings of Eurospeech 1997* (1997).

[468] VELDHUIS, R. Consistent pitch marking. In *Proceedings of the International Conference on Speech and Language Processing 2000* (2000).

[469] VEPA, J., AND KING, S. Kalman-filter based join cost for unit-selection speech synthesis. In *Proceedings of Eurospeech 2003* (2003).

[470] VEPA, J., AND KING, S. Subjective evaluation of join cost functions used in unit selection speech synthesis. In *Proceedings of the International Conference on Speech and Language Processing 2004* (2004).

[471] VEPA, J., KING, S., AND TAYLOR, P. Objective distance measures for spectral discontinuities in concatenative speech synthesis. In *Proceedings of the International Conference on Speech and Language Processing 2002* (2002).

[472] VEREECKEN, H., MARTENS, J., GROVER, C., FACKRELL, J., AND VAN COILE, B. Automatic prosodic labeling of 6 languages. In *Proceedings of the International Conference on Speech and Language Processing 1998* (1998), pp. 1399–1402.

[473] VERHELST, W., COMPERNOLLE, D. V., AND WAMBACQ, P. A unified view on synchronized overlap–add methods for prosodic modification of speech. In *Proceedings of the International Conference on Spoken Language Processing 2000* (2000), vol. 2, pp. 63–66.

[474] VÉRONIS, J., DI CRISTO, P., COURTOIS, F., AND CHAUMETTE, C. A stochastic model of intonation for text-to-speech synthesis. *Speech Communication* **26**, 4 (1998), 233–244.

[475] VINCENT, D., ROSEC, O., AND CHONAVEL, T. Estimation of LF glottal source parameters based on an ARX model. In *Proceedings of Eurospeech, Interspeech 2005* (2005), pp. 333–336.

[476] VITERBI, A. J. Error bounds for convolutional codes and an asymptotically optimum decoding algorithm. *IEEE Transactions on Information Theory* **13**, 2 (1967), 260–269.

[477] VON KEMPELEN, W. *Mechanismus der menschlichen Sprache nebst Beschreibung einer sprechenden Maschine* and *Le Mécanisme de la parole, suivi de la description d'une machine parlante*. Vienna: J. V. Degen (1791).

[478] VONWILLER, J. P., KING, R. W., STEVENS, K., AND LATIMER, C. R. Comprehension of prosody in synthesized speech. In *Proceedings of the Third International Australian Conference on Speech Science and Technology* (1990).

[479] VOSNIDIS, C., AND DIGALAKIS, V. Use of clustering information for coarticulation compensation in speech synthesis by word concatenation. In *Proceedings of Eurospeech 2001* (2001).

[480] WANG, D., AND NARAYANAN, S. Piecewise linear stylization of pitch via wavelet analysis. In *Proceedings of Interspeech 2005* (2005).

[481] WANG, L., ZHAO, Y., CHU, M., ZHOU, J., AND CAO, Z. Refining segmental boundaries for TTS database using fine contextual-dependent boundary models. In *Proceedings of the International Conference on Acoustics Speech and Signal Processing 2004* (2004).

[482] WEBSTER, G. Improving letter-to-pronunciation accuracy with automatic morphologically-based stress prediction. In *Proceedings of Interspeech 2004* (2004).

[483] WELLS, J. C. *Longman Pronunciation Dictionary*. London: Pearson Education Limited (2000).

[484] WER, B. R., LEROUX, A., DELBROUCK, H. P., AND LECLERCS, J. A new parametric speech analysis and synthesis technique in the frequency domain. In *Proceedings of Eurospeech 1995* (1995).

[485] WIGGINS, R. An integrated circuit for speech synthesis. *Proceedings of the International Conference on Acoustics, Speech, and Signal Processing 1980* (1980), pp. 398–401.

[486] WIGHTMAN, C. ToBI or not ToBI. *Speech Prosody* (2002), 25–29.

[487] WIGHTMAN, C. W., AND ROSE, R. C. Evaluation of an efficient prosody labeling system for spontaneous speech utterances. In *Proceedings of the IEEE Automatic Speech Recognition and Understanding Workshop 1* (1999), pp. 333–336.

[488] WIKIPEDIA. Main page – Wikipedia, the free encyclopedia (2007). [Online; accessed 30 January 2007].

[489] WILHELMS-TRICARICO, R. Physiological modeling of speech production: Methods for modeling soft-tissue articulators. *Journal of the Acoustical Society of America* **97**, 5 (1995), 3085–3098.

[490] WILHELMS-TRICARICO, R. A biomechanical and physiologically-based vocal tract model and its control. *Journal of Phonetics* **24** (1996), 23–28.

[491] WILHELMS-TRICARICO, R., AND PERKELL, J. S. Biomechanical and physiologically based speech modeling. In *Proceedings of the Second ESCA/IEEE Workshop on Speech Synthesis* (1994), pp. 17–20.

[492] WILHELMS-TRICARICO, R., AND PERKELL, J. S. Towards a physiological model of speech production. In *Proceedings of the XIVth International Congress of Phonetic Science 1999* (1999), pp. 1753–1756.

[493] WILLEMS, N. J. A model of standard English intonation patterns. *IPO Annual Progress Report* (1983).

[494] WOODLAND, P. C., AND POVEY, D. Large scale discriminative training of hidden Markov models for speech recognition. *Computer Speech and Language* **16** (2002), 25–47.

[495] WOUTERS, J., AND MACON, M. W. A perceptual evaluation of distance measures for concatenative speech. In *Proceedings of the International Conference on Speech and Language Processing 1998* (1998).

[496] WOUTERS, J., AND MACON, M. W. Unit fusion for concatenative speech synthesis. In *Proceedings of the International Conference on Spoken Language Processing 2000* (2000).

[497] WRENCH, A. A. An investigation of sagittal velar movement and its correlation with lip, tongue and jaw movement. In *Proceedings of the International Congress of Phonetic Sciences* (1999), pp. 435–438.

[498] WU, Y., KAWAI, H., NI, J., AND WANG, R.-H. Minimum segmentation error based discriminative training for speech synthesis application. In *Proceedings of the International Conference on Acoustics, Speech, and Signal Processing 2004* (2004).

[499] XU, J., CHOY, T., DONG, M., GUAN, C., AND LI, H. On unit analysis for Cantonese corpus-based TTS. In *Proceedings of Eurospeech 2003* (2003).

[500] XU, Y. Speech melody as articulatory implemented communicative functions. *Speech Communication* **46** (2005), 220–251.

[501] XU, Y. Speech prosody as articulated communicative functions. In *Proceedings of Speech Prosody 2006* (2006).

[502] SAGISAKA, Y. KAIKI, N. I., AND MIMURA, K. ATR – ν-TALK speech synthesis system. In *Proceedings of the International Conference on Speech and Language Processing 1992* (1992), vol. 1, pp. 483–486.

[503] YAMAGISHI, J., MASUKO, T., AND KOBAYASHI, T. MLLR adaptation for hidden semi-Markov model based speech synthesis. In *Proceedings of the 8th International Conference on Spoken Language Processing* (2004).

[504] YAMAGISHI, J., ONISHI, K., MASUKO, T., AND KOBAYASHI, T. Modeling of various speaking styles and emotions for HMM-based speech synthesis. In *Proceedings of Eurospeech 2003* (2003).

[505] YAROWSKY, D. Homograph disambiguation in speech synthesis. In *Second ESCA/IEEE Workshop on Speech Synthesis* (1994).

[506] YAROWSKY, D. Homograph disambiguation in text-to-speech synthesis. In *Computer Speech and Language* (1996).

[507] YOON, K. A prosodic phasing model for a Korean text-to-speech synthesis system. In *Proceedings of the International Conference on Speech and Language Processing 2004* (2004).

[508] YOSHIMURA, T., TOKUDA, K., MASUKO, T., KOBAYASHI, T., AND KITAMURA, T. Simultaneous modelling of spectrum, pitch and duration in HMM-based speech synthesis. In *Proceedings of Eurospeech 1999* (1999).

[509] YOSHIMURA, T., TOKUDA, K., MASUKO, T., KOBAYASHI, T., AND KITAMURA, T. Mixed excitation for HMM-based speech synthesis. In *Proceedings of the European Conference on Speech Communication and Technology 2001*, vol. 3 (2001), pp. 2259–2262.

[510] YOUNG, S. J., EVERMANN, G., KERSHAW, D. ET AL. *The HTK Book* (1995–2006).

[511] YU, Z.-L., WANG, K.-Z., ZU, Y.-Q., YUE, D.-J., AND CHEN, G.-L. Data pruning approach to unit selection for inventory generation of concatenative embeddable Chinese TTS systems. In *Proceedings of Interspeech 2004* (2004).

[512] YUAN, J., BRENIER, J., AND JURAFSKY, D. Pitch accent prediction: Effects of genre and speaker. In *Proceedings of Interspeech* (2005).

[513] ZEN, H., AND TODA, T. An overview of Nitech HMM-based speech synthesis system for Blizzard Challenge 2005. In *Proceedings of Interspeech 2005* (2005).

[514] ZEN, H., TOKUDA, K., MASUKO, T., KOBAYASHI, T., AND KITAMURA, T. Hidden semi-Markov model based speech synthesis. In *Proceedings of the 8th International Conference on Spoken Language Processing, Interspeech 2004* (2004).

[515] ZERVAS, P., MARAGOUDAKIS, M., FAKOTAKIS, N., AND KOKKINAKIS, G. Bayesian induction of intonational phrase breaks. In *Proceedings of Eurospeech 2003* (2003).

[516] ZHANG, J., DONG, S., AND YU, G. Total quality evaluation of speech synthesis systems. In *International Conference on Speech and Language Processing* (1998).

[517] ZHAO, Y., WANG, L., CHU, M., SOONG, F. K., AND CAO, Z. Refining phoneme segmentations using speaker-adaptive context dependent boundary models. In *Proceedings of Interspeech 2005* (2005).

[518] ZHENG, M., SHI, Q., ZHANG, W., AND CAI, L. Grapheme-to-phoneme conversion based on a fast TBL algorithm in Mandarin TTS systems. *Lecture Notes in Computer Science*. Berlin: Springer-Verlag (2005), p. 600.

[519] ZOLFAGHARI, P., NAKATANI, T., AND IRINO, T. Glottal closure instant synchronous sinusoidal model for high quality speech analysis/synthesis. In *Proceedings of Eurospeech 2003* (2003).

[520] ZOVATO, E., SANDRI, S., QUAZZA, S., AND BADINO, L. Prosodic analysis of a multi-style corpus in the perspective of emotional speech synthesis. In *Proceedings of Interspeech 2004* (2004).

[521] ZWICKER, E., AND FASTL, H. *Psychoacoustics, Facts and Models*. Berlin: Springer-Verlag (1990).

Index

abbreviations, decoding, 98
Abjab writing system, 34–5
acoustic models of speech production, 309–40
   about the acoustic models, 309, 339–40
   assumptions discussion, 336–40
   components of the model, 309–11
   models with vocal-tract losses, 335–6
   nasal cavity modelling, 333–5
   oral cavity sound source positions, 335
   radiation models, 330
   source and radiation effects, 336
   see also glottis/glottal source; sound, physics of; vowel-tube models
acoustic representations, 156–9
   spectogram, 157–9
   spectral analysis/frequency domain analysis, 156
   spectral envelope, 156–7
acoustic-space formulation (ASF), 485, 493–7
acoustic theory see sound, physics of; vowel-tube model
acoustic waves, 316–18
acronyms, decoding, 99
adapting systems if TTS, 50
addition paradigm, 71
Advanced Telecommunications Research (ATR), 512–14
affective communication, 8–9
affective prosody, 17, 123–4
affricates, 153, 201
air-flow measurement (mouth air flow), 155
algorithms and features, 79–82
   hand written algorithms, 80
   see also text-classification algorithms
allophones, 162
allophonic variation, 166–7
all-pole modelling, assumptions, 337–8
alphabetic writing, 34
alternative spellings, decoding, 98
alveolar, 154
ambiguity issues, 22
   different words, same form, 54
   homograph ambiguity, 22

AM model see autosegmental–metrical (AM) intonation model
analogue signals, 262–78
   aperiodic signals, 262
   complex exponential sinusoid, 266–9
   complex numbers, 268
   conjugate symmetric complex amplitudes, 268–9
   Euler's formula, 266–8
   Fourier series/synthesis/analysis, 265–6, 269–70
   frequency, 264–5
   frequency domain, 270–5
   frequency range, 274
   fundamental frequency (F0), 148, 265
   harmonic frequency, 265
   periodic signals, 262–9, 305–7
   phase shift, 264
   quasi-periodic signals, 262
   sinusoid signals, 263–9
   time domain, 270
   waveforms, 262
   see also Fourier transform
APL (Anderson, Pierrehumbert and Liberman) synthesis scheme, 246
applications, future of, 538
approximants, 154, 201
arbitrariness, 15
architectures for TTS, 71–5
   addition paradigm, 71
   associative arrays (maps), 72
   atomic values, 72–3
   autosegmental phonology of data structures, 73
   Delta formulation/structure, 74
   dictionaries, 72
   finite partial functions, 72
   heterogeneous relation graph (HRG) formalism, 72–5
   list/tree/ladder relations, 73
   lookup tables, 72
   overwrite paradigm, 71
   utterance structure, 71
articulatory gestures, 406
articulatory phonetics see speech production/articulatory phonetics

articulatory phonology, 183, 406
articulatory physiology, 406
articulatory synthesis, 405–7
ASCII encoding, 70
aspiration, 168
assimilation effect, 167
associative arrays (maps), 72
assumed intent for prosody, 49–50
atomic values, 72–3
audio-visual speech synthesis, 527–9
   about audiovisual synthesis, 406–7, 527–8
   and speech control, 528–9
   texture mapping, 528
   visemes, 528
auditory scales, 351–2
augmentative prosody, 18, 125–6
autocorrelation function, for pitch detection, 381
autocorrelation source separation method, 360–1
automatic labelling, 521
autosegmental-metrical (AM) intonation model, 227, 237–9
   analysis with, 248
   APL synthesis scheme, 246
   data-driven synthesis, 247–8
   deterministic synthesis, 246–7
   prediction of labels from text, 246
   synthesis with, 245–8
   and the ToBI scheme, 247, 248
autosegmental phonology, 73, 183
auxiliary generation for prosody, 49–50

bag-of-features approach, 84
Baum–Welch algorithm, 449
Bayes' rule, 86–7, 545
beam pruning, 509
bilabial constriction, 154
Blizzard Challenge testing, 526
boundary accents/tones, 121, 236, 238
braille, 27
break index concept, 115
British English MRPA phoneme inventory, 554
British intonation school, 227, 236–7

Campbell timing model, 258
canned speech, 43–4
cepstra
   linear-prediction cepstra, 369
   mel-frequency cepstral coefficient (MFCC), 370
cepstral coefficients, synthesis from, 429–31
cepstrum speech analysis, 353–7
   as deconvolution, 355–6
   definition, 353
   discussion, 356–7
   the magnitude spectrum as a signal, 353–5
   for pitch detection, 379
chain rule, 546

channel/medium (means of conversion), 13
character
   character-to-phoneme conversion, 55
   definition, 54
   encoding schemes, 69–70
CHATR system, 513
Cholskey decomposition technique, 359
Chomskian field, 534
classical linear-prediction (LP) synthesis, 399–405
   about LP synthesis, 399
   a complete synthesiser, 403–4
   formant synthesis comparison, 399–400
   impulse/noise source model, 400–1
   LP diphone-concatenative synthesis, 401–3
   source modelling, 378
   source problems, 404–5
classification *see* text-classification algorithms
classifiers, F0 models, 228
clitics, 60–1
closed-phase analysis, 374–7
   instants of glottal closure points, 374
   pre-emphasis, 375
cluster impurity, 88
coarticulation, 168
Cocke–Younger–Kasami (CYK) algorithm, 104
collocation rule, 84
colouring effect, 167
common-form model of TTS, 5–6, 38
communication processes, 18–23
   ambiguity issues, 22
   common ground issues, 20
   dialogue turns, 18
   effectiveness factor, 19–20
   efficiency factor, 19–20
   encoding/decoding, 18–19, 21–2
   Grice's maxims, 20
   homograph ambiguity, 22
   information–theoretic approach, 23
   message generation, 18–21
   messages, 18
   semiotics, 23
   speech, redundancy in, 21
   text decoding/analysis, 22
   understanding, 19, 22–3
communication, types of, 8–13
   about communication, 8, 23–5
   affective communication, 8–9
   iconic communication, 9–10
   interpreted communication, 8
   meaning/form/signal, 12–13
   signals, 13
   symbolic communication, 10–12
   see also human communication
comparison tests, 524
competitive evaluations, 526
complex numbers, 268

Index

component/unit testing, 525–6
compound-noun phrases, 116–17
compound prosodic domains theory, 114
comprehension tests, 523
compression, lossless and lossy, 215
computational phonology, 184
concatenative synthesis, 401–3
   issues, 431–2
   macro-concatenation, 431, 497
   micro-concatenation, 431
   optimal coupling, 432
   phase mismatch issues, 431–2
concept-to-speech systems, 42–3
   future of, 537
conditional probability, 545
conjugate symmetric complex amplitudes, 268–9
consonants, 153–5
   affricates, 153
   alveolar, 154
   approximants, 154
   bilabial, 154
   difficult consonants, 199–200
   fricatives, 153
   glides, 154–5
   IPA charts, 555
   labiodental, 153
   nasal stops, 153
   obstruent, 154
   oral stops, 153
context-free grammars (CFGs), 102–4
context-orientated-clustering, 495
context-sensitive modelling, 451–4
context-sensitive rewrite rule, 83–4
context-sensitive rules, 182
context-sensitive synthesis models, 461–3
continuants, 150
contractions, 60
convolution sum, 292
correlation coefficient, 544
covariance matrix, 439
covariance method, 358–60
coverage (in unit-selection synthesis), 510
cumulative density functions, 549–10
curse of dimensionality, 81, 534

databases, 517–22
   automatic labelling, 521
   avoiding explicit labels, 521–2
   hand labelling, 519–21
   and labelling, 519–21
   prosody databases, 518–19
   text materials, 518
   unit-selection databases, 517–18
data-driven intonation models, 250–4
   about data-driven models, 250–1
   dynamic-system models, 252–3

   functional models, 254
   HMM models, 253–4
   SFC model, 254
   unit-selection synthesis, 251–2
data-driven synthesis, 247–8, 435, 470–1
   see also hidden Markov model (HMM)
data sparsity problem, 81, 193
decision lists, 85–6
decision trees, 87–8, 221, 452–5
   clustering, 494–6
decoding/encoding messages, 18–19, 21–2
   text decoding/analysis, 22
   see also text decoding/analysis
delta formulation/structure, 72
delta/velocity coefficients, 438–9
dental constriction, 154
dependency phonology, 183
deterministic acoustic models, synthesis with, 248–50
   Fujisaki superimpositional models, 249
   Tilt model, 249–50
deterministic phrasing prediction, 130–1
   deterministic content function (DCF), 130
   deterministic content function punctuation (DCFP), 131
   deterministic punctuation (DP), 130
   verb-balancing rule, 131
deterministic synthesis models, 246–7
dialogue turns, 18
dictionaries, 72
digital filters, 288–94, 308
   about digital filters, 288–9
   convolution sum, 292, 293–4
   difference equations, 289
   FIR filter, 289
   IIR filter, 289
   impulse response, 289–91
   linearity principle, 289
   linear time-invariant (LTI) filter, 288
   recursive filters, 289
   scaling, 288–9
   superposition, 289
   third-order filters, 289
   transfer function, 293–4
   and the z-transform, 293–4
digital filters, analysis/design, 294–305
   about digital filter design, 304–5
   anti-resonances, 303
   characteristics, 298–304
   complex-conjugate pairs of poles, 300–2
   polynomial analysis (poles and zeros), 294–7
   practical properties, 304–6
   resonance/resonators, 300
   skirts of poles, 302
   and the z-domain transfer function, 297–8

digital signals, 278–84, 307
  digital representations, 280
  digital waveforms, 279
  discrete Fourier transform (DFT), 281–2
  discrete-time Fourier transform (DTFT), 280–1
  and the frequency domain, 283–4
  Laplace transform, 283
  Nyquist frequency, 279
  sample rate/frequency, 279
  z-transform, 282–3
diphone-concatenative synthesis, 401–3
diphone inventories, 414
diphones from speech, 414–15
diphone unit-selection system, 505
diphthongs, 153, 201
discourse-neutral renderings, 116
discrete Fourier transform (DFT), 281–2
discreteness, 15–16
discrete random variables, 540–1
discrete-time Fourier transform (DTFT), 280–1
discrete-tube model, assumptions, 337
distinctiveness of speech issues, 171–2
downdrift/declination, 230–3
duality principle, 15
duration synthesis modelling, 463–4
Dutch intonation school, 237
dynamic-system synthesis models, 252–3
dynamic-time-warping (DTW) technique, 219, 469

ease of data acquisition, and synthesis with vocal-tract models, 407
egressive pulmonic air stream, 147
eigenface model, 528
electoglottography/laryngography, 155, 383
electromagnetic articulography (EMA), 156
electropalatography, 155
emotional speech synthesis, 529–31
  describing emotion, 529
  with HMM techniques, 531
  with prosody control, 529–30
  with unit selection, 531
  with voice transformation, 530
emotion axes, 123
emphasis, 118–19
encoding/decoding messages, 18–19, 21–2
engineering approach to TTS, 4
engine/rule separation, 83
entropy, 546–7, 552
epoch detection, 381–4
  electroglottograph, 383
  epoch-detection algorithm (EDA), 381
  instant of glottal closure (IGC), 382–3
  laryngograph/laryngograph signals (Lx signals), 383–4
  pitch-synchronous analysis, 381

epoch manipulation for TD-PSOLA, 417–20
equivalent rectangular bandwidth (ERB) auditory scale, 352
Euclidean distance, 486
Euler's formula, 266–8
evaluation, 522–6
  about evaluation, 522–3
  see also tests/testing
exceptions dictionaries, 208
expressive speech see emotional speech synthesis

feature geometry, 183
features and algorithms, 79–82
filter-bank speech analysis, 352–3
filters see digital filters
finite-impulse-response (FIR) filter, 289
finite partial functions, 72
first-generation synthesis see vocal-tract models, synthesis with
forced alignment, 468
formants (speech resonance), 159–60
formant synthesis, 388–99
  about formant synthesis, 388–9
  consonant synthesising, 392–4
  copy synthesis technique, 394–6
  Klatt synthesiser, 394–5
  lumped-parameter speech generation model, 389
  parallel synthesisers, 392
  phonetic input, 394–7
  quality issues, 397–9
  serial/cascade synthesisers, 391–2
  single formant synthesis, 390–1
  sound sources, 389–90
formant tracking, 370–2
form/message-to-speech synthesis, 42
Fourier series/synthesis/analysis, 265–6, 269–71
Fourier transform, 275–8
  discrete Fourier transform (DFT), 281–2
  discrete-time Fourier transform (DTFT), 280–1
  duality principle, 278
  inverse Fourier transform, 277–8
  scaling property, 277
  sinc function, 277
frame shift in speech analysis, 346–7
frequency, 264
  angular frequency, 265
frequency domain, 270–5, 307
  analysis/spectral analysis, 156
  for digital signals, 283–4
  for pitch detection, 381
fricatives, 153
Fujisaki intonation model, 227, 239–42
Fujisaki superimpositional models
  analysis with, 250
  synthesis with, 249

fundamental frequency (F0), 148, 265
  and pitch, 225
  see also pitch detection/tracking
fundamental frequency (F0) contour models, 227–9
  acoustic model, 228
  classifiers, 228
  regression algorithms, 228
  target points, 228

Gaussian mixture models, 469
Gaussian/normal distribution/bell curve, 436–8, 549
general partial-synthesis functions, 496–7
generative models, 89–90
glides, 154–5
  off-glides, 155
  on-glides, 155
glottis/glottal source, 148, 330–3
  assumptions, 338–9
  glottal-flow derivative, 333
  Lijencrants-Fant model, 332
  open/return/closed phases, 330–1
  parameterisation of glottal-flow signals, 379
government phonology, 183
graphemes, 28
  definition, 54–5
  TTS models, 39
grapheme-to-phoneme (G2P) conversion, 55, 218–22
  with decision trees, 221
  dynamic time warping (DTW), 219
  G2P algorithms, 208, 218
  G2P alignment, 219
  memory-based learning, 220–1
  NetTalk algorithm, 219–20
  neural networks, 219–20
  pronunciation by analogy, 220–1
  rule-based techniques, 218–19
  rule ordering, 219
  statistical techniques, 221–2
  with support-vector machines, 221
Grice's maxims, 20

hand labelling, 519–21
hand written algorithms, 80
harmonic/noise models (HNMs), 426–9
harmonics, 148–9
Harvard sentences, 523
Haskins sentences, 523
heterogeneous relation graph (HRG) formalism, 72–5
hidden Markov model (HMM)
  about the HMM, 89–91, 435, 471–3
  and intonation synthesis, 253–4
  and phrasing prediction, 133–5
hidden Markov model (HMM) formalism, 435–56
  about HMM formalism, 435–6

acoustic representations, 439–40
backoff techniques, 444
Baum–Welch algorithm, 449
context-sensitive modelling, 451–4
covariance matrix, 439
decision trees, 452–5
delta delta/ acceleration coefficients, 439
delta/velocity coefficients, 438–9
diagonal covariance, 439
discrete state problems, 454–5
forced-alignment mode, 448
forward–backward algorithm, 449–50
as generative models, 440–3
generative nature issues, 455–6
independence of observations issues, 454
language models, 444
linearity problems, 455
recognising with HMMs, 440–3
self-transition probability, 440
smoothing techniques, 444
states of phone models, 440
training HMMs, 448–51
transition probabilities, 440
triphone models, 451
Viterbi algorithm, 444–8
see also observations for HMMs
hidden Markov models (HMMs), labelling databases with, 465–8
  about labelling, 465
  alignments quality measurement, 470
  dynamic-time-warping (DTW) technique, 469
  forced alignment, 468
  Gaussian mixture models, 469
  phone boundaries determination, 468–70
  phone sequence determination, 467–8
  word sequence determination, 467
hidden Markov models (HMMs), synthesis from, 456–64, 514
  about synthesis from HMMs, 456–7
  acoustic representations, 460–1
  context-sensitive models, 461–3
  duration modelling, 463–4
  example systems, 464
  likeliest observations for a given state sequence, 457–60
hidden semi-Markov model (HSMM), 464
homographs, 56
  abbreviation homographs, 54
  accidental homographs, 54
  ambiguity issues, 22, 46
  decoding, 98
  disambiguation, 79, 99–101
  homograph disambiguation, 56
  part-of-speech homographs, 54
  resolution of, 53
  true homographs, 54

homonyms, 58
  pure homonyms, 58
homophones, 56–7
human communication, 13–18
  about human communication, 13–14
  affective prosody, 17
  augmentative prosody, 18
  see also linguistic levels; verbal communication
Hunt and Black algorithm, 477–9, 504

iconic communication, 9–10
impulse/noise models
  classical LP prediction, 378
  classical LP synthesis, 400–1
independence concept, 543
independent feature formulation (IFF), 485
infinite-impulse–response (IIR) filter, 289
information–theoretic approach, 23
inside–outside algorithms, 105
instant of glottal closure (IGC) points, 374, 382–3
integrated systems, future of, 536
intelligibility issues, 3, 48–9, 510, 523
International Phonetic Association (IPA)
  alphabet, 163–5
  consonant chart, 555
  symbol set (IPA alphabet), 163–5
interpreted communication, 8
interpreting characters, 69–71
intonational phonology, 121
intonational phrases, 114
intonation behaviour, 229–36
  boundary tones, 236
  downdrift/declination, 230–3
  nuclear accents, 230
  pitch accents, 230, 234–6
  pitch range, 233–4
  tune, 229–30
intonation synthesis, 225–9
  about intonation, 225, 259–61
  F0 and pitch, 226
  F0 synthesis, 229
  intonational form, 226–7
  intonational synthesis, 225
  micro-prosody, 229
  pitch-accent languages, 227
  tone languages, 227
intonation theories and models, 236–45, 250–4
  about data-driven models, 250–1
  autosegmental–metrical (AM) model, 237–9
  British school, 227, 236–7
  data driven models, 250–4
  Dutch school, 237
  F0 contour models, 227–9
  Fujisaki model, 227, 239–42, 250
  intonational phonology, 237
  INTSINT model, 239
  phonological versus phonetic versus acoustic, 244–5
  purpose, 244
  superimpositional models, 242
  superimpositional versus linear, 245
  Tilt model, 227, 242–4
  ToBI scheme, 237
  tones versus shapes, 245
  traditional model, 236–7
  see also autosegmental–metrical (AM) intonation model; data-driven intonation models; deterministic acoustic models, synthesis with
intonation and tune, 121–2
  prediction issues, 139
INTSINT intonation model, 239
inverse filtering, 372
IPA *see* International Phonetic Association (IPA)
ISO 8859, 70

java speech markup language, 69
join functions, 497–504
  about joining units, 497–8
  acoustic-distance join costs, 499–500
  categorical and acoustic join costs, 500–1
  join classifiers, 497, 502–4
  join costs, 497–8
  join detectability, 498
  join probability, 497
  macro-concatenation issue, 497
  phone-class join costs, 498–9
  probabilistic and sequence join function, 501–2
  sequence join classifier, 503
  singular-value decomposition (SVD), 502
  splicing costs, 499

Kalman filter, 252
Klatt deterministic rules, 256–7
Klatt synthesiser, 394–5
Kullback–Leibler distance, 552

labelling databases, 519
  automatic labelling, 521
  avoiding explicit labels, 521–2
  hand labelling, 519–21
  see also hidden Markov models (HMMs), labelling databases with
labiodental constriction, 153
language models, 444
  N-gram language model, 444
language origin, and pronunciation, 223
Laplace transform, 283
laryngograph/laryngograph signals (Lx signals), 383–4
larynx, 148
Laureate system, 513
least modification principle, 477

Index

letter sequences, decoding, 99
Levinson–Durbin recursion source-filter separation technique, 361–2, 367
lexemes, inflected forms, 59
lexical phonology/word formation, 179–81
lexical stress, 116, 186–9
lexicons, 63, 207–18
   compression, lossless and lossy, 215
   computer lexicons, 207
   exceptions dictionaries, 208
   formats, 210–12
   grapheme-to-phoneme algorithms, 208
   language lexicons, 207
   memorising the data, 209
   offline lexicon, 213–14
   orthographic and pronunciation variants, 210–12
   orthography–pronunciation lexicons, 207
   over-fitting data, 209
   quality of, 215–16
   as a relational database, 210–11
   rules for, 208–10
   simple dictionary formats, 210
   speaker's lexicons, 207
   system lexicon, 214–15
   unknown word problems, 216–18
Lijencrants–Fant model for glottal flow, 332, 374, 376
limited-domain synthesis systems, 44
linear filters, assumptions concerning, 337
linear-prediction cepstra, 369
linear-prediction (LP) PSOLA, 423–4
linear-prediction (LP) speech analysis, 357–65
   about linear prediction, 357–8
   autocorrelation method, 360–1
   Cholskey decomposition method, 359
   covariance method for finding coefficients, 358–60
   Levinson-Durbin recursion technique, 361–2
   perceptual linear prediction, 370
   spectra for, 362–5
   Toeplitz matrix, 361
linear-prediction (LP) synthesis *see* classical linear-prediction (LP) synthesis; residual-excited linear prediction
linear time-invariant (LTI) filters, 288, 310
   for nasalised vowels, 333–4
line-spectrum frequencies (LSFs), 367–9
linguistic-analysis TTS models, 39
linguistic levels, 16–17
   morphemes/morphology, 16
   phonetics/phonology, 16
   pragmatics, 17
   semantics, 16–17
   speech acoustics, 16
   syntax, 16

linguistics/speech technology relationship, 533–6
   future of, 537–8
log area ratios, 367
logographic writing, 34
logotomes/nonsense words, 415
log power spectrum, 343–4
lookup tables, 72
lossless tube, assumptions, 338
LP *see* linear-prediction (LP) ...
lumped-parameter speech generation model, 389

machine-readable phonetic alphabet (MRPA) phoneme inventory, 204–5
machine translation, 1
macro-concatenation, 497
magnetic resonance imaging (MRI), 156
Manhattan distance, 486
marginal distributions, 543
markup languages, 68–9
   java speech markup language, 69
   speech synthesis markup language (SSML), 69
   spoken text markup language, 69
   VoiceXML, 69
maximal onset principle, 185
MBROLA technique, 429
meaning/form/signal, and communication, 12–13
meaning-to-speech system, 42–3
mean opinion score, 524
medium (means of conversion), 13
mel-frequency cepstral coefficients (MFCCs), 370, 429–31, 439
mel-scale, 351
memorising the data (machine learning), 209
memory-based learning, 220–1
message/form-to-speech synthesis, 42
messages, 18
   message generation, 20–1
metrical phonology, 114, 120, 183
metrical stress, 188
micro-prosody, 229
minimal pair principle/analysis, 163, 197–9, 204
mis-spellings, decoding, 98
model effectiveness, and synthesis with vocal-tract models, 407
models of TTS, 37–41
   common-form model, 38
   comparisons, 40–1
   complete prosody generation, 40
   full linguistic-analysis, 39–40
   grapheme form, 39
   phoneme form, 39
   pipelined, 39
   prosody from the text, 40
   signal-to-signal, 39
   text-as-language, 39
modified rhyme test (MRT), 523, 524

modified timit ascii character set, 166
modularity, and synthesis with vocal-tract models, 407
moments of a PMF, 542
monophthongs, 153
morphemes/morphology, 16
morphology, 222–3
  derivational, 59, 222
  inflectional, 222
  morphological decomposition, 222–3
  and scope, 59
MRPA phoneme inventory, 204–5
multi-band-excitation (MBE), 427
multi-centroid analysis, 371
multi-pass searching, 509

naive Bayes' classifier, 86–7
names, pronunciation, 223
nasal cavity modelling, 333–5
nasalisation colouring, 167
nasal and oral sounds, 150
nasal stops, 153
natural-language parsing, 102–5
  Cocke–Younger–Kasami (CYK) algorithm, 104
  context-free grammars (CFGs), 102–4
  probabilistic parsers, 104–5
  statistical parsing, 105
natural-language text decoding, 46–7, 97–101
  about natural-language text, 97–8
  acronyms, 99
  homograph disambiguation, 99–101
  letter sequences, 99
  non-homographs, 101
naturalness issues/tests, 3, 47–8, 510, 523, 524
  mean opinion score, 524
natural phonology, 183
NetTalk algorithm, 219–20
neural networks, and G2P algorithms, 219–20
neutral vowel sound, 152–3
NextGen system (AT&T), 513–14
n-gram model, 91
non-linear phonology, 183
non-linguistic issues, 32–3
non-natural-language text decoding, 92–7
  about non-natural-language text, 92
  parsing, 95
  semiotic classification, 92–4
  semiotic decoding, 95
  verbalisation, 95–7
nonsense words/logotomes, 415
non-standard words (NSWs), 106
non-uniform unit synthesis, 480
nuclear accents, 230
null/neutral prosody, 18
number-communication systems, 33
Nyquist frequency, 279

observations for HMMs, 436–8
  covariance matrix, 437
  Gaussian/normal distribution/bell curve, 436–8
  multivariate Gaussian, 437
  probabilistic models, 436
  probability density functions (pdfs), 436
  standard deviation, 436
  variance, 436
obstruent consonants, 154
offline lexicon, 213–14
open-phase analysis, 377–8
optimal coupling, 432
optimality theory, 183
oral cavity, 152
  sound source positions, 335
oral and nasal sounds, 150
oral stops, 153
over-fitting data, 209
overwrite paradigm, 71

palatalisation, 180
parameterisation of glottal-flow signals, 379
parsing/parsers, 53, 95, 103
  probabilistic parsers, 104–5
  statistical parsing, 105
  see also natural-language parsing
partial-synthesis function, 493
part-of-speech (POS) tagging, 82, 88–92
  generative models, 89–90
  hidden Markov model (HMM), 89–91
  n-gram model, 91
  observation probabilities, 90
  POS homographs, 88
  syntactic homonyms, 88–9
  transition probabilities, 90
  Viterbi algorithm, 92
perceptual linear prediction, 370
perceptual substitutability principle, 485
periodic signals, 262–9, 305–7
phase mismatch issues, 431–2
phase shift, 264
phase-splicing systems, 44
phone-class join costs, 498–9
phoneme inventories, 204–5
  British English MRPA, 554
  modified TIMIT for General American, 553
phonemes
  and graphemes, 28
  and verbal communication, 14, 16, 161–4
phoneme TTS models, 39
phones
  about phones, 162–4
  definitions, 553–5
phonetic similarity principle, 197–9

phonetics/phonology, 16
    phonetic context, 167
    phonetic variants, 57
    see also phonological theories; phonology; phonotactics; speech production/articulatory phonetics
phonological theories, 181–4
    articulatory phonology, 183
    autosegmental phonology, 183
    computational phonology, 184
    context sensitive rules, 182
    dependency phonology, 183
    feature geometry, 183
    government phonology, 183
    metrical phonology, 183
    natural phonology, 183
    non-linear phonology, 183
    optimality theory, 183
    The Sound Pattern of English (SPE), 181–2
phonology, 172–89
    about phonology, 172, 189–91
    lexical stress, 186–9
    maximal onset principle, 185
    metrical phonology, 114
    palatalisation, 180
    phonological phrases, 114
    syllabic consonants, 184
    syllables, 184–6
    word formation/lexical phonology, 179–81
    see also phonological theories; phonotactics
phonotactics, 172–9
    distinctive features, 174
    feature structure, 174–9
    phonotactic grammar, 172–7, 207
    primitives issues, 176
    syllable structures, 176–7
phrasing prediction, 129–36
    classifier approaches, 132–3
    deterministic approaches, 130–1
    experimental formulation, 129–30
    HMM approaches, 133–5
    hybrid approaches, 135–6
    precision and recall scheme, 130
phrasing/prosodic phasing, 112–15
    about phrasing, 112–13
    phasing models, 113–15
pictographic writing, 34
pipelined TTS models, 39
pitch-accent languages, 121, 227
pitch accents, 230, 234–6
    alignment factors, 235–6, 534–5
    height factors, 235
pitch detection/tracking, 379–81
    pitch-detection algorithms (PDAs), 379
pitch-marking, 381
pitch range, 233–4

pitch-synchronous overlap and add (PSOLA) techniques, 415–21
    about PSOLA, 415–16, 421
    epoch manipulation, 417–20
    time-domain PSOLA (TD-PSOLA), 416–17
pitch-synchronous speech analysis, 347, 381
polynomial analysis (poles and zeros), 294–7
post-lexical processing, 223–4
pragmatics, 17
pre-processing, 52
pre-recorded prompt systems, 43
probabilistic models, 436
probabilistic parsers, 104–5
probabilistic and sequence join function, 501–2
probability density functions (pdfs), 436
probability mass functions (PMFs), 541
probability theory, continuous random variables, 547–50
    cumulative density functions, 549–50
    expected values, 548–9
    Gaussian (normal) distribution, 549
    uniform distribution, 549
probability theory, discrete probabilities, 540–42
    discrete random variables, 540–1
    expected values, 541–2
    moments of a PMF, 542
    probability mass functions (PMFs), 541
probability theory, pairs of continuous random variables, 550–2
    entropy for, 552
    independent versus uncorrelated, 551
    Kullback–Leibler distance, 552
    sum of two, 551–2
probability theory, pairs of discrete random variables, 542–7
    Baye's rule, 545
    chain rule, 546
    conditional probability, 545
    correlation, 544
    entropy, 546–7
    expected values, 543
    higher-order moments and covariance, 544
    independence, 543
    marginal distributions, 543
    moments of a joint distribution, 544
    sum of random variables, 545–6
problems in text-to-speech, 44–50
    adapting systems, 50
    assumed intent for prosody, 49–50
    auxiliary generation for prosody, 49–50
    homograph ambiguity, 46
    intelligibility issues, 48–9
    natural language text decoding, 46–7
    naturalness, 47–8
    syntactic ambiguity, 46–7
    text classification/semiotic systems, 44–6

Processing documents, 68–71
  character encoding schemes, 69–70
  interpreting characters, 69–71
  see also markup languages
productiveness property, 15
prominence, 115–21
  data and labelling, 119–21
  discourse prominence patterns, 118–19
  emphatic prominence, 119
  nuclear prominence, 116
  prominence shift, 117–18
  prominence systems, 119–21
  syntactic prominence patterns, 116–18
prominence prediction, 136–9
  compound-noun phrases, 136–7
  data-driven approaches, 138–9
  function-word prominence, 138
pronunciation, 192–224
  about pronunciation, 192–3, 224
  abstract phonological representations, 196–7
  by analogy, 220–1
  data sparsity problem, 193
  language origin issues, 223
  morphology, 222–3
  names, 223
  phonemic and phonetic input, 193–4
  phonetic input problem, 194–5
  post-lexical processing, 223–4
  structured phonemic representation, 195–6
  see also grapheme-to-phoneme (G2P) conversion; lexicons
pronunciation, a phonological system for, 197–207
  about phonological system development, 197
  affricates, 201
  approximants/vowel combinations, 201–2
  difficult consonants, 199–200
  diphthongs, 201
  glides, 201–2
  inventory defining, 203–4
  minimal pairs principle/analysis, 197–9, 204
  MRPA phoneme inventory, 204–5
  phoneme inventory, 204–5
  phoneme names, 204–6
  phonetic similarity principle, 197–9
  phonotactics/phonotactic grammar, 207
  rhotic/non-rhotic accents, 202
  simple consonants and vowels, 197–9
  syllabic issues, 206–7
  syllable boundaries, 206
  TIMIT phoneme inventory, 203–4
prosodic hierarchy, 114
prosodic interaction, 146
prosodic meaning and function, 122–7
prosodic style, 127
prosodic and verbal content, 30–1

prosody
  about prosody, 14
  affective, 17
  augmentative, 18
  null/neutral, 18
  in reading aloud, 36–7
prosody, determination from text, 127–9
  augmentative prosody control, 128
  prosody and human reading, 127–9
  prosody and synthesis techniques, 128–9
  TTS models, 40
prosody, prediction from text, 53, 111–45
  about prosody, 111–12, 144–5
  affective prosody, 123–4
  augmentative prosody, 125–6, 142
  intonational-tune prediction, 139
  intonation and tune, 121–2
  labelling schemes/accuracy, 139–41
  linguistic theories/prosody, 141–2
  phrasing, 112–15
  prosodic meaning and function, 122–7
  prosodic phase structures, 113
  prosodic style, 127
  real dialogues, 143–4
  speaker choice/variability, 142–3
  suprasegmentality, 124
  symbolic communication, 126–7
  underspecified text, 142
  see also phrasing prediction; prominence; prominence prediction; prosody determination
pruning methods, 508–9
  beam pruning, 509
PSOLA *see* pitch-synchronous overlap and add (PSOLA) techniques
punctuation
  status markers, 65
  and tokenisation, 65–6
  underlying punctuation, 66
pure unit selection, 477

quality improvements, future of, 537

radiation models for sound, 330
  assumptions, 338
Reading aloud, 35–7
  prosody in, 36–7
  and silent reading, 35–6
  style issues, 37
  verbal content, 37
RealSpeak system, 514
reduced stress, 187
redundancy, in speech, 21
reflection coefficients, 366–7
regression algorithms, F0 contour models, 228
resequencing algorithms, 477

residual analysis, for pitch detection, 381
residual-excited linear prediction, 421–4
  about residual excited LP, 421–3
  linear-prediction PSOLA, 423–4
  residual manipulation, 423
residual manipulation, 423
residual speech signals, 372–4
  error signals, 372
  inverse filtering, 372
resonance, 159
  formants, 159–60
  resonant systems, 311–13
rhotic/non-rhotic accents, 202
rVoice system, 514

scope and morphology, 59
secondary stress, 187
second-generation synthesis systems, 412–34
  about second-generation systems, 412–13, 433–4
  cepstral coefficients, synthesis from, 429–31
  concatenation issues, 431–2
  diphone inventory creation, 414
  diphones from speech, 414–15
  MBROLA technique, 429
  speech units in, 413–15
  see also pitch-synchronous overlap and add (PSOLA) techniques; residual-excited linear prediction; sinusoidal models techniques
segments, 162
semantics, 16–17
semiotic classification, 45, 79, 92–4
  open-class rules, 93
  specialist sub-classifiers, 93
  and translation, 45–6
semiotic decoding, 95
semiotics, 23
semiotic systems, 33–4
sentences, 14, 62–3
  sentence-final prosody, 67
  sentence splitting, 53, 67–8
  style manuals, 68
sentential stress, 186–7
sequence join classifier, 503
signal processing and unit-selection, 511
signals
  and communication, 13
  see also analogue signals; digital signals; transforms
signal-to-signal TTS models, 39
sinc function, 277
singular-value decomposition (SVD), 502
sinusoidal models techniques, 424–9
  about sinusoidal models, 424–5
  harmonic/noise models (HNMs), 426–9
  multi-band-excitation (MBE), 427
  pure sinusoidal models, 425–6

sinusoid signals, 263–5
Sound Pattern of English, The (SPE), 181–2
sound, physics of, 311–19
  acoustic capacitance, 317
  acoustic impedance, 317
  acoustic inductance, 317
  acoustic reflection, 318
  acoustic resistance, 317
  acoustic waves, 316–18
  boundary conditions, 315
  lossless tubes, 317
  resonant systems, 311–13
  sound propagation, 317
  speed of sound, 316
  standing waves, 314
  travelling waves, 313–15
  see also vowel-tube model
sound sources see speech production/articulatory phonetics
source/filter model of speech, 151
source-filter separation see cepstrum speech analysis; filter-bank speech analysis; linear-prediction (LP) speech analysis
source-filter separation assumptions, 338
source signal representations, 372–9
  closed-phase analysis, 374–7
  impulse/noise models, 378
  open-phase analysis, 377–8
  parameterisation of glotta-flow signals, 379
  residual signals, 372–4
speaker choice/variability, 142–3
spectral analysis/frequency domain analysis, 156
spectral-envelope, 156–7
  and vocal-tract representations, 362–72
spectral representations of speech, short term, 343–5
  envelopes, 345
spectrograms, 157–9
  in speech analysis, 348–51
speech, disfluences in, 30
speech acoustics, 16
speech, communicative use principles, 160–72
  about communicating with speech, 161–2
  allophones, 162
  allophonic variation, 166–7
  aspiration, 168
  assimilation effect, 167
  coarticulation, 168
  colouring effect, 167
  continuous nature issues, 169–70
  distinctiveness issues, 171–2
  IPA alphabet, 163–5
  minimal pair, 163
  modified timit ascii character set, 166
  nasalisation colouring, 167
  phonemes, 161–4
  phones, 162–4

speech, communicative (cont.)
  phonetic context, 167
  segments, 162
  targets, 168–9
  transcriptions, 170–1
speech production/articulatory phonetics, 146–56
  about speech production, 146–7
  consonants, 153–5
  continuants, 150
  egressive pulmonic air stream, 147
  examining speech production, 155–6
  fundamental frequency (F0), 148
  harmonics, 148–9
  larynx, 148
  neutral vowel sound, 152–3
  oral cavity, 152
  oral and nasal sounds, 150
  source/filter model of speech, 151
  stop sounds, 150
  timbre, 149
  unvoiced sounds, 150
  velum, 150
  vocal folds, 148
  vocal organs, 147
  vocal-tract filter, 150–1
  vowels, 151–3
  see also acoustic models of speech production; glottis/glottal source
speech recognition, 1, 22
speech, redundancy in, 21
speech signals analysis, 341–86
  about speech analysis, 341, 384–6
  spectral-envelope and vocal-tract representations, 362–72
  see also cepstrum speech analysis; epoch detection; filter-bank speech analysis; linear-prediction (LP) speech analysis; pitch detection; source signal representations
speech signals analysis, short term, 341–52
  auditory scales, 351–2
  envelopes, 345
  equivalent rectangular bandwidth (ERB) scale, 352
  frame lengths and shifts, 345–9
  mel-scale, 351
  pitch-asynchronous analysis, 347
  pitch-synchronous analysis, 347
  spectral representations, 343–5
  spectrograms, 348–51
  time-frequency tradeoff, 346
  windowing, 342–5
Speech synthesis markup language (SSML), 69
speech technology/linguistics relationship, 533–6
  future of, 537–8
speech/writing comparisons, 26–35
  component balance, 31–2
  form comparisons, 28–9

non-linguistic contents, 32–3
number-communication systems, 33
physical natures of, 27–8
prosodic and verbal contents, 30–1
semiotic systems, 33–4
speech spontaneity, 30
usage of each, 29–30
speed of sound, 316
SPE (*The Sound Pattern of English*), 181–2
splicing costs, 499
spoken text markup language, 69
standard deviation, 436
standing waves, 314
statistical parsing, 105
status markers, 65
stochastic signals, 288
stop sounds, 150
stress in speech, 116
  lexical stress, 116, 186–9
  metrical stress, 188
  reduced stress, 187
  secondary stress, 187
  sentential stress, 186–7
strict layer hypothesis, 114
style manuals, 68
sum of random variables, 545–6
sums-of-products model, 257–8
superimpositional intonation models, 242
superposition of functional contours (SFC) model, 254
support-vector machines, 221
suprasegmentality, 124
syllables, 184–6
  boundaries, 206
  syllabic consonants, 184
  syllabic writing, 34
symbolic language/communication, 10–12, 126–7
  combinations of symbols, 11–12
synonyms, 57–8
syntactic ambiguity, 46–7
syntactic analysis, 102
syntactic homonyms, 88–9
syntactic prominence patterns, 116–18
syntactic trees, 102
syntax, 16
  syntactic hierarchy, 16
  syntactic phrases, 16
synthesis
  articulatory synthesis, 405–7
  synthesis algorithms, future of, 536–7
  synthesis specification, 387–8
  see also classical linear-prediction (LP) synthesis; formant synthesis; hidden Markov models (HMMs), synthesis from; second-generation synthesis systems; vocal-tract models, synthesis with

synthesis of prosody *see* autosegmental-metrical (AM) intonation model; data-driven intonation models; deterministic acoustic models, synthesis with; intonation ...; timing issues
system lexicon, 214–15
system testing, 523

tagging, 82–3
talking-head synthesis, 406–7, 527
  see also audio-visual speech synthesis
targets, 168–9
tests/testing, 523–6
  Blizzard Challenge testing, 526
  comparison tests, 524
  competitive evaluations, 526
  Harvard sentences, 523
  Haskins sentences, 523
  modified rhyme test (MRT), 523, 524
  naturalness tests, 524
  semantically unpredictable sentences, 523
  system testing, 523
  test data, 525
  unit/component testing, 525–6
  word-recognition tests, 523–4
text analysis, future of, 536
text anomalies, 105
text-as-language TTS models, 39
text-classification algorithms, 79–92
  ad-hoc approaches, 83
  bag-of-features approach, 84
  cluster impurity, 88
  collocation rule, 84
  context-sensitive rewrite rule, 83–4
  curse of dimensionality, 81, 534
  data driven approach, 80
  decision lists, 85–6
  decision trees, 87–8
  deterministic rule approaches, 83
  engine/rule separation, 83
  features and algorithms, 79–82
  hidden Markov model (HMM), 89–91
  naive Bayes' classifier, 86–7
  part-of-speech (POS) tagging, 82, 88–92
  probabilistic approach, 80
  statistical approach, 80
  tagging, 82–3
  trigger tokens, 84–5
  unsupervised approach, 80
  word-sense disambiguation (WSB), 82–3
text decoding/analysis, 22, 52–3, 78–110
  about text decoding, 78–9, 105–10
  see also natural-language parsing; non-natural-language text decoding; text-classification algorithms
text materials, 518
text normalisation, 44, 106

text segmentation and organisation, 63–8
  about text segmentation, 52–3, 75–7
  sentence splitting, 67–8
  tokenisation, 64–7
  see also architectures for TTS; processing documents; sentences; words
text-to-speech (TTS)
  about text-to-speech, 1 2, 26, 50–1
  basic principles, 41
  common-form model, 5–6
  development goals, 3
  engineering approach, 4
  intelligibility issues, 3
  naturalness issues, 3
  purposes, 2
  see also models of TTS; problems in text-to-speech
texture mapping, 528
third-generation techniques *see* hidden Markov model (HMM); unit-selection synthesis
Tilt intonation model
  analysis with, 250
  synthesis with, 227, 242–4, 249–50
time-domain PSOLA (TD-PSOLA), 416–17
  pitch-scale modification, 416–17
  time-scale modification, 416
time-frequency tradeoff, in speech analysis, 346
time invariance, assumptions concerning, 337
timing issues, 254–9
  about timing, 254–5
  Campbell model, 258
  durations, 254
  Klatt rules, 256–7
  nature of timing, 255–6
  phase-final lengthening, 256
  sums-of-products model, 257–8
TIMIT phoneme inventory, 203–6, 553
  modified timit ascii character set, 166
ToBI intonation scheme, 237, 247, 248
Toeplitz matrix, 361
token, definition, 54
tokenisation, 53, 64–7
  and punctuation, 65–6
  tokenisation algorithms, 66–7
tone languages, 124, 227
tonemes, 121
transcriptions, 170–1
transfer-function poles, 364–5
transforms, 284–8, 307
  about transforms, 284
  analytical analysis, 287
  convolution, 287
  duality for time and frequency, 284–5
  frequency shift, 286
  impulse properties, 285
  Laplace transform, 283

transforms (*cont.*)
  linearity, 284
  modulation, 286
  numerical analysis, 287
  scaling, 285
  stochastic signals, 288
  time delay, 286
  z-transform, 282–3
  see also Fourier transform
translation from semiotic classification, 45–6
tree-banks, 105
trigger tokens, 84–5
triphone models, 451
tune and intonation, 121–2

understanding, 19, 22–3
uniform distribution, 549
unit back-off searching, 505–8
unit/component testing, 525–6
unit-selection databases, 517–18
  speaker choice issues, 518
unit-selection synthesis, 474–516
  about unit selection synthesis, 251–2, 474–9, 510–11, 515–16
  ATR family contribution, 512–14
  CHATR system, 513
  concatenation of units, 477
  coverage, 510
  extending from concatenative synthesis, 475–7
  features, cost and perception, 511–12
  HMM system, 514
  Hunt and Black algorithm, 477–9
  Laureate system, 513
  NextGen system (AT&T), 513–14
  principle of least modification, 477
  pure unit selection, 477
  RealSpeak system, 514
  resequencing algorithms, 477
  rVoice system, 514
  signal processing issues, 511
  see also join functions
unit-selection synthesis, features, 479–84
  base types, 479–10
  dimensionality reduction/accuracy tradeoff, 483
  feature choosing, 481–2
  feature combination structures, 481
  feature types, 482–4
  hand labelling technique, 480–1
  heterogeneous systems, 480
  homogeneous systems, 480
  intelligibility issue, 510
  join feature structure, 481
  left/right join feature structure, 481
  linguistic and acoustic features, 480–1
  naturalness issues, 510
  non-uniform unit synthesis, 480
  original/derived features, 481
  partial synthesis, 482
  script technique, 480
  target feature structure, 481
unit-selection synthesis, searching, 504–9
  about searching, 504–5
  beam pruning, 509
  diphone unit-selection system, 505
  half-phone solution, 506–7
  Hunt and Black algorithm, 504
  multi-pass searching, 509
  pre-selection, 508–9
  pruning methods, 508–9
  unit back-off solution, 505–6
  Viterbi algorithm/search, 504–5, 508
unit-selection synthesis, target function formulation, 484–93
  about the target function, 484–5
  acoustic-space formulation (ASF), 485, 493–7
  context-orientated-clustering, 495
  decision-tree clustering, 494–6
  disruption issues, 485
  distance/cost issues, 484
  equal-error-rate approach to learning, 491
  Euclidean distance, 486
  feature axis scaling, 488
  full set of candidates, 484
  general partial-synthesis functions, 496–7
  hand tuning, 491
  independent feature formulation (IFF), 485–8
  independent-feature formulation limitations, 491–3
  Manhattan distance, 486
  perceptual approaches, 490–1
  perceptual space formulating/defining, 485–6
  perceptual substitutability principle, 485
  search candidates set, 484
  target weights setting, 488–91
unknown words
  decoding, 98
  problems with, 216–18
unvoiced sounds, 150
UTF-8, 71
UTF-16, 71
utterance structure, 71

variance, 436
velum, 150
verbal communication, 14–16
  arbitrariness, 15
  discreteness, 15–16
  duality, 15
  phonemes, 14
  productiveness, 15
  sentences, 14
  words, 14

verbalisation, 95–7
verb-balancing rule, 131
visemes, 528
Viterbi algorithm, 92, 444–8, 504–5, 508
vocal organs, 147
vocal-tract
 filter, 150–1
 sound loss models, 335–6
 straight tube assumptions, 337
 transfer function, 310–11
vocal-tract models, synthesis with, 387–411
 about synthesis with vocal-tract models, 387, 407–11
 ease of data acquisition, 407
 effectiveness of models, 407
 modularity issues, 407
 synthesis specification, 387–8
 see also articulatory synthesis; classical linear-prediction (LP) synthesis; formant synthesis; residual-excited linear prediction; vowel-tube models
vocal-tract and spectral-envelope representations, 362–72
voice transformation, and synthesizing emotion, 530
VoiceXML, 69
vowel sounds, 151–3
 diphthongs, 153
 monophthongs, 153
 neutral vowel, 152
vowel-tube models, 319–30
 about the vowel tube, 319
 all-pole resonator model, 329–30

discrete time and distance, 320
junction special cases, 322–3
junction of two tubes, 320–2
multi-tube vocal-tract model, 327–9
reflection coefficient, 322
single-tube vocal-tract model, 325–7
transmission coefficient, 322
two-tube vocal-tract model, 323–5

windowing, 342–5
word formation/lexical phonology, 179–81
word-recognition tests, 523–4
words, 14
 ambiguity issues, 54
 defining in TTS, 55–9
 definitions/terminology, 54–5
 form issues, 53–4
 hyphenated forms, 61–2
 shortened forms, 61
 slang forms, 61
 word variants, 57
word-sense disambiguation (WSB), 82–3
writing *see* speech/writing comparisons
writing systems, 34–5
 Abjab, 34–5
 alphabetic, 34
 logographic, 34
 pictographic, 34
 syllabic, 34

z-transform, 282–3
 and digital filters, 293–4, 297–8